CHEMICAL PHYSICS

ELECTRONS AND EXCITATIONS

CHEMICAL PHYSICS

ELECTRONS AND EXCITATIONS

Sven Larsson

CRC Press
Taylor & Francis Group
Boca Raton London New York

CRC Press is an imprint of the
Taylor & Francis Group, an **informa** business

CRC Press
Taylor & Francis Group
6000 Broken Sound Parkway NW, Suite 300
Boca Raton, FL 33487-2742

First issued in paperback 2019

© 2012 by Taylor & Francis Group, LLC
CRC Press is an imprint of Taylor & Francis Group, an Informa business

No claim to original U.S. Government works

ISBN-13: 978-1-4398-2251-7 (hbk)
ISBN-13: 978-0-367-38169-1 (pbk)

Library of Congress Cataloging-in-Publication Data

Larsson, Sven, 1941-
    Chemical physics : electrons and excitations / Sven Larsson.
        p. cm.
    Includes bibliographical references and index.
    ISBN 978-1-4398-2251-7 (hardback)
    1. Chemistry, Physical and theoretical--Textbooks. I. Title.

QD453.3.L37 2012
541'.2--dc23
                                                                2011051332

Visit the Taylor & Francis Web site at
http://www.taylorandfrancis.com

and the CRC Press Web site at
http://www.crcpress.com

Contents

Preface

In the second half of the nineteenth century, *physical chemistry* developed as a well-defined subject, consisting of thermodynamics, kinetics, and transport processes, and mainly dealing with bulk properties and continuum models. When quantum mechanics was discovered in 1925, paving the way for modern molecular physics, this subject was less well received by the chemists. Partly this was due to the morass of equations and calculations one sinks into, just to get insight into such a simple concept as the chemical bond. Application of quantum methods in chemistry was pioneered by people like Henry Eyring, Linus Pauling, Robert Mulliken, Per-Olov Löwdin, Björn Roos, and many others. Most of these scientists called themselves "quantum chemists."

The calculation of electronic energies and wave functions made impressive progress during the second half of the twentieth century, a development that ran parallel to the improvement of computer technology. At the same time, the new subject of *quantum chemistry* became standardized and automatized. There is now an army of users of ready programs who calculate bond lengths and reaction barriers with ever increasing accuracy. Unfortunately, there appears to be some hesitation to tackle more difficult problems, such as transfer of protons, electrons, and excitations. These fields were pioneered by, for example, Rudy Marcus, Norman Sutin, Noel Hush, Joshua Jortner, John R. Miller, and Ahmed Zewail. It has become customary to refer to physics and chemistry based on the quantum mechanical behavior of the elementary particles, atoms and molecules, as *chemical physics*, and this explains the main title of this book.

Chemical physics has received enormous attention in recent years, while at the same time there are quite few text-books available. The intention with the present book is to summarize the necessary theoretical background for research in electrochemistry, biochemistry, photophysics and photochemistry, natural and artificial photosynthesis, conducting organic crystals, and conducting polymers. The book is an extension of my courses for physicists as well as chemists at Chalmers University of Technology. In my experience, physics students usually have a good training in mathematics, while their knowledge of molecules is limited. Chemistry students are less well equipped in mathematics and quantum mechanics, but their background in chemistry and molecules is adequate. I therefore included some basic quantum mechanics and atomic physics that the physicists already know. At the same time, I have included some basic knowledge of molecules that the chemists already are experts on. The first nine chapters therefore contain quantum mechanics, based on calculation of wave functions, molecular orbital theory, statistical mechanics, kinetics, and proton transfer. One idea with the book is to make it possible for physicists and chemists to meet on a common ground. Individuals lacking a background in physics and chemistry often question the difference between physics and chemistry. A simple answer is that atoms and metals belong to physics, molecules to chemistry. This is not a good answer, however, since chemists must understand the periodic table of the elements and physicists need to understand chemical bonds and spectra of molecules.

While I was writing the book, I became aware of another difference in approach that actually has poisoned the relations between chemists and physicists for a long time. Physicists prefer to treat processes with delocalized electrons, where the reorganization of the structure is negligible. This is necessary to be able to work in momentum space. Unfortunately, some important problems cannot be treated if reorganization is neglected. Conducting polymers belong to the latter problems, and the spectra of molecules and localization belong to the problems connected to "Hubbard U." Molecular physicists are, of course, aware of the fact that excited states generally have equilibrium geometries different from the ground state. From this follows the importance of structural reorganization (via the Stokes law; the Stokes shift is equal to the reorganization energy).

Chapter 10 is a large chapter on electron transfer. Since electron transfer is of central importance in biology, biological electron transfer is treated in a separate Chapter (Chapter 11). Chapters 12 through 15 deal with excited states, photo-induced electron transfer, and excitation energy transfer. Chapter 16 deals with metals and includes solar cells of the original semiconductor kind. Chapters 17 and 18 contain applications to conductivity, where structural reorganization cannot be ignored or hidden behind an effective mass.

Consistent with my background as a quantum chemist, almost all figures have been made using quantum chemical methods. I have used the *HyperChem* programs, developed by Neil Ostlund and others (Hypercube, Inc.; originally Waterloo, Ontario, Canada; now Gainesville, Florida, USA). The reason is the unique simplicity in using *HyperChem*. The calculated structures may be less than perfect in some cases. Expert quantum chemists are cordially invited to improve and finish up the calculations using methods that include electronic correlation. To me it was necessary to get a quick result. The PC on my desk did all calculations while I was writing. In this way my impossible dreams, as a student of quantum chemistry in the 1960s, materialized. The *HyperChem* programs are also excellent as a pedagogical means, particularly for students who are not quantum chemists. The readers of this book are recommended to use them.

Sven Larsson

Acknowledgments

There are many people to thank. First of all, I would like to thank my students Anders Broo, Bruno Källebring, Manuel Braga (deceased), Lucia Rodrigues-Monge, Agris Klimkāns, and Fredrik Blomgren for doing the hard, but necessary background work. Exam workers Erik Johansson and Michael Ahl should also be thanked. I have had many other collaborators. When it comes to the subject of the book, particularly Nikolaj Ivashin, Edgars Silinsh (deceased), Gregorius Dusčesas, Örjan Hansson, Tõnu Pullerits, Igor Filatov, and Johan Olofsson should be mentioned. I have been happy to benefit from lunch discussions with Sture Nordholm, Gunnar Nyman, Gunnar Svensson, Bengt Nordén, Nicola Markovic, Joakim Andréasson, Staffan Wall, and Björn Åkerman. Last but not least, Bo Albinsson and his research group in the femtolab should be mentioned for increasing my knowledge on photochemistry and fast reactions.

1 Quantum Theory

1.1 INTRODUCTION

Toward the end of the nineteenth century, it became clear that some experiments cannot be explained on the basis of known physics, now referred to as classical physics. In 1900, Max Planck was able to correctly describe the distribution of black-body radiation energy as a function of wavelength λ or frequency $\nu = c/\lambda$. He had to accept the existence of energy packets (quanta) with energy $E = h\nu$. A few years later, Albert Einstein found that this unexpected relation between energy and frequency is also necessary in order to understand photoemission from metallic surfaces. A decade later, Niels Bohr arrived at the correct equation for the electronic spectrum of the hydrogen atom by applying "quantum conditions" on the motion of the electron around the nucleus and the Planck's relation $E = h\nu$ on the energy difference between the energy levels (1916). A decade later, Heisenberg, Born, and Jordan introduced "quantum mechanics" and de Broglie and Schrödinger (1925–1926) introduced "wave mechanics."

Full understanding of modern chemistry would be impossible without quantum theory. Chemistry existed as its own, phenomenological science long before the year 1900, and has a number of features that defy explanation in terms of the classical laws of physics, for example, chemical bonds, reaction barriers in chemical reactions, and spectra. After the year 1900, a number of chemical phenomena have been described using quantum mechanics. Chemical bonds can now be accurately calculated with the help of a personal computer. Electron transfer can be understood, as can excitation energy transfer and other phenomena in photochemistry and photophysics. Chemistry has become a branch of physics: chemical physics.

1.2 ELECTROMAGNETIC RADIATION

The fundamental, classical theory for electricity and magnetism is contained in the Maxwell equations and the Lorentz force law. The Maxwell equations allow for electromagnetic (EM) radiation to spread with the velocity of light. In 1888, H. R. Hertz produced EM radiation experimentally and showed that it has the same reflection and refraction properties as light. It was therefore concluded that light is EM radiation with wavelengths between 400 and 700 nm. In due course, microwave, infrared (IR), ultraviolet (UV), and x-ray radiations were placed on the frequency scale of EM radiation. In the case of EM waves, there are two oscillating fields, one electric field and one magnetic field perpendicular to it. Both are orthogonal to the direction of propagation.

1.2.1   POLARIZATION OF EM RADIATION

The equations for the $\vec{E}$ and $\vec{B}$ fields of EM radiation propagating in the direction $\vec{k}$ are as follows:

$$\vec{E}(\vec{r},t) = \vec{E}_0 \exp\left(i\vec{k}\cdot\vec{r} - i\omega t\right)$$

$$\vec{B}(\vec{r},t) = \vec{B}_0 \exp\left(i\vec{k}\cdot\vec{r} - i\omega t\right). \tag{1.1}$$

Equation 1.1 describes a *plane wave*. The direction of propagation is $\vec{k}$. Vectors $\vec{r}$ with a given value of $\vec{k}\cdot\vec{r}$ are in a plane perpendicular to $\vec{k}$. For constant time t, the complex wave is a function of the perpendicular distance $\vec{k}\cdot\vec{r}$ to the origin. As time increases for constant $\vec{k}\cdot\vec{r}$, complex waves pass the plane with frequency equal to ω.

$\vec{E}_0$ and $\vec{B}_0$ are perpendicular to the propagation direction and determine polarization:

$$\vec{E}_0 = \begin{pmatrix} \cos\theta\exp(i\alpha_x) \\ \sin\theta\exp(i\alpha_y) \end{pmatrix} \quad \vec{B}_0 = \begin{pmatrix} -\sin\theta\exp(i\alpha_x) \\ \cos\theta\exp(i\alpha_y) \end{pmatrix}. \tag{1.2}$$

For linearly polarized light, $\alpha_x = \alpha_y$. For circularly polarized light, $\alpha_y = \alpha_x \pm \pi/2$, where $+$ or $-$ are used for the two possible polarizations (left or right).

The speed of propagation is $\omega/k = c$, where c is the velocity of light. Two waves may be added and undergo interference, whereby the ordinary vector summation rules are obeyed.

1.2.2   PLANCK'S LAW

The interaction between EM radiation and matter includes experiences of everyday life. If a piece of iron is heated in the blacksmith's shop, it will turn red after a while. With further heating, the color changes to yellow. Everything heated to 6000°C will be as white as sunlight. The frequency distribution and color of the heated body depend only on its temperature. Wien had arrived at his "displacement law" and even derived an equation for the energy distribution, but it did not fit perfectly to very accurate measurements of energy distribution as a function of frequency.

In the case of light radiation from a heated black body the oscillators present at the surface are the source of the radiation. Max Planck found that the distribution of energy on different wavelengths for a certain temperature of a black piece of metal agrees with the experimental one only if one assumes that the energy of the radiation comes in packets whose energy depends on the frequency, $\nu = c/\lambda$, where c is the velocity of light and λ the wavelength, in the following simple way:

$$E = h\nu. \tag{1.3}$$

Equation 1.3 cannot be derived from classical physics and might be regarded as the first quantum law. With the help of Equation 1.3, Planck was able to derive the following law:

$$dE = \frac{8\pi hc}{\lambda^5}\left[\exp\left(\frac{hc}{\lambda kT}\right) - 1\right]^{-1} d\lambda, \qquad (1.4)$$

giving the radiation energy within a certain wavelength gap $d\lambda$ as a function of λ and temperature T of the radiating black body, in perfect agreement with the experiments. Boltzmann's constant is k. His work was published in 1901 (Figure 1.1). By comparing Equation 1.4 to experimental measurements, Planck obtained quite accurate values of both h, now called Planck's constant, and the Boltzmann constant k.

To Planck, Equation 1.3 was little more than a necessary mathematical equation. In the later quantum mechanical picture, the excitations of the black body are quantized as *outgoing* waves. Equation 1.3 may be derived using time-dependent quantum mechanics for the excitation process. The possible frequencies depend on the material in the case of an insulator. Planck assumed that he had a "black body," meaning that all frequencies are available. $\Delta E = h\nu$ has turned out to be a general equation in quantum mechanics, holding for excitations, electrons, vibrations, fields, etc.

The unique importance of Planck's law was not clear until 4 years later (1905) when Albert Einstein noted it and used it in connection with his derivation of the photoelectric effect. Einstein needed quanta of energy in the *incoming* radiation to be able to explain photoelectron radiation.

1.2.3 PHOTOELECTRIC EFFECT

Usually, it is possible to decrease the span of frequencies in a ray to a single frequency ν (for all practical purposes) and this radiation is called *monochromatic*

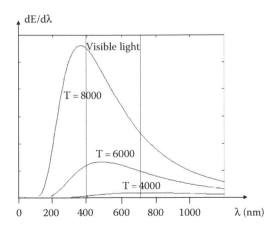

FIGURE 1.1 Black-body radiation at different temperatures.

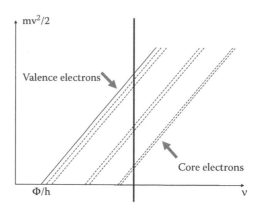

FIGURE 1.2 Photoelectric effect. The vertical full-drawn line corresponds to the v-axis in the XPS experiment. The fine structure depends on how the atom is bound in the molecule.

radiation. When monochromatic radiation is absorbed on a metal surface, electrons are emitted if the incoming radiation has a sufficiently high frequency. The kinetic energy of the electrons, $mv^2/2$, can be measured in electric and magnetic fields and plotted against the frequency v (Figure 1.2). This experiment was developed into UV and x-ray photoelectron spectroscopy (UPS and XPS, respectively). Another name for XPS is ESCA (electron spectroscopy for chemical analysis).

The following equation holds:

$$hv = \Phi + \frac{mv^2}{2} \Rightarrow \frac{mv^2}{2} = hv - \Phi, \qquad (1.5)$$

where h is the Planck's constant and Φ is a material constant. Since no electrons are emitted for $hv < \Phi$, Φ may be interpreted as the binding energy of the loosest bound electron. When $hv > \Phi$, electrons are emitted with maximum kinetic energy, $mv^2/2$. If the intensity of the incoming radiation is increased, there is no change in the energy, only an increase in the density of the emitted electrons. If, on the other hand, v is increased, the emitted electrons have a higher kinetic energy and new electrons that previously remained fixed to the atom can be emitted. In summary, the energy of the photoelectrons depends only on hv.

Einstein was the first person to interpret the photoelectric effect and conclude that the incoming monochromatic EM radiation consists of particles (later called *photons*) with the Planck's energy $E = hv$, despite the fact that light clearly diffracts and shows interference in the same way as waves. The energy of the incoming photon leads to emission of a *single electron*. The energy $E = hv$ is absorbed from the radiation field, in one way or another. The wave–particle "duality" of light was going to be a puzzle in physics for a long time to come.

1.2.4 X-RAYS

X-rays were serendipitously discovered 2 years earlier than electrons. On a dark November night in 1895, Wilhelm Conrad Röntgen at the University of Würzburg had wrapped up his x-ray tube in black cardboard for a routine experiment. In his laboratory there happened to be some pieces of paper, drained in a fluorescing solution. When Röntgen switched on the power to obtain a discharge through the tube, there was a flush from the paper pieces, although they were several meters away and the tube was wrapped up. Apparently, the radiation was not absorbed by the black cardboard. Röntgen had no idea of the nature of the penetrating rays and therefore simply called them "x-rays."

After several years, Max von Laue proved (1912) that the radiation emanating from the tube is EM radiation with a wavelength so small that he could use the lattice of a crystal to obtain a diffraction pattern. X-rays turned out to be very useful in medicine, to examine bone fractures and to localize bullets. Their importance in chemistry is immense, since the bonding structure of a crystallized chemical compound can be determined by x-ray diffraction. The structure of proteins with many subunits can be obtained, including photosynthetic reaction centers and antenna systems.

In 1914, H. G. J. Moseley used x-ray absorption spectroscopy (XAS) to show the relationship between atomic number (Z) and the frequency ($\nu = c/\lambda$) of the absorption peaks, and showed that the atomic structure depends on the charge of the nucleus rather than on its mass. The shell structure of the atoms was revealed. Hardest to eject are the K electrons, which therefore have to be located closest to the nucleus. Moseley also found the less energetic L series and M series, apparently emanating from shells further out. At about the same time, Bohr introduced the principal quantum number n of the electrons. It was easy to conclude that $n = 1$ for the K-shell, $n = 2$ for the L-shell, and $n = 3$ for the M-shell.

If an electron is missing in an inner shell of an atom, the hole may be filled by an electron from a higher energy level and radiation corresponding to $\Delta E = h\nu$ is emitted. This is the basis for x-ray emission spectroscopy (XES). The energy of the emission line depends on the atomic number and the subshell that is filled.

1.3 ELECTRONS

Before the electron was discovered in 1897, oxidation and reduction reactions were already well-known concepts and so were chemical bonding and conductivity. It was suspected that a particle, the "electron," was behind these phenomena. We now know that *oxidation* of a component of a system is the same as removal of electrons and that *reduction* is the same as addition of electrons. A chemical compound can only be oxidized if something else is reduced at the same time. In an electrochemical cell, ions from the electrolyte are reduced and deposited on the cathode. Corresponding oxidation reactions take place at the anode. The time-integrated current was originally measured in *equivalents* of charge. One equivalent corresponds to one mole of deposited metal, thus it is equal to Avogadro's number of electrons.

In the SI system, one ampere (1 A) is defined on the basis of magnetic interactions. The SI unit for charge, one coulomb (1 C or 1 As), is equal to 1 A during 1 sec, which is equal to 6.2415096×10^{18} electron charges. Since Avogadro's constant is equal to $6.02214179 \times 10^{-23}$, one equivalent (or one mole) is equal to 96485.34 C (Faraday's constant).

1.3.1 DISCOVERY OF THE ELECTRON

In 1897, Thomson discovered the electron in radiation from vacuum tubes (Figure 1.3). A similar tube is used in a television set. The ray was deflected in a way that suggested that electrons are small, negatively charged particles (called corpuscles by Thomson) with charge e and mass m. The ratio e/m can be determined in the experiment and is independent of the (very low) pressure in the tube and the nature of the thin gas in it.

At the time of its discovery, the electron was considered a particle and this is still a popular opinion. The reason was that the electron has a well-defined charge and mass as other particles. For most practical purposes, the electron behaves as a particle, thus it is easy to assume that it *is* a particle. The same applies to the photon.

As we will see next, electrons are quantum mechanical *wave packets*. Several wavelengths are present in the wave packet. The wave packet can be proven to have velocity and acceleration consistent with electronic mass and charge, measured under circumstances where the particle characteristics of the electron are unquestioned (such as the Millikan oil drop experiment).

1.3.2 QUANTUM CONDITIONS IN THE ATOM

Thus, the electron was first known as a negatively charged particle. The work of Rutherford shows that it is localized in the outer parts of the atom and is 1841 times lighter than the hydrogen atom nucleus. The latter carries the positive charge and most of the atomic weight.

The spectrum of a particular atom is unique. At the beginning of the last century, the atom was frequently thought to consist of a number of oscillators, one for each spectral line. Ritz found that sometimes two frequencies in a spectrum for a given atom sum up to another frequency of the same atom. It was also considered as very remarkable that the frequencies of the hydrogen atom seemed to have a strange integer relationship, known as Balmer's rule.

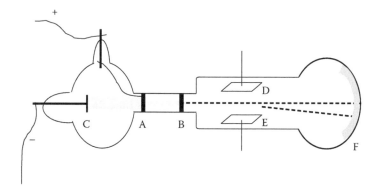

FIGURE 1.3 Apparatus used by Thomson in his historic experiment. A, anode with a slit; B, slit to sharpen the cathode ray further; C, cathode; D and E, two insulated plates connected with the opposite poles of a high voltage battery; F, recording equipment. The magnetic field is perpendicular to the electric field.

A rough understanding of the spectrum and structure of atoms was first achieved by Niels Bohr in Copenhagen. He realized that a classical treatment of the hydrogen atom, similar to planet motion, would not give rise to a discrete spectrum, not even a stable atom. Therefore, he introduced *stationary states* for the first time. This was a hint of what was to come in the form of quantum mechanics about 10 years later. His principal assumptions are worth citing, since the main differences between classical theory and what is required by a novel theory are clearly stated:
Assume:

1. That the dynamical equilibrium of the systems in the stationary states can be discussed by help of the ordinary mechanics, while the passing of the systems between different stationary states cannot be treated on that basis.
2. That the latter process is followed by the emission of a homogeneous radiation, for which the relation between the frequency and the amount of energy emitted is the one given by Planck's theory.

Thus, to get the frequency of a discrete spectral line, he applied Planck's theory to the energy *difference* between the stationary states of the oscillating electrons, not to the energy of the individual electrons themselves:

$$h\nu = E_n - E_m. \tag{1.6}$$

This is a general law that may be applied to other electronic systems as well.

1.3.3 OLD QUANTUM THEORY

Bohr had looked behind the curtain hiding the new quantum world. His theory for the hydrogen atom is consistent with the shell structure that Henry Moseley had just found to be generally true for atoms, using XAS.

The Bohr theory was extended in various ways, particularly by Arnold Sommerfeld. This "old quantum theory" is still used as a tool to "understand" quantum mechanics and to obtain approximate solutions. The old quantum theory deals with, for example, the Bohr–Sommerfeld quantum condition (quantization of action along a path); EBK theory of A. Einstein, L. Brillouin, and J. B. Keller; and JWKB theory of H. Jeffreys, G. Wenzel, H. A. Kramers, and L. Brillouin. Unfortunately, there is no possibility of giving justice to the various aspects of old quantum theory in this book.

Einstein insisted on the particle character of light. In the long wavelength limit of radio waves and alternating currents, the wave interpretation seemed to be the only possibility though. Concerning other types of EM radiation, it is easy to accept that γ-rays are particles like α and β particles. In 1923, the American physicist Arthur Compton showed experimentally that electrons are scattered by x-rays. This experiment is very easy to interpret, if it is assumed that x-rays consist of particles. An interpretation in terms of waves is, in fact, also possible, but it had to wait for several decades. It was now concluded that Einstein was right in his proposal that x-rays consist of particles. Photon was accepted as a name for this elementary particle.

Of course, there is ample evidence that light consists of waves (diffraction and interference experiments), and this must also have been clear to Einstein. How is it possible, then, that light *appears to be* at the same time a point-like particle and a wave extended over space? In the photoelectric effect and the Planck's equation, EM radiation appears as particles with energy, hν, but, in reality, they are dealing with absorption and emission events. We do not really know whether EM radiation after emission and before reabsorption consists of particles.

At the beginning of the 1920s, things were confused and they were going to get worse. A young French nobleman was preparing for a most daring and radical proposal: electrons are waves.

1.3.4   MATTER WAVES

p = E/c, where E is the energy density, is the momentum density for EM radiation, and causes a radiation pressure, for example, on comet tails. If this equation is applied on the Planck's wave packet with energy E = hν, the momentum p = hν/c = h/λ is obtained.

Louis de Broglie proposed that electrons, known as particles or corpuscles with mass and negative charge, can be considered as waves. He derived the wavelength λ for a particle by applying Einstein's theory of relativity on a wave packet. His conclusion was that λ should be related to the momentum p as

$$\lambda = h/p, \qquad (1.7)$$

that is, the same relation as for radiation pressure. In analogy with the EM wave, according to de Broglie, the wave associated with a particle may be written as

$$A \sin(kx - \omega t) = A \sin(2\pi x/\lambda - 2\pi \nu t). \qquad (1.8)$$

This is the sinus form for wave expansion in one dimension. According to Equation 1.8, at a certain point (x = constant), waves with time period $2\pi/\omega$ will pass. The frequency will be $\nu = \omega/2\pi$. On the other hand, if the time is kept constant, waves with wavelength $\lambda = 2\pi/k$ will be distributed on the x-axis.

In the mind of de Broglie, the wave accompanies the particle. He also talked about a *pilot wave* that guided the particle in its path. It is unclear in the theory of de Broglie what the physical meaning of the matter wave is. Is it a longitudinal wave like sound waves? How then could waves appear in a vacuum? The meaning of the waves was not cleared up until the interpretation of Max Born some years later.

The wavelength may be calculated as follows. Let us assume that the electron is moving in a negative potential V. The energy of the electron is increased by $p^2/2m = e \cdot V$. The momentum is $p = (2meV)^{1/2}$. According to Equation 1.7, the de Broglie wavelength is

$$\lambda = \frac{h}{\sqrt{2meV}} = \sqrt{\frac{150}{e \cdot V}} \text{ Å}. \qquad (1.9)$$

Thus, if the electron has achieved a kinetic energy of $|e| \cdot V = 1\,eV$ (in atomic units), the de Broglie wavelength is roughly 12 Å (1.2 nm).

Equation 1.8 may be generalized to a complex three-dimensional function:

$$\psi(x,y,z) = e^{i(\vec{k}\cdot\vec{r}-\omega t)} = \exp\left[i\left(k_x x + k_y y + k_z z - \omega t\right)\right], \quad \text{where } |k| = 2\pi/\lambda. \quad (1.10)$$

The vector $\vec{k} = (k_x, k_y, k_z)$ with the components k_x, k_y, and k_z is called the wave vector. All points $\vec{r} = (x, y, z)$, for which the scalar product $\vec{k} \cdot \vec{r} = k_x x + k_y y + k_z z - \omega t = C = $ constant, have the same value, $\exp(iC)$, of the wave function. These points form a plane since the equation of the plane is $k_x x + k_y y + k_z z = C + \omega t = $ constant, if t is constant. When time lapses and t is increased, the plane is moved forward, parallel to the original plane. The wave that is described by Equation 1.10 is therefore a plane *matter wave*. The vector (k_x, k_y, k_z) is orthogonal to the plane and defines the direction of propagation. The same value of the wave function appears for $C = C + 2\pi n$, since $\exp(2\pi i n) = 1$ for $n = 1, 2, \ldots$.

If de Broglie was right, electrons should provide a diffraction pattern if passed through a double slit (Figure 1.4). The double slit experiment for light is well known and was first performed by Thomas Young in 1803. G. P. Thomson, son of the discoverer of the electron, first obtained the corresponding diffraction pattern for electrons.

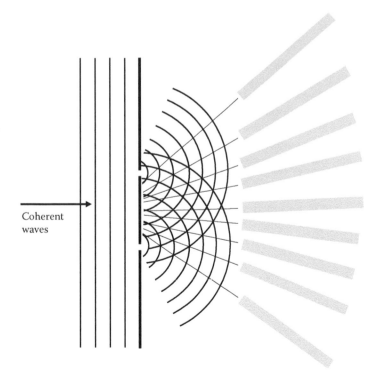

Coherent
waves

FIGURE 1.4 Interference pattern in double slit experiment. Radiation is EM radiation or electron waves.

Davisson and Germer (1927) used a metal lattice to show that interference between matter waves takes place. De Broglie's theory was confirmed.

Thus, it can be shown that both light and electrons consist of waves. An important question is the meaning of the diffraction pattern in the case of electrons. In the Davisson–Germer experiment, a local electric current is measured, and high intensity of the diffracted electron wave shows where the density is high. However, the wave function Ψ can be both positive and negative. How can one explain that a particle density can be both positive and negative? An explanation was given by Born, who interpreted $|\Psi|^2$, not Ψ, as an electronic probability density.

Having accepted the wave description, the next question is how we are going to describe mathematically that electrons (or neutrons) sometimes appear as particles. Two or more waves may be added to give interference (Figure 1.5). From acoustics, it is well known that two waves with close frequencies give a group wave modulation of the old wave (beats). This is illustrated in Figure 1.5, where waves with 24 and 22 wave maxima are added to give a wave with three main wave maxima, called "beats." If we superpose two wave functions, ψ_1 and ψ_2, with wave vectors $(k - \delta k)$ and $(k + \delta k)$ and angular frequencies $(\omega - \delta\omega)$ and $(\omega + \delta\omega)$, respectively, we obtain after some calculations:

$$\Psi = \psi_1 + \psi_2 = e^{i\left[(k-\delta k)x - (\omega - \delta\omega)t\right]} + e^{i\left[(k+\delta k)x - (\omega + \delta\omega)t\right]}$$

$$= 2e^{i(kx - \omega t)}\cos(\delta k \cdot x - \delta\omega \cdot t). \tag{1.11}$$

The exponential in the right member of Equation 1.11 is a complex function of magnitude 1. The cosine factor forms the beat as a contour for the function ψ, defining what we call a wave packet. The velocity of the wave packet is given by $= \delta\omega/\delta k = d\omega/dk$. For a free electron, $E = h\nu = \hbar\omega = p^2/2m = (\hbar k)^2/2m$, thus $\omega = \hbar \cdot k^2/2m$. We obtain $d\omega/dk = \hbar k/m = p/m = v$. The velocity of the beat (= the group velocity) for a matter wave is thus equal to the velocity of the particle. One may show that a wave packet also moves with the same acceleration as the particle.

The general three-dimensional expression for the matter waves is as follows: $g(p_x, p_y, p_z)$ is a weight factor function; ω is replaced by E according to $E = h\nu = \hbar\omega$; and k is replaced by p according to the de Broglie relation (1.7): $|k| = 2\pi/\lambda = p/\hbar$. The following equation is obtained for the wave packet:

FIGURE 1.5 Interference between two one-dimensional matter waves of different wavelengths. Real wave functions have been multiplied by an exponential damping factor.

$$\psi(x,y,z,t) = \frac{1}{h^{3/2}} \iiint_{E^3} g(p_x, p_y, p_z) e^{(i/\hbar)(p_x x + p_y y + p_z z - Et)} dp_x dp_y dp_z. \qquad (1.12)$$

Equation 1.12 is a Fourier transform of the function g to the function ψ, where ψ is a function of the Cartesian coordinates x, y, and z, and time t. The function g may be obtained in a corresponding Fourier transformation of Ψ; Ψ consists of an unimportant complex exponential and a contour function that defines the wave packet, moving with the velocity and acceleration of the particle.

Waves may be assumed to be *coherent* in the enlightened area. Coherent means that the same wave, complex exponential or sine wave, with a certain frequency, forms the wave function. The *coherence length* is the length of a region over which light is coherent.

1.4   TIME-INDEPENDENT SCHRÖDINGER EQUATION

Erwin Schrödinger evaluated the theory of de Broglie and became convinced that matter waves, as many other waves, are functions in three dimensions and solutions of a differential equation. Due to the boundary conditions, there are eigenvalues. It would be reasonable if there was a correspondence between these eigenvalues and the Bohr quantum numbers.

1.4.1   SCHRÖDINGER'S STANDING WAVES

Schrödinger started out from classical mechanics in the way it was derived by the mathematician W. R. Hamilton (1805–1865). The Hamilton function in this theory is simply the sum of kinetic energy, expressed as a function of the momentum p and potential energy V, expressed as a function of the position of the particle. According to Schrödinger, physical observables must be replaced by *operators*.

An operator transforms a function to another one. The operator may be a function that multiplies the function it "acts" on. For example, the operator x transforms the function sin(x) into the new function x · sin(x). Alternatively, the action of the operator may be to perform a derivation. The derivative operator transforms the function sin(x) to the function cos(x).

Other physical variables than kinetic and potential energy also correspond to an operator. The operator for momentum is a derivative multiplied by Planck's constant and (−i), where i is the imaginary unit. If we let that operator act on (one-dimensional) matter waves, we obtain

$$-i\hbar \frac{\partial}{\partial x} e^{i(k \cdot x - \omega t)} = k\hbar e^{i(k \cdot x - \omega t)} = p e^{i(k \cdot x - \omega t)}. \qquad (1.13)$$

The momentum operator acting on the matter wave thus gives the product of the momentum and the matter wave. The kinetic energy $p^2/2m$ may thus be replaced by a second derivative:

$$-\frac{1}{2m}\hbar^2\frac{\partial^2}{\partial x^2}e^{i(k \cdot x-\omega t)}=\frac{k^2}{2m}\hbar^2 e^{i(k \cdot x-\omega t)}=\frac{p^2}{2m}e^{i(k \cdot x-\omega t)}. \qquad (1.14)$$

Thus, this operator always represents the kinetic energy of a single electron.

The potential terms depend on the system. In the case of the hydrogen atom, the potential comprises the attraction between electron and nucleus, proportional to 1/r. In this case, the operator is simply multiplication by 1/r and a constant.

When the Hamilton function had been replaced by the *Hamiltonian operator*, Schrödinger was able to write down a *time-independent (Schrödinger) equation* (SE) for standing waves as a function of the electron coordinates:

$$-\frac{\hbar^2}{2m}\frac{d^2\psi}{dx^2}+V\psi=E\psi. \qquad (1.15)$$

The boundary conditions depend on the system and give the eigenvalues spectrum.

Schrödinger also wrote down a wave equation that includes time and may be regarded as the wave equation that replaces classical equations of motion. That equation is referred to as the *wave equation* or the *Schrödinger equation*. To solve the wave equation (see Chapter 5), we use the solutions of the time-independent equation. With boundary conditions, we obtain the motion of the wave packet.

1.4.2 Particle in a Box

It is instructive to solve the time-independent SE in a very simple case: a particle in a one-dimensional box with infinite walls (Figure 1.6). The particle may move around inside the walls, where V(x) = 0. The electrons of a piece of metal are freely mobile inside the metal, that is, the box. The same can be assumed to be the case for

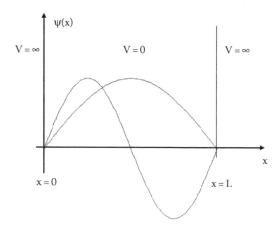

FIGURE 1.6 One-dimensional box. Wave functions for n = 1 and n = 2 are given. Note the boundary conditions.

the so-called π electrons in an aromatic molecule, where the molecule is the box. All molecules, freely mobile in a container, is a third example.

Equation 1.15 may have a number of solutions, for example, $\psi(x) = \exp(kx)$ and $\exp(-kx)$; $\psi(x) = \exp(ikx)$ and $\psi(x) = \exp(-ikx)$. The latter two may be combined into $\psi(x) = \sin(kx)$ and $\psi(x) = \cos(kx)$. If we insert these wave functions into SE, we find that the energy eigenvalue $E = -\hbar^2k^2/2m$ in the first two cases and $E = \hbar^2k^2/2m$ in the latter two cases.

$$k = n\pi/L. \tag{1.16}$$

We must now take the boundary conditions into account. It generally holds that ψ must have a derivative at all points where the potential $V(x)$ is continuous. If $V(x)$ is discontinuous and infinite, as in Figure 1.6 for $x = 0$ and $x = L$, it is sufficient if ψ is continuous. Since $\psi = 0$ in those points where $V = \infty$, we thus require that $\psi(0) = 0$ and $\psi(L) = 0$. The only one of the given functions that satisfies this condition for $x = 0$ is $\psi = \sin kx$, where k is a constant. By choosing k, we may fit these functions in such a way that the boundary conditions are satisfied. The remaining functions are of no use. We thus choose k so that $\psi(L) = 0$. Since $\sin(kx) = 0$ for $kx = n\pi$, where n is an integer number, we must have $k = n\pi/L$.

Since $\sin(-kx) = -\sin(kx)$, we obtain the same wave function since the sign is of no interest. Furthermore, $n = 0$ implies $\psi = 0$. We therefore ignore all solutions where n is the negative integer number or equal to zero. The solutions of SE in the case of "particle in the box" are thus

$$E = E_n = \frac{\hbar^2k^2}{2m} = \frac{\hbar^2\pi^2}{2mL^2}n^2, \quad n = 1,2,\dots \tag{1.17}$$

$$\psi_n = \sin(n\pi x/L). \tag{1.18}$$

Equations 1.17 and 1.18 define *stationary states* and standing waves. The ground state corresponds to $n = 1$ and the excited states to $n > 1$.

The states in Equation 1.17 are one-electron states. The time-independent SE can also be an equation for a wave function that depends on the position of several electron coordinates or nuclear coordinates. If the particles are electrons (or fermions in general), they obey the *Pauli principle*, according to which at most two electrons, with different spins, are allowed to have the same spatial wave function.

The functions in Equation 1.18 should have a magnitude equal to unity and be *normalized*:

$$\int \psi^*\psi dx = 1. \tag{1.19}$$

Thus, set $\psi = \psi_n$ and calculate Equation 1.20:

$$\int_0^L \psi^* \psi dx = \int_0^L \sin^2(n\pi x/L)dx = \int_0^L \frac{1-\cos(2n\pi x/L)}{2}dx = \left|\frac{x}{2}\right|_0^1 = \frac{L}{2}. \quad (1.20)$$

Then, multiply the wave functions of Equation 1.18 by the square root of the reciprocal of the result [$\sqrt{(2/L)}$], and Equation 1.19 will be satisfied. These functions are shown in Figure 1.6 for $n = 1$ and $n = 2$.

If the particle in a box model is used in two dimensions for electrons that are assumed to be freely mobile in the π system of a planar aromatic molecule, the excitation energies ($E_m - E_n$) are obtained, which agree quite well with the energy of spectral lines in the visible or UV part of the spectrum. It is certainly not to be expected that a perfect agreement is obtained, since $V = 0$ can only hold in a very approximate way. Anyway, this free-electron model is a rather good model for π systems. The electrons are not really free, but bound to the molecule, since $V = \infty$ outside.

If the box model is applied to molecules instead of electrons and the extension of the box (L) has macroscopic dimensions, the energy differences between the lower energy levels become very small in Equation 1.17. It is then enough with a very small energy, for example, in the form of heat, to excite the electrons to higher levels. For heavy particles, the quantization is thus without any importance since almost any energy is possible. Quantization is most important for small particles such as electrons.

Summarizing, there are many solutions to a time-independent SE, but only those that satisfy the boundary conditions correspond to reality. The boundary conditions lead to energy quantization. This has far-reaching consequences in chemistry. For example, a chemical compound has a well-defined free energy, which may be calculated from the eigenvalues. The free energy for a molecule in a confinement is always quantized. Very often, however, the boundary conditions are automatically included. This is the case in quantum chemical calculations with basis functions. Only sums and products of exponential functions that automatically satisfy the boundary conditions are used.

1.4.3 Finite Walls, Tunneling

In classical mechanics, the particle moves between the walls with constant probability. It is assumed to bounce back at the wall, retaining the same velocity in the opposite direction. The classical particle can only reach regions where its kinetic energy is higher than the potential energy $V(x) = V_0(x)$. If the potential $V_0(x)$ is finite in a certain region (Figure 1.7), Equation 1.15 in that region may be written as

$$\frac{\hbar^2}{2m}\frac{d^2\psi}{dx^2} = (V_0 - E)\psi \Rightarrow \psi = \exp(\pm kx) \quad \text{where } k = \sqrt{2m(V_0 - E)}/\hbar. \quad (1.21)$$

If $E < V_0$, this is the only allowed solution to the time-independent SE in the right region of Figure 1.7. For $E > V_0$, Ψ is a complex exponential function, typical for currents.

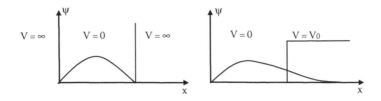

FIGURE 1.7   Particle in a well with infinite walls and one finite wall.

In the left part of Figure 1.7, the solution is one of those corresponding to $V = 0$ obtained above. In the right part of Figure 1.7, the first derivative is continuous at the right turning point, since V is not infinite. This condition allows us to obtain a solution. The wave function cannot be zero for $E < V_0$, but tends smoothly to zero with increasing x.

It is easy to realize that if there is another allowed region to the right of the barrier, the wave function continues there with the value of ψ in the barrier. In the quantum mechanical description, a particle may thus penetrate a forbidden region.

This unexpected quantum mechanical passing of the energy barrier is referred to as *tunneling* or the *tunneling effect*. The term was introduced by the Russian physicist Gamow, who was known to have a good sense of humour. The reason for the name "tunneling" is that if the energy of the particle is insufficient to go above the barrier, there is still a possibility to penetrate through the barrier in a "tunnel." Gamow was particularly interested in nuclear decay, where α-, β-, or γ-particles are emitted. An unstable isotope has a limited life time, that is, if we wait long enough it will decay by tunneling of an α-, β-, or γ-particle through the barrier.

In chemistry, there are a great number of processes where the tunneling phenomenon is important. It may be necessary to use a more accurate treatment than the standard tunneling model, however.

1.4.4   Interpretation of the Wave Function

EM waves passing through two parallel slits interfere. This is also the case for matter waves; but what is interfering? Matter waves are positive in some regions and negative in others, but matter itself cannot be negative, of course. So what is the meaning of matter waves and the quantum mechanical wave function ψ? It was Max Born, head of the famous theoretical physics group in Göttingen, Germany, who provided the now accepted interpretation. According to him, it is the absolute square of the wave function:

$$\rho(\vec{r}) = |\psi|^2 = \psi^*\psi, \tag{1.22}$$

which means the charge density of the electron. At the same time, the meaning of Equation 1.22 is a *probability density* of the electron. Consequently, $\psi^*\psi$ integrated over the area A,

$$\int_A \psi^*\psi \, dx \, dy \, dz, \tag{1.23}$$

means the probability to find the electron within the region A. If A is the whole space, this integral has to be equal to unity since the total probability to find the electron is equal to unity. The probability density $|\psi|^2$ is always positive. The wave function ψ itself is complex and shows interference with other wave functions.

It follows that the contribution of electrons to the dipole moment $\bar\mu$ is

$$\mu_x = e\int_A x\rho(\bar r)\,dx\,dy\,dz; \quad \mu_y = e\int_A y\rho(\bar r)\,dx\,dy\,dz; \quad \mu_z = e\int_A z\rho(\bar r)\,dx\,dy\,dz, \tag{1.24}$$

where A is the whole space (R^3). The nuclei may be assumed to be point-like. Their contribution to the dipole moment is

$$\mu_N = |e|\sum_I Z_I \bar R_I. \tag{1.25}$$

The wave function is thus probability *amplitude*. The remarkable thing is that by superposing plane waves (adding probability amplitudes), we can obtain a new wave function whose absolute square also means a probability density.

If plane waves are added, they add up in some regions and cancel in others:

$$\psi(x) = \sum_n c_n \exp(p_n 2\pi i x/h),$$

where $p_n = p + n\delta p$; Assume

$$c_n \neq 0 \quad \text{for} -\frac{\Delta p}{2\delta p} < n < \frac{\Delta p}{2\delta p}. \tag{1.26}$$

One may show that if $n \to \infty$, ψ is concentrated in a region roughly of magnitude $\Delta x = \hbar/\Delta p$. Δp is the *uncertainty* in p, since the sum in Equation 1.26 contains terms with $c_n \neq 0$ in an interval Δp. It is possible to show the *uncertainty relation* of Heisenberg:

$$|\Delta x||\Delta p| \geq \hbar/2, \tag{1.27}$$

where Δx means the uncertainty in position and Δp means the uncertainty in momentum. Δp has an exact definition (Ψ is normalized):

$$\Delta p = \sqrt{\int \Psi^* p^2 \Psi dV - \left(\int \Psi^* p \Psi dV\right)^2}. \tag{1.28}$$

In the extreme case of a plane wave, it is easily shown that $\Delta p = 0$ (insert the expression for a plane wave in Equation 1.10 and subsequently the latter in Equation 1.28). For $\psi = \exp(2\pi ix/\lambda)$, $\psi^*\psi = 1$ everywhere. The interpretation is that the electron has a well-defined momentum but a completely undetermined position.

In the other extreme case, the electron has a well-defined position, say the origin. We have to add plane waves over an infinite range of $|p|$ to obtain a sharp peak.

An electron in a molecule cannot leave the molecule without adding energy. In a Fourier expansion, we need as many plane waves with a positive k as with a negative k. The second term under the square root sign in Equation 1.28 is again equal to zero.

The expression for the kinetic energy, $p^2/2m = (hn)^2/(2L)^2/2m$, of the particle in the box may be used to derive the uncertainty in momentum; p may be positive or negative. We find $\Delta p = 2|p| = hn/L$. Δx is simply equal to the width of the box, L. Finally, we obtain

$$|\Delta x||\Delta p| = L\frac{hn}{L} = hn > \hbar/2. \qquad (1.29)$$

Equation 1.29 shows that an electron in a molecule satisfies the Heisenberg uncertainty relation if its wave function satisfies time-independent SE with suitable boundary conditions. The product of uncertainties, $|\Box x||\Box p|$, is larger in the excited states. Equation 1.29 may be given a more general proof, but for this we refer to textbooks in quantum mechanics.

The Schrödinger wave thus explains the Heisenberg uncertainty relations. Particle properties are found if the wave packet has a small extension in space. $|\Box p|$ then has to be large. On the other hand, it may be shown that the average momentum and average acceleration are that of the classical particle.

If $|\Box p|$ is small, the wave properties are dominant and the wavelength is quite well defined. Interference is possible. This is shown in the double slit experiment (Figure 1.4). A classical particle would pass through either of the two slits and the probability distribution would not be affected if one of the slits were covered. A wave, on the other hand, will go through both slits. An interference pattern appears, as if two waves simultaneously spread from both slits. Since this experiment can be done with electrons, electrons are not true particles. In the same sense, it is incorrect to say that light consists of particles (phonons). As we will see in Chapter 12, the laser experiment, where a time resolution of 1 fs or less can be achieved, is dependent on the fact that the "photon splits" and can reach the target along two different pathways. This is impossible for a classical particle, but not at all strange in the case of waves.

Many (N) electrons *cannot* be described by normalizing to N instead of to unity. Two electrons are described by a function of the form $f(x_1, x_2)$, where x_1 and x_2 are the independent variables for the two electrons. From this, it follows that the particles can always be counted using integer numbers.

A mathematical description for electrons, phonons (vibrational quanta), and photons exists and is referred to as "second quantization." "First quantization" amounts to the ordinary quantization for a single particle, discussed so far in this chapter, which is a consequence of the boundary conditions in a given experiment. The EM field is simply described as a classical field in the sense of Maxwell's equations. Second quantization is a formalism to "keep track" of the number of particles. Excitation or de-excitation of the field is quantized and therefore the field is quantized. The particles of second quantization are referred to as "almost particles" (*quasi-particles*).

1.5 MATHEMATICAL BACKGROUND

1.5.1 EIGENVALUES AND EIGENFUNCTIONS

Summarizing, observable physical entities (observables) are represented by an operator. For example, momentum in three-dimensional space corresponds to the operator:

$$\hat{\mathbf{p}} = -i\hbar \left(\frac{\partial}{\partial x}, \frac{\partial}{\partial y}, \frac{\partial}{\partial y} \right). \tag{1.30}$$

The "roof" symbol is sometimes used to denote an operator, but it may also be left out. If the result of the action of the operator is just multiplication of the function by a constant ω, that is,

$$\hat{\Omega}\Psi = \omega\Psi, \tag{1.31}$$

we say that the wave function ψ is an *eigenfunction* of the operator Ω with the *eigenvalue* ω. Ψ may be normalized, and then Ψ is determined up to a complex exponential, $\exp(i\phi)$.

It is also possible that two different functions, Ψ_1 and Ψ_2, have the same eigenvalue even if they are not proportional. Ψ_1 and Ψ_2 are said to be *degenerate*.

ω is said to be a sharp value of the observable, if Equation 1.31 is satisfied. Comparing with Equation 1.13, we see that the operator $-i\hbar(d/dx)$ has the eigenfunction $\exp(2\pi i x/\lambda)$ with the eigenvalue $p = h/\lambda$. Thus, h/λ is the sharp value of the momentum in this case (the de Broglie case).

The function $\exp(2\pi i x/\lambda)$ is not an eigenfunction of the position operator x, since $x \cdot \exp(2\pi i x/\lambda)$ is not the same function as $\exp(2\pi i x/\lambda)$. This is consistent with the fact that the operators of momentum and position do not commute:

$$\left[-i\hbar \frac{\partial}{\partial x}, x \right] = -i\hbar \left(\frac{\partial}{\partial x} x - x \frac{\partial}{\partial x} \right) = -i\hbar. \tag{1.32}$$

If the two operators have common eigenfunctions, the multiplication order does not matter, and hence their commutator has to be equal to zero. This is not the case for the momentum and position operators in Equation 1.32.

If Ω corresponds to kinetic energy, the eigenvalue equation is given by Equation 1.14. The operator for potential energy is defined by simply multiplying V(x), thus $\Omega\psi(x) = V(x)\psi(x)$. In this case, Equation 1.31 is fulfilled only if V(x) is a constant.

The total Hamiltonian operator is H = T + V, or (in one dimension):

$$\hat{H} = \hat{T} + \hat{V} = -\frac{\hbar^2}{2m}\frac{d^2}{dx^2} + V(x). \tag{1.33}$$

The eigenvalue equation of the Hamiltonian is identical to the time-independent SE:

$$\hat{H}\psi = E\psi. \tag{1.34}$$

The eigenvalue E is equal to the total energy (kinetic energy plus potential energy) of the system, thermodynamically to be compared with the enthalpy. Since ψ has to satisfy the boundary conditions, only discrete energy values are allowed (energy quantization).

1.5.2 HERMITEAN OPERATORS

A quantum mechanical operator has to satisfy certain conditions to represent an observable. For superposition to make sense, it must be a linear operator:

$$\hat{\Omega}(C_1\psi_1 + C_2\psi_2) = C_1\hat{\Omega}\psi_1 + C_2\hat{\Omega}\psi_2. \tag{1.35}$$

Furthermore, for the operator $\hat{\Omega}$ to have real eigenvalues, it has to be a Hermitean operator with the property

$$\int \Psi^* \hat{\Omega}\Phi dv = \int (\hat{\Omega}\Psi)^* \Phi dv. \tag{1.36}$$

The integration may be written using the braket $\langle\ \rangle$ symbol:

$$\langle \Psi | \hat{\Omega} | \Phi \rangle = \int \Psi^* \hat{\Omega}\Phi dv. \tag{1.37}$$

The integration region has been left out in Equations 1.36 and 1.37. It is understood that the integration is carried out over the region of space where the wave functions are different from zero.

1.5.3 EXPECTATION VALUE

If electronic wave packets are going to behave as particles, their wave packets must have properties like velocity and acceleration. A wave packet is not necessarily an eigenfunction to the operator related to a certain property or observable, for example, angular momentum or kinetic energy. A distribution of values is observed. The *expectation value* is defined as a kind of average value for the region of space where the electron can be found:

$$\langle \hat{\Omega} \rangle = \frac{\langle \psi | \hat{\Omega} | \psi \rangle}{\langle \psi | \psi \rangle} = \frac{\int \psi^* \hat{\Omega} \psi dx}{\int \psi^* \psi dx}. \tag{1.38}$$

The number calculated is a real value, called the expectation value of $\hat{\Omega}$, since $\hat{\Omega}$ is a Hermitean operator. If Ψ is an eigenfunction of $\hat{\Omega}$, the expectation value is equal to the eigenvalue, as is easily seen in Equation 1.38.

If the function wave Ψ is normalized, the denominator in Equation 1.38 may be left out. Plane waves extended over the whole space cannot be normalized, of course.

A set of functions $\{\Psi_i\}$ is said to be *orthogonal* or *non-overlapping* if

$$\langle \Psi_i | \Psi_j \rangle = \int \Psi_i^* \Psi_j dx = 0 \quad \text{for } i \neq j. \tag{1.39}$$

The overlap between two functions, Ψ_i and Ψ_j, is called the overlap integral. If the integral of Equation 1.39 is $= 1$ for $i = j$ and $= 0$ for $i \neq j$, the set of functions $\{\Psi_i\}$ is said to be *orthonormal*.

1.5.4 SEPARATION OF VARIABLES

A frequent case is that the wave function depends on two independent variables, x_1 and x_2, and that the Hamiltonian can be written as a sum:

$$H = H_1 + H_2. \tag{1.40}$$

Quite generally, a function of two variables may be expanded in a series of the type

$$\psi(x_1, x_2) = \sum_{i,j} C_{ij} \varphi_i(x_1) \chi_j(x_2). \tag{1.41}$$

We now try to factorize ψ, that is we try to include only one term in Equation 1.41 and set $\psi(x_1, x_2) = \varphi(x_1)\chi(x_2)$. If ψ were an eigenfunction of H, we would have

$$(H_1 + H_2)\varphi(x_1)\chi(x_2) = [H_1\varphi(x_1)]\chi(x_2) + \varphi(x_1)[H_2\chi(x_2)] = E\varphi(x_1)\chi(x_2). \tag{1.42}$$

In the operation $H_1\varphi(x_1)\chi(x_2)$, $\chi(x_2)$ is a constant. Similarly, in $H_2\varphi(x_1)\chi(x_2)$, $\varphi(x_1)$ is a constant. Hence, there are only two terms in the central member.

If Equation 1.42 is divided by $\varphi(x_1)\chi(x_2)$, we obtain, after some calculation

$$\frac{H_1\varphi(x_1)}{\varphi(x_1)} = -\frac{H_2\chi(x_2)}{\chi(x_2)} + E. \tag{1.43}$$

The left member of Equation 1.43 only depends on x_1, while the right member only depends on x_2. The right and the left members may thus be varied independently of each other; therefore Equation 1.43 is valid only if both members are equal to a constant K. Hence,

$$\frac{H_1\varphi(x_1)}{\varphi(x_1)} = K; \quad -\frac{H_2\chi(x_2)}{\chi(x_2)} + E = K,$$

which may be rewritten as

$$H_1\varphi(x_1) = K\varphi(x_1);\dagger \quad H_2\chi(x_2) = (E-K)\chi(x_2). \tag{1.44}$$

We have shown that both φ and χ are eigenfunctions of H_1 and H_2, respectively.

It is easy to show the reverse theorem. If φ and χ are eigenfunctions of H_1 and H_2, respectively, $\varphi(x_1)\chi(x_2)$ is an eigenfunction of $H_1 + H_2$, as is shown by operating with $H_1 + H_2$ on $\varphi(x_1)\chi(x_2)$.

The theorem may be generalized to more than two variables, that is, if the Hamiltonian is a sum of three or more terms, each acting on functions of a single variable. Variable 1 is separated first and subsequently variable 2, from the others.

An important example of separation of variables is the three-dimensional box. We already know the solution of the one-dimensional box. The extension of the box is $\{L_x, L_y, L_z\}$. The Hamiltonian separates in x, y, and z. The time-independent SE is

$$-\frac{\hbar^2}{2m}\left(\frac{d^2\psi}{dx^2} + \frac{d^2\psi}{dy^2} + \frac{d^2\psi}{dy^2}\right) = -\frac{\hbar^2}{2m}\Delta\Psi = E\Psi, \tag{1.45}$$

(the sum of the second derivative, the operator Δ, is called the Laplace operator). E is the sum:

$$E = E_x + E_y + E_z, \tag{1.46}$$

where, for example, E_x is a solution of the one-dimensional SE:

$$-\frac{\hbar^2}{2m}\left(\frac{d^2\psi}{dx^2}\right) = E_x\psi. \tag{1.47}$$

The wave function may be written as

$$\psi(x,y,z) = \left(\frac{8}{L_xL_yL_z}\right)^{1/2}\sin\frac{n_x\pi x}{L_x}\cdot\sin\frac{n_y\pi y}{L_y}\cdot\sin\frac{n_z\pi z}{L_z}. \tag{1.48}$$

The normalized eigenfunction is a product of the normalized one-dimensional eigenfunctions.

1.6   VARIATION PRINCIPLE: LINEAR EXPANSION

The variation principle is used to get the "best possible" approximation of a function. In almost all applications of "quantum chemistry," the coefficients are optimized in a *linear* expansion. In Chapter 2, we will use the variation principle to derive the Hartree–Fock equation.

1.6.1   ENERGY EXPECTATION VALUE IS $\geq E_0$, THE LOWEST EIGENVALUE OF H

Assume that we know the eigenfunctions $\{\Psi_i\}$ and eigenvalues $\{E_i\}$ of the Hamiltonian H. We also assume that an arbitrary function Φ can be expanded in terms of these eigenfunctions (in principle):

$$\Phi = \sum_i C_i \Psi_i. \tag{1.49}$$

The set $\{\Psi_i\}$ is assumed to be orthonormal. The overlapping case is treated in Appendix 5. We require from Φ that it is normalized:

$$\langle \Phi | \Phi \rangle = \sum_i |C_i|^2 = 1. \tag{1.50}$$

We will only show that the energy difference δE between the Hamiltonian expectation value for Φ and the lowest energy eigenvalue of H is always positive. The difference may be written as

$$\delta E = \int \Phi^* H \Phi d\tau - E_0. \tag{1.51}$$

Using Equations 1.49 through 1.51, we obtain

$$\delta E = \left\langle \sum_i C_i \Psi_i \middle| (H - E_0) \sum_j C_j \Psi_j \right\rangle = \sum_{i,j} C_i^* C_j (E_j - E_0) \langle \Psi_i | \Psi_j \rangle = \sum_i |C_i|^2 (E_i - E_0). \tag{1.52}$$

Since $E_i \geq E_0$, we obtain from Equation 1.52 that

$$\delta E \geq 0. \tag{1.53}$$

This proves the theorem and defines the term "best possible." In optimization, we can never end up with an energy expectation value *smaller* than the lowest exact eigenvalue of H. The expectation value is always an *upper bound* to the true eigenvalue and this is the first part of the variation principle.

The second part of the variation principle demonstrates that the error in the energy goes to zero as the square of the error in the wave function. The proof is left out.

1.6.2 Linear Expansion

The coefficients in Equation 1.49 are calculated in Appendix 5. Here, we will use a simplified derivation. Φ in Equation 1.49 satisfies SE if $H\Phi = E\Phi$. In reality, the best Φ in Equation 1.49 gives

$$H\Phi(x) = \sum_i C_i H\Psi_i(x) \approx E\Phi. \tag{1.54}$$

If we manage to find a good linear combination of N functions, we may straighten out $\approx$, and consider Φ an eigenfunction of H. We multiply Equation 1.54 with Ψ_j^* and integrate

$$\sum_i \int \Psi_j^* H\Phi dx = E\sum_i C_i \int \Psi_j^* \Psi_i dx = E\sum_i C_i \delta_{ij} = EC_j. \tag{1.55}$$

If we call $H_{ij} = \int \Psi_j^* H\Psi_i dx$, we find that Equation 1.55 is a matrix eigenvalue problem:

$$\begin{pmatrix} H_{11} & H_{12} & \cdots & H_{1N} \\ H_{21} & H_{22} & \cdots & H_{2N} \\ \vdots & \vdots & \vdots & \vdots \\ H_{N1} & H_{N2} & \cdots & H_{NN} \end{pmatrix} \begin{pmatrix} C_1 \\ C_2 \\ \vdots \\ C_N \end{pmatrix} = E \begin{pmatrix} C_1 \\ C_2 \\ \vdots \\ C_N \end{pmatrix}. \tag{1.56}$$

The matrix H with the matrix elements H_{ij} is called the *Hamiltonian matrix*. The coefficient matrix is obtained by diagonalization.

A nontrivial solution (not all $C_i = 0$) requires that the determinant of $H - E$, where E is the diagonal eigenvalue matrix, is equal to zero. Evaluation of this matrix gives a polynomial P(E). The *secular equation* P(E) = 0 gives the eigenvalues and guarantees a nontrivial solution. The eigenvalues E_i satisfy the secular equation.

1.7 SPIN

In 1924, Pauli suggested that a fourth quantum number is necessary to explain the properties of the atoms in the periodic table. In the original version of his *exclusion principle*, no two electrons can have the same set of quantum numbers. At this time, Pauli did not want to connect the fourth quantum number to a spin of the particle since a spinning particle should have extension in space and this would lead to further inconsistencies. On the other hand, the mere existence of the spin seemed to explain many features of atomic spectra, as was discovered but not published by Kronig (since Kronig happened to be Pauli's assistant). Instead, two other Dutchmen, Samuel Goudsmit and George Uhlenbeck, are considered as the discoverers of electron spin.

Goudsmit and Uhlenbeck explained multiplets in atomic spectra with the help of the spin of the electron. For example, the well-known splitting of the yellow 3s–3p line in the spectrum of Na was explained by the existence of two quantum states for 3p. The state energies depend on whether the electron spin acts with or against the orbital spin. The magnetic moment of the spinning electron and the magnetic moment of the electron in its orbit interact, and this interaction is referred to as *spin-orbit coupling*.

1.7.1   SPIN OF A SINGLE ELECTRON

The solution to the spin problem followed when Dirac generalized the SE to a relativistically invariant equation (1928). In the Dirac equation, the spin-orbit coupling appears as a natural part of the formalism. The Dirac wave function ψ has four components, but only two of them refer to electrons. The other two coordinates are positron coordinates. It is reasonable to assume that our particle is 100% an electron, with no probability density for being a positron.

The electronic components correspond to electronic spin. We may assume that the spin is either up or down:

$$\psi = \begin{pmatrix} \varphi^\uparrow \\ \varphi^\downarrow \end{pmatrix} = \varphi^\uparrow \begin{pmatrix} 1 \\ 0 \end{pmatrix} + \varphi^\downarrow \begin{pmatrix} 0 \\ 1 \end{pmatrix}. \tag{1.57}$$

$\varphi^\uparrow$ and $\varphi^\downarrow$ are ordinary spatial orbitals, that is, functions of a point in space $\vec{r} = (x, y, z)$. The arrows are just indices in the Dirac equation, suggestive of the role of $\varphi^\uparrow$ and $\varphi^\downarrow$ as the spatial part of the wave functions for "spin-up" or "spin-down" electrons. An orbital with spin that can have spin-up or spin-down in this way is called a *spin orbital*.

It is important to check that the new object in Equation 1.57 has the same mathematical properties as an ordinary wave function. In the scalar product $\langle \psi_1 | \psi_2 \rangle$, we need to replace the complex conjugate of the simple function ψ_1 by the "transposed complex conjugate" of the vector representing ψ_1:

$$\langle \psi_1 | \psi_2 \rangle = \int \begin{pmatrix} \varphi_1^{\uparrow *} & \varphi_1^{\downarrow *} \end{pmatrix} \begin{pmatrix} \varphi_2^\uparrow \\ \varphi_2^\downarrow \end{pmatrix} dv = \langle \varphi_1^\uparrow | \varphi_2^\uparrow \rangle + \langle \varphi_1^\downarrow | \varphi_2^\downarrow \rangle. \tag{1.58}$$

From this equation, we may conclude that spin-up and spin-down wave function components never mix.

We now need operators that operate on row vectors. An operator for a row vector is a 2×2 square matrix. In addition to the unit matrix e, Dirac used the *Pauli spin matrices*:

$$e = \begin{pmatrix} 1 & 0 \\ 0 & 1 \end{pmatrix}; \quad \sigma_x = \begin{pmatrix} 0 & 1 \\ 1 & 0 \end{pmatrix}; \quad \sigma_y = \begin{pmatrix} 0 & -i \\ i & 0 \end{pmatrix}; \quad \sigma_z = \begin{pmatrix} 1 & 0 \\ 0 & -1 \end{pmatrix}. \quad (1.59)$$

These matrices (operators) play an important role in the solution of the Dirac equation. Notice that the squares of all four operators are all equal to the unit operator. σ_z operates in the following way:

$$\sigma_z \begin{pmatrix} 1 \\ 0 \end{pmatrix} = \begin{pmatrix} 1 & 0 \\ 0 & -1 \end{pmatrix} \begin{pmatrix} 1 \\ 0 \end{pmatrix} = \begin{pmatrix} 1 \\ 0 \end{pmatrix} \quad \sigma_z \begin{pmatrix} 0 \\ 1 \end{pmatrix} = \begin{pmatrix} 1 & 0 \\ 0 & -1 \end{pmatrix} \begin{pmatrix} 0 \\ 1 \end{pmatrix} = (-1) \begin{pmatrix} 0 \\ 1 \end{pmatrix}. \quad (1.60)$$

Apparently, the two unit 1×2 row operators are eigenvectors of σ_z with eigenvalues 1 and -1, respectively.

It is easy to show that the following commutation relations hold:

$$\sigma_x \sigma_y - \sigma_y \sigma_x = 2i\sigma_z,$$

$$\sigma_y \sigma_z - \sigma_z \sigma_y = 2i\sigma_x, \quad (1.61)$$

$$\sigma_z \sigma_x - \sigma_x \sigma_z = 2i\sigma_y.$$

The Pauli spin operators ($\vec{s}$) may be defined as

$$\vec{s} = (s_x, s_y, s_z) = \frac{\hbar}{2} (\sigma_x, \sigma_y, \sigma_z). \quad (1.62)$$

Inserting this in Equation 1.61, we obtain

$$s_x s_y - s_y s_x = i\hbar s_z,$$

$$s_y s_z - s_z s_y = i\hbar s_x, \quad (1.63)$$

$$s_z s_x - s_x s_z = i\hbar s_y.$$

These commutation relations are the same as for angular momentum (Appendix 1).

Next, we will use atomic units where $\hbar = 1$.

1.7.2 Properties of Spin Functions

There is great advantage in writing a spin orbital as a function rather than a row vector. We therefore introduce spin functions α and β:

$$\phi(\vec{r}) \begin{pmatrix} 1 \\ 0 \end{pmatrix} \leftrightarrow \phi(\vec{r})\alpha(\varsigma); \quad \phi(\vec{r}) \begin{pmatrix} 0 \\ 1 \end{pmatrix} \leftrightarrow \phi(\vec{r})\beta(\varsigma). \quad (1.64)$$

The function value need not be specified, since α and β components should not be added as functions anyway. However, the spin functions have to be orthonormalized and assuming that the spatial function φ is already orthonormalized, we must have

$$\int \alpha(\varsigma)^* \alpha(\varsigma)d\varsigma = 1; \quad \int \beta(\varsigma)^* \beta(\varsigma)d\varsigma = 1; \quad \int \alpha(\varsigma)^* \beta(\varsigma)d\varsigma = 0; \quad \int \beta(\varsigma)^* \alpha(\varsigma)d\varsigma = 0.$$

(1.65)

If these rules are obeyed, nothing bad will happen.

From Equations 1.60 and 1.62, we obtain

$$s_z \alpha = \frac{1}{2}\alpha; \quad s_z \beta = -\frac{1}{2}\beta.$$

(1.66)

The projection of spin along the z-axis, as is in fact the meaning of s_z, and the square of the angular momentum $(\vec{s} \cdot \vec{s})$ (we write this as $\vec{s}^2$) correspond to conserved observables. We thus have to construct spin functions that are eigenfunctions also to the square of angular momentum.

We now introduce the auxiliary operators:

$$s^{\pm} = s_x \pm i s_y.$$

(1.67)

According to our definition of the spin operators (Equation 1.62), we have

$$s^+ \alpha = 0; \quad s^+ \beta = \alpha; \quad s^- \alpha = \beta; \quad s^- \beta = 0.$$

(1.68)

Equation 1.68 is easy to prove if the matrices for s^+, s^-, α, and β are written out. Because of Equation 1.68, $s^{\pm}$ are referred to as *step-up* and *step-down* operators, respectively.

Using Equations 1.66 through 1.68, we define an operator $\hat{s}^2$ as follows and obtain

$$\hat{s}^2 = s_x^2 + s_y^2 + s_z^2 = s^- s^+ + s_z^2 + s_z \quad \text{and} \quad \hat{s}^2 = s^+ s^- + s_z^2 - s_z.$$

(1.69)

Hence,

$$\hat{s}^2 \alpha = s^- s^+ \alpha + s_z^2 \alpha + s_z \alpha = 0 + \frac{1}{2} \cdot \frac{1}{2}\alpha + \frac{1}{2}\alpha = \frac{3}{4}\alpha,$$

(1.70)

and similarly for β. Hence, α and β are eigenfunctions of both s_z and s^2.

In summary, there are two spin functions: α for $m_s = 1/2$ and β for $m_s = -1/2$. Both are eigenfunctions of $\hat{s}^2$ with the eigenvalue 3/4. If we introduce the spin quantum number s (at this stage, there is an operator $\hat{s}$, a vector $\vec{s}$, and a number s), we find that the eigenvalue in Equation 1.70 is $s(s + 1) = 1/2(1/2 + 1) = 3/4$. The eigenvalue always turns out to be $s(s + 1)$.

The spin functions may be generalized to many-electron spin functions; $\alpha(\zeta_1)$ $\alpha(\zeta_2)$, $\alpha(\zeta_1)\beta(\zeta_2)$, $\beta(\zeta_1)\alpha(\zeta_2)$, and $\beta(\zeta_1)\beta(\zeta_2)$ are all spin functions for two electrons. In a product, the independent variables are assumed to come in the order: first 1 and then 2. The eigenvalues of the operator $S_z = s_{z1} + s_{z2}$ are obviously 1, 0, 0, and −1, for the spin functions $\alpha\alpha$, $\alpha\beta$, $\beta\alpha$, and $\beta\beta$, respectively. For two electrons, we have: $S^+ = s_1^+ + s_2^+$ and $S^- = s_1^- + s_2^-$. It is customary to write $\vec{s} = (s_x, s_y, s_z)$ for one single electron and with capitals $S(S_x, S_y, S_z)$ in the case of many electrons. Since for N electrons, $S_z = s_{z1} + s_{z2} + \cdots + s_{zN}$, M is equal to the number of α spin functions minus the number of β spin functions, multiplied by 1/2.

The square of the operator $\vec{S}$ may now be obtained as

$$\hat{S}^2 = S_x^2 + S_y^2 + S_z^2 = S^-S^+ + S_z^2 + S_z = S^+S^- + S_z^2 - S_z. \qquad (1.71)$$

For two electrons, the spin functions are $\alpha\alpha$, $\alpha\beta$, $\beta\alpha$, and $\beta\beta$. Starting from $\beta\beta$, we obtain

$$S^+\beta\beta = (\alpha\beta + \beta\alpha). \qquad (1.72)$$

Acting on Equation 1.72 by S^-, we obtain

$$(s_1^- + s_2^-)(\alpha\beta + \beta\alpha) = \left[s_1^-(\alpha\beta + \beta\alpha) + s_2^+(\alpha\beta + \beta\alpha) \right] = \alpha\alpha + \alpha\alpha = 2\alpha\alpha. \quad (1.73)$$

Finally, we act by the operator S^2, defined in Equation 1.71, on $\alpha\alpha$, $\alpha\beta + \beta\alpha$, and $\beta\beta$, and find that all these functions are eigenfunctions with the eigenvalue equal to $2 = 1(1 + 1)$.

We notice that the eigenvalues in Equation 1.70 for a single electron and Equation 1.73 for two electrons in both cases can be written as $S(S + 1)$, where $S = 1/2$ in the first case and $S = 1$ in the second case. Both these numbers are equal to the largest possible M_S quantum number in the respective cases and this conforms to the general result for angular momentum.

Hence, after normalization we obtain the spin functions for $S = 1$:

$$S = 1; \ M_S = 1: \quad \alpha\alpha$$

$$S = 1; \ M_S = 0: \quad (\alpha\beta + \beta\alpha)/\sqrt{2} \qquad (1.74)$$

$$S = 1; \ M_S = -1: \quad \beta\beta.$$

The spin functions $\alpha\beta$ and $\beta\alpha$ may also form a linear combination $\alpha\beta - \beta\alpha$, which is orthogonal to the previous one, $\alpha\beta + \beta\alpha$. It is obvious that the eigenvalue of S_z is equal to zero for the former spin function. Changing all + signs to − signs in Equation 1.73, we find that $\alpha\beta - \beta\alpha$ is an eigenfunction to $\hat{S}^2$ with the eigenvalue 0 (equal to $S(S + 1)$ if $S = 0$). Hence, the two-electron spin function for $S = 0$ is

$$S = 0; \quad M_S = 0: \quad (\alpha\beta - \beta\alpha)/\sqrt{2}. \tag{1.75}$$

For three electrons, the total number of spin functions is obviously $2^3 = 8$. The highest possible M_S is equal to 3/2 and the eigenfunction $\alpha\alpha\alpha$. There are four spin functions with $S = 3/2$. The remaining spin functions correspond to the lower S-value M_S with equal to 1/2 and $-1/2$. Apparently, there are two spin functions with $S = 1/2$ and $M_S = 1/2$. One of them can be chosen as $\alpha\beta\alpha - \beta\alpha\alpha$, since this function is obviously orthogonal to $\alpha\beta\alpha + \beta\alpha\alpha + \alpha\alpha\beta$. It also fits well to the Li atom consisting of a K-shell with $\alpha\beta - \beta\alpha$, multiplied by α from the right. The third spin function for $S = 1/2$ and $M_S = 1/2$ is found if we choose one that is orthogonal to the other two: $2\alpha\alpha\beta - \beta\alpha\alpha - \alpha\beta\alpha$. It is easy to check that the eigenvalue of S^2 for the latter two spin functions is $S = 1/2$. In summary, for three electrons we obtain

$$
\begin{aligned}
S &= 3/2; \quad M_S = 3/2: \quad \alpha\alpha\alpha \\
S &= 3/2; \quad M_S = 1/2: \quad \alpha\alpha\beta + \alpha\beta\alpha + \beta\alpha\alpha \\
S &= 3/2; \quad M_S = -1/2: \quad \alpha\beta\beta + \beta\alpha\beta + \beta\beta\alpha \\
S &= 3/2; \quad M_S = -3/2: \quad \beta\beta\beta
\end{aligned}
\tag{1.76}
$$

and

$$
\begin{aligned}
S &= 1/2; \quad M_S = 1/2: \quad \alpha\beta\alpha - \beta\alpha\alpha, \ 2\alpha\alpha\beta - \alpha\beta\alpha - \beta\alpha\alpha \\
S &= 1/2; \quad M_S = -1/2: \quad \alpha\beta\beta - \beta\alpha\beta, \ 2\beta\beta\alpha - \beta\alpha\beta - \alpha\beta\beta.
\end{aligned}
\tag{1.77}
$$

1.7.3 Spin Multiplicity

The total number of spin functions for N electrons is 2^N. We will make use of the well-known binomial expansion of $(a + b)^N$:

$$(a+b)^N = \binom{N}{0}a^N + \binom{N}{1}a^{N-1}b + \binom{N}{2}a^{N-2}b^2 + \cdots + \binom{N}{k}a^{N-k}b^k + \cdots + \binom{N}{N}b^N, \tag{1.78}$$

where the *binomial coefficients* are given by (if $k \le N$ and $0! = 1$)

$$\binom{N}{k} = \frac{N!}{(N-k)!k!}; \quad \binom{N}{0} = 1; \quad \binom{N}{1} = N. \tag{1.79}$$

If we set $a = 1$ and $b = 1$ in Equation 1.78, we find that

$$2^N = \sum_{k=0}^{N} \binom{N}{k}. \tag{1.80}$$

2^N is equal to the total number of spin functions for N electrons. For each of the two possibilities for the first electron, there are two for the spin of the second electron.

It is easy to prove that

$$\binom{N+1}{k} = \binom{N}{k-1} + \binom{N}{k}. \tag{1.81}$$

If the numbers under the summation sign of Equation 1.80 are ordered in rows, starting from $k = 0$, we obtain *Pascal's triangle* (Figure 1.8). Owing to Equation 1.81, each number in this triangle is the sum of the two closest numbers in the row above. Now let [1 1] in the line with $N = 1$ represent that α and β are the possible spin functions for $N = 1$ ($S = 1/2$). For $N = 2$ in the line below, we get $\alpha\alpha$ from α by multiplying by α; $\alpha\beta$ from α by multiplying by β; $\beta\alpha$ from β by multiplying α; and $\beta\beta$ from β by multiplying by β. These spin functions have M_S values -1, 0, 0, and 1, respectively, and are thus represented by the line [1 2 1].

For two electrons, there are one singlet, [0 1 0], and one triplet, [1 1 1], spin function possible. This case is very common, since we may apply it not only when the number of electrons is equal to two, but also in the case with two "active" electrons. For example, we may excite one electron from one molecular orbital to another. There are now two different orbitals for the spatial functions with one electron in each; however, for the spin functions, all possibilities have to be considered. They combine to the one singlet and three triplet functions given in Equations 1.74 and 1.75.

The next line corresponds to three electrons, decomposed as [1 3 3 1] = [1 1 1 1] + 2[0 1 1 0], which means that for three electrons there are one spin quartet and two spin doublets. Assuming that the electrons are in three different spatial orbitals, for example, 1s, 2s, and 3s for the Li atom, there are eight states possible in the spectrum, ordered on three different energy levels in the absence of a magnetic field.

```
                1
             1     1
          1     2     1
       1     3     3     1
    1     4     6     4     1
 1     5    10    10     5     1
1    6    15    20    15     6    1
1   7   21   35   35   21   7   1
1  8  28  56  70  56  28  8  1
```

FIGURE 1.8 Pascal's triangle. Each number is obtained by summing the two numbers above.

If spin-orbit coupling can be ignored, each spin function that is an eigenfunction of S^2 and S_z corresponds to a quantum mechanical eigenfunction. In a magnetic field, there will be $2S + 1$ nondegenerate components. For $S = 0$, there is a singlet; for $S = 1/2$, there is a doublet; for $S = 1$, there is a triplet; in general, thus, a $(2S + 1)$-tet. From Figure 1.8 it follows that for $N = 1$, there is one doublet; for $N = 2$ one singlet and one triplet; for $N = 3$ one quartet and two doublets; and for $N = 4$ one pentet, three triplets, and two singlets. The notation for the state is ^{2S+1}X.

This can be applied to electrons with spatial angular momentum too. The configuration $1s^2 2s^2 2p^2$ for the carbon atom with the same radial function consists of 15 different functions on three energy levels: nine in a ^3P state (ground state), five in a ^1D state, and one in a ^1S state.

1.8 MANY-ELECTRON THEORY

The wave function for a single electron, $\psi(\vec{r}_1, \zeta_1)$, is a function of the position coordinates of the electron: $\vec{r}_1 = (x_1, y_1, z_1)$ and the spin coordinate ζ_1. To simplify the notation, we write this as: $\psi(1) \equiv \psi(\vec{r}_1, \zeta_1)$. The absolute square of function ψ, $\psi^*(1)\psi(1) = |\psi(1)|^2$, is the probability density of the electron. As already stated, a function of position in space and spin is called a *spin orbital*. A function of position only is called an *orbital*. If we integrate over a volume A, we obtain the probability to find the electron in A. ψ itself can be added to (or superposed on) other wave functions to form a new wave function. Physically, this is equivalent to interference. According to Born, ψ is *probability amplitude*.

1.8.1 WAVE FUNCTION FOR MANY ELECTRONS

A wave function for many (N) electrons (or other particles) is interpreted in the same way. The wave function is a function of position and spin of each of the individual electrons:

$$\Psi = \Psi(1, 2, \ldots, N). \tag{1.82}$$

The absolute square of Ψ gives the probability density that electron 1 is at $\vec{r}_1$ with spin ζ_1, electron 2 at $\vec{r}_2$ with spin ζ_2, etc. Such a function can be described in many different ways. A correct although not very useful way would be to make a table of Ψ in possible positions and spins of N electrons. We would need to do this for a great number of points for each electron, let us say for 100 values of $\vec{r}_1$. For each choice of $\vec{r}_1$, 100 values of $\vec{r}_2$ are necessary and for each pair $(\vec{r}_1, \vec{r}_2)$, 100 values of $\vec{r}_3$, etc. In the case of the iron atom with $N = 26$, we need to give Ψ in 10^{52} different positions of the electrons! Of course, this is impossible even using all the compact disks and USB sticks in the world.

A simpler description is to assume that the electrons move independently of each other. The electronic motion is separable and the wave function is a product of orbitals (Section 1.5.4). Therefore, we try to approximate as an orbital product:

$$\Psi(1,2,\ldots,N) \approx \psi_1(1) \cdot \psi_2(2) \cdot \ldots \cdot \psi_N(N). \qquad (1.83)$$

To describe the orbital product in Equation 1.83, we may plot 100 orbital values for the positions of each of the $N = 26$ electrons, altogether 2600 points in the case of the iron atom. Alternatively, one may use a linear expansion of the orbital, in which case only the coefficients have to be listed.

Expansion in an orbital product is thus necessary, at least as a first approximation. The problem is that such an expansion cannot be exact in general. Equation 1.83 distributes electron 1 according to ψ_1, irrespective of the coordinates of the other electrons. This is an approximation, since the probability amplitude for electron 1 depends on the coordinates of the other electrons. If electron 2 moves a bit, this should influence the way electron 1 is distributed. The positions are, in fact, *correlated*, and this correlation cannot be correctly represented by just a product of orbitals.

Another problem has to do with symmetry. An electron system must be antisymmetric in the electron coordinates. We will first make the orbital product antisymmetric, which will automatically introduce correlation between electrons with the same spin, but not, unfortunately, between electrons with different spins.

1.8.2 PAULI EXCLUSION PRINCIPLE

The results of Moseley (1914) and Bohr (1916) showed that the electrons of the atoms in the periodic table are ordered in shells with $2N^2$ electrons. The first shell has two electrons, the second eight, etc. Within each shell, there are subshells.

Each spin orbital has one unique set of quantum numbers. Helium has two electrons in the first s-shell, since an s-shell can take only two electrons, one for each spin quantum number ($s_z = 1/2$ and $s_z = -1/2$). In the next element, lithium, the third electron enters the second shell ($n = 2$). This *Aufbau principle* explains the periodic table (see Chapter 2). The two-rule mystified the physicists and was the reason why Pauli, in 1924, introduced a fourth quantum number, later to be identified with spin. The first version of the Pauli exclusion principle holds that: *any two electrons cannot have the same set of quantum numbers*.

After the discovery of quantum mechanics, particularly the relativistic quantum theory of Dirac, the Pauli exclusion principle was reformulated as follows: a one-electron spatial function (orbital) can be occupied by at most two electrons with different spins. An alternative way of saying this is: *a spin orbital can be occupied by only one single electron*.

The two-rule of the chemical bond could now be explained. Methane (CH_4) has four bonds but eight electrons. A chemical bond is formed by two electrons in the Lewis–Langmuir bond model. Each bonding orbital can hold two electrons, one with spin up (α) and one with spin down (β).

The modern version of the Pauli exclusion principle is derived from another principle: the *principle of indistinguishability*. All electrons are identical. Thus, the total electron density must be invariant if the coordinates of two electrons are interchanged. Thus, if 1 and 2 are interchanged:

$$\Psi(2,1,3,\ldots,N)\Psi^*(2,1,3,\ldots,N) = \Psi(1,2,3,\ldots,N)\Psi^*(1,2,3,\ldots,N). \quad (1.84)$$

Equation 1.84 is satisfied if the wave function itself remains the same, or changes its sign:

$$\Psi(2,1,3,\ldots,N) = \pm\Psi(1,2,3,\ldots,N). \quad (1.85)$$

Electrons belong to a greater group of particles called *fermions*, which change sign in Equation 1.85 for all possible interchanges of particle coordinates. The group of fermions also includes neutrons and protons. In the case of fermions, from Equation 1.85 we obtain

$$\Psi(1,1,3,\ldots,N) = -\Psi(1,1,3,\ldots,N) = 0, \quad (1.86)$$

implying that if the coordinates of two electrons are the same ($1 = 2$), the probability amplitude is equal to zero. In other words, two fermions with the same spin cannot be at the same point in space at the same time. This is the most general formulation of the Pauli exclusion principle.

If the wave function does not change sign at the interchange of two particles, we are dealing with *bosons*.

The Pauli exclusion principle gives structure to matter. It explains why atoms and molecules cannot penetrate each other. The electrons of two equal atoms are subject to the Pauli exclusion principle. If the atoms are distant, orbitals in different atoms may be regarded as different, so that each may be occupied by two electrons. If the two atoms come closer to each other, the pair of overlapping outer orbitals need to be redefined into one bonding and one antibonding (molecular) orbital. The new bonding orbital increases the stability of the molecule when the atoms start to penetrate each other. However, the occupation of the antibonding orbital decreases the stability more than the occupation of the bonding orbital increases it. Hence, if all orbitals are filled, the whole molecule becomes unstable. This explains the repulsion between molecules with closed shells and the repulsion between noble gas atoms that are close to each other.

Chemical bonding appears when the electron count is such that the antibonding orbital need not be occupied by electrons. Examples are diatomic molecules, such as C_2, N_2, O_2, and F_2. In the second row, all are stable except Ne_2 (see Chapter 3).

1.8.3 INDEPENDENT ELECTRON MODEL

The spin orbital product (Equation 1.83) is a many-electron wave function, but it is neither symmetric nor antisymmetric. Antisymmetry is taken into account as follows:

$$\Phi(1,2,3,\ldots,N)=\frac{1}{\sqrt{N!}}\begin{vmatrix} \psi_1(1)\psi_1(2)\psi_1(3) & \cdots & \psi_1(N) \\ \psi_2(1)\psi_2(2)\psi_2(3) & \cdots & \psi_2(N) \\ \vdots & \cdots & \vdots \\ \vdots & \cdots & \vdots \\ \psi_N(1)\psi_N(2)\psi_N(3) & \cdots & \psi_N(N) \end{vmatrix}=\left|\psi_1\psi_2\cdots\psi_N\right|. \quad (1.87)$$

Instead of an orbital product, we use a determinant function. This way of writing the wave function Φ was introduced by the American physicist J. C. Slater and is therefore called the *Slater determinant*. The factor before the determinant is a normalization factor. The rightmost member of Equation 1.87 is an abbreviation.

Interchanging $1 \leftrightarrow 2$ (for example, in Equation 1.87 is equivalent to interchanging the first two rows in the determinant. The sign of the determinant is then changed, and this is exactly what is required for an electronic wave function. Thus, the Pauli principle is automatically satisfied. In this model, the electrons in a certain sense still move independently of each other, and the model is therefore referred to as the *independent electron model*.

Formally, Equation 1.87 may be written with the help of a projection operator called the *antisymmetrizer*, A_N:

$$\Psi(1,2,3,\ldots,N)=\frac{1}{\sqrt{N!}}\sum_P(-1)^p P\left[\phi_1(1)\phi_2(2)\ldots\phi_N(N)\right]=A_N\left[\phi_1(1)\phi_2(2)\ldots\phi_N(N)\right].$$

$$(1.88)$$

In Equation 1.88, P denotes a permutation of the electrons and p denotes the parity of this permutation. If p is even, $(-1)^p = 1$, and if p is odd, $(-1)^p = -1$. A_N thus projects out the antisymmetric component from the orbital product and normalizes it by the factor $1/\sqrt{N!}$

Orbitals may be transformed into each other without changing the Slater determinant. Assume that we have two orbitals with the same spin, ϕ_1 and ϕ_2. Their Slater determinant is $\phi_1(1)\,\phi_2(2) - \phi_1(2)\,\phi_2(1)$. We now perform the transformation as follows:

$$\psi_1 = (\cos\theta)\phi_1 + (\sin\theta)\phi_2$$
$$\psi_2 = (-\sin\theta)\phi_1 + (\cos\theta)\phi_2. \qquad (1.89)$$

We have used $\cos\theta$ and $\sin\theta$ as coefficients since it is automatically guaranteed that the new orbitals ψ_1 and ψ_2 in Equation 1.89 are orthonormal, provided ϕ_1 and ϕ_2 are orthonormal. Equation 1.89 is actually the simplest example of a unitary transformation.

Now the Slater determinant of ψ_1 and ψ_2 is calculated. The result may be written as

$$
\begin{vmatrix} \psi_1(1) & \psi_1(2) \\ \psi_2(1) & \psi_2(2) \end{vmatrix} = \begin{vmatrix} \cos\theta & \sin\theta \\ -\sin\theta & \cos\theta \end{vmatrix} \begin{pmatrix} \phi_1(1) & \phi_1(2) \\ \phi_2(1) & \phi_2(2) \end{pmatrix} = \begin{vmatrix} \phi_1(1) & \phi_1(2) \\ \phi_2(1) & \phi_2(2) \end{vmatrix},
\tag{1.90}
$$

proving that the 2×2 Slater determinant is invariant under a unitary transformation. In Equation 1.90 we have used the rule that a determinant of a product of matrices is equal to the product of determinants.

1.8.4 CORRELATION HOLE

In Equation 1.86, we found that if we try to put two electrons with the same spin at the same point, the wave function is equal to zero. It is quite easy to see in Equation 1.87 that if the two electrons approach each other, the determinant wave function tends to zero and is proportional to the distance between them, δ. (Set $1 = \vec{r}_1\alpha$ and $2 = \vec{r}_2\alpha = (\vec{r}_1 + \vec{\delta})\alpha$ in Equation 1.87 and use the Taylor expansion to get $\psi_\mu(\vec{r}_2) = \psi_\mu(\vec{r}_1 + \vec{\delta})$. The result is a sum of two Slater determinants where one has two columns equal and the other one is proportional to δ.) This means that the density of electrons with the same spin, that is, the absolute square of the wave function, tends to zero as δ^2. If the position of an electron is assumed fixed, the probability density of electrons with the same spin tends to zero near to the fixed electron. The excluded probability density amounts to a full electron, as will be proven for a Slater determinant in Chapter 2. This "hole" is called the *exchange hole*. Electrons with the same spin are thus *correlated* in a Slater determinant. The correlation problem is the problem of accounting for a correlated motion between the electrons.

The Slater determinant does not account for any correlation between electrons with different spins, since spin orbitals with different spins are always orthogonal and cannot be equal. As we will see, electrons with different spins also try to avoid each other and create a hole; in this case, referred to as a *Coulomb hole*. The Coulomb hole is usually smaller than the exchange hole. The probability density of two electrons with different spins to be at the same point is, in fact, different from zero. The two holes, exchange hole and Coulomb hole, together form the *correlation hole*.

As an example, we may assume that two electrons from the 1s shell are at the same distance from the nucleus but at different distances from each other, as in Figure 1.9. In the first case, they are close ($r_{12} = |\vec{r}_1 - \vec{r}_2|$ small) and in the second case, distant (r_{12} large). The probability density is expected to be higher at a large distance than at a small distance since the electrons repel each other. Despite the fact that they are at the same distance from the nucleus, the probability density of the positions in the right figure is higher than that of the positions in the left figure. In the wave function, the two orbitals may be identical 1s orbitals whose value depends only on the distance from the nucleus. Therefore, the orbital product wave function *incorrectly* gives the same probability amplitude for the two geometrical arrangements of Figure 1.9.

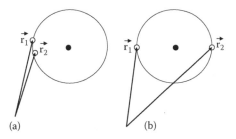

FIGURE 1.9 Electrons 1 and 2 at the same distance from the nucleus, but (a) close to each other and (b) far from each other.

To improve the description, we may choose the following type of wave function:

$$\Psi(\vec{r}_1,\vec{r}_2) \approx \psi(\vec{r}_1)\cdot\psi(\vec{r}_2)\cdot(1+K\cdot r_{12}), \qquad (1.91)$$

where K is a positive constant. Clearly, the probability amplitude ψ is higher if r_{12} is large compared to if r_{12} is small or zero. Equation 1.91 thus introduces a correlation between the positions of the electrons with different spins. The best possible K may be obtained using the variation principle and minimizing $\langle\Psi|H|\Psi\rangle/\langle\Psi|\Psi\rangle$. It turns out that adding a basis function of the type in Equation 1.91 gives an important variational energy stabilization to the wave function.

Electrons with different spins are thus correlated in their motion. This correlation cannot be accounted for by a wave function in the form of a single Slater determinant, since Equation 1.91 cannot be written as a product of two orbitals. The exact properties of the Coulomb hole can be derived from the Hamiltonian operator. One may show that there is cusp of the following type when two electrons come close to each other:

$$\Psi(\vec{r}_1\alpha,\vec{r}_2\beta,\ldots) = \text{const.}\left(1+\frac{r_{12}}{2}+\cdots\right), \quad \text{for } \vec{r}_1 = \vec{r}_2 + \vec{r}_{12} \text{ and } r_{12} \rightarrow 0. \qquad (1.92)$$

The Coulomb hole is smaller in its extension than the exchange hole and integrates to zero. The charge of opposite spin around the first electron is thus removed to some extent and placed somewhere else in the system.

1.8.5 CORRELATION ENERGY

The difference between the exact ground state energy eigenvalue of the SE, E_0, and the best (in the sense of the variation principle) energy expectation value $\langle H\rangle$ over the Slater determinant, is due to the neglect of correlation between electrons in their positions, particularly between electrons with different spins, in the Slater determinant. This was first pointed out by the Swedish quantum chemist Per-Olov Löwdin. He referred to the error of the Slater determinant as *correlation energy*:

$$E_{corr} = \langle H\rangle - E_0. \qquad (1.93)$$

The correlation energy of the He atom is about 0.04 H (≈ 1 eV ≈ 100 kJ/mol). The correlation energy for an electron pair in general is roughly of this order of magnitude. This is a large value, comparable with bond energies. However, in many applications, the correlation error is still unimportant. For example, in the calculation of chemical reaction barriers, the correlation energies for reactants and products are often approximately the same. Therefore, the independent electron model serves well as a first approximation of the electronic wave function.

1.8.6 CONFIGURATION INTERACTION

The total wave function can be much improved if it is written with the help of two or many configurations, each corresponding to a Slater determinant wave function. The most general way of writing a wave function of this type is

$$\Psi = C_0\Phi_0 + \sum_{i,a} C_i^a \Phi_i^a + \sum_{\substack{i,j \\ a,b}} C_{ij}^{ab} \Phi_{ij}^{ab} + \cdots . \tag{1.94}$$

In Φ_i^a, spin orbital i has been replaced by spin orbital a; in Φ_{ij}^{ab}, spin orbitals i and j are replaced by spin orbitals a and b, etc.

Another way to derive Equation 1.94 is to make a straightforward expansion of the total N-electron wave function. Let D_i be expansion coefficients:

$$\Psi(1,2,\ldots,N) = \sum_i D_i \Psi_i(1) \Psi_i^N(2,3,\ldots,N)$$

$$= \sum_i D_i \psi_i(1) \sum_j D_j \psi_j(2) \Psi_j^{N-1}(3,4,\ldots,N) \tag{1.95}$$

$$= \sum_{i,j,k,\ldots} D_i D_j \ldots D_q \psi_i(1) \psi_j(2) \psi_k(3) \ldots \psi_p(N-1) \psi_q(N).$$

If the antisymmetrizer is applied to Equation 1.95, we obtain an expansion of the same type as in Equation 1.94.

In the following examples, we will show that configuration interaction (CI) solves the correlation problem, in principle. Unfortunately, the computational problem rapidly increases with the size of the molecule and the required accuracy of the expansion.

1.8.7 ELECTRONIC DENSITY MATRIX

The density matrix corresponding to an orbital product is defined as

$$\gamma_0(1;1') = \sum_{i=1}^N \phi_i(1)\phi_i^*(1'). \tag{1.96}$$

For $1 = 1'$, we obtain $\rho(1) = \gamma_0(1;1)$, equal to the electronic density. In Equation 1.96, 1 and $1'$ are different in order to be able to operate on only the right function, as is required when an expectation value is calculated.

Generally, $\gamma(1;1')$ may be calculated from an arbitrary wave function using the definition (Löwdin 1955):

$$\gamma(1,1') = N \int \Psi^* (1,2,\ldots,N) \Psi(1',2,\ldots,N) d2 d3 \ldots dN, \qquad (1.97)$$

provided that the wave function Ψ is antisymmetrized; γ is called the *one-electron reduced density matrix* or simply *one-matrix*. To calculate the expectation value of, for example, the kinetic energy, from γ, we have to take the second derivative of the function that contains $1'$, then set $1 = 1'$ and integrate over 1. The same is true for all one-electron operators. All one-electron expectation values can be obtained from $\gamma(1;1')$. In this sense, the full wave function contains "unnecessary information" (for $N > 1$) in the form of dependence on all electron coordinates.

To be able to calculate two-electron expectation values, knowledge of $\gamma(1;1')$ is insufficient. To obtain the same simplification as for the one-electron expectation values, the two-matrix, a function of 1, $1'$, 2, and $2'$, might be introduced, but this is not considered here. The two-matrix contains all necessary information, but unfortunately the two-matrix cannot easily be obtained without first calculating the wave function (the N-representability problem).

If the wave function is a Slater determinant:

$$\Psi(1,2,3,\ldots,N) = \frac{1}{\sqrt{N!}} \sum_P (-1)^P \left[\phi_1(1)\phi_2(2)\ldots\phi_N(N) \right], \qquad (1.98)$$

the first-order reduced γ has the same expansion as the one given in Equation 1.96:

$$\gamma_0(1;1') = \sum_{i=1}^{N} \phi_i(1)\phi_i^*(1'), \qquad (1.99)$$

which is quite easy to show by going through all permutation types.

We note that the following relation holds in the case when Ψ is a Slater determinant:

$$\int \gamma_0(1;1')\phi_i^*(1')d1' = \phi_i(1). \qquad (1.100)$$

Equation 1.100 means that ϕ_i is an eigenfunction of the integral operator γ_0 with the eigenvalue equal to unity.

In the general case, instead of Equation 1.100, we obtain

$$\int \gamma(1;1')\chi_i^*(1')d1' = v_i\chi_i(1), \qquad (1.101)$$

where χ_i is an eigenfunction of γ with the eigenvalue v_i. In this case, the eigenvalues are usually smaller than unity:

$$0 \leq v_i \leq 1. \tag{1.102}$$

The function χ_i is called a *natural spin orbital* (NSO). The eigenvalue v_i is the *occupation number* of χ_i. It may be shown that the sum of occupation numbers is equal to N. There is generally an infinite set of NSOs, except when the wave function is approximated by a finite number of Slater determinants.

Let us first assume that the exact expression for the wave function is dominated by a single Slater determinant. If the number of electrons is N, the first N spin orbitals have an occupation number quite close to unity. In the case of the helium atom ground state, the largest two eigenvalues $v_1 = v_2 \approx 0.98$. These NSOs are said to be *strongly occupied*. An NSO is usually considered to be strongly occupied if its occupation number is larger than, say, 0.1. In the helium atom, the sum of the remaining occupation numbers is ≈ 0.04, since the sum of all occupation numbers has to be equal to $N = 2$.

Electronic states, where the number of strongly occupied spin orbitals is larger than the number of electrons, are common. In a diatomic molecule near the dissociation limit between the atoms, all spin orbitals, which are strongly occupied for the individual atoms in their ground states, or their linear combinations, are also strongly occupied for the diatomic molecule in its ground state. In the hydrogen molecule, for example, there are four strongly occupied NSOs near the dissociation limit. In this type of situation, the electronic states are nearly degenerate. The single Slater determinant, no matter the properties of its occupied spin orbitals, is a bad approximation of the true wave function.

The first-order reduced density matrix of the exact wave function also has a number of "weakly occupied" orbitals for which $i > N$. These orbitals never appear in a single Slater determinant method, but are important for a correct description of the correlation hole.

From Equation 1.101 we obtain

$$\gamma(1;1') = \sum_{i=1}^{\infty} v_i \chi_i(1) \chi_i^*(1'). \tag{1.103}$$

Notice that a correct representation of γ requires an infinite set of spin orbitals.

1.8.8 CORRELATION ERROR

The following treatment of correlation is best suited to the nondegenerate type of wave function. If N NSOs with the highest occupation numbers are occupied in a single Slater determinant, we obtain

$$\gamma_{NSO}(1;1') = \sum_{i=1}^{N} \chi_i(1) \chi_i^*(1'). \tag{1.104}$$

Löwdin showed that γ_{NSO} is the best possible approximation to the exact density (Equation 1.103). The one-matrix $\gamma_0(1;1')$ of the trial wave function corresponding to a single Slater determinant with orbitals $\{\phi_i\}$ may be expanded as follows:

$$\gamma_0(1;1') = \sum_{i=1}^{\infty} \phi_i(1)\phi_i^*(1').$$ (1.105)

Let us compare this to the exact expansion (γ) in terms of NSO χ_i (Equation 1.103). Adding and subtracting γ_0 and using Equation 1.103, we obtain

$$\gamma(1;1') = \gamma(1;1') + \gamma_0(1;1') - \gamma_0(1;1')$$

$$= \sum_{i=1}^{N} \phi_i(1)\phi_i^*(1') + \left(\sum_{i=1}^{N} \chi_i(1)\chi_i^*(1') - \sum_{i=1}^{N} \phi_i(1)\phi_i^*(1') \right)$$

$$+ \sum_{i=1}^{N} (v_i - 1)\chi_i(1)\chi_i^*(1') + \sum_{i=N+1}^{\infty} v_i\chi_i(1)\chi_i^*(1')$$

$$= \gamma_0(1,1') + \varepsilon_{orbital}(1,1') + \varepsilon_{CI}(1,1').$$ (1.106)

This expansion shows that there are two errors in any independent electron model. The first error in the wave function, $\varepsilon_{orbital}$, depends on the fact that the spin orbitals calculated are not NSO. The last two terms in the second member of Equation 1.106, together forming ε_{CI}, are due to the fact that the occupation numbers are not equal to unity or zero in the first and second part of ε_{CI}, respectively. This error appears since the true wave function is an expansion in terms of many Slater determinants and cannot be accurately approximated by a single Slater determinant.

2 Atoms

2.1 ATOMIC UNITS

It is sometimes advantageous to use special units for atoms and molecules. In the system of *atomic units*, the unit of length is called 1 Bohr and is equal to the radius of the hydrogen atom. Other atoms are roughly the same size. The atomic unit of energy is called Hartree and is an order of magnitude larger than that of the excitation energies in the visible spectrum. The atomic unit of mass is the rest mass of the electron. Atomic units are obtained as follows:

Define:

1. Atomic unit for action (energy × time) or moment of momentum: Planck's constant h divided by 2π ($\hbar = 1$).
2. Atomic unit for charge: absolute value of charge of the electron ($|e| = 1$).
3. Atomic unit for mass: rest mass of the electron (m = 1).
4. Atomic unit for permittivity: 4π times the permittivity of vacuum ($4\pi\varepsilon_0 = 1$).

Derive:

1. Atomic unit for length: 1 Bohr (a_0) = $4\pi\varepsilon_0 h^2/me^2$ = 0.529178 Å.
2. Atomic unit for energy: 1 Hartree (H or E_H) = $e^2/(4\pi\varepsilon_0 a_0)$ = 27.2113 eV.
3. Atomic unit for time: $\hbar/E$, where E = 1 H; $\hbar/E = 4\pi\varepsilon_0 a_0 \hbar/e^2$ = 2.41889 × 10^{-17} sec.

Atomic units, except H and a_0, are abbreviated a.u.

A problem with atomic units is that natural constants are "lost" in equations, along with their values in ordinary units. A good advice is to write all equations in ordinary units also. Atomic units are good for rough estimations in the atomic world, however. It is interesting that the mean velocity for an electron in the innermost orbital, the 1s orbital, is quite easily calculated and is equal to Z a.u. (Bohr/time-a.u.); thus for the uranium atom, the average velocity of the innermost electrons is 92. The velocity of light is only 137 Bohr/time-a.u. Thus, 1s electrons move around the uranium nucleus with a mean velocity close to the velocity of light. It is likely that there will be relativistic corrections to the wave functions and energies for the heavy atoms.

Since the proton and neutron are 1836 times as heavy as the electron, the atomic nuclei move quite slowly (on the femtosecond scale) compared with the electrons. It is possible to measure the motion of nuclei in fast chemical reactions with the help of laser light pulses. This rather new science is sometimes called "femtochemistry."

2.2 HYDROGEN ATOM

It was particularly the efforts to understand the spectrum of the hydrogen atom that led to the discovery of quantum mechanics. The wave functions of the hydrogen atom function as model orbitals in other atoms and molecules.

2.2.1 TIME-INDEPENDENT SCHRÖDINGER EQUATION FOR THE HYDROGEN ATOM

Bohr obtained the correct energy levels of the hydrogen atom by using two "quantum conditions" (see Section 1.3.2). It was impossible to derive these two conditions from more fundamental principles. Bohr obtained a perfect discrete spectrum (line spectrum) of the hydrogen atom. For larger atoms, there appeared to be no simple solution, although the appearance of line spectra was now less of a mystery. Bohr's theory also permitted an adequate understanding of the periodic table. For a moment, it seemed that the theory of de Broglie could explain the Bohr orbits as standing waves, but this picture was less than consistent and actually the reason why Schrödinger tried a three-dimensional differential equation. The quantum mechanics of Heisenberg and Born also did not seem to provide a simple way out.

Stimulated by the wave theory of de Broglie, but well aware of the pitfalls and limitations of this theory, Erwin Schrödinger set up an equation for the hydrogen atom. The time-independent Schrödinger equation (SE) for a single electron and a nucleus with atomic number Z may be written using SI units and atomic units, respectively:

$$H\Psi = (T+V)\psi = -\frac{\hbar^2}{2m}\Delta\psi - \frac{Z|e|}{4\pi\varepsilon_0 r}\psi = E\psi \rightarrow H\Psi = -\frac{1}{2}\Delta\psi - \frac{Z}{r}\psi = E\psi. \quad (2.1)$$

The electron mass should be replaced by the reduced mass, in the case of hydrogen equal to $\mu = m(1 - 1/1836)$, since the nucleus has a finite mass equal to 1836m. The right part of Equation 2.1, where atomic units are used, assumes that the nucleus is infinitely heavy. The attraction between the electron and the nucleus is given by $V(r) = -Z/r$. Since V only depends on the distance r to the nucleus (central field), a polar coordinate system is used (Figure 2.1). The complete expression in polar coordinates for the Hamiltonian operator is given by

$$H = -\frac{1}{2}\Delta - \frac{Z}{r} = -\frac{1}{2}\frac{1}{r^2}\left[\frac{\partial}{\partial r}\left(r^2\frac{\partial}{\partial r}\right) + \frac{1}{\sin\theta}\frac{\partial}{\partial\theta}\left(\sin\theta\frac{\partial}{\partial\theta}\right) + \frac{1}{\sin^2\theta}\frac{\partial^2}{\partial\varphi^2}\right] - \frac{Z}{r}. \quad (2.2)$$

r, θ, and φ may be separated and the wave function written as a product of a radial function and an angular function:

$$\psi(r,\theta,\varphi) = R(r)Y(\theta,\phi). \quad (2.3)$$

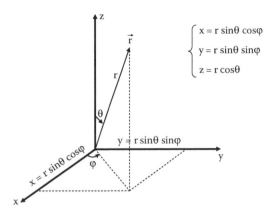

FIGURE 2.1   Polar coordinates.

ψ is an eigenfunction of H if Y and R are solutions to Equations 2.4 and 2.5, respectively,

$$-\frac{1}{2}\left[\frac{1}{\sin\theta}\frac{\partial}{\partial\theta}\left(\sin\theta\frac{\partial}{\partial\theta}\right)+\frac{1}{\sin^2\theta}\frac{\partial^2}{\partial\varphi^2}\right]Y_{\ell m}\left(\theta,\varphi\right)=\frac{\ell\left(\ell+1\right)}{2}Y_{\ell m}\left(\theta,\varphi\right). \quad (2.4)$$

The functions $Y_{\ell m}$ are the well-known spherical harmonic functions: l is a quantum number called the azimuthal quantum number; m (or m_ℓ) is the magnetic quantum number; and K = l(l + 1)/2 is the eigenvalue and if it is inserted in Equation 2.5, we obtain

$$\left[-\frac{1}{2}\frac{1}{r^2}\frac{d}{dr}\left(r^2\frac{d}{dr}\right)-\frac{Z}{r}+\frac{1}{2}\frac{\ell\left(\ell+1\right)}{r^2}\right]R=ER. \quad (2.5)$$

This equation will be solved in Section 2.2.3.

2.2.2   ANGULAR FUNCTION

The spherical harmonic functions Y(θ,φ) are eigenfunctions of the operator for the square of angular momentum ($\hat{l}^2$) for a particle in a central field, according to the equation

$$\hat{l}^2Y_{\ell m}=\ell\left(\ell+1\right)Y_{\ell m}; \quad \ell=1,2,\dots. \quad (2.6)$$

The separation of the variables θ and φ leads to a product of a function that depends only on θ and another function that depends only on φ:

$$Y_{\ell m}\left(\theta,\varphi\right)=\Theta_{\ell m}\left(\theta\right)\Phi_m\left(\varphi\right) \quad (2.7)$$

The first few associated Legendre functions (unnormalized) are

$$\ell = 0 \quad \Theta_{00} = 1$$

$$\ell = 1 \quad \Theta_{10} = \cos\theta$$

$$\Theta_{11} = \Theta_{1-1} = \sin\theta$$

$$\ell = 2 \quad \Theta_{20} = 3\cos^2\theta - 1 \qquad (2.8)$$

$$\Theta_{21} = \Theta_{2-1} = \cos\theta\sin\theta$$

$$\Theta_{22} = \Theta_{2-2} = \sin^2\theta.$$

For l, we use the following letter code:

$$\ell = 0 \quad \ell = 1 \quad \ell = 2 \quad \ell = 3 \quad \ell = 4 \quad \ell = 5 \quad \ell = 6 \quad \ell = 7$$

$$\text{s} \qquad \text{p} \qquad \text{d} \qquad \text{f} \qquad \text{g} \qquad \text{h} \qquad \text{i} \qquad \text{j}$$

The notations s, p, and d originate in atomic spectroscopy.

Φ_m satisfies the equation

$$\frac{d^2\Phi_m}{d\varphi^2} = -m^2\Phi_m \qquad (2.9)$$

which, with the boundary condition on Φ, $\Phi(\varphi) = \Phi(\varphi + 2\pi)$, leads to

$$\Phi_m(\varphi) = \exp(im\varphi); \quad m \text{ integer number} \qquad (2.10)$$

m (or m_ℓ) is called the magnetic quantum number. For $\Theta_{\ell m}(\theta)$, the following condition is valid: $m \le |\ell|$, thus for all l: $m = -1, -1 + 1, ..., 1 - 1, 1$.

The functions $Y_{\ell m}(\theta,\varphi) = \Theta_{\ell m}(\theta)\Phi_m(\varphi)$ are normalized over a spherical surface.

A wave function that depends on just the coordinates of a single electron [(r,θ,φ) or (x,y,z)] in an atom is called an *atomic orbital* (AO) and its energy is called *orbital energy*. The atoms of the periodic table have possess s, p, d, or f orbitals. In the absence of a magnetic field, all AOs with the same l and the same radial function are degenerate (have the same energy). Linear combinations of the functions $\Phi_m(\varphi) = \exp(im\,\varphi)$ may be used to obtain real AOs:

$$\exp(im\varphi) + \exp(-im\varphi) = 2\cos(m\varphi)$$

$$\exp(im\varphi) - \exp(-im\varphi) = 2i \cdot \sin(m\varphi). \qquad (2.11)$$

Numbers with absolute value equal to unity multiplying the wave functions are without any importance in quantum mechanics, thus the imaginary unit can be left out in the second equation. The resulting products, $\Theta(\theta) \cdot \Phi(\varphi)$, forming the angular functions for AO have a specific directional dependence. For example, $p_x = \sin\theta \cdot \cos\varphi = x/r$ and $p_y = \sin\theta \cdot \sin\varphi = y/r$ (unnormalized) are directed along the x- and y-axes, respectively. In the same way, the d and f orbitals are calculated. For example, d_{yz} and d_{zx} are formed from the $l = 2$ and $m = \pm1$ functions.

Thus, we obtain the following normalized spherical harmonics functions:

$$s \quad Y_{00} = \frac{1}{\sqrt{4\pi}},$$

$$p(z) \quad Y_{10} = \sqrt{\frac{3}{4\pi}}\cos\theta = \sqrt{\frac{3}{4\pi}}\cdot\frac{z}{r},$$

$$p(x) \quad \frac{Y_{11}+Y_{1-1}}{\sqrt{2}} = \sqrt{\frac{3}{4\pi}}\sin\theta\cos\varphi = \sqrt{\frac{3}{4\pi}}\cdot\frac{x}{r},$$

$$p(y) \quad \frac{Y_{11}-Y_{1-1}}{\sqrt{2}} = \sqrt{\frac{3}{4\pi}}\sin\theta\cdot\sin\varphi = \sqrt{\frac{3}{4\pi}}\cdot\frac{y}{r},$$

$$d(z^2) \quad Y_{20} = \frac{1}{2}\sqrt{\frac{5}{4\pi}}(3\cos^2\theta-1) = \frac{1}{2}\sqrt{\frac{5}{4\pi}}\cdot\frac{2z^2-x^2-y^2}{r^2},$$

$$d(zx) \quad \frac{Y_{21}+Y_{2-1}}{\sqrt{2}} = \sqrt{\frac{15}{4\pi}}\cos\theta\cdot\sin\theta\cdot\cos\varphi = \sqrt{\frac{15}{4\pi}}\cdot\frac{zx}{r^2},$$

$$d(yz) \quad -i\frac{Y_{21}-Y_{2-1}}{\sqrt{2}} = \sqrt{\frac{15}{4\pi}}\cos\theta\cdot\sin\theta\cdot\sin\varphi = \sqrt{\frac{15}{4\pi}}\cdot\frac{yz}{r^2},$$

$$d(x^2-y^2) \quad \frac{Y_{22}+Y_{2-2}}{\sqrt{2}} = \frac{1}{2}\sqrt{\frac{15}{4\pi}}\sin^2\theta\cdot\cos2\varphi = \frac{1}{2}\sqrt{\frac{15}{4\pi}}\cdot\frac{x^2-y^2}{r^2},$$

$$d(xy) \quad -i\frac{Y_{22}-Y_{2-2}}{\sqrt{2}} = \frac{1}{2}\sqrt{\frac{15}{4\pi}}\sin^2\theta\cdot\sin2\varphi = \sqrt{\frac{15}{4\pi}}\cdot\frac{xy}{r^2}. \quad (2.12)$$

Since s orbitals, according to Equation 2.12, do not depend on θ and φ, they are spherically symmetric. There are three p orbitals along the coordinate axes; five d orbitals and seven f orbitals. Orbitals with $m = 0$ are independent of φ and extend around the z-axis, where $\cos^2\theta$ have large values.

2.2.3 Radial Function

The radial function R(r) in Equation 2.3 is determined from the radial time-independent SE with the boundary conditions that it is finite or zero at $r = 0$ and tends exponentially to zero as $r \to \infty$:

$$-\frac{1}{2}\frac{1}{r^2}\frac{d}{dr}\left(r^2\frac{dR}{dr}\right)+\left(\frac{\ell(\ell+1)}{2r^2}-\frac{Z}{r}\right)R(r)=ER(r).\tag{2.13}$$

This equation does not depend on the m quantum number since the single electron is assumed to move in a central field. The radial functions are thus the same for orbitals, which are different only in m. A number of known sets of functions satisfy the mentioned boundary conditions. The Laguerre functions are known to be solutions of Equation 2.13. R(0) has to be finite at the nucleus and $R(r) \to 0$ exponentially when $r \to \infty$, and this is, in fact, satisfied by Laguerre functions that are of the following type:

$$R(r)=\left(C_p r^p + C_{p-1}r^{p-1}+\cdots\right)\exp(-\alpha r).\tag{2.14}$$

All powers of the Laguerre polynomial multiplying the exponential have to be nonnegative integer numbers. If there is only one term in the polynomial, one may show, by inserting Equation 2.14 into Equation 2.13, that $p = l$. We define a principal quantum number $n = p + 1$. Thus, $n = l + 1$ in this case. In the case of one term (no nodes for $r > 0$), we have the following AO: 1s, 2p, 3d, 4f, …, with powers $p = 0$, 1, 2, …, respectively.

If there are two terms in the Laguerre polynomial in Equation 2.14, one may show that $p = l + 1$ and, since $n = p + 1$, in this case $n = l + 2$. We have the following AO with two terms in the polynomial: 2s, 3p, 4d, 5f, etc. Generally, one may show that $n = l +$ (number of terms in the polynomial).

The lowest power in a Laguerre polynomial is $n -$ (number of terms) $= l$. Thus, all s orbitals have a constant term in the polynomial, that is $R(0) \neq 0$; all p orbitals are linear in r for small r; all d orbitals are quadratic in r for small r, etc. AO with $l = 1$ cannot have a constant term in the polynomial in Equation 2.14, since Equation 2.13 is then unsatisfied. AO with $l = 2$ also cannot have a linear term. The radial functions are thus summarized in Table 2.1.

The unnormalized radial functions $R_{n\ell}(r)$ may be written as

$$(\mathbf{1s}) \sim \exp(-Zr);\, (\mathbf{2s}) \sim (2-3r)\exp(-Zr/2);\, (\mathbf{2p}) \sim r\exp(-Zr/2)$$

$$(\mathbf{3s}) \sim \left(3-2Zr+2Z^2r^2/9\right)\exp(-Zr/3);\, (\mathbf{3p}) \sim r\left(2-Zr/3\right)\exp(-Zr/2)\tag{2.15}$$

$$(\mathbf{3d}) \sim r^2\exp(-Zr/3).$$

For atoms with many electrons, the atomic state in question is not always spherically symmetric. The central field approximation is applied, meaning that V(r) is an average over θ and φ. Variable separation can still be made in Equation 2.2.

TABLE 2.1
Radial Functions

n	p				nl			
0	1	1s	2p	3d	4f	5g	6h	7i
1	1 + 1		2s	3p	4d	5f	6g	7h
2	1 + 2			3s	4p	5d	6f	7g
3	1 + 3				4s	5p	6d	7f
4	1 + 4					5s	6p	7d
5	1 + 5						6s	7p
6	1 + 6							7s

n = number of nodes; p = highest power. Bold texts are used to denote orbitals used for Z ≤ 92.

2.2.4 ENERGY SPECTRUM

One may show that the exponential factor α in Equation 2.14 can be expressed as $\alpha = Z/n$, and that the orbital energies E are given by

$$E_n = -\frac{\alpha^2}{2} = -\frac{Z^2}{2}\frac{1}{n^2} = -R_\infty hc\frac{Z^2}{n^2}, \quad n = 1, 2, 3, \ldots. \tag{2.16}$$

If atomic units are used in Equation 2.16 and R_∞ is expressed in reciprocal centimeter (cm^{-1}) and the velocity of light c in cm/s, the fourth member will be expressed in joule (J) if h is expressed in Js. Equation 2.16 happens to be the same equation that was derived by Bohr by using two quantum conditions for the electron motion according to classical mechanics. Equation 2.16 is in perfect agreement with experimental spectra. R_∞, valid for infinitely heavy mass, should be replaced by $R = R_\infty(1 - m/M)$ for a nucleus with mass M. The energy of an atom with a single electron (hydrogen-like atoms), according to Equation 2.16, is independent of the quantum numbers l (accidental degeneracy). The emission spectrum of the hydrogen atom is given in Figure 2.2.

The *Rydberg constant*, named after the Swedish atomic physicist Johannes (Janne) Rydberg, is the constant R_∞ that multiplies $1/n^2$ in Equation 2.16. It has the value 0.5 H for an infinitely heavy nucleus. In reciprocal centimeter (remove hc in Equation 2.16), it has the value 109737,315 cm^{-1}. In the Lyman series, where the final state is 1s, we obtain the wave numbers 82,303, 97,544, 102,879, …, 109,737 cm^{-1}, where the last value is the limit for ionization (n = ∞). The Balmer series, partly in the visible region, corresponds to the wave numbers 15,241, 20,576, 23,045, …, 27,434 cm^{-1}. The Paschen series with wave numbers 5,334, 7,804, 9,145, …, 12,193 cm^{-1}, is entirely in the infrared region.

The grey region in Figure 2.2 represents a continuous spectrum for E > 0. The electron is no longer bound to the atomic nucleus but can be regarded as free.

The sun emits almost continuous "black-body" radiation according to Planck's radiation law. The exceptions are some sharp black lines in the visible spectrum that were discovered by Fraunhofer in 1814. The wavelengths of these black lines agree

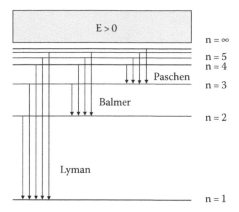

FIGURE 2.2 Emission spectrum of hydrogen.

with that of the hydrogen discrete lines Equation 2.16. The explanation is straight-forward. On its way from the sun, the electromagnetic (EM) radiation has to pass a region with mainly atomic hydrogen. The black lines correspond to absorption of radiation in this region. Other black lines are also found, suggesting that the atoms they belong to are also present in the outer parts of the sun.

2.2.5 SIZE OF ORBITALS

The probability to find the electron at r, θ, and φ in the volume element $r^2 dr \cdot \sin \theta \, d\theta \cdot d\varphi$ is given by

$$\left| \Psi_{n\ell m} \left(r, \theta, \varphi \right) \right|^2 r^2 dr \cdot \sin \theta \, d\theta \cdot d\varphi. \tag{2.17}$$

Here, it is assumed that $R_{n\ell}$, $\Theta_{\ell m}$, and Φ_m are normalized. The probability of finding the electron in a spherical shell of width dr is then D(r)dr, where

$$D\left(r \right) = r^2 R_{n\ell}^2 \left(r \right) \iint\limits_{\substack{0 < \theta < \pi \\ 0 < \varphi < 2\pi}} \sin \theta \, d\theta \, d\varphi \cdot \Theta_{\ell m}^* \Theta_{\ell m} \Phi_m^* \Phi_m = r^2 R_{n\ell}^2 \left(r \right), \tag{2.18}$$

where D(r) is called the radial distribution. For the angle-independent s orbitals, the angular integration leads to

$$D\left(r \right) = 4\pi r^2 \psi^2. \tag{2.19}$$

Since $R_{n\ell}$ is normalized according to

$$\int\limits_0^\infty R_{n\ell}^2 \left(r \right) r^2 dr = 1, \tag{2.20}$$

it follows from the definition of D in Equation 2.19 that

$$\int_0^\infty D(r)\,dr = 1. \qquad (2.21)$$

For large r, D(r) behaves as $\sim r^{2n}\exp(-2Zr/n)$. This function has the following maximum:

$$r_{max} = n^2/Z, \qquad (2.22)$$

where r_{max} is a measure of the size of the orbital expressed in atomic units. The hydrogen 1s orbital has $r_{max} = 1$ Bohr (=the Bohr radius = 0.5292 Å). Most of the electronic density is distributed close to r_{max} and within $2r_{max}$. The size of the hydrogen-like atom is thus directly proportional to n^2, where n is the quantum number of the highest occupied orbital, and inversely proportional to the nuclear charge (Z).

The least bound electrons engage in chemical bonds with other atoms and are therefore called *valence electrons*. The inner orbitals are called *core* orbitals.

2.2.6 SLATER'S SCREENING RULES: SIZE OF ATOMS

The core electrons screen the positive charge of the atomic nucleus in its action on the valence electrons. It is possible to define an effective charge of the nucleus, Z_{eff}, which is smaller than the actual charge, Z. This effective nuclear charge, Z_{eff}, substitutes Z in the time-independent SE for the electron:

$$-\frac{1}{2}\Delta\psi - \frac{Z_{eff}}{r}\psi = E\psi. \qquad (2.23)$$

We assume that a part of each occupied AO contributes to a screening factor S of the nucleus. According to electrostatics, charges inside the radius of the test charge contribute to S. If we know S, Z_{eff} may be calculated as

$$Z_{eff} = (Z-S). \qquad (2.24)$$

The orbitals obtained from Equation 2.23 decrease more slowly than the hydrogen-like ones. The number of nodes and the angular dependence remain the same.

The screening factor S gets its contributions from the inner electronic shells and the shell where the electron belongs to. Nothing comes from shells with higher principal quantum number n than the orbital of the electron. According to *Slater's rules*, S should be calculated in the following ways:

Contribution to S from the same shell: 0.35 from every other electron

Contribution to S from near shell below: 0.85 per electron

Contribution to S from deeper shells: 1.0 per electron. (2.25)

The contributions are summed and used in Equation 2.24 to calculate Z_{eff}.

As an example, we choose the carbon atom with configuration $1s^2 2s^2 2p^2$ and derive α for the 2s and 2p orbitals: contribution from the same shell: $3 \times 0.35 = 1.05$; contribution to S from inner shells: $2 \times 0.85 = 1.7$; sum: $S = 2.75$. Thus, $Z_{eff} = 6 - 2.75 = 3.25$. The 2s and 2p orbitals should then be described by $r \exp(-3.25r/2)$. The constant term in the Laguerre polynomial for 2s is neglected. Thus, α in Equation 2.14 is equal to 1.625.

Slater derived the screening rules from empirical data and these rules work well, down to the transition metal series. The hydrogen-like 3d and 4f orbitals are described by a single term in the Laguerre polynomial, but this simple description is not favorable when penetration and screening are included. The 3d and 4f orbitals have an inner part where screening is less important than in the outer parts. At least two exponential functions are needed for a relevant description.

At increasing value of Z, it may be energetically favorable to start filling a sub-shell with a higher principal quantum number (for example, 4s) before the first shell (3d) is started. This holds true for neutral atoms, but not necessarily for positive ions.

The rare gas atoms are the smallest in each period, since the screening is not complete in the outer shell (the *valence shell*); therefore, the effective charge is larger the higher the number of electrons. The smallest atom is the He atom with a radius only 60% of that of the hydrogen atom. The neon atom is about the same size as the hydrogen atom. In the heavier rare gas atoms, the increase of the factor n^2 is more important than the increase of the effective charge of the nucleus due to less screening. The size of the rare gas atoms thus increases with the atomic number.

The largest atom is the caesium atom. The reason is a high principal quantum number for the valence orbital, while at the same time, screening is high since all other electrons are located in the inner shells.

2.3   EQUATION OF MOTION FOR SINGLE ELECTRONS

An improvement of Slater's screening rules would be to make the screening potential consistent with the actual charge density of the occupied orbitals. The English physicist Douglas R. Hartree advanced and developed this idea.

As we saw in the previous chapter, one-electron representation of the motion of electrons can never be exact. The best many-electron wave function is obtained if we put the spin orbitals into a Slater determinant. To obtain a good many-electron wave function, it is necessary to include other configurations, as demanded by the variation principle. We will first review the most popular one-electron methods, in particular the Hartree–Fock method. This method was used primarily for atoms for more than 30 years, until 1950.

2.3.1   HARTREE'S SELF-CONSISTENT FIELD (SCF) METHOD

Hartree set forth to calculate wave functions for atoms with many electrons soon after the discovery of the SE of the hydrogen atom. His idea was to write down the repulsive average potential for the electrons, which is easily done using electrostatic theory. Then, he solved the SEs for the electrons in the potential of the nucleus and the screening potential.

The problem is that the screening potential can only be calculated if the orbitals are already known. Therefore, Hartree had to find an *iterative* solution. He made an initial guess as to the extension of the orbitals in space and thus calculated a screening potential. The SE for the orbitals was solved and the screening potential updated. The procedure was repeated and new orbitals were calculated, until the orbitals no longer changed from one iteration to the next. The final result was said to be *self-consistent*. The screening potential now corresponds exactly equal to the repulsion between the electrons of the occupied orbitals.

Hartree did not have access to computers, of course, but he constructed the so-called differential analyzer with the help of a Meccano kit. He managed to calculate accurate AOs already in 1928.

2.3.2 HARTREE–FOCK

Later, J. C. Slater showed that the Hartree equations can be obtained if the variation principle is applied to a product of spin orbitals. The Russian theoretical physicist Vladimir A. Fock pointed out that certain symmetry conditions are not obeyed in the Hartree method, of which the most important one is the antisymmetric property of the total wave function. The variation principle was now applied to an antisymmetrized product of spin orbitals, that is, a Slater determinant. This is a fundamental method in electronic structure calculations and is referred to as the *Hartree–Fock method* or simply "Hartree–Fock."

In atomic units and assuming a fixed nucleus, we may write the Hamiltonian for an atom with N electrons as

$$H = \sum_{i=1}^{N} \left[-\frac{1}{2} \Delta_i - \frac{Z}{r_i} \right] + \sum_{i<j}^{N} \frac{1}{r_{ij}}; \quad r_{ij} = |\vec{r}_i - \vec{r}_j|, \tag{2.26}$$

where Z is the atom number (the nuclear charge is $|e|Z = Z$ a.u.). The first summation contains kinetic energy and nuclear attraction. The second summation is the operator for the Coulomb repulsion between the electrons. Due to the mathematical form of this operator, we cannot write the solution exactly as an antisymmetrized orbital product. Instead, we use the variation principle to find the lowest possible expectation value of the energy over a Slater determinant Φ, constructed from the spin orbitals $\{\psi_i\}_{i=1,N}$. The expectation value is

$$\langle \Phi | H | \Phi \rangle = \sum_{i=1}^{N} \left\langle \psi_i \left| -\frac{1}{2} \Delta - \frac{Z}{r} + \frac{1}{2} \sum_{j=1}^{N} (J_j - K_j) \right| \psi_i \right\rangle. \tag{2.27}$$

The derivation of Equation 2.27 is mathematically simple and straightforward, but lengthy, so it is left out here. Notice that capitals are used for many-electron wave functions, while lowercase symbols are used for spin orbitals.

The *Coulomb operator* J_j corresponds to the Coulomb repulsion at $\bar{r}_i$ from the electron in spin orbital ψ_j and is defined by

$$J_j\psi_i(x_1) = \int \frac{\psi_j^*(x_2)\psi_j(x_2)}{r_{12}} dx_2\psi_i(x_1). \tag{2.28}$$

The *exchange operator* K_j is defined by

$$K_j\psi_i(x_1) = \int \frac{\psi_j^*(x_2)\psi_i(x_2)}{r_{12}} dx_2\psi_j(x_1). \tag{2.29}$$

The integral operator K_j is almost negligible, except when it acts on its own spin orbital ψ_j. Since $K_j\psi_j(x) = J_j\psi_j(x)$, $J_j\psi_j(x)$ and $K_j\psi_j(x)$ cancel in Equation 2.27. Physically, this means that the electron in spin orbital ψ_j does not repel itself. There is no *self-energy*.

Equation 2.27 is the total energy if all electrons are obtained from the following differential equation, called the *Fock equation*:

$$-\frac{1}{2}\Delta\psi - \frac{Z}{r}\psi + \sum_{j=1}^{N}\left(J_j - K_j\right)\psi = \hat{F}\psi = \varepsilon\psi, \tag{2.30}$$

where $\hat{F}$ acts on one-electron functions and is called the *Fock operator*. Equation 2.30 has a very simple meaning. The kinetic energy operator and the attraction to the nucleus are the same as in any one-electron equation for an atom. The screening terms consist of the repulsion between all electrons, compensated, as expected, by the exchange terms.

The Fock Equation 2.30 has solutions that are occupied or unoccupied, meaning that they are included or not included, respectively, in J and K. The unoccupied orbitals, called *virtual orbitals*, are not subject to the self-energy correction, since an unoccupied orbital is not included in J and K.

We do not yet know if the spin orbitals that are the solution to Equation 2.30 are also the looked-for spin orbitals, minimizing the energy expectation value in Equation 2.27. The proof will be carried out in steps. First, we have to construct a many-electron operator that corresponds to the Fock one-electron operator. This is easy to do if we subtract a constant term to compensate for the fact that electron repulsion is counted twice. We obtain the many-electron Hamiltonian as

$$H_0 = \sum_{i=1}^{N}\left[-\frac{1}{2}\Delta_i + \frac{Z}{r_i} + \sum_{j=1}^{N}\left(J_j - K_j\right)\right] - \frac{1}{2}\sum_{i=1}^{N}\sum_{j=1}^{N}\left\langle\psi_i\left|J_j - K_j\right|\psi_i\right\rangle. \tag{2.31}$$

It is now easy to check by insertion that the Slater determinant is an eigenfunction of the Hamiltonian operator in Equation 2.31 with the eigenvalue equal to the expectation value of Equation 2.27. It may also be shown (see Section 2.3.3) that the spin orbitals satisfying Equation 2.30, actually give the lowest total energy.

The occupied spin orbitals included in the operators J and K have to be the solutions of Equation 2.30. An iterative method is used, where the successive solutions of Equation 2.30 define J and K. This procedure is repeated until the energy eigenvalues are within a stipulated convergence limit. This solution is called the self-consistent field (SCF) solution. Physically, SCF means that screening and penetration effects are taken into account in the best possible way within the one-electron approximation.

The Hartree–Fock approximation works well regarding electron density (see Section 2.3.6) for both atoms and molecules. Even the spin polarization, which is a consequence of an unequal number of up- and down-spin electrons, agrees well with experimental data.

The Hartree–Fock energy is always related to an error, the correlation energy, but serves well as the point of departure for further improvement of the wave function.

Orbitals that are the solutions of Equation 2.30 are called *canonical* Hartree–Fock orbitals. Please remember that a unitary transformation can be applied to the orbitals of a Slater determinant without changing the total (Slater determinant) wave function.

2.3.3 Brillouin's Theorem

The Hartree–Fock method is related to a number of theorems, for example, the *Brillouin theorem* (Equation 2.32). We form a new Slater determinant by substituting one of the occupied spin orbitals, ψ_i, by an unoccupied one, ψ_a, to obtain a new Slater determinant, called Φ_i^a; ψ_a is orthogonal to all occupied ψ_i. We calculate $\left\langle \Phi_i^a \middle| H \middle| \Phi \right\rangle$ and find

$$\left\langle \Phi_i^a \middle| H \middle| \Phi \right\rangle = \left\langle \psi_a \middle| -\frac{1}{2}\Delta + \frac{Z}{r} + \sum_{j=1}^{N} (J_j - K_j) \middle| \psi_i \right\rangle = \varepsilon_i \left\langle \psi_a \middle| \psi_i \right\rangle = 0. \quad (2.32)$$

The proof is nontrivial, but is left out. We may now show that the occupied orbitals, which satisfy Equation 2.30, are those that give the best expectation value of the total energy according to the variation principle. Let $\delta\Phi$ be a variation in the wave function Φ and consider the change $\delta\left\langle \Phi \middle| H \middle| \Phi \right\rangle$. H cannot be changed, so the variation is in the Slater determinant Φ and is equal to $\delta\Phi$. If $\delta\Phi$ is composed of orbitals that are already occupied, applying $\delta\Phi$ amounts to forming new linear combinations of the same orbitals. It is quite easy to show that the variation has to be in a single spin orbital, say ψ_i, to which a part of a previously unoccupied spin orbital ψ_a is added, say $\delta x \cdot \psi_a$. According to Equation 2.32, we then have

$$\left\langle \delta\Phi \middle| H \middle| \Phi \right\rangle = \delta x \cdot \left\langle \Phi_i^a \middle| H \middle| \Phi \right\rangle = 0. \quad (2.33)$$

This is Brillouin's theorem.

The conclusion is that the spin orbitals that satisfy Equation 2.30, and thereby Equation 2.33, are the best possible ones from the point of view of total energy and thus the desired optimal Hartree–Fock spin orbitals.

It also follows from Brillouin's theorem that the total energy cannot be improved in a configuration interaction (CI) calculation that involves only the ground state

Slater determinant and the singly substituted Slater determinants. The off-diagonal matrix elements in the CI calculation are equal to zero by Equation 2.33. However, it is meaningful to involve all singly substituted Slater determinants in a calculation of the excited states.

If the number of electrons is even, the spatial functions will be the same for one spin-up and one spin-down spin orbital, since Equation 2.30 is satisfied for both spins. Whether different orbitals for different spins give a better energy is an old problem in quantum chemistry, which we will not discuss here.

If the number of electrons is odd, Equation 2.30 will not be the same for spin up and spin down. One example is the ground state of the atom Li ($1s^2 2s$). In this case, the 1s orbital is occupied with α and β, but the 2s orbital only with α. The $1s\alpha$ orbital will have an attractive exchange term with $2s\alpha$, and is therefore more attracted to the nucleus than the $1s\beta$ orbital that has no such exchange interaction. The difference between the orbitals is hardly visible in a diagram, but some expectation values, for example, the spin density at the nucleus, have a quite large contribution from the difference between the $1s\alpha$ and $1s\beta$ orbitals.

Unfortunately, differences in the spatial orbitals for different spins may be impractical if CI calculations are performed afterward to improve the Hartree–Fock calculation. Therefore, the variation is restricted, to keep the same spatial function for spin up and spin down. This method is called restricted Hartree–Fock (RHF). The original method with unlimited variation of the orbital functions is called unrestricted Hartree–Fock (UHF). Only the latter satisfies the Brillouin theorem.

Application of the Hartree–Fock method shows that the electrons to some extent penetrate into inner shells where they are less screened. The accidental degeneracy between orbitals with different l quantum numbers disappears. For example, 2s is more strongly bound than 2p. A lower l quantum number has a lower energy for the same principal quantum number.

2.3.4 Ionization Energy, Electron Affinity, and Disproportionation Energy

The *ionization energy* for atoms and molecules may be determined by studying the ionization limit in a monochromatic EM field. The ionization limit is the first ionization energy, the lowest frequency necessary to ionize the atom or molecule (equal to $v = \Phi/h$ in Figure 1.2). The second, third, etc., ionization energies refer to consecutive ionization of the ions.

In principle, *oxidation* is the same phenomenon as ionization. However, oxidation refers to a solid or liquid medium where a number of corrections have to be applied. In vacuum, the first ionization energy is defined as

$$I = E_{N-1} - E_N, \tag{2.34}$$

where E_{N-1} is the ground state energy of the ionized system and E_N is the ground state energy of the neutral system. The second ionization energy is defined as the first ionization energy of the positive ion.

The *electron affinity* (~gas *phase reduction potential*) is defined as

$$A = E_N - E_{N+1}. \tag{2.35}$$

The electron affinity is thus positive if it is possible to lower the total energy by attaching an electron. A negative ion usually has itself a negative electron affinity, meaning that doubly negative ions do not exist except in a few cases (C_{60}^{2-} is one of these exceptions).

If Equation 2.35 is subtracted from Equation 2.34, we obtain the *disproportionation energy* U:

$$U = I - A = E_{N-1} + E_{N+1} - 2E_N, \tag{2.36}$$

where U refers to the same conditions as I and A. Under vacuum conditions (or gas phase), the process is one of moving one electron from one atom to the next. U is positive and ranges from a few to several tens of electron volts. In physics, U is referred to as "Hubbard U," named after the English physicist John Hubbard.

In chemistry, the American chemist Ralph G. Pearson (1963) introduced the concept of hardness. Originally, the definition referred to hard and soft acids and bases. The hard-soft acid-base (HSAB) principle states that hard "Lewis acids" prefer hard "Lewis bases" and conversely that soft Lewis acids prefer soft Lewis bases. We will not dig deeply into the chemical background here. What is important is that Pearson (~1980) later found that hardness may simply be defined as

$$\eta = \frac{I - A}{2}, \tag{2.37}$$

which is formally equal to U/2. Hardness is used in typical chemical situations, for example, to predict that an aqueous solution of NaCl and AgF form NaF and AgCl. The Pearson model is a qualitative model where one works with atomic quantities.

In a molecule as well as in solids, the thermodynamic disproportionation energy is an *adiabatic* energy difference, meaning that the equilibrium distances for the nuclei have been used in the definition. When one electron is moved from one site to the next, the equilibrium distances change. Hubbard U, on the other hand, refers to constant positions of the nuclei.

2.3.5 KOOPMANS' THEOREM

One may define ionization energies from the inner shells if the monochromatic radiation has sufficient energy, $h\nu$, to remove an electron from an inner subshell, say subshell i, and leave the atom or molecule in the state $E_{i,N-1}$ (which is an "almost" discrete state in a continuum of states). The velocity (v), and thereby kinetic energy ($mv^2/2$), of the ejected electron is measured. We have

$$h\nu - mv^2/2 = I_i = E_{i,N-1} - E_N. \tag{2.38}$$

The higher the ionization energy, the lower is the velocity (v) of the emitted electron.

In Equation 2.38, the ionization energy from the inner shells is slightly modified by the chemical environment. This energy correction is measured by a technique called ESCA (electron spectroscopy for chemical analysis). It is a useful technique, particularly to study the surface and parts of the bulk near the surface. If UV light is used instead of x-rays, the valence shell may be analyzed in greater detail.

In a hydrogen-like atom, the ionization energy is equal to the negative of the orbital energy of the electron. In a many-electron system, one might suspect that the negative of the orbital energy in the Hartree–Fock model is related to the ionization energy. This relationship will be studied next.

Koopmans' theorem states that the negative of the orbital energy is approximately equal to the ionization energy. To derive Koopmans' theorem, we calculate E_N and $E_{i,N-1}$ using the Hartree–Fock approximation. The assumption is made that the orbitals of the neutral system can also be used to calculate a reasonably accurate expectation value for the ionized system. Φ^k denotes a Slater determinant missing spin orbital ψ^k. Using Equation 2.38,

$$I_k = E_{k,N-1} - E_N \approx \left\langle \Phi^k \left| H_{N-1} \right| \Phi^k \right\rangle - \left\langle \Phi \left| H \right| \Phi \right\rangle$$

$$= \sum_{i=1;i\neq k}^{N} \left\langle \psi_i \left| -\frac{1}{2}\Delta_1 + V(\vec{r}_1) + \frac{1}{2}\sum_{j=1;j\neq k}^{N} \left(J_j^1 - K_j^1 \right) \right| \psi_i \right\rangle$$

$$- \sum_{i=1}^{N} \left\langle \psi_i \left| -\frac{1}{2}\Delta_1 + V(\vec{r}_1) + \frac{1}{2}\sum_{j=1}^{N} \left(J_j^1 - K_j^1 \right) \right| \psi_i \right\rangle \qquad (2.39)$$

$$= -\left\langle \psi_k \left| -\frac{1}{2}\Delta_1 + V(\vec{r}_1) + \sum_{j=1}^{N} \left(J_j^1 - K_j^1 \right) \right| \psi_k \right\rangle = -\left\langle \psi_k \left| \varepsilon_k \right| \psi_k \right\rangle = -\varepsilon_k.$$

In this equation, we have used $V(\vec{r}_1)$ for the attraction of the electrons to all nuclei. We have also used that

$$\left\langle \psi_k \left| \left(J_j - K_j \right) \right| \psi_k \right\rangle = \left\langle \psi_j \left| \left(J_k - K_k \right) \right| \psi_j \right\rangle. \qquad (2.40)$$

This completes the proof of Koopmans' theorem. The conclusion is that the orbital energy ε_k in the Hartree–Fock method may be interpreted as (approximately) the negative of the ionization energy.

Note that Koopmans' theorem does not state that the negative of the orbital energy is *exactly* equal to the ionization energy. The correct definition of ionization energy is given in Equation 2.38: $I_k = E_{k,N-1} - E_N$. $E_{k,N-1}$ has to be obtained with the correct wave function for the ionized system, but in the proof we use the orbitals of the neutral system. Using the relaxed orbitals of the ionized system would make $E_{k,N-1}$ more negative and thus I_k smaller. However, there is a compensating correction, the correlation energy correction. The latter depends to a large extent on how many electron pairs are correlated. There is thus a larger correlation energy correction on E_N than on $E_{k,N-1}$, since the neutral system contains more electron pairs. This correction thus lowers E_N more than $E_{k,N-1}$, hence the correlation error compensates approximately for the relaxation error.

It can be proven in about the same way that electron affinity is approximately the negative of the orbital energy of an unoccupied orbital. In this case, the error tends to be larger since the above mentioned compensation between relaxation and correlation does not happen. Instead, the errors add up. The experimental electron affinity is often a few electron volts higher than the calculated one, using Koopmans' theorem. The result is improved if E_N and E_{N+1} are calculated from many-electron energies and accurate wave functions.

Hartree–Fock theory plays a great role in quantum chemistry. In molecules, the orbitals are referred to as *molecular orbitals* (MO). The orbital energy of the highest occupied MO (HOMO) is approximately equal to the ionization energy according to Koopmans' theorem. The orbital energy of the lowest unoccupied MO (LUMO) is approximately equal to the electron affinity. By Equation 2.36, the energy difference between the HOMO and LUMO approximates Hubbard U, according to Koopmans' theorem.

A common misunderstanding is that the HOMO–LUMO energy difference is an excitation energy (=the band gap in a solid). In calculations for atoms, molecules, or solids, the HOMO–LUMO gap is *not* equal to the excitation energy. This is consistent with the fact that the virtual orbital energy is uncorrected for self-energy (see Section 2.3.2). An excited electron, of course, has the same number of electrons around it as an electron in the ground state. The true excitation energy can only be approximated if two different calculations are carried out, one for the ground state and one for the excited state, and the total energy difference is taken.

2.3.6 Møller–Plesset (MP) Theorem

The density and spin density provided by the UHF method are generally quite accurate. The theorem of Møller and Plesset (MP) states that the UHF density and spin density are correct to second order in a perturbation theory where the second-order energy contribution is nonzero, more precisely of the order of the correlation energy.

The perturbation theory is set up as follows (Appendix 3). The "unperturbed" state is supposed to be the eigenstate of the many-electron operator of Equation 2.31, with the UHF Slater determinant as the zero-order function. The perturbation V is simply the difference between the exact Hamiltonian operator H and the approximation H_0 of Equation 2.31. The excited states of the unperturbed Hamiltonian H_0 are singly, doubly, and higher substituted Slater determinants. The substituting orbitals are the virtual orbitals ϕ_a of the UHF method.

In standard perturbation theory, the first-order correction to the energy is

$$E_0^1 = \left\langle \Psi_0^0 \middle| V \middle| \Psi_0^0 \right\rangle. \tag{2.41}$$

It is quite easy to show that this contribution is equal to zero (if the repulsion energy had not been subtracted in Equation 2.31, it would have appeared as the first-order correction). The second-order contribution to the energy is

$$E_0^2 = \sum_{i=1}^{\infty} \frac{\left|\left\langle \Psi_0^0 \left| V \right| \Psi_i^0 \right\rangle\right|^2}{E_0^0 - E_i^0}. \tag{2.42}$$

This energy is of the order of the correlation energy and thus very significant. The largest contribution here comes from the doubly substituted Slater determinants.

The first-order contribution to the wave function is (Appendix 3)

$$\Psi_0^1 = \sum_{i=1}^{\infty} \frac{\left|\left\langle \Psi_0^0 \left| V \right| \Psi_i^0 \right\rangle\right|}{E_0^0 - E_i^0} \Psi_i^0. \tag{2.43}$$

It is easy to show that if the excited state, Ψ_i^0, is a singly substituted Slater determinant, the matrix element is equal to zero. Since $V = H - H_0$ and the functions appearing are eigenfunctions of H_0, the integrals with H_0 vanish because of orthogonality between the eigenfunctions. The H-integrals vanish for singly substituted determinants because of Brillouin's theorem. We conclude that none of the Ψ_i^0 appearing in the perturbation expansion are singly substituted. All are, at least, doubly substituted.

We thus obtain for the density:

$$\rho = \int \left(\Psi_0^* + \Psi_0^{1*}\right)\left(\Psi_0 + \Psi_0^1\right) d\tau_2 \ldots d\tau_N = \int \Psi_0^* \Psi_0 d\tau_2 \ldots d\tau_N + \int \Psi_0^1 \Psi_0^1 d\tau_2 \ldots d\tau_N. \tag{2.44}$$

The cross terms between Ψ_0 and Ψ_i^0 vanish since the determinants are different in at least two orbitals. The remaining term is a second order term since Ψ_i^0 is of first order. Their Hartree–Fock density (or density matrix for that matter) is thus correct to second order.

2.3.7 BEST OVERLAP ORBITALS

We have derived the Hartree–Fock method from the requirement that the orbitals should be those that give the lowest energy expectation value of a Slater determinant. In a CI expansion where only singly substituted Slater determinants are used, all coefficients for the latter are equal to zero according to Brillouin's theorem. However, if doubly substituted Slater determinants (and higher) are introduced in the total wave function, the coefficients for the singly substituted Slater determinants are no longer equal to zero.

It is possible to choose spin orbitals where the coefficients of the singly substituted determinant are equal to zero, even if doubly and higher substitutions are introduced. In a CI type expansion using these orbitals, the Slater determinant with the latter spin orbitals has the maximum possible overlap with the true wave function. The new spin orbitals are therefore called *best overlap orbitals*. In a Slater determinant Φ, these orbitals minimize

$$\int |\Psi - \Phi|^2 \, d1d2...dN, \tag{2.45}$$

where Ψ is the exact wave function. Since Φ and Ψ are normalized, the overlap integral

$$\int |\Psi\Phi| \, d1d2...dN \tag{2.46}$$

is maximized. The best overlap orbitals may be determined from an equation like Equation 2.46, where the integration does not include integration over 1 (see paper by Larsson). Notice that best overlap orbitals are determined only up to a unitary transformation of the orbitals.

The best overlap orbitals are not the same as the natural spin orbital, but are probably very close (choosing the best unitary transformation of the best overlap orbitals). In a CI expansion, the singly substituted determinants disappear if the best overlap orbitals of the same CI problem are used. The best overlap orbitals are also called *Brueckner orbitals* (BO), after the American physicist Keith Brueckner, or are referred to as *correlation corrected orbitals*.

The problem with the best overlap orbitals is that the total wave function has to be calculated before the correlation corrected orbitals can be obtained. This problem may be circumvented by calculating correlation potentials for each correlated pair of electrons and subsequently sum the correlation potentials for all pairs.

2.3.8 EXCHANGE HOLE

The sum of J terms in the Fock operator Equation 2.27 permits a very simple representation:

$$J = \sum_{j=1}^{N} J_j = \int \frac{\sum_{j=1}^{N} \psi_j^*(x_2)\psi_j(x_2)}{r_{12}} dx_2 = \int \frac{\rho(\vec{r}_2)}{r_{12}} dv_2, \tag{2.47}$$

where $\rho(\vec{r})$ is the total electronic density at the point $\vec{r}$. Working with the density instead of the orbitals is a great simplification in the Hartree equations. Unfortunately, Equation 2.47 includes repulsion between all electrons including the electron whose equation we are writing down. The self-energy cannot be subtracted as in Equation 2.27. If it is subtracted by just removing one of the terms in the summation in Equation 2.47, there will be different equations for different orbitals. There is no longer any guarantee that the final orbitals are orthogonal. In the UHF and RHF methods, these problems are resolved, but the exchange potentials lead to rather complicated equations. Slater and others therefore set up approximation schemes for the exchange operator K in terms of the density ρ.

The sum of all exchange operators, acting on an orbital ϕ, may be written as

$$\hat{K}\phi = \sum_{j=1}^{N} K_j\phi = \sum_{j=1}^{N} \int \frac{\phi_j^*(x_2)\phi(x_2)}{r_{12}} dx_2 \phi_j(x_1).$$ (2.48)

To get insight into the action of K, Equation 2.48 is rewritten as

$$\hat{K}\phi = \sum_{j=1}^{N} K_j\phi = \sum_{j=1}^{N} \int \frac{1}{r_{12}} \frac{\phi_j^*(x_2)\phi^*(x_1)\phi_j(x_1)\phi(x_2)}{\phi^*(x_1)\phi(x_1)} dx_2 \phi(x_1).$$ (2.49)

Equation 2.49 looks similar to Equation 2.47, except that instead of the charge density $\rho(\vec{r})$, we have an *exchange charge density*:

$$\rho_{ex}(x_1,x_2) = \sum_{j=1}^{N} \frac{\phi_j^*(x_2)\phi^*(x_1)\phi_j(x_1)\phi(x_2)}{\phi^*(x_1)\phi(x_1)}.$$ (2.50)

Where is the exchange charge density ρ_{ex} located, when the exchange operator acts on an occupied spin orbital? We assume that electron 1 is fixed in its position with spin up. Set $\phi = \phi_k$ and calculate ρ_{ex} at the point $\vec{r}_2$. The resulting function is still complicated. First, we find the value of ρ_{ex} at the point $\vec{r}_1$ where the first electron is located:

$$\rho_{ex}(x_1,x_1) = \sum_{j=1}^{N} \frac{\phi_j^*(x_1)\phi_k^*(x_1)\phi_j(x_1)\phi_k(x_1)}{\phi_k^*(x_1)\phi_k(x_1)} = \sum_{j=1}^{N} \phi_j^*(x_1)\phi_j(x_1).$$ (2.51)

Obviously, if the exchange operator $\hat{K}$ acts on a spin-up electron, the operator removes the full charge of all spin-up electrons at the point $\vec{r}_1$. The corresponding is true if it works on a spin-down electron. No spin-down charge is removed in the equation for the spin-up electron, and vice versa.

To find out exactly how much charge is removed around point $\vec{r}_1$, we integrate Equation 2.50 over the coordinates of the second electron:

$$\int \rho_{ex}(x_1,x_2) dv_2 = \int \sum_{j=1}^{N} \frac{\phi_j^*(x_2)\phi_k^*(x_1)\phi_j(x_1)\phi_k(x_2)}{\phi_k^*(x_1)\phi_k(x_1)} dv_2$$

$$= \sum_{j=1}^{N} \frac{\phi_k^*(x_1)\phi_j(x_1)}{\phi_k^*(x_1)\phi_k(x_1)} \delta_{jk} = 1.$$ (2.52)

It is not too difficult to show that the charge removed tends to the full charge in a quadratic way close to $\vec{r}_1$. It is as if the electron in its motion is surrounded by a hole of missing charge. This hole is called the *exchange hole*. In this way, the Hartree–Fock method accounts for correlation between electrons with the same spin.

2.3.9 LOCAL EXCHANGE AND DENSITY FUNCTIONAL THEORY

The earliest density-based method, dating back to 1927, is due to Thomas and Fermi. An addition by Dirac to account for the exchange hole resulted in the Thomas–Fermi–Dirac model. The theory was much improved by Slater, Gombás, and Gáspár and Kohn, Hohenberg, and Sham. The Slater X_α method is an interpolation scheme between different $\rho^{1/3}$ exchange methods. The Hohenberger–Kohn–Sham (HKS) method was interpreted to mean that the exact energy can be obtained in one-electron equations if the potential term is a certain functional of the electronic density. The HKS method was improved in several stages. Generally, these methods give results for bond lengths, which are in better agreement with experimental numbers, than the Hartree–Fock method.

Unfortunately, there are several misconceptions about HKS theories. We have to remember that all wave functions derived from density functional theory (DFT) are single Slater determinants and therefore represent a function that does not possess enough flexibility to mathematically represent the true wave function.

1. *Slater's derivation*: Previously we found that in the Hartree–Fock equation, the electron moves in the repulsive field of all electrons, including itself, but with density subtracted corresponding to one missing electron around it. We assume that the density is constant. Slater derived the radius of the hole (r_s) and then calculated the decrease in the repulsive potential because of the hole. The following relation must hold:

$$\frac{4\pi}{3} r_s^3 \rho = 1 \Rightarrow r_s = \left(\frac{3}{4\pi\rho}\right)^{1/3}. \tag{2.53}$$

We now integrate the potential field due to the missing electron. This is easy, since we have assumed a constant density

$$V_{exch} = -\int_0^{r_s} \frac{\rho}{r} 4\pi r^2 dr = -\frac{4\pi r_s^2}{2}\rho = -\left(4.5\pi\right)^{1/3}\rho^{1/3}. \tag{2.54}$$

V_{exch} has to replace the K operator in the Hartree–Fock equations. In the resulting equation, the Hamiltonian is a function of the density of α or β spin density. The Slater exchange has the correct dependence of the density with power of 1/3, but it seems to overcorrect for some inadequacies in the results of the Hartree–Fock method.

2. *Gáspár's derivation*: In the original Thomas–Fermi model, one derives that the kinetic energy density is proportional to the 5/3 power of the electronic density. The total kinetic energy is then given by

$$T = const \int \rho(r)^{5/3} dv. \tag{2.55}$$

Using a novel derivation, Gáspár found that the exchange operator may be written as proportional to the 1/3 power of the density, but with a proportionality factor different from Slater's. At this stage of development, the total energy may be calculated as a functional of the electronic density.

Still, the approximation for the kinetic energy in Equation 2.55 is the main problem in the theory of Gáspár. In later improvements, Slater arrived at the Xα method, where orbitals are calculated and the one matrix (Section 1.8) is obtained. The expression for the kinetic energy is derived from this one matrix. In this method, the parameter α multiplies the local exchange expression, and this makes it possible to use empirical data to interpolate between α = 2/3 (Gáspár) and α = 1 (original Slater).

3. *The Hohenberg–Kohn and Kohn–Sham theorems* simply state that the total energy can be obtained by applying the variation principle to the total energy density functional and suggest that the one-electron equations obtained in this way also account for electronic correlation. HKS derive another theorem that the total energy is uniquely determined by the density.

2.3.10 DFT Method as a Practical Calculation Method

The Kohn–Sham methodology leads to a one-electron equation, which in some cases appeared to give better results than the Hartree–Fock method. Nowadays, the local exchange methods are referred to as density functional theories (DFT). The DFT method is claimed to account for electronic correlation. It is commonly believed that a density functional, including exchange, can be found, which will give the correct total energy of the system. However, this functional has never been found and it remains questionable whether it can be defined without first solving the exact many-electron SE.

There is an ongoing effort to improve the density functionals. Becke interpolated in linear sums between DFT and Hartree–Fock exchange potentials to agree with the experiments and accurate calculations. The most popular of these potentials is a three-parameter hybridization between Hartree–Fock and local exchange by Becke (B3). The potential also includes correlation potentials based on correlation theory of Lee, Yang, and Parr (LYP). This method is thus referred to as B3LYP.

The DFT method has been very useful in the calculation of bond lengths, vibrational spectra, and a number of other properties of molecules. Contrary to the case with the Hartree–Fock method, which can be improved using various forms of CI, there is no straightforward way to improve the DFT methods further. One cannot use CI in DFT since the method is already supposed to contain correlation effects. A common opinion appears to be that no further refinement is necessary for the ground state.

The following criticism of the DFT method appears well-founded:

1. It is true that with the correct density functional, we may obtain the correct energy. This is hardly of any interest, however, since no "correct" density functional has ever been proposed. The correct functional does not have to be a simple mathematical expression. It is not even known what kind of

functional would be needed to obtain the correct kinetic energy from the density. In most work, the kinetic energy is obtained from the Slater determinant wave function in the normal way.

2. The correlation potential cannot be obtained in any simple way. In some applications, correlation is independent of density. For example, choose a series of two-electron atoms (He, Li$^+$, Be^{2+}, ..., U^{90+}). The density is increasing with Z, since the size of the 1s orbital is inversely proportional to 1/Z. However, as shown by J. Linderberg and H. Shull, the correlation energy is almost constant at 0.04 eV.

3. A number of expectation values cannot be obtained from the density, but require the one-matrix or the two-matrix. To obtain the one-matrix or two-matrix, one has to first define a wave function. Normally, the Slater determinant for the N lowest energy orbitals is used, but a single Slater determinant cannot possibly be the correct wave function. As has been shown at the end of Chapter 1, the correct one-matrix contains weakly occupied NSOs, because the correct wave function is a superposition of many Slater determinants. It is unthinkable that the DFT orbitals would give correct results for all expectation values, when the nonzero occupation numbers of the one-matrix are incorrectly equal to unity.

In this book, a number of cases where correlation effects are important will be presented. Multiconfigurational expansion of the wave function cannot be neglected. Unfortunately, trying to account for correlation accurately on the basis of DFT is a dead end.

2.4 CORRELATION AND MULTIPLET THEORY

It is very common that the ground and excited states cannot be described with a single Slater determinant. The configuration (Slater determinant) $1s\alpha 2s\beta$ for the helium atom corresponds to two states, ^1S $[(1s2s + 2s1s)(\alpha\beta - \beta\alpha)]$ and ^3S $[(1s2s - 2s1s)(\alpha\beta + \beta\alpha)]$, with a substantial energy difference between them. The main reason is that the Slater determinant is not an eigenfunction of the square of the angular momentum operator. There are large correlation effects also in cases when the state is a spin singlet state and a single Slater determinant is a reasonable approximation. In this section, we will be dealing with methods that go beyond the Hartree–Fock (or DFT) methods, with the aim to describe the exact solution for the many-electron system.

2.4.1 HYLLERAAS' METHOD

The Hamilton operator for the atom He (using atomic units and fixed nucleus) consists of the kinetic energy for two electrons and all possible Coulomb interactions between the positively charged nuclei and the negatively charged electrons:

$$H\Psi = -\frac{1}{2}(\Delta_1 + \Delta_2)\Psi + \left(-\frac{2}{r_1} - \frac{2}{r_2} + \frac{1}{r_{12}}\right)\Psi = E\Psi. \qquad (2.56)$$

The Coulomb repulsion between the two electrons is $1/r_{12}$, where r_{12} is the distance between the electrons. We are going to use the variation principle on a linear expansion. All basis functions contain one (or more) exponential function where the effective nuclear charge is obtained from screening theory (Section 2.2.6). Since there is one other electron in the same shell, the screening factor is equal to $S = 0.35$. A rather good function to approximate the 1s orbital is thus $\exp[-(2 - 0.35)r] = \exp(-1.65r)$.

This solution may be improved in several ways. Using the variation principle, it is found that $\alpha = 1.7$ is a better value than 1.65, but the improvement is quite negligible.

A more significant improvement is obtained by using a linear expansion where some of the terms contain the two-electron function r_{12}:

$$\Psi = \sum_{i=1}^{M} C_i \Phi_i. \tag{2.57}$$

The coefficients C_i are determined variationally in an eigenvalue problem. The Norwegian physicist Egil Hylleraas was the first to calculate an accurate wave function for the He atom. He used six basis functions of the following type:

$$\Phi_1 = \exp(-1.7r_1)\exp(-1.7r_2)$$

$$\Phi_2 = (r_1 + r_2)\exp(-1.7r_1)\exp(-1.7r_2)$$

$$\Phi_3 = r_1 r_2 \exp(-1.7r_1)\exp(-1.7r_2)$$

$$\Phi_4 = (r_1 + r_2)^2 \exp(-1.7r_1)\exp(-1.7r_2) \tag{2.58}$$

$$\Phi_5 = r_{12} \exp(-1.7r_1)\exp(-1.7r_2)$$

$$\Phi_6 = r_{12}^2 \exp(-1.7r_1)\exp(-1.7r_2).$$

Notice that Φ_5 and Φ_6 are two-electron functions, which cannot be factorized into one-electron functions. By calculating all matrix elements and solving the 6×6 eigenvalue problem, Hylleraas, in 1928, obtained, without comparison, the best description of the helium atom with the energy -2.903329 H, compared to the earlier best value of -2.86 H. With the help of modern computers, it was recently possible to determine the ground state energy with more than accurate 20 decimal places (-2.903724 H) using essentially the Hylleraas method.

The functions Φ_5 and Φ_6 are particularly important since they give the electrons a possibility to *correlate* their positions. The values of Φ_5 and Φ_6 are equal to zero when the electrons are located at the same point in space.

The Hylleraas method gives a very good description of the electronic states, and is the only method that reproduces the cusp for $\vec{r}_2 \rightarrow \vec{r}_1$. The model needs great computer resources and can only be used for small molecules and atoms. Werner Kutzelnigg, the German quantum chemist, has carried out particularly accurate and important work.

2.4.2 CENTRAL FIELD APPROXIMATION

The Hartree–Fock method usually works well only for the ground state and for the lowest state of each symmetry. For example, the 1s, 2s, and 2p orbitals may be part of many possible states:

$$1s^2 \, {}^1S \, (\text{ground state for He})$$

$$1s2s \, {}^1S, {}^3S \, (\text{excited states for He})$$

$$1s2p \, {}^1P, {}^3P \, (\text{excited states for He}) \tag{2.59}$$

$$(2s)^2 \, {}^1S \, (\text{ground state for Be; excited state for He})$$

$$(2p)^2 \, {}^3P, {}^1S, {}^1D \, (\text{ground state for C; excited states for He and C}).$$

Lowercase is used for orbitals and capitals for the total electronic state. Capital S means that the total orbital momentum is $S = 0$. In Equation 2.59, this follows if all orbitals are s orbitals and also if both orbitals are p orbitals. Capital P means that $L = 1$, which follows for the orbital combination (s,p) and (p,p). Upper left index in the total state designation is the spin multiplicity: 1 for a singlet and 3 for a triplet. *Singlet state* means that there is a single state where for every α spin orbital there is a β spin orbital. *Triplet state* appears when there is a possibility for three degenerate states, for which the spins always point in the same direction. This degeneracy is split in a magnetic field.

How are the many-electron states obtained from the orbitals? If we have two orbitals, 1s and 2s, we may create Slater determinants with different spin functions:

$$\Phi_1 = |1s\alpha 2s\alpha|; \, \Phi_2 = |1s\alpha 2s\beta|; \, \Phi_3 = |1s\beta 2s\alpha|; \, \Phi_4 = |1s\beta 2s\beta|. \tag{2.60}$$

Linear combination of $|1s\alpha 2s\beta|$ and $|1s\beta 2s\alpha|$ gives two wave functions:

$$\Psi_+ = |1s\alpha 2s\beta| + |1s\beta 2s\alpha| = |1s2s|(\alpha\beta + \beta\alpha)$$

$$\Psi_- = |1s\alpha 2s\beta| - |1s\beta 2s\alpha| = |1s2s|(\alpha\beta - \beta\alpha), \tag{2.61}$$

where Ψ_+ and Ψ_- are normalized if divided by $\sqrt{2}$. If the linear problem is solved as in Section 1.6.2 after the necessary matrix elements have been calculated, it turns out that Φ_1, Φ_4, and Ψ_+ have the same energy, while Ψ_- has a slightly higher energy. The triplet of states with degenerate eigenvalues are all included in the *multiplet* ${}^3S$. The singlet state, ${}^1S$, is represented by Ψ_-.

If we instead use the orbitals 1s and 2p, there are more Slater determinants. Every orbital has either α or β spin function. The orbital 2p may have different quantum numbers m. Altogether there are 12 different functions:

$$\left|1s\alpha 2p_+\alpha\right|, \quad \left|1s\alpha 2p_0\alpha\right|, \quad \left|1s\alpha 2p_-\alpha\right|.$$

$$\left|1s\alpha 2p_+\beta\right|, \quad \left|1s\alpha 2p_0\beta\right|, \quad \left|1s\alpha 2p_-\beta\right|,$$

$$\left|1s\beta 2p_+\alpha\right|, \quad \left|1s\beta 2p_0\alpha\right|, \quad \left|1s\beta 2p_-\alpha\right|, \qquad (2.62)$$

$$\left|1s\beta 2p_+\beta\right|, \quad \left|1s\beta 2p_0\beta\right|, \quad \left|1s\beta 2p_-\beta\right|.$$

These two functions combine into two electronic multiplet states: ^3P and ^1P. ^3P is ninefold degenerate and ^1P threefold degenerate. The spin-triplet has the lowest energy. This rule is called *Hund's rule* (after Friedrich Hund who started his career as an assistant of Werner Heisenberg in Göttingen). The rule is not absolute, but if some approximations are made, it may be proven mathematically.

As a further example of multiplets, we may obtain the ground state of the carbon atom with the configuration $1s^22s^22p^2$. The 1s and 2s subshells are fully occupied (closed subshells), while the 2p subshell is only occupied by two electrons of the possible six (open subshell). We only need to treat the open $2p^2$ subshell.

The Slater determinants may be ordered as in Table 2.2. M_S is the sum of all m_s quantum numbers ($m_s = 1/2$ for α and $m_s = -1/2$ for β). M_L is the sum of all m quantum numbers. The Slater determinants where two rows are the same, such as $\left|p_+\alpha p_+\alpha\right|$ and $\left|p_+\beta p_+\beta\right|$, are equal to zero.

According to Section 1.6.2, we should make a linear combination of all possible Slater determinants, calculate the matrix elements, and determine the coefficients in an eigenvalue problem. We will choose a simpler way, however. It is sufficient to note that certain combinations of (M_L,M_S) quantum numbers correspond to one, two, or three Slater determinants. Wave functions with different quantum numbers cannot be mixed. For a given linear combination, the number of states in the boxes in Table 2.2 are the same. Each (L,S)-multiplet must be represented by a single Slater determinant for a given (M_L,M_S) – quantum number. Thus, the following scheme must be possible:

$$
\begin{array}{c}
1 \quad\quad 1 \\[4pt]
1\,2\,1 \quad 1 \quad 1\,1\,1 \\[4pt]
2\,3\,2 = 1 \;+\; 1\,1\,1 \;+\; 1 = {}^1D + {}^3P + {}^1S \qquad (2.63) \\[4pt]
1\,2\,1 \quad 1 \quad 1\,1\,1 \\[4pt]
1 \quad\quad 1
\end{array}
$$

The ground configuration of the carbon atom thus gives rise to three different multiplets: ^1D, ^3P, and ^1S. We may use Hund's rule to establish the energy order. The first and most important of these rules is that multiplets with the highest spin S have the lowest energy; hence, the ^3P state is the ground state. Another rule of Hund says that if the spins are the same, the multiplet with the highest L quantum number has the lowest energy. The multiplet with the highest L quantum number, ^1D, is thus the second state

TABLE 2.2
Possible Slater Determinants for the 2p² Configuration

M_L	$M_S = 1$	$M_S = 0$	$M_S = -1$
2		$\lvert p_+\alpha\ p_+\beta \rvert$	
1	$\lvert p_+\alpha\ p_0\alpha \rvert$	$\lvert p_+\alpha\ p_0\beta \rvert , \lvert p_0\alpha\ p_+\beta \rvert$	$\lvert p_+\beta\ p_0\beta \rvert$
0	$\lvert p_0\alpha\ p_0\alpha \rvert , \lvert p_+\alpha\ p_-\alpha \rvert$	$\lvert p_+\alpha\ p_-\beta \rvert , \lvert p_0\alpha\ p_0\beta \rvert , \lvert p_-\alpha\ p_+\beta \rvert$	$\lvert p_0\beta\ p_0\beta \rvert , \lvert p_+\beta\ p_-\beta \rvert$
−1	$\lvert p_0\alpha\ p_-\alpha \rvert$	$\lvert p_0\alpha\ p_-\beta \rvert , \lvert p_-\alpha\ p_0\beta \rvert$	$\lvert p_0\beta\ p_-\beta \rvert$
−2		$\lvert p_-\alpha\ p_-\beta \rvert$	

in energy, while ¹S has the highest energy of all the multiplets arising from the 2p² configuration. This is in good agreement with the measured atomic spectrum (Figure 2.3).

2.4.3 CORRELATION

In calculations on large molecules or chemical reactions, the Hylleraas model can hardly be used. CI methods are more useful, particularly when there is one dominating first approximation in the form of a Slater determinant. In the case when several Slater determinants have the same (or approximately the same) energy, a multiconfigurational SCF approach is useful. The latter includes reoptimization of the orbitals in an *active space*, meaning a set of orbitals involved with rather strong occupancy. The best known of these methods is the complete active space–self-consistent field (CASSCF) method, which was developed particularly by the Swedish quantum chemist Björn Roos. This method is primarily a method for molecules, however, and uses Gaussian orbital basis sets. Exponential-type basis sets lead to faster convergence, but the calculation of the integrals needs considerably greater efforts than if Gaussian functions are used, in the case of molecules. A problem with Gaussian basis sets is that they are flat at the origin, while the true wave function can

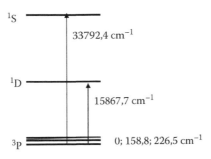

FIGURE 2.3 Electronic multiplet states belonging to the carbon atom in the ground configuration [He]2s²2p².

be proven to have a cusp (the derivative of the wave function at r = 0 with respect to r is nonzero).

A good approximation in the case of atoms is to use exponential basis functions in a multiconfigurational approach, in combination with Hylleraas functions to account for the cusp in the wave function when two electrons of different spin approach each other.

2.5   ATOMS IN CHEMISTRY

The concepts of oxidation and reduction go back to discoveries made at the end of the eighteenth century, particularly by the French chemist Lavoisier. He made a number of discoveries and improvements during the French revolution (1789–1799). For example, he discovered oxygen and introduced the SI unit for length (1 m). Unfortunately, he was executed by the guillotine by the revolutionary government.

Chemists, for example, the Swede Jöns Jakob Berzelius, began to determine atomic masses and valence numbers. Well before the end of the nineteenth century (1869), the Russian chemist Dmitri Mendeleev established the periodic table.

2.5.1   Periodic Table of the Elements

The periodic table can be explained essentially with the help of Slater's screening rules, the angular and spin degeneracy, and the necessary antisymmetry of the electronic wave function (the Pauli principle).

From one atom to the next, the screening factor is increased by 0.35. The increase in nuclear charge is thus not compensated and hence the atoms become smaller for increasing Z, according to Equation 2.22. At the end of a shell where all possible orbitals with the same principal quantum number are filled, we find the small and inert noble (or rare) gas elements helium, neon, argon, krypton, xenon, and radon. They belong to group 18 of the periodic table and do not react very well with other elements. They were not known at the time of Mendeleev, but finally obtained by liquidification of air in the early 1900s.

In the atom succeeding a noble gas atom, the electron has to be put into the next shell with the principal quantum number n increased by one unit. To that electron the inner shell electrons now have a much increased screening effect. The single valence electron is very reactive. The pure element forms a metal of low density. These are the alkali metals (Li, Na, K, Rb, Cs, Fr) found in group 1. The alkali elements easily form *atomic ions*. The valence electron transfers easily to another atom, leaving a positively charged ion behind. Alkali ions are noble in the sense that they have a closed shell. The only distinguishing property among the alkali ions is the size.

The alkaline earth metals (Be, Mg, Ca, Sr, Ba, Ra) share some of the properties of the alkalis, but are less reactive and harder. All are metals and all form doubly positive ions.

The d and f series are exceptions to the rule that the elements become progressively smaller through a period. The reason is that the new electron already belongs to an inner shell and the screening almost compensates the increase of nuclear charge.

When molecules are formed, the number of electrons in the valence shell is important. Phosphorous has the same number of electrons in the p subshell as nitrogen. Thus, we find compounds such as KNO_2 and KPO_2, where the bonding conditions are the same and the valence state is +3 for both elements. We also find KNO_3 and KPO_3 with valence state +5. The different size of the atoms N and P causes some differences. KPO_3 prefers to add a water molecule to form KH_2PO_4.

The oxidation state +4 appears to be missing in group 5a, which contains the elements N, P, As, Sb, and Bi. The solids $BaSbO_3$ and $BaBiO_3$ do exist, but it turns out from the BiO distances that there are alternant sites with Sb^{3+}/Sb^{5+} and Bi^{3+}/Bi^{5+}, respectively. The reason is found if we consider the atomic occupancies of the ions. The Sb^{5+} ions are stripped of electrons down to the Kr core and the Bi^{5+} ions down to a Xe core. The Sb^{3+} and Bi^{3+} ions contain in addition a $5s^2$ pair and a $6s^2$ pair, respectively, that is, a filled subshell. A 5s and 6s subshell with a single electron in the valence shell is not feasible, since this electron demands almost the same space as two electrons. It is better to add another electron in the same subshell for the "same price."

In the same way, the valence states are –2, 0, 2, 4, and 6 in the chalcogen group (O, S, Se, Te, Po) and –1, 0, 1, 3, and 5 in the halogen group (F, Cl, Br, I, At). The number of valence shell electrons is the same for all elements in a group. The *number* of electrons in a shell or subshell is most important to determine the chemical properties. Missing valence numbers are of interest in electron transfer since single electrons cannot easily be traded between ions of the same element. In this case, the electrons appear in pairs and behave as pairs.

2.5.2 Hydrogen Atom

In 1781, Henry Cavendish first identified hydrogen as an elemental compound. Lavoisier gave it the name hydrogen (French: *hydrogène* meaning "generator of water"). Hydrogen gas (H_2) is not a very common compound on the earth, although it is produced in large amounts industrially and used mostly as a fuel. When the hydrogen molecule is formed from two hydrogen atoms, 4.5 eV of energy is released. Hydrogen in the form of atoms is much rarer and exists almost exclusively in the laboratory. In organic chemistry, hydrogen atoms are only known to be dissolved in some very strong acids. For the same reason, *transfer* of hydrogen atoms is almost nonexistent, while transfer of electrons and protons separately is very common.

Lavoisier found that when metal is obtained from metal oxide, using hydrogen gas or carbon in a *reduction* reaction, the metal weighs less than the amount of oxide it originates from. Conversely, a metal, when treated with air, may go through an *oxidation* reaction and regain the lost weight. At the same time, the very special properties of the metal are lost, of course. Metal oxides are chemical compounds where metal binds to oxygen.

Oxygen also oxidizes organic compounds. Ethyl alcohol (C_2H_5OH) first forms acetaldehyde (CH_3CHO), which at further oxidation forms acetic acid (CH_3COOH). The name *oxygen* means literally to *generate acid*.

Acids have high proton-donating capacity and *bases* have high proton-accepting capacity. Protons do not occur in free form. In biosystems, protons are mostly bound to oxygen (−OH) or nitrogen (−NH₂). In water solvent, the acid is the hydronium ion (H_3O^+) and the base is the hydroxyl ion (OH^-).

In biosystems, proton gradients are built up because of electron transfer, which in turn is driven by free-energy gradients. The *proton motive force* resulting from proton gradients is used to drive the life processes. This will be discussed further, particularly in Chapters 9 and 11.

The understanding of acidity as due to hydrogen ions (H^+) and basicity as due to OH^- ions emanated from the theory of dissociation into ions in water by the Swedish physical chemist Svante Arrhenius. Later, in 1919, Ernest Rutherford recognized the hydrogen nucleus as an elementary particle and gave the name *proton*.

Hydrogen thus exists only strongly bound to itself (H_2) or to other elements (H_2O, CH_4, etc.) or, in gas phase, as positive or negative ions.

2.5.3 OXIDATION STATES AND OXIDATION POTENTIALS

The discovery of the electron simplified electrochemistry. Electrons are provided at one electrode (the cathode) or absorbed at the other electrode (the anode). The concept of "equivalents" was no longer needed.

During the nineteenth century, it was found that some elements can be oxidized in steps, where each step is characterized by an *oxidation state*. A way to describe this was gradually developed. In oxidation of the hydrogen atom to the hydrogen positive ion (proton), the oxidation state is raised by one unit, from zero to +1. The hydrogen atom is capable of reducing a number of oxides to the pure element, and in this process the hydrogen atom is reduced by one unit, to the positive hydrogen ion. Water was found to be a molecule formed by two hydrogen atoms and one oxygen atom. The only possibility for the oxidation state of the oxygen atom was equal to 0 (as an element) or equal to −2 in water and acids.

When finally the electron was discovered in 1897, it was realized that oxidation is equivalent to *removal of electrons* and reduction is equivalent to *addition of electrons* to the chemical system in question. It remained customary to associate *oxidation state* with atoms. For example, in the compounds H_2S, S_2, SO, SO_2, and H_2SO_4, sulfur has the oxidation states −2, 0, +2, +4, and +6, respectively. This is the *formal* oxidation state of an atom, which is important when the electrons in the chemical reaction are counted to find the stoichiometric coefficients in the reaction equation. The size of an added or removed electron is usually too large, due to the uncertainty relations, to be associated with a single atom. LiH, for example, can be regarded as Li^+H^-, LiH, or Li^-H^+, but there is no unique way to find out from the calculated density which case applies. The bonding valence electrons are distributed over the whole molecule.

Oxidation and reduction should be equivalent to electron transfer. However, for atoms in a covalently bonded molecule, the latter concept is quite meaningless. Not even in the case of a transition metal ion is it reasonable to assign electrons to the metal ion, as is sometimes done. It is reasonable to assign electrons only to larger units such as a metal complex or in the whole π-system. This is obvious when, for example, C_{60} is reduced. The new electron in C_{60}^- belongs to the whole molecule, not to a particular carbon atom. Measured reduction potentials are properties of the whole molecule. In the same way, the assignment of oxidation state +2 to copper in $CuCl_6^{4-}$ does not mean that the Cu 3d orbital is occupied by exactly nine electrons.

2.5.4 HYBRIDIZATION OF ATOMIC ORBITALS

The American physical chemist Gilbert N. Lewis (1875–1946) was the first person to consider the electron pair as the origin of the chemical bond. He described chemical bonding long before quantum mechanics was known. He is the one behind the concept of Lewis acids and Lewis bases, which was briefly mentioned above. More surprising perhaps is that it was G. N. Lewis who coined the name "photon" for Planck's quanta.

G. N. Lewis is probably most known for the theoretical concept of a "covalent bond." This bond is nonpolar and formed by an electron pair. In the molecules H_2 and Cl_2, for example, the atoms are connected by a covalent bond. We write this bonding as H–H, Cl–Cl, or H:H, Cl:Cl. Each dot means an electron and each bar a pair of electrons of opposite spin. The oxygen molecule has a double bond (O=O) and the nitrogen molecule a triple bond. Methane (CH_4) may also be written with bars or dots between the carbon atom and a hydrogen atom.

The Lewis bonding model may be rationalized using quantum mechanics if the ordinary C2s and C2p orbitals, for example, are mixed in such a way that they point into the corners of a tetrahedron with the carbon atom at its center. Generally, AOs on the *same* atom mix to form directed orbitals. This is referred to as (atomic) *hybridization*. Hybridization is important for the interpretation of the wave function, but of little interest computationally, since it just amounts to a unitary transformation of the orbitals, and Slater determinants are invariant under unitary transformations.

If one s orbital and one p_x orbital (or p_y or p_z) from the same shell are superposed, two new asymmetric AOs are formed, which point along the x-axis in the positive and negative directions. One s and two p orbitals from the same shell (p_x and p_y, for example) form three new, asymmetric orbitals in the xy plane, directed with 120° between them. Finally, if one s orbital and all three p orbitals are combined, four new AOs are formed, pointing into the corners of a tetrahedron.

The AOs, χ_{2s}, χ_{2px}, χ_{2py}, and χ_{2pz} are nonoverlapping and normalized. The directed AOs are formed as

$$\chi_1 = \left(\chi_{2s} + \chi_{2pz}\right)/\sqrt{2}. \tag{2.64}$$

It is easily seen that $\chi_{2s} + \chi_{2pz}$ is an AO that has large positive values on the positive z-axis ($\theta = 0$), where both 2s and 2p are positive, but close to zero on the negative z-axis ($\theta = 180$). This type of hybridization is called *sp-hybridization*. The linear combination $\chi_{2s} - \chi_{2pz}$, on the other hand, is directed toward the negative z-axis.

In planar organic molecules, each carbon atom binds to three other atoms. One may combine 2s, $2p_x$, and $2p_y$ orbitals to form hybridized orbitals:

$$\chi_1 = \left(\chi_{2s} + \sqrt{2}\chi_{2py}\right)\Big/\sqrt{3}$$

$$\chi_2 = \left(\chi_{2s} + \sqrt{\frac{3}{2}}\chi_{2px} - \sqrt{\frac{1}{2}}\chi_{2py}\right)\Big/\sqrt{3} \qquad\qquad (2.65)$$

$$\chi_3 = \left(\chi_{2s} - \sqrt{\frac{3}{2}}\chi_{2px} - \sqrt{\frac{1}{2}}\chi_{2py}\right)\sqrt{3}.$$

In this case, we have *sp²-hybridized* orbitals in the xy-plane, which may be used to form bonds in the plane, called σ orbitals. The new orbitals are pointing at the angles of 0°, 120°, and 240°. The $2p_z$ orbitals perpendicular to the plane form MOs, which are called π orbitals. The gap between the occupied and unoccupied π orbitals is smaller than the corresponding gap for σ orbitals.

If all p orbitals and s orbitals are used, it is possible to form bond orbitals, which point to the corners of a tetrahedron:

$$\chi_1 = \left(\chi_{2s} + \chi_{2px} + \chi_{2py} + \chi_{2pz}\right)/2$$

$$\chi_2 = \left(\chi_{2s} - \chi_{2px} - \chi_{2py} + \chi_{2pz}\right)/2$$

$$\chi_3 = \left(\chi_{2s} - \chi_{2px} + \chi_{2py} - \chi_{2pz}\right)/2 \qquad\qquad (2.66)$$

$$\chi_4 = \left(\chi_{2s} + \chi_{2px} - \chi_{2py} - \chi_{2pz}\right)/2.$$

It is quite easy to see that the new orbitals in Equation 2.66 are directed. For example, χ_1 has a larger value in the point (x, y, z) = (1, 1, 1) than in the point (−1, −1, −1). The other orbitals are directed in the (−1, −1, 1), (−1, 1, −1), and (1, −1, −1) directions, that is, toward the corners of a regular tetrahedron (or every second corner of a cube). This is the quantum mechanical description of the Lewis electron pairs. In time-independent quantum mechanics, we are interested in probability distributions. The electrons require a region of a certain size. sp³-hybridization shows that if the 2s and 2p AOs are accessible, four bonds can be found with as little overlapping as possible. The new orbitals no longer have the nucleus as an inversion center.

If all directed orbitals are fully occupied at sp³-hybridization, the density is the same as if all s and p orbitals are fully occupied. The 2s orbital forms itself a spherically symmetric charge distribution. The density of the 2p orbitals is the sum of the squares of x, y, and z orbitals, thus of type $(x^2 + y^2 + z^2)F(r)$. Filled octets thus means, seen from the atomic nucleus, a spherical charge distribution as in the inert gases. This is, in fact, a good explanation for why the octet rule works.

The transformations given in Equations 2.64 through 2.66 are all unitary. The wave function, if it can be written in the form of a Slater determinant, is not changed, provided that all orbital combinations are occupied.

The differences between the physicist's model (Bohr's spherical electron orbits) and the chemist's model (the electrons are located in the corners of a tetrahedron) became apparent in the years around 1920. This obvious disagreement seemed to be just another proof that physicists and chemists belong to different species. However, the differences between the Bohr model and the Lewis model disappear in light of quantum mechanics. Bohr's orbits and Lewis' fixed electron pairs are approximations to equally justified electronic orbits, blown up to orbitals to satisfy the uncertainty principle of Heisenberg.

There are also other types of hybridization for the case of d and f orbitals. Almost all simple bonding types seem to be explicable in terms of hybridized orbitals. Since the latter are constructed from realistic AOs, hybridization may be understood as a guarantee that the electrons get the space demanded by the uncertainty relations.

On the other hand, the hybridization concept is less useful, the less the radial overlap between the participating AOs. This is often the case when d and f orbitals are involved. This also means that the rules for chemical bonding become increasingly complicated further down in the periodic table.

3 Molecules

3.1 INTRODUCTION

There are well over one million known organic molecules, not counting macromolecules such as proteins. Quite a few are useful as participants in electron transfer (ET) or excitation energy transfer (EET) reactions. The ET rate difference between otherwise quite similar molecules varies over several orders of magnitude. In the case of ET between transition metal complexes, the variation in rate spans over more than 10 orders of magnitude. These differences are due to variation in structure when the electrons or excitations are given off or received. The less the structural change, the faster is the reaction.

In this chapter, chemical bonds will be discussed. One goal is to understand what happens to the structure when electrons are added or removed or when electronic excitation takes place.

3.2 CHEMICAL BONDING

One consequence of the discovery of the structure of atoms through the work of Rutherford was that the concept of chemical bonding became a subject of theoretical consideration. In Bohr's theory, the most loosely bound electrons, the *valence* electrons, are responsible for chemical bonding. Before quantum mechanics was completed, Kossel, Lewis, and Langmuir invented a phenomenological electronic concept, the *electron pair bond*, corresponding to the dash used previously to indicate a covalent bond between atoms.

In quantum mechanics, chemical bonding depends on the fact that two electrons occupy the lower energy-quantized molecular orbitals (MO), while the upper MO is empty, thereby gaining stabilization energy for the molecule. If the molecule has many atoms, the bonding contribution from a given MO is usually distributed over many bonds.

The electronic density in a molecule is close to a superposition of the atomic electron densities. The bonding electron orbitals are superpositions of atomic valence orbitals.

3.2.1 HYDROGEN MOLECULE, H_2

The time-independent Schrödinger equation (SE) for a diatomic molecule with N electrons and fixed nuclei A and B, with nuclear charges Z_A and Z_B, respectively, may be written (atomic units) as

$$H = \sum_{i=1}^{N} \left[-\frac{1}{2}\Delta_i - \frac{Z_A}{\left|\vec{R}_A - \vec{r}_i\right|} - \frac{Z_B}{\left|\vec{R}_B - \vec{r}_i\right|} \right] + \sum_{i<j}^{N} \frac{1}{r_{ij}} + \frac{Z_A Z_B}{\left|\vec{R}_A - \vec{R}_B\right|}. \tag{3.1}$$

For the hydrogen molecule (H_2), we have $Z_A = Z_B = 1$ and the number of electrons $N = 2$. The interelectronic repulsion is represented by a single term, $1/r_{12}$. The nuclei are considered as fixed at a distance of

$$R = \left| \vec{R}_A - \vec{R}_B \right|. \tag{3.2}$$

The solution of the time-independent SE can be achieved using large-scale ("quantum chemical") calculations. The boundary conditions are automatically satisfied by the chosen basis functions, which are atomic orbitals (AOs). In most cases, the Hartree–Fock or density functional theory (DFT) method is used to obtain bond lengths and energies. Unfortunately, a description in terms of a single Slater determinant is not possible in general if the dissociation energy of a molecule is desired. It is then necessary to use more advanced multiconfigurational methods (CI, Coupled Cluster, CASSCF, etc.).

The dissociation energy for the hydrogen molecule is 4.52 eV (436 kJ/mol). The bond distance is 0.74 Å. The bond between the two hydrogen atoms is thus very strong, which is partly due to electron correlation effects. The electronic structure will now be described.

For simplicity, we consider the atomic 1s orbitals a and b, on the hydrogen atoms A and B, respectively. A good approximation of an MO is a superposition of AOs. In the symmetric case, the two MOs are (a + b) and (a − b) (not normalized); (a + b) is the bonding orbital, doubly occupied in the ground state. The spatial part of the two-electron wave function in the form of an orbital product (corresponding to Equation 1.83) is the product of MO wave functions:

$$\Psi_{MO}\left(\vec{r}_1,\vec{r}_2\right) = \left[a\left(\vec{r}_1\right) + b\left(\vec{r}_1\right)\right]\left[a\left(\vec{r}_2\right) + b\left(\vec{r}_2\right)\right]. \tag{3.3}$$

The ground state of a hydrogen molecule, like most ground states, is a spin singlet state. In the case of two electrons, the spin part of the wave function is $\alpha\beta - \beta\alpha$ (left out in Equation 3.3). In MO a + b, there is an increase in the electronic density between the atoms, compared with two superposed atomic densities. In the unoccupied MO a − b, on the other hand, there is a decrease in the density. There is a nodal surface perpendicular to the molecular axis. In both cases, the cusp conditions at the atomic nuclei are the same as in the atoms. A general rule is that the greater the number of nodal surfaces, the higher is the orbital energy.

Let us assume that electron 1 is fixed near nucleus A, but far away from nucleus B. Since b is an exponential function located at B, the probability amplitude in position 1 of AO b is very small, so we put

$$b\left(\vec{r}_1\right) = 0. \tag{3.4}$$

The MO wave function for the ground state (Equation 3.3) may now be written as

$$\Psi_{MO}\left(\vec{r}_1,\vec{r}_2\right) = C\left[a\left(\vec{r}_2\right) + b\left(\vec{r}_2\right)\right]. \tag{3.5}$$

$C = a(\vec{r}_1)$ is a constant since electron 1 is fixed. We find that electron 2 is described by the MO (a + b), delocalized over both atoms, A and B, independent of the location of particle 1. This is a situation typical for the independent electron model.

The same AOs, a and b, may be used in a wave function that contains electron correlation. This is the *valence bond* (VB) wave function, first derived by Heitler and London. In unnormalized form, it is given by

$$\Psi_{VB}(\vec{r}_1, \vec{r}_2) = a(\vec{r}_1)b(\vec{r}_2) + b(\vec{r}_1)a(\vec{r}_2). \tag{3.6}$$

Let us now look at the VB wave function when electron 1 is fixed near nucleus A, so that $b(\vec{r}_1) = 0$. The second term in the right member can be neglected, and we obtain

$$\Psi_{VB}(\vec{r}_1, \vec{r}_2) = a(\vec{r}_1)b(\vec{r}_2) = C \cdot b(\vec{r}_2). \tag{3.7}$$

Electron 2 is thus described by a 1s function on atom B, with no contribution from atom A. In the total wave function Ψ_{VB}, electron 2 is thus near H_B if electron 1 is near H_A. In the same way, we may show that if electron 2 is near H_A, electron 1 has to be near H_B. The two electrons are thus correlated in their motion. The VB wave function converges to the correct atomic form when the nuclei are separated. In Ψ_{MO}, on the other hand, there is no correlation and the separation of atoms is incorrectly described.

Unfortunately, VB wave functions are hard to use for molecules larger than H_2. Slater determinants are easier to handle and generally give a rather good approximation at the equilibrium positions of the nuclei. We therefore primarily use the MO approximation as a basis for our discussions.

What is the difference between the MO and VB wave functions? We may write

$$\Psi_{MO} = (a+b)(a+b) = (ab+ba) + aa + bb = \Psi_{VB} + aa + bb. \tag{3.8}$$

In the case of the hydrogen molecule, we thus find that the MO wave function includes the ionic terms aa and bb. A wave function aa + bb means that two electrons are either on A, corresponding to $H^- + H^+$, or on B, corresponding to $H^+ + H^-$. H^- is a well-studied system that has a slightly lower energy than the neutral H atom (-0.52 H). The attraction energy between H^+ and H^- is $-1/R$ H, where R is given in Bohr, and thus the energy of the aa + bb state is $(-0.52 - 1/R)$ H, tending to -0.52 H as $R \to \infty$. The correct ab + ba wave function, on the other hand, has the energy $2 \times (-0.5) = -1$ H for $R \to \infty$.

The ionic states correspond to a charge disproportionation. From the above, it follows that if one electron is moved from one atom to the other, the energy is $(-0.52 - (-1)) = 0.48 \text{ H} = 13 \text{ eV}$:

$$H_A + H_B \to H_A^+ + H_B^- \quad \Delta G^0 = U = 13 \text{ eV}. \tag{3.9}$$

In the following, U will be frequently used to denote the disproportionation energy.

The VB wave function may be written as a configuration interaction (CI) expansion:

$$\Psi_{VB}(1,2) = (ab + ba) = \left[(a+b)(a+b) - (a-b)(a-b)\right]/2. \qquad (3.10)$$

In the second term of the CI expansion, the orbital $(a + b)$ is substituted with $(a - b)$. Thus, the VB model is equivalent to a CI model, where the bonding MO $(a + b)$ is substituted with the antibonding orbital $(a - b)$ in two spin orbitals.

3.2.2 REPRESENTATION OF MO

In a number of applications, it is sufficient to know the coefficients of the orbitals. However, it may be instructive to represent MO also with pictures.

An AO $\phi(x,y,z)$ is a function in three dimensions and more difficult to represent than an ordinary one-dimensional function $\phi(x)$ in one dimension. To get an idea about the extension of MOs in space, one may choose two plot values, say $+0.01$ and -0.01. A set of points that have one of these two values of the wave function are found. The surface with the same value is called an isosurface, which is represented by chicken wire. The three-dimensional isosurface is projected onto a two-dimensional picture. This type of representation has been used in all orbital figures of this chapter.

3.2.3 HOMONUCLEAR DIATOMIC MOLECULES

In diatomic molecules of the second and third periods, both s and p orbitals (2s, 2p and 3s, 3p, respectively) come into play, for example, in N_2, O_2, Na_2, F_2, and Cl_2. MOs have to satisfy certain symmetry conditions, since the Hamiltonian is invariant under rotations, reflections, and other possible symmetry operations that leave the molecule invariant.

Let the x-axis be the molecular axis. We abbreviate, for example, ψ_{1sA} as $1s_A$. If the MO has the same value in all directions counted from the bond axis, it is called a σ *orbital* (Table 3.1). An MO that changes sign in the same way as a p orbital at $180°$ rotation around the bond axis is called a π *orbital* (π_y or π_z). Finally, δ *orbitals* change sign at a turn of $90°$, in the same way as a d orbital (δ_{yz} or δ_{y2-z2}).

In homonuclear diatomic molecules, the center of the molecule forms a center of inversion. A given MO is either symmetric or antisymmetric under inversion. Symmetric MOs are indexed by g (from German: *gerade* = even). Antisymmetric MOs are indexed by u (German: *ungerade* = odd) (Table 3.1).

Understanding of orbital mixing for homonuclear diatomic molecules may be simplified if we assume sp-hybridization. A hybridized AO of type $2s_A + 2p_{xA}$ is directed along the positive x-axis and is called σ_+; $2s_A - 2p_{xA}$ is directed along the negative x-axis and is called σ_-. Figure 3.1 shows that the higher the energy of the MO, the more nodal surfaces (set of points where the wave function is equal to zero) are present in a given MO.

(N_2) The hybridization model is a good model if the contributing orbitals have approximately the same energy and radial function of the orbitals. MOs for N_2 may

---

TABLE 3.1

Unnormalized MOs for Homonuclear Diatomic Molecules

Molecules	MO	Notation
H_2–He_2	$1s_A + 1s_B$	σ_g
	$1s_A - 1s_B$	σ_u
Li_2–Ne_2	$2s_A + 2s_B$	σ_g
	$2s_A - 2s_B$	σ_u
	$2p_{xA} + 2p_{xB}$	σ_u
	$2p_{xA} - 2p_{xB}$	σ_g
	$2p_{yA} + 2p_{yB}, 2p_{zA} + 2p_{zB}$	π_{uy}, π_{uz}
	$2p_{yA} - 2p_{yB}, 2p_{zA} - 2p_{zB}$	π_{gy}, π_{gz}

Note: The x-axis is the interatomic axis.

---

be understood as strongly hybridized between 2s and 2p on each of the atoms. N_2 has 10 valence electrons. Since the π orbitals are doubly degenerate and the bonding π_u MO allows occupation by four electrons, we only have to occupy the MOs that have a negative orbital energy, up to π_u.

Valence orbitals of the same symmetry are counted from below and the number (if not 1) is inserted just before the symmetry label. The orbital σ_g has the lowest energy and is strongly bonding. sp-hybridized atomic orbitals overlap strongly in the region between the atoms. It is reasonable to count a single bond between the

Orbital	ε_i (eV)	Symmetry	Hybridization	Character
	16,1	$2\sigma_u$	$\sigma_{a+} - \sigma_{b-}$	antibonding
	4,9	π_g	$\pi_a - \pi_b$	π–antibonding
	−16,9	π_u	$\pi_a + \pi_b$	π–bonding
	−17,2	$2\sigma_g$	$\sigma_{a-} + \sigma_{b+}$	lone pair
	−21,0	σ_u	$\sigma_{a-} - \sigma_{b+}$	lone pair
	−40,5	σ_g	$\sigma_{a+} + \sigma_{b-}$	bonding

FIGURE 3.1 MOs for the molecule N_2. Left atom = a; right atom = b.

N atoms (N:N) due to the σ_g orbital, but this bond contribution is partly cancelled by σ_u. There is a bonding contribution also from $2\sigma_g$. The sum of contributions to the bond order from the σ orbitals is close to unity and corresponds to an ordinary single bond.

Two lone pair of MOs are $2\sigma_g$ and σ_u, but at the same time are weakly bonding and antibonding. These two MOs together give only a small contribution to the bond order. Largely, the four electrons that occupy $2\sigma_g$ and σ_u correspond to the electron pairs in N_2 (:N N:).

π_u is doubly degenerate (y and z) and is occupied by four electrons. This MO is bonding, since there is no nodal surface in the bonding region. The MO energy is decreased compared with the AOs and the density is increased between the atoms. π_u contributes with two perpendicular π bonds. The total bond order is $1 + 2 = 3$ (triple bond).

(O$_2$): In the O_2 molecule, the two new electrons appear in the antibonding π_g. Therefore, one of the π bonds coming from π_u is cancelled, but one σ and one π bond remain. The bond order is ≈ 2. O_2 has a double bond.

Since π_g is doubly degenerate, it can be occupied by two electrons of the same spin. According to Hund's main rule, the state with equally directed spins are the lowest in energy. Hence, the oxygen molecule ground state has the total spin $S = 1$ $(2S + 1 = 3)$ corresponding to a triplet state. This occupation is consistent with the paramagnetic ground state of oxygen gas.

Capital letters are used for the total symmetry of a state. The sign change under operations is the product of sign changes of the orbitals. The two antibonding electrons in O_2 form the spatial ground state wave function $\pi_{gx}(\vec{r}_1)\pi_{gy}(\vec{r}_2) - \pi_{gy}(\vec{r}_1)\pi_{gx}(\vec{r}_2)$. The spin function is the triplet $\alpha\alpha$ (or $\alpha\beta + \beta\alpha$, or $\beta\beta$). The spatial function is antisymmetric and the spin function is symmetric under exchange of electrons 1 and 2, thereby satisfying the Pauli principle. If each MO is rotated $180°$, the sign is changed, but since there is a product of two such orbitals in the total wave function, the symmetry is Σ. All MOs are gerade, thus the state is $^3\Sigma_g$.

There now appears a final symmetry property for the total wave function: reflection in a plane containing the intermolecular axis. Let us choose the xz-plane. π_{gx} is symmetric in this plane, while π_{gy} is antisymmetric. The product $\pi_{gx}(\vec{r}_1)\pi_{gy}(\vec{r}_2)$ is thus antisymmetric and the same is true for $\pi_{gy}(\vec{r}_1)\pi_{gx}(\vec{r}_2)$. The symmetry of the total wave function is therefore antisymmetric at reflection in the xz-plane. A minus sign is added in the symmetry notation for the state. The full ground state of the oxygen molecule is thus $^3\Sigma_g^-$.

3.2.4 HETERONUCLEAR DIATOMIC MOLECULES

In the heteronuclear diatomic molecules, the MOs resemble those of Figure 3.1, but the distribution on the two atoms is different because the energy and size of the AOs are different. Indices g and u are excluded since the inversion center is missing. The final charge distribution is polar to some extent. In LiF, it is reasonable to describe the bond as an ionic bond of the type Li^+F^-, formed by first transferring the Li 2s electron to the F 2p subshell, followed by strong interaction between the ions. The measured dipole moment agrees with this picture. In molecules with nearly the same nuclear charge, for example, CO, the dipole moment is numerically small and its direction is hard to predict.

In this connection, it may be of interest to mention the electronegativity scale proposed by Mulliken. The *electronegativity* of an atom (χ) is the arithmetic mean of the first ionization energy and the electron affinity:

$$\chi = \left(I + A\right)/2. \tag{3.11}$$

χ is a measure of how "willingly" the atom receives electrons; Cs is the least electronegative atom and F is the most electronegative atom.

The central teaching of the Lewis model is that the electrons are localized in the corners of a tetrahedron, where they are shared by two atoms. The Lewis model is, to a large extent, inconsistent with physical principles, since charged particles cannot be fixed in space, according to classical mechanics. However, the Lewis model may be justified on the basis of quantum mechanics. This is expressed in the natural bond orbital (NBO) model of Frank Weinhold. The MOs are also expressed in terms of hybridized AOs. The hybridized electron "clouds" replace the Lewis dots. In most cases, the NBO model shows that the time-independent wave function is in agreement with the Lewis model.

3.2.5 IONIC BONDS

At a large interatomic distance, an alkali atom and a halogen atom are more stable than their ions. This follows the fact that the ionization energies for alkali atoms are all higher than the electron affinity for halogens. Let us consider LiF. The energy gained by transferring one electron to the F atom (A = 3.40 eV) is not sufficient to ionize Na, which requires I = 5.14 eV. Among the two candidates to form the ground state, the neutral atoms F + Na are thus the winner and the ionic state F^- + Na^+ is an excited state 5.14 − 3.40 = 1.74 eV higher in energy (at a large distance between the nuclei) than the ground state.

As the atoms start to approach each other, the energy of the first-mentioned atomic state is hardly changed, since the atoms are uncharged. The ionic state, on the other hand, lowers its energy, depending on the attraction between the different charges of the ions. It is easy to calculate the Coulomb energy:

$$U = -\frac{1 \cdot 1}{R \cdot 1.89}\left(a.u\right) = -\frac{1 \cdot 1}{R \cdot 1.89} \, 27.2 \left(eV\right). \tag{3.12}$$

We have used atomic units in Equation 3.1. The unit for the charge is the absolute value of the electron charge $|e|$, and since the charges of the ions are ± 1, we get -1×1 in the numerator, where R is expressed in Å (=0.1 nm) and hence we have to multiply by 1.89 to get the distance in B. Finally, we multiply by 27.21 to obtain the answer in electron volts (Table 3.2).

Thus, the total energy difference between the ionic state and the neutral state is

TABLE 3.2

Ionization Energies and Electron Affinities (eV) for Alkali Atoms and Halogens

Atom	I	A	χ
Li	5.39	0.62	3.01
Na	5.14	0.55	2.85
K	4.34	0.50	2.42
Rb	4.18	0.49	2.34
Cs	3.89	0.47	2.18
F	17.4	3.40	10.4
Cl	13.0	3.62	8.31
Br	11.8	3.37	7.59
I	10.5	3.06	6.78

$$E = I - A - \frac{1}{R \cdot 1.89} \cdot 27.2. \tag{3.13}$$

The states have the same energy if

$$I - A = \frac{1}{R \cdot 1.89} \cdot 27.2. \tag{3.14}$$

If we insert the values for Na and F, we will find that this happens at the inter-atomic distance of 8.7 Å. For Cs and F, the energy is the same at 29.4 Å and for I and Li at 6.2 Å.

What happens with the atoms when the neutral state is no longer the one with the lowest energy, when the atoms move toward each other? There will be ET from the alkali atom to the halogen atom. The probability depends on the distance between the atoms and on their speed relative to one another. After the leap, the atoms are ions with opposite charge and will immediately attract each other. This can be seen in some scattering experiments and is referred to as the harpoon effect.

3.2.6 BOND DISTANCE DEPENDS ON OCCUPATION

In the neutral molecule H_2, the interatomic distance is 0.74 Å, considerably shorter than the bond in H_2^+ (1.06 Å). The smaller bond length in H_2 is due to the fact that the bonding orbital (a + b) is occupied by two electrons in H_2 but only a single electron in H_2^+.

(C_2): Watts and Barlett have calculated the C_2 molecule. The bond corresponds closely to a double bond. The calculated atomic distance is 1.24 Å, in agreement with the experimental result. In both cases, the π_u orbitals are fully occupied, while $2\sigma_g$ or σ_u is empty. There are two important configurations, one where σ_u is fully occupied and $2\sigma_g$ empty, and another where σ_u is empty and $2\sigma_g$ fully occupied.

For C_2^- the calculated bond distance is 1.27 Å in the ground state, again in very good agreement with the experimental result. Normally, one expects the negative molecule ion to be more unstable and have a larger interatomic distance. In the ground state C_2^-, the slightly bonding $2\sigma_g$ is occupied by a single electron and hence the bond length is only slightly increased.

In the case of C_2^+, finally, the electron is taken from the π_u MO in the lowest spin singlet state. Since this is a bonding orbital, the bond length is larger than for the neutral molecule, or 1.30 Å. There is a lower energy spin quartet ground state with a bond length of 1.40 Å. In the latter case, there are only two electrons in the π_u MO.

(O_2): In the oxygen molecule, the highest occupied MO (HOMO) is antibonding. Therefore, the positive ion, O_2^+, has a stronger bond than the neutral molecule. The bond length, 1.12 Å, is considerably shorter than in O_2 (1.21 Å). In O_2^- (superoxide), another electron has entered the antibonding π_g orbital and hence the bond distance is longer than in O_2 (1.34 Å). In peroxides, corresponding to O_2^{2-}, with another two electrons in the antibonding MO compared with O_2, the distance is even longer, 1.49 Å.

The ET reaction:

$$O_2 + e^- \rightarrow O_2^-, \qquad (3.15)$$

is a rare case among diatomic molecules where ET is important. In natural photosynthesis, oxygen molecules are formed from H_2O via O_2^-. Respiration is the reverse process, where oxygen molecules from breathing are used to form water via O_2^-. As we will see in later chapters, the reaction is slow due to the large distance change. Enzymatic reactions have evolved in order to speed up the reaction (see Chapter 11).

3.3   POLYATOMIC MOLECULES

One would hope that MO theory would be capable of predicting the structure without large-scale calculations. This is true only to a limited extent. For example, from MO theory, it is easy to predict why ozone (O_3) is bent while CO_2 is linear and NO_2 tends to form a dimer (N_2O_4). Next, we will concentrate on two important groups of compounds: aliphatic and aromatic hydrocarbons. In particular, aromatic molecules (planar π-systems) are very often involved in ET and excitation transfer reactions. The bonding capacity of the π orbitals is spread out over many bonds, and this leads to minimal reaction barriers in this type of ET reaction. The excitation energies are low and often in the visible region. The same is true for other planar π-systems, which contain heteroatoms (particularly N and O).

Aliphatic molecules have only single bonds and low reaction barriers for rotation around the bonds. In aromatic molecules, the reaction barrier for rotation around a bond is large in most cases. Aromatic molecules are stiff due to the additional bond strength of the π electrons perpendicular to the plane.

3.3.1 WATER MOLECULE

The water molecule H_2O has a special role to play in ET systems, not as a carrier of electrons—it is one of the smallest *insulating* molecules available—but rather as a carrier of protons. Adding electrons (reduction) or removing electrons (oxidation) is related to very large energies and simply does not take place under normal circumstances, particularly not in biological systems. The important role of the water molecule is as a carrier of protons in the molecular ion H_3O^+ (*hydronium* ion), or as a donor of protons to form the ion OH^- (*hydroxyl* ion; *hydroxide* in compounds).

Although the number of protons around the oxygen atom may vary, the number of electrons and the electron orbitals remain essentially the same. The valence shell of the oxygen atom is short of two electrons for a complete shell and the hydrogen atoms have one electron each. The latter two electrons are donated to fill the oxygen valence shell. The bonding scheme is quite simple, with four lone-pair electrons and two electrons in each bond (Figure 3.2).

The bonding may be thought of as due to hybridized AOs or as a modification of the AOs of the oxygen atom. In any case, since the molecule is planar, one MO is perpendicular to the plane and corresponds very closely to an atomic 2p orbital. This orbital also forms part of the two oxygen lone pairs. The resulting hybridization scheme would be sp^2 with a bond angle of 120°. The energies of the 2s and 2p orbitals are too different to admit efficient atomic hybridization, however. The 2s orbital forms a strong bond together with the hydrogen 1s orbital (a_1), while the 2p AO contributes to $2a_1$. The remaining bonding character is in the b_2 MO, antisymmetric in a plane through the oxygen atom, perpendicular to the molecular plane. The only atomic character possible in b_2 is H1s and O2p. O2s does not mix. The best bonding between O2p and H1s is when the overlap is maximized, which requires that the two OH bonds are perpendicular to each other. This is not the case, however, and may be due to the fact that the H nuclei are positive and repel each other. Thus, the bond angle opens up and is larger than 90°, particularly when the bonds are short, as in the

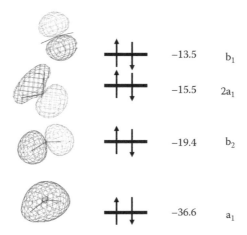

FIGURE 3.2   Molecular orbitals for the water molecule.

H_2O molecule. The HOH angle is 104.45° in the water molecule and closer to 90° in H_2S, H_2Se, and H_2Te.

The final configuration (not counting the oxygen 1s orbital) is $(a_1^2b_2^22a_1^2b_1^2)$; a_1 is similar to O2s, while the other three MOs (b_2, $2a_1$, and b_1) look like O2p orbitals. The unoccupied $3a_1$ and $2b_2$ orbitals are antibonding and high in energy, due to the short OH bond distance of 0.9584 Å. The water molecule has a negative electron affinity and comparatively high ionization energy, so it does not take part in ET reactions. Protons, on the other hand, may quite easily be added to the lone pair and also quite easily removed.

In the linear molecule OH⁻, a_1 becomes σ while $2a_1$ becomes 2σ. The former looks like an O2s orbital and the latter like an O2p orbital; b_2 and b_1 become π orbitals obtained from O2p orbitals.

In the case of H_3O^+, the molecule is a trigonal pyramid like isoelectronic NH_3. The reason is that the interaction between O2p and H1s orbitals is greater if the molecule is nonplanar. The bonding energy to O2s is not greatly diminished by pyramidalization. In this case, as well as in the case of H_2O and OH⁻, one MO is similar to 2s and the three other occupied MOs are similar to O2p. Unoccupied MOs are high in energy.

3.3.2 SATURATED HYDROCARBONS

Hydrocarbons contain only hydrogen and carbon atoms. *Aliphatic* hydrocarbons contain only single bonds, which may be thought of as sp^3-hybridized AOs. The smallest aliphatic hydrocarbon is methane (CH_4). If one of the hydrogen atoms is replaced by a carbon atom, a maximum of three new hydrogen atoms can be added, corresponding to H_3C-CH_3 (C_2H_6, ethane). If further hydrogen atoms are replaced by CH_3 (methyl) groups, we obtain $H_3C-CH_2-CH_3$ (C_3H_8, propane), $H_3C-CH_2-CH_2-CH_3$ (C_4H_{10}, buthane), $H_3C-CH_2-CH_2-CH_2-CH_3$ (C_5H_{12}, penthane), $H_3C-CH_2-CH_2-CH_2-CH_2-CH_3$ (C_6H_{14}, hexane), etc. The general formula for noncyclic, aliphatic hydrocarbons is C_nH_{2n+2}, where methane has n = 1. The number of bonds is four for each carbon atom and one for each hydrogen atom; thus, the total number of bonds is equal to $(4 \times n + 2n + 2)/2 = 3n + 1$.

Since carbon has four valence electrons and hydrogen has one, the number of valence electrons is obviously $4 \times n + (2n + 2) = 6n + 2$. Thus, there are two electrons per bond, which is considered the optimum bonding situation for the Lewis electron pairs. In all these cases, there are no cyclic structures.

Cyclic structures may be formed by removing two hydrogen atoms and connecting two carbon atoms (Figure 3.3). The number of hydrogen atoms decreases by two, so for a single cycle the general formula is C_nH_{2n}. One example is cyclohexane (C_6H_{12}). The number of electrons decreases by two (one for each hydrogen) and the number of bonds decreases by one. The number of bonds is still the same as the number of valence electron pairs. Each bar corresponds to an electron pair. This means that the electron pairs fit precisely into the structure. Aliphatic hydrocarbons are therefore said to be *saturated*.

In an aliphatic hydrocarbon, a CH unit may be replaced by a nitrogen or a phosphorus atom without much change of the electron structure. A CH_2 unit may be replaced by an oxygen or sulfur atom in the same way.

(a)

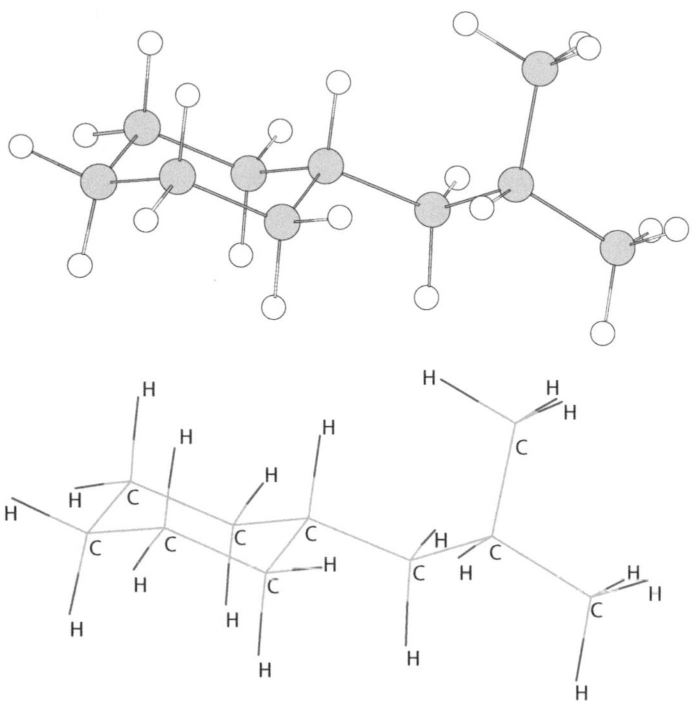

(b)

FIGURE 3.3   Aliphatic hydrocarbon $C_{10}H_{20}$ with a cyclic part (cyclohexane).

3.3.3   AROMATIC (UNSATURATED) HYDROCARBONS

In *aromatic* hydrocarbons, the number of hydrogen atoms is lower than for the corresponding aliphatic hydrocarbons with the same number of carbon atoms. Therefore, these molecules are referred to as unsaturated hydrocarbons. For example, if two hydrogen atoms are removed from $C_{10}H_{20}$ in Figure 3.3, we obtain Figure 3.4a ($C_{10}H_{20}$). All carbon atoms, except two, have four neighbors. The two carbon atoms with only three neighbors are connected by another bond bar. This bond is now referred to as a double bond.

Any carbon atom with a double bond has a planar structure around it. Three of its electrons go to sp^2-hybridized single bonds. The third C2p orbital is a $C2p_z$ orbital perpendicular to the xy plane. All $C2p_z$ orbitals are parallel and form π orbitals.

We may continue to remove hydrogen atoms and obtain 2, 3, or 4 double bonds as in Figures 3.4b through 3.4d. If there are two double bonds with just one single bond in between, we are referring to *conjugated* double bonds. A single double bond behaves as the ethylene (C_2H_4) double bond. Conjugated double bonds interact quantum mechanically. In Figure 3.4b, there are two conjugated double bonds and in Figure 3.4c, there are three double bonds that form a phenyl group.

In Figure 3.4d, there appears to be only three conjugated double bonds, since the last double bond forms a plane by itself. The reason for this is that in a fully planar

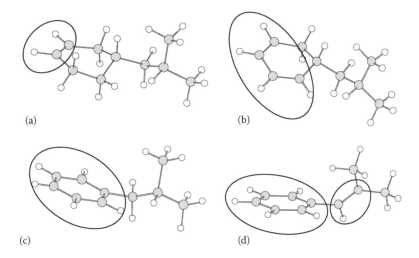

(a)   (b)

(c)   (d)

FIGURE 3.4 Unsaturated hydrocarbons obtained by removing hydrogens from the molecule in Figure 3.3. The structures are reoptimized: (a) $C_{10}H_{18}$ with one double bond, (b) $C_{10}H_{16}$ with two, (c) $C_{10}H_{14}$ with three, and (d) $C_{10}H_{12}$ with four double bonds. The conjugated parts of the system are encircled.

molecule there is *steric hindrance* between the methyl group and the phenyl group. The hydrogen atoms would be too close if the molecule were planar. Vinylbenzene (C_6H_5–C_2H_3) is fully planar with four conjugated double bonds.

Unsaturated hydrocarbons thus solve the problem of not being saturated with hydrogen by forming double bonds. Around each double bond, the molecule prefers to be planar. The single bonds in the xy-plane are formed from sp²-hybridized AOs. The almost free rotation in single bonds disappears if a CC bond has some double bond character. In ethylene (C_2H_4), the barrier of rotation is very large, ≈ 3 eV.

3.4 HÜCKEL MODEL FOR AROMATIC HYDROCARBONS

In 1930, the German theoretical chemist Erich Hückel proposed a quantum mechanical model that can be applied to the aromatic π electrons. Aromatic systems are also referred to as π-systems. π electrons in a system of conjugated double bonds interact in a very special way. The Hückel model is a very simple model. Nevertheless, it is behind a lot of photophysics and ET. The results have proven to be surprisingly accurate. The model is perfect for understanding, but nowadays quantum chemical calculations provide more accurate results, particularly for spectra.

3.4.1 Hückel Model

The $2p_z$-AOs constitute the basis set $\{\chi_\mu\}$ and are numbered in the same way as the carbon atoms (Figure 3.5).

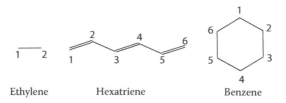

FIGURE 3.5 Carbon skeleton numbering in ethylene (ethene), hexatriene, and benzene.

A matrix eigenvalue problem is derived in the ordinary way: $(H - \varepsilon)c = 0$, where H is the Hamiltonian matrix and c is the eigenvector. The following simplifications are done in the Hamilton matrix:

$$H_{\mu\mu} = \int \chi_\mu{}^* H \chi_\mu dv = \alpha$$

$$H_{\mu v} = \int \chi_\mu{}^* H \chi_v dv = \beta < 0, \text{ if } \mu \text{ and } v \text{ are neighboring carbon atoms}$$

$$H_{\mu v} = 0, \text{ if carbon atoms } \mu \text{ and } v \text{ are not neighbors}$$

$$S_{\mu v} = \int \chi_\mu{}^* \chi_v dv = \delta_{\mu v} \left(= 1 \text{ if } \mu = v; = 0 \text{ if } \mu \neq v\right). \tag{3.16}$$

As an example, we may consider ethylene (or ethene, C_2H_4) and benzene (C_6H_6) with the following $(H - \varepsilon)$ matrices:

$$\begin{pmatrix} \alpha - \varepsilon & \beta \\ \beta & \alpha - \varepsilon \end{pmatrix} \text{ and } \begin{pmatrix} \alpha - \varepsilon & \beta & 0 & 0 & 0 & \beta \\ \beta & \alpha - \varepsilon & \beta & 0 & 0 & 0 \\ 0 & \beta & \alpha - \varepsilon & \beta & 0 & 0 \\ 0 & 0 & \beta & \alpha - \varepsilon & \beta & 0 \\ 0 & 0 & 0 & \beta & \alpha - \varepsilon & \beta \\ \beta & 0 & 0 & 0 & \beta & \alpha - \varepsilon \end{pmatrix}. \tag{3.17}$$

If the carbon atoms are inequivalent, we still choose the same α and β. In the case of ethene, the eigenvalue problem may be written as

$$\begin{pmatrix} \alpha - \varepsilon & \beta \\ \beta & \alpha - \varepsilon \end{pmatrix} \begin{pmatrix} c_1 \\ c_2 \end{pmatrix} = \begin{pmatrix} 0 \\ 0 \end{pmatrix}. \tag{3.18}$$

This system of equations has a nontrivial solution if the determinant of the matrix $= 0$. The secular equation is

$$\begin{vmatrix} \alpha - \varepsilon & \beta \\ \beta & \alpha - \varepsilon \end{vmatrix} = (\alpha - \varepsilon)^2 - \beta^2 = 0, \tag{3.19}$$

which yields

$$\varepsilon = \alpha \pm \beta. \tag{3.20}$$

The eigenfunctions are calculated by inserting the eigenvalues one by one into the eigenvalue problem, Equation 3.18, for example, $\varepsilon_+ = \alpha + \beta$ gives

$$\begin{pmatrix} \alpha - (\alpha+\beta) & \beta \\ \beta & \alpha - (\alpha+\beta) \end{pmatrix}\begin{pmatrix} c_1 \\ c_2 \end{pmatrix} = \begin{pmatrix} 0 \\ 0 \end{pmatrix} \Leftrightarrow \begin{pmatrix} -\beta & \beta \\ \beta & -\beta \end{pmatrix}\begin{pmatrix} c_1 \\ c_2 \end{pmatrix} = \begin{pmatrix} 0 \\ 0 \end{pmatrix}. \tag{3.21}$$

The coefficients may be determined from this equation, disregarding the normalization factor. In the present case, Equation 3.21 yields $c_1 = c_2$ and after normalization: $c_1 = 1/\sqrt{2}$ and $c_2 = 1/\sqrt{2}$. The eigenvector for $\varepsilon_- = \alpha - \beta$ is, in the same way, $c_1 = 1/\sqrt{2}$ and $c_2 = -1/\sqrt{2}$.

Since $\beta < 0$, ε_+ is the lowest energy solution. Corresponding eigenfunctions have the same sign of the two AOs, while the higher energy solution (ε_-) has different signs. This agrees with the general rule: the more nodal surfaces, the higher is the energy.

The solutions for the benzene molecule are given in Figure 3.6. Since the three lower MOs are occupied by two electrons each, they gain together $2 \cdot 2|\beta| + 4|\beta| = 8|\beta|$ by π bonding. In general, we define the energy gain that we obtain for N π electrons (π *binding energy*) as (n_i is the occupation number of orbital i)

$$N\alpha - \sum_{i=1}^{N} n_i \varepsilon_i. \tag{3.22}$$

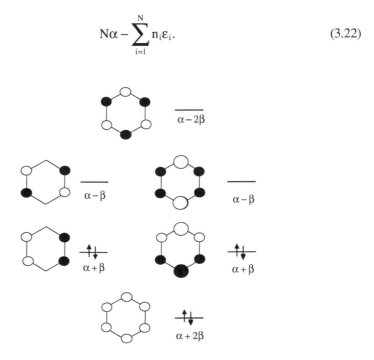

FIGURE 3.6 π MOs for benzene, seen from above the molecular plane.

The π binding energy for ethylene is thus $2\alpha - 2(\alpha + \beta) = -2\beta \approx 4$ eV. This, of course, is a very rough estimation. Anyway, if the same estimation were used for three isolated double bonds (with more than one single bond in between), the gain in energy would have been $6|\beta|$. This should be compared with $8|\beta|$ for three conjugated double bonds, as we obtain from Figure 3.6. The additional gain in energy obtained by letting all π electrons interact is called the π *delocalization energy*. The latter is thus the difference between π binding energy and the energy we would have obtained, had there been only local double bonds:

$$N\alpha - \sum_{i=1}^{N} n_i \varepsilon_i + N\beta. \tag{3.23}$$

The Hückel model gives a simple way to determine bond lengths without performing a full geometry optimization by minimizing the total energy (see Section 3.5.2). One first defines the *charge–bond order matrix* ($P_{\mu\nu}$):

$$P_{\mu\nu} = \sum_{i=1}^{N} n_i c_{\mu i}^* c_{\nu i}. \tag{3.24}$$

We remind ourselves that n_i is the occupation number of orbital i. The diagonal element $P_{\mu\mu}$ gives the (fractional) total occupation of π electrons on atom μ. The off-diagonal elements in $P_{\mu\nu}$ (Equation 3.24) may be interpreted as π bond orders for the π bond between the carbon atoms μ and ν. For ethylene, we have $P_{12} = 2 \times 1/\sqrt{2} \times 1/\sqrt{2} = 1$, which means that the π electrons contribute one bond in addition to the already present single bond. Together with the σ bond, we obtain in total a double bond. Had the upper orbital also been occupied by two electrons, there would have been contributions to P_{12} equal to -1, that is, the total contribution equal to zero from the π electrons, meaning that if all π orbitals are fully occupied there is no π bond.

The bond length may be obtained using an empirical relation. A typical CC single bond in a saturated hydrocarbon is 1.54 Å, while the CC bond in ethylene is 1.34 Å. We may interpolate between these limit values to obtain the bond length $R_{\mu\nu}$ between the carbon atoms μ and ν:

$$R_{\mu\nu} = -0.2 \cdot P_{\mu\nu} + 1.54. \tag{3.25}$$

This simple equation gives CC bond lengths in very good agreement with the experiments. The Hückel model thus gives a simple picture of the bond character in aromatic molecules.

The change in bond length when the molecule is oxidized or reduced depends on the change in $P_{\mu\nu}$ in Equation 3.25. In Equation 3.24, the change in $P_{\mu\nu}$ depends on the change in occupation number, since in the ordinary Hückel model the orbitals and eigenvalues are independent of occupation. The same is the case at excitation. The change in bond length thus depends on the lowest unoccupied MO (LUMO) when an electron is added and on the HOMO when an electron is removed. At excitation, both the HOMO and the LUMO are important.

By choosing a suitable value of β (for example, $\beta = -2$ eV), the Hückel model may be used to estimate ionization energies and electron affinities. We remind ourselves that $\varepsilon_{LUMO} - \varepsilon_{HOMO}$ is not a valid approximation for excitation energies in the Hartree–Fock model. In the Hückel model, with a chosen β, $\varepsilon_{LUMO} - \varepsilon_{HOMO}$ may be used to compare the spectra between different molecules.

3.4.2   BOND-LENGTH-DEPENDENT COUPLINGS

An improvement of the Hückel model may be obtained if it is realized that the shorter the bond, the stronger is the coupling. Longuet–Higgins suggested the following equation:

$$\beta = \beta_0 \exp\left[-\left(R_{\mu\nu} - R_0\right)/a\right]. \tag{3.26}$$

For $R_{\mu\nu} = R_0$, the coupling (β_0) is the same for all bonds. Set $R_0 = 1.4$ and $a = 0.3106$ Å. A calculation is started by assuming that all bond lengths are equal. Equations 3.25 and 3.26 are used to update the $R_{\mu\nu}$ and β values of the different bonds. The iterative solution converges quite fast.

3.4.3   CYCLIC π-SYSTEMS

In a cyclic π-system, we assume that N atoms are ordered on a circle. A turn of $2\pi/N$ should give an identical system (Figure 3.7). The wave functions of interest are π orbitals perpendicular to the plane of the circle. The basis functions are called χ_μ, where μ is the site index.

The wave function must be invariant under rotation, disregarding a constant of absolute value equal to unity. We have reason to try the following expression:

$$\phi = \sum_{\mu=0}^{N-1} N^{-\frac{1}{2}} \chi_\mu \cdot \exp(\mu i\theta), \tag{3.27}$$

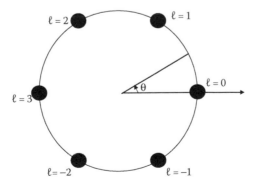

FIGURE 3.7   Cyclic π-system (cyclohexatriene = benzene).

where θ has to be determined by satisfying the boundary condition:

$$\exp(N\theta i) = 1 \Rightarrow \theta = \frac{2\pi}{N}\ell; \quad \ell = \text{integer}, \tag{3.28}$$

which implies

$$
\begin{aligned}
\text{for N even:} \quad & \ell = 0, \pm 1, \ldots, \pm N/2 \\
\text{for N odd:} \quad & \ell = 0, \pm 1, \ldots, \pm(N-1)/2.
\end{aligned} \tag{3.29}
$$

For different values of l, different values of θ are obtained in Equation 3.28 and different orbitals in Equation 3.27. Other values of l do not lead to any new orbitals.

It is practical to divide all matrix elements in Equation 3.17 by $\beta \neq 0$ and set $(\alpha - \varepsilon)/\beta = x$. The secular equation for the cyclic system of benzene may then be written as

$$
\begin{vmatrix}
x & 1 & 0 & 0 & 0 & 1 \\
1 & x & 1 & 0 & 0 & 0 \\
0 & 1 & x & 1 & 0 & 0 \\
0 & 0 & 1 & x & 1 & 0 \\
0 & 0 & 0 & 1 & x & 1 \\
1 & 0 & 0 & 0 & 1 & x
\end{vmatrix} = 0. \tag{3.30}
$$

Equation 3.30 defines a polynomial equation, $P(x) = 0$.

Let us check that the solution for $l = 0$ satisfies the eigenvalue problem in the case of N atoms:

$$
\begin{pmatrix}
x & 1 & 0 & \cdots & 1 \\
1 & x & 1 & \cdots & 0 \\
0 & 1 & x & 1\ldots & 0 \\
\vdots & \vdots & \vdots & \cdots & \vdots \\
1 & 0 & 0 & \ldots 1 & x
\end{pmatrix}
\begin{pmatrix}
1 \\
\exp(i\theta) \\
\exp(2i\theta) \\
\vdots \\
\exp[(N-1)i\theta]
\end{pmatrix} =
\begin{pmatrix}
0 \\
0 \\
0 \\
0 \\
0
\end{pmatrix}. \tag{3.31}
$$

If $N > 2$, all N equations may be written as

$$\exp(-i\theta) + x + \exp(i\theta) = 0, \tag{3.32}$$

which, in the case of even N, is satisfied for

$$x_\ell = -2\cos\theta = -2\cos\frac{2\pi\ell}{N}; \quad \ell = 0, \pm 1, \ldots, \pm N/2, \tag{3.33}$$

or, equivalently,

$$\varepsilon_\ell = \alpha + 2\beta\cos\frac{2\pi\ell}{N}; \quad \ell = 0,\pm 1,\ldots,\pm N/2, \tag{3.34}$$

and correspondingly for odd N.

The eigenfunctions corresponding to Equation 3.34 are

$$\phi_\ell = \sum_{\mu=0}^{N-1} N^{-\frac{1}{2}}\chi_\mu \cdot \exp\left(\mu i\,\frac{2\pi\ell}{N}\right). \tag{3.35}$$

For benzene (N = 6), there are two degeneracies. The degenerate orbitals may be added and subtracted (in general, transformed unitarily). Figure 3.6 is obtained.

For cyclobutadiene, we obtain the coefficients of Figure 3.8. Cyclobutadiene (C_4H_4) is not stabilized according to Hückel's rule. In fact, the delocalization energy Equation 3.22 is equal to zero. The ground state is degenerate, and consequently, according to Hund's rule, the ground state is a triplet state. This is also obtained in quantum chemical (*ab initio*) calculations, provided the bond lengths (and couplings) are forced to be the same. If bond length-independent couplings are obtained, the solution is two double bonds and two single bonds. Apparently, there is no energy to gain by delocalization in the case of cyclobutadiene. In the

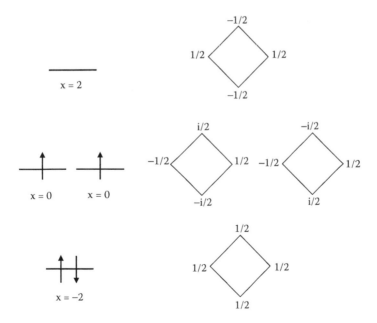

FIGURE 3.8 Hückel x-values and eigenfunction coefficients for cyclobutadiene in the ground state.

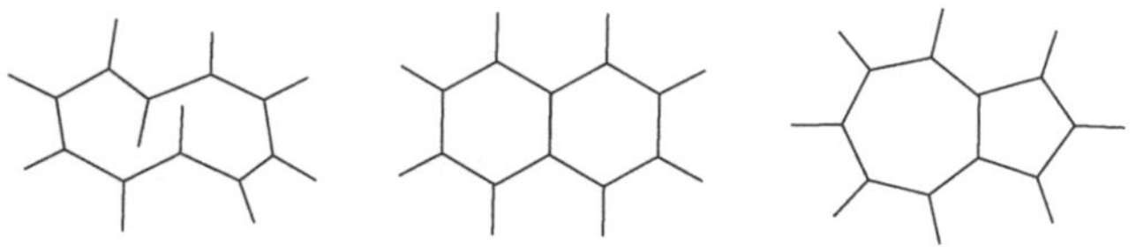

Cyclodecapentaene   Naphthalene   Azulene

FIGURE 3.9   Cyclodecapentaene, naphthalene, and azulene.

same way, the coefficients for benzene are obtained. After transformation, the real orbitals of Figure 3.6 are obtained.

From Equation 3.33 it follows that cyclic systems, for which the number of electrons can be written as $N = 4n + 2$, where n is an integer, are especially stable. For $n = 1$ ($N = 6$), we have benzene and for $n = 2$ ($N = 10$) (cyclodeca—pentaene = "quasi-naphthalene") (Figure 3.9). Naphthalene itself ($C_{10}H_8$) has a bridge across the middle of the molecule, which makes this molecule even more stable. Also, azulene is closely related to cyclodecapentaene. For larger cyclic systems, the stabilization is less pronounced. This $(4n + 2)$ rule is called *Hückel's rule*.

If bond length-dependent β is used, cyclobutadiene will obtain a rectangular shape. The two longer bonds are ordinary single bonds and the two shorter bonds are double bonds. Cyclobutadiene does not fulfill Hückel's rule, and does not gain energy by conjugation. Benzene, on the other hand, fulfills the $4n + 2$ rule, has equal bond lengths, and a delocalization energy of $2|\beta|$. The next cyclic with an even number of carbon atoms is cyclooctatetraene with four short and four long bonds. In this case, the molecule is nonplanar. If Hückel's rule is not fulfilled, bonding effects not included in the Hückel model become important. Cyclodecapentaene appears to be stable in the planar structure seen in Figure 3.9. For an increasing number of carbon atoms, the Hückel rule is less important.

3.4.4   LINEAR π-SYSTEMS

The secular equation for a linear system may be written as in Equation 3.30, but since the linear molecule has no connection between atoms 1 and N, the matrix is missing β values in the corners. The x matrix has x in the diagonal, surrounded by 1's, while the rest of the matrix elements are equal to zero. For hexatriene (C_6H_8) with three conjugate double bonds, the secular equation is

$$\begin{vmatrix} x & 1 & & & & \\ 1 & x & 1 & & & \\ & 1 & x & 1 & & \\ & & 1 & x & 1 & \\ & & & 1 & x & 1 \\ & & & & 1 & x \end{vmatrix} = 0. \tag{3.36}$$

The matrix elements of the empty space in the corners are all equal to zero. The general matrix for a linear π-system may be written as

$$
D_N(x) = \begin{vmatrix}
x & 1 & & & & & \\
1 & x & 1 & & & & \\
& 1 & x & 1 & & & \\
& & & \cdot & \cdot & \cdot & \\
& & & & \cdot & \cdot & \cdot \\
& & & & & 1 & x & 1 \\
& & & & & & 1 & x
\end{vmatrix}
$$

$$
= x \cdot \begin{vmatrix}
x & 1 & & & \\
1 & x & 1 & & \\
& & \cdot & \cdot & \cdot \\
& & & \cdot & \cdot & \cdot \\
& & & 1 & x & 1 \\
& & & & 1 & x
\end{vmatrix}
-
\begin{vmatrix}
1 & 1 & & & \\
& x & 1 & & \\
& & \cdot & \cdot & \cdot \\
& & & \cdot & \cdot & \cdot \\
& & & 1 & x & 1 \\
& & & & 1 & x
\end{vmatrix}
\tag{3.37}
$$

$$
= x \cdot D_{N-1}(x) -
\begin{vmatrix}
x & 1 & & \\
1 & x & 1 & \\
& \cdot & \cdot & \cdot \\
& & 1 & x & 1 \\
& & & 1 & x
\end{vmatrix}
= x \cdot D_{N-1}(x) - D_{N-2}(x).
$$

We obtain a recursion formula from which we may calculate the secular determinant for any linear system. It turns out that there is a general solution:

$$
x_n = 2 \cos\left(\frac{\pi}{N+1} n\right); \quad \varepsilon_n = \alpha + 2\beta \cos\left(\frac{\pi}{N+1} n\right); \quad n = 1, 2, \ldots, N, \tag{3.38}
$$

with the corresponding orbitals:

$$
\phi_n = \sum_{\mu=1}^{N} \chi_\mu c_{\mu n} = \sqrt{\frac{2}{N+1}} \sum_{\mu=1}^{N} \chi_\mu \sin\left(\frac{\pi}{N+1} n\mu\right) \tag{3.39}
$$

As the length of the chain is increased, the splitting between the eigenvalues becomes smaller. From Equation 3.38, we derive that the HOMO–LUMO gap is

$$
4|\beta| \sin\left(\frac{\pi}{2(N+1)}\right). \tag{3.40}
$$

As $N \to \infty$, the HOMO–LUMO gap becomes the *Fermi gap* and tends to zero. Molecules of the type $H_2C=CH-CH=\ldots=CH-CH=CH_2$ (polyenes) have a finite HOMO–LUMO gap, however, so this is in disagreement with the experiments. The lowest excitation in polyenes is about 1.5 eV. If, on the other hand, the method of bond length-dependent β is used (Equation 3.26), there is a finite band gap.

According to Equations 3.38 and 3.39, the lowest eigenvalue for a linear polyene with an even number (N) of carbon atoms (n = 1) has a wave function with all coefficients positive and the largest values in the central part of the molecule. For n = 2, the first half of the coefficients are positive and the second half are negative. In the case of butadiene (N = 4), the charge-bond order matrix (Equation 3.24) for the first and last bond are both large, which means a short bond. The central bond in butadiene has a small bond order and is therefore a long bond.

For N = 6, HOMO (n = 3) has the signs $+ + - - + +$. Reminding ourselves about the sign pattern for n = 1 and n = 2, we find a bond length alternation of the type short–long–short–long–short–. Every second bond is a short bond. The same pattern appears in accurate calculations.

3.4.5 ALTERNANT SYSTEMS

In an alternant aromatic molecule, the atoms may be divided in two sets—starred and unstarred—so that two members of the same set are never bonded to each other (Figure 3.10).

The orbital energy of an alternant aromatic system shows an important pairing property. If the energy spectrum contains the energy $\alpha + x_n\beta$, it also contains the energy $\alpha - x_n\beta$. The corresponding orbitals are

$$\phi_n = \sum_{\mu}^{*} \chi_\mu c_{\mu n} + \sum_{\nu}^{o} \chi_\nu c_{\nu n}$$

$$\phi_n' = \sum_{\mu}^{*} \chi_\mu c_{\mu n} - \sum_{\nu}^{o} \chi_\nu c_{\nu n}.$$

(3.41)

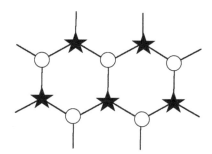

FIGURE 3.10 Alternate system (naphthalene).

Thus, to get an antibonding orbital from a bonding one, we just have to change the sign of the components of the unstarred atoms.

The proof is simple if we realize that the eigenvalue problem may be written as

$$
\begin{pmatrix}
x & 1 & 0 & . & 0 & 1 \\
1 & x & 1 & 0 & . & 0 \\
0 & 1 & x & 1 & 0 & . \\
\vdots & \vdots & \vdots & \vdots & \vdots & \vdots \\
0 & . & 0 & 1 & x & 1 \\
1 & 0 & . & 0 & 1 & x
\end{pmatrix}
\begin{pmatrix}
c_{1n} \\
c_{2n} \\
c_{3n} \\
\vdots \\
c_{(N-1)n} \\
c_{Nn}
\end{pmatrix}
=
\begin{pmatrix}
0 \\
0 \\
0 \\
\vdots \\
0 \\
0
\end{pmatrix}. \tag{3.42}
$$

Here, the first, third, etc., atoms belong to the starred set. The first row and every second row in the $N \times N$ matrix in Equation 3.42, and correspondingly for columns, belong to the starred set. Alternatively to Equation 3.42, we may thus write

$$
x_n c_{\mu n} + \sum_{\nu}^{o} c_{\nu n} = 0, \tag{3.43}
$$

where only those ν terms are included under the summation, which are bonded to μ. If μ belongs to the starred set, ν must belong to the unstarred set. Conversely, if μ belongs to the unstarred set, ν must belong to the starred set. All eigenvalue equations can be written as Equation 3.43. But if Equation 3.43 is satisfied by the eigenvalue x_n and the eigenvector coefficients c_i, which are included in Equation 3.43, there is also an equation where x_n has changed sign, $c_{\mu n}$ has kept its sign, and $c_{\nu n}$ have changed their sign. If x_n is negative and $c_{\mu n}$ is positive, then $c_{\nu n}$ included in Equation 3.43, corresponding to the coefficients bonded to μ, must be positive. If, on the other hand, x_n is positive and $c_{\mu n}$ is positive, then $c_{\nu n}$ included in Equation 3.43 must be negative. This means that Equation 3.41 is satisfied. We also find that for each negative eigenvalue x_n, there is a positive eigenvalue $|x_n|$.

An important corollary to this theorem for alternant π-systems is the Coulson–Rushbrooke theorem, which states that the electronic charge on all atoms in an alternant system is equal to $-|e|$. This means that all atoms are uncharged, since one electron was originally removed from each atom to form the π-system. Consider the coefficient matrix

$$
C =
\begin{pmatrix}
c_{11} & c_{12} & \cdots & c_{1N} \\
c_{21} & c_{22} & \cdots & c_{2N} \\
\vdots & \vdots & \vdots & \vdots \\
c_{N1} & c_{N2} & \cdots & c_{NN}
\end{pmatrix}, \tag{3.44}
$$

where the eigenvectors are ordered in columns. The second index is an eigenvector index. Since the eigenvectors form an orthonormal set, the matrix C is a unitary matrix. It is easy to check that the matrix product

$$P_{\mu\nu} = \left(C^{-1}C\right)_{\mu\nu} = \left(C^{t*}C\right)_{\mu\nu} = \sum_{i=1}^{N} c_{\mu i}^{*} c_{\nu i} = \delta_{\mu\nu}. \qquad (3.45)$$

The transposed conjugate matrix is the inverse matrix for unitary matrices. P is thus a unit matrix with 1 in the diagonal and otherwise zeroes.

We now set $\mu = \nu$ in Equation 3.45. $P_{\mu\mu}$ gives the contribution to the electronic charge on atom μ, added over all orbitals i, as if all MOs had been occupied by a single electron (occupation numbers $n_i = 1$ in Equation 3.24). To obtain the ground state charge, we should set $n_i = 2$ for the lowest half of the MOs (lowest in energy) and $n_i = 0$ for the other half of the MOs. In the case of alternant systems, the same result is obtained, however, since any upper MO gives the same contribution to $P_{\mu\mu}$ as the lower one with the corresponding eigenvalue on the lower, doubly occupied half (same absolute value, different sign). Therefore, $P_{\mu\mu}$ of Equation 3.45, multiplied by $-|e|$, is equal to $P_{\mu\mu}$ of Equation 3.24 for the ground state. This proves the Coulson–Rushbrooke theorem.

3.4.6 FULLERENES

Aromatic molecules may consist of only carbon. This is the case with *graphene*, which consists of a planar lattice of carbon hexagons. Stacked layers form *graphite*, which is a mineral found on earth. Graphene and graphite are electrically conducting. If calculations are performed on the finite structures of graphene, the end carbon atoms have to be saturated by hydrogen atoms. As the system gets larger, the energy between HOMO and LUMO tends to zero.

Replacement of a hexagon by a pentagon in a graphene layer produces curvature in the planar graphene sheet. Polyhedra may be created by replacing further hexagons by pentagons. Such molecules, C_n, are called *fullerenes*. C_{60} is particularly stable. Although C_{60} exists in large amounts in "carbon black," the molecule, along with other fullerenes, was not discovered until 1985 in mass spectra.

A polyhedron has to obey a formula by Euler, which states that in a three-dimensional convex polyhedron, the number of corners (C) plus the number of faces (F) minus the number of edges (E) is equal to two. This is quite easy to see if we try to draw an ordinary polyhedron. We begin by one face (F = 1). Obviously, C = E, so C + F − E = 1. When we continue to add a face, the number of corners is always increased by one less than the number of edges. Thus, $\Delta C + \Delta F - \Delta E = 0$. When we finally close the polyhedron, $\Delta F = 1$; however, since all corners and edges already exist, $\Delta C = \Delta F = 0$; hence, $\Delta C + \Delta F - \Delta E = 1$. Adding up the original numbers and all the increments, we obtain

$$C + F - E = 2. \qquad (3.46)$$

In the case of fullerenes, all faces are hexagons and pentagons, say n_5 pentagons and n_6 hexagons. Thus, if a fullerene C_C can be drawn in this way, we obviously have

$$C = 5n_5/3 + 6n_6/3; \quad F = n_5 + n_6; \quad E = 5n_5/2 + 6n_6/2. \quad (3.47)$$

Using Equation 3.46, we thus find that

$$C + F - E = \left(5n_5/3 + 6n_6/3\right) + \left(n_5 + n_6\right) - \left(5n_5/2 + 6n_6/2\right) = 2. \quad (3.48)$$

From this equation follows

$$n_5/6 = 2. \quad (3.49)$$

Hence, the number of pentagons is $n_5 = 12$ and the number of hexagons is undetermined. All existing polyenes with only hexagons and pentagons contain 12 pentagon faces and an arbitrary number of hexagon faces.

The smallest fullerene, C_{20}, contains no hexagon face. C_{60} contains 20 hexagons. Thus, it is the 12 pentagons that cause the curvature of a fullerene (Figure 3.11).

The MOs of fullerenes may be calculated exactly using the Hückel model (Figure 3.11). Unfortunately, neither MOs in general, nor the Hückel model as such, can precisely reproduce anything experimentally measurable. A minor problem is that the system is only "locally" planar.

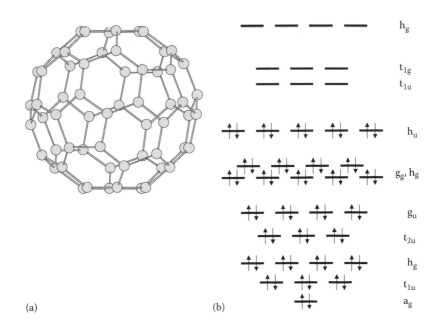

(a) (b)

FIGURE 3.11 C_{60} and its molecular orbitals (Hückel).

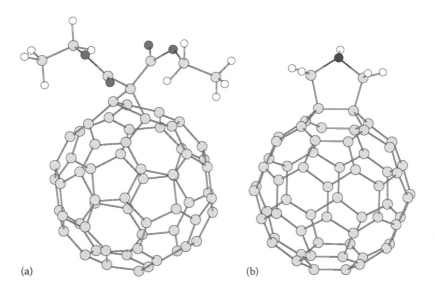

(a)   (b)

FIGURE 3.12   (**See color insert**) C_{60} molecule with groups covalently bonded to C_{60}.

For C_{60}, one may alternatively use a model for free electrons on a surface of a sphere. The orbital energy structure is similar to that of the Hückel model for the deep, occupied orbitals. Deviations occur for the valence orbitals. The Hückel MOs should be considered as more correct than the free-electron MOs.

The lowest MO in Figure 3.11 has symmetry a_g, meaning that all $2p_z$ MOs have the same sign, resembling an atomic s orbital. Next, there is a threefold degenerate t_{1u} orbital. It consists, like all other MOs shown in Figure 3.11, of $2p_z$ AOs. The gross angular distribution of t_{1u} is that of the three atomic p orbitals. Next, the h_g MOs resemble atomic d orbitals. Further up, g_u and t_{2u} correspond to atomic f orbitals but the icosahedral symmetry of C_{60} does not permit degeneracy. On the other hand the occupied g_g and $2h_g$ are (accidentally) degenerate.

HOMO is a fivefold degenerate h_u orbital. LUMO is t_{1u}. At a slightly higher energy, we find a threefold degenerate t_{1g} MO. At higher energy, we find $3h_g$ and another 19 MOs not shown in Figure 3.11. Figure 3.11 also does not show σ orbitals, since only π orbitals are obtained in Hückel.

Organic chemists have managed to bind organic groups to the fullerene atoms. Two examples are shown in Figure 3.12, both synthesized by C. Bingel in 1993. The so-called Bingel reaction allows the introduction of useful organic groups of different kinds on the fullerene surface.

Nanotubes are fullerenes that are cylindrical except at the end points where six pentagons are involved at each end to close the structure.

3.5   EXCITED STATES

Calculation of accurate excited states of molecules is an order of magnitude more difficult than calculations of the ground state. The Hartree–Fock model can be used to obtain the ground state solution for a given symmetry (irreducible representation). Most interesting molecules are not very symmetric, however, so representation group theory is not of much use here. If symmetry cannot be used to distinguish the states, one may replace the occupied spin orbitals in turn by virtual spin orbitals and make a CI calculation. The number of excited states obtained is equal to the number of substituted occupied orbitals multiplied by the number of virtual orbitals used for substitution. There is still a long way to go before at least a few can be obtained that accurately approximate the true states.

Simpler one-electron models, such as the Hückel model, may be used to obtain virtual orbitals and orbital energies, but not many-electron states. In the case of planar π-systems, a model founded on the Hartree–Fock model and developed by Pariser, Parr, and Pople (PPP) during the 1960s, still appears as a simple and useful approach to obtain the excited states of $\pi\pi^*$ type.

In large-scale calculations, the CASSCF *ab initio* model has proven very useful for the excited states. Another useful approach is the TD-DFT (time-dependent DFT method). The name refers to a technique to carry out special summations in perturbation theory that is based on time-dependent theory. The wave function itself is time independent.

3.5.1 Diatomic Molecules

Most problems concerning excited states already appear for diatomic molecules. For diatomic molecules, it becomes clear that we want to have a good representation of the electronic states for all interatomic distances. We assume for the time being that the nuclei are held fixed in the calculation of the excited states. Close to the equilibrium position, we may be able to use singly substituted Slater determinants to approximate the lowest excited states. Particularly for large interatomic distances, it is necessary to include electronic correlation to get the correct separation between the excited states.

The physics of the excited states is very rich in phenomena and includes many cases of intersection of electronic states. The character of the state is identified by the properties of the wave function, but the energy order of the states may be different at the equilibrium distance and at other interatomic distances. Each excited state has its own equilibrium distance. State crossing also occurs frequently for ionic molecules, both in the ground and the excited states.

Diatomic molecules have the advantage that the orbital energies or total energies can be plotted along the R-axis, where R is the interatomic distance. For polyatomic molecules, we have to imagine a one-dimensional parameter as the independent variable.

3.5.2 Aromatic Molecules

The electronic transitions of organic molecules with the π-system are located in the UV region and sometimes also in the visible region with transition energies below 3 eV. The photophysics of π-systems is of great general interest. For example, natural photosynthesis is built around molecules related to porphyrin (see Chapters 14 and 15).

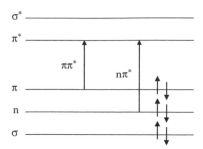

FIGURE 3.13 Orbital energy diagram for π-systems and types of excitations.

The excitations may be of different type (Figure 3.13). The π orbitals form HOMO and LUMO and therefore, in most cases, the lowest excited state is a $\pi\pi^*$ transition from an occupied π orbital to a virtual π^* orbital; n denotes a lone pair. The orbital energy is usually higher than that of the bonding σ orbitals.

3.5.3 TRANSITION MOMENT

Figure 3.14 shows the spectrum of an organic molecule (naphthalene). The first excitation energy is at about 4 eV (300 nm). For organic molecules in general, the excitation energy is well above 1 eV, which means that only the ground state is populated before the radiation is switched on. All excitations are thus from the ground state to the excited states.

The intensity of the absorption line depends on the oscillator strength, which in turn depends on the transition moment squared (see Chapter 7). The electronic part of the transition moment for x-polarized light may be written as

$$\mu_x = \int \phi_i^* \cdot x \cdot \phi_a dx dy dz. \tag{3.50}$$

For y- or z-polarized light, x is replaced by y or z, respectively, in Equation 3.50.

Notice that according to Equation 3.50, $\pi \to \pi^*$ transitions are forbidden for z-polarized light. Both ϕ_i and ϕ_a are antisymmetric at reflection in the xy-plane, and so is the function z. The product $\phi_i \cdot \phi_a$ is thus symmetric, the integrand $\phi_i \cdot z \cdot \phi_a$ is antisymmetric, and the integral is equal to zero. This means that the transition is forbidden.

3.5.4 SPECTRA OF CYCLIC AND LINEAR π-SYSTEMS

The spectrum for naphthalene ($C_{10}H_8$) is shown in Figure 3.14. Spectra of other cyclic aromatic systems resemble the naphthalene spectrum, and this has stimulated development of theories, for example, those of Platt and Gouterman.

The expressions for the orbital energy and orbitals for cyclic systems were derived in Equations 3.34 and 3.35. The orbitals are given by

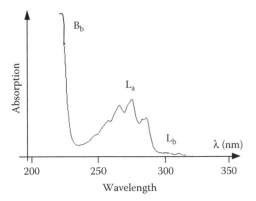

FIGURE 3.14   Spectrum of naphthalene.

$$\phi_\ell = \sum_{\mu=0}^{N-1} N^{-\frac{1}{2}} \chi_\mu \cdot \exp\left(\mu i \frac{2\pi\ell}{N} \right). \tag{3.51}$$

We may interpret l as a quantum number; l(HOMO) = (N/2 − 1)/2. We first add and subtract degenerate orbitals with the same |l| and obtain real wave functions:

$$\phi_+ = \sum_{\mu=0}^{N-1} (2/N)^{\frac{1}{2}} \chi_\mu \cdot \cos\left(\frac{2\pi\ell\mu}{N} \right)$$

$$\phi_- = \sum_{\mu=0}^{N-1} (2/N)^{\frac{1}{2}} \chi_\mu \cdot \sin\left(\frac{2\pi\ell\mu}{N} \right). \tag{3.52}$$

If an electron is excited from l to l + 1 or from −l to −(l + 1), the quantum number is changed by one unit. If, on the other hand, an electron is excited from l to − (l + 1) or from −l to (l + 1), the quantum number is changed by 2l + 1 = N/2 units. Since the sum of the l values in the ground state is equal to zero, it is equal to ±1 or ±N/2 for the excited states. In the notation of Platt, the states with l = 0, ±1, ±2, ±3, ... are called A, B, C, ... states, while the excited states with l = ±N/2, ±(N/2 + 1), ±(N/2 + 2), ... are called L, M, N, ... states. A is the index for the ground state. We are mainly concerned with the lowest excited states, which are the B and L states.

Total excited state wave functions interact if the value of |l| is the same. For each |l|, there are two wave functions:

$$\Psi_a = \left(\Phi_\ell + \Phi_{-\ell} \right) / \sqrt{2} \text{ and } \Psi_b = \left(\Phi_\ell - \Phi_{-\ell} \right) / \sqrt{2}. \tag{3.53}$$

L_a and L_b are the lowest states and the B_a and B_b states are at considerably higher energy. The proof is left out here.

The Platt model just described may also be used for systems that do not have full cyclic symmetry, for example, naphthalene ($C_{10}H_8$). The ϕ_+ and ϕ_- orbitals of Equation 3.52 are no longer degenerate, but recognizable as cosine and sine functions. The L_a, L_b and B_a, B_b functions may still be constructed and are energy ordered in the same way.

The transition density between the ground state and a singly substituted Slater determinant may be calculated for the B_a, B_b and L_a, L_b states, respectively:

$$A \rightarrow B_a: \quad \gamma_a = \left(\phi_\ell^* \phi_{\ell+1} + \phi_{-\ell}^* \phi_{-\ell-1} \right) = \left(2\sqrt{2}/N \right) \sum_{\mu=0}^{N-1} \chi_\mu^2 \cdot \cos\left(\frac{2\pi\mu}{N} \right)$$

$$A \rightarrow B_b: \quad \gamma_b = \left(\phi_\ell^* \phi_{\ell+1} - \phi_{-\ell}^* \phi_{-\ell-1} \right) = \left(2\sqrt{2}/N \right) \sum_{\mu=0}^{N-1} \chi_\mu^2 \cdot \sin\left(\frac{2\pi\mu}{N} \right)$$

$$(3.54)$$

$$A \rightarrow L_a: \quad \gamma_a = \left(\phi_{-\ell}^* \phi_{\ell+1} + \phi_\ell^* \phi_{-\ell-1} \right) = \left(2\sqrt{2}/N \right) \sum_{\mu=0}^{N-1} \chi_\mu^2 \cdot \cos\left(\pi\mu \right)$$

$$A \rightarrow L_b: \quad \gamma_b = \left(\phi_{-\ell}^* \phi_{\ell+1} - \phi_\ell^* \phi_{-\ell-1} \right) = \left(2\sqrt{2}/N \right) \sum_{\mu=0}^{N-1} \chi_\mu^2 \cdot \sin\left(\pi\mu \right).$$

In Equation 3.54, it is rather easy to see that the transition density for the L states consists of a series of canceling terms. Transition densities for the B states, on the other hand, have a single nodal plane either along the y-axis (B_a) or along the x-axis (B_b). According to Equation 3.54, this means high intensity for both states, for x- and y-polarized light, respectively. This agrees well with the experimental absorption spectrum (Figure 3.14).

For linear systems, the situation is simple. There are no degeneracies (Equation 3.38). Equation (3.39) shows that the eigenfunction with the lowest energy is symmetric at reflection in a plane through the center of the molecule, perpendicular to the molecular axis. The next are alternatingly antisymmetric and symmetric. The transition density for the HOMO → LUMO transition has to be antisymmetric. If the molecular axis is the x-axis, the transition is strongly allowed for x-polarized light, but not for other polarizations. The strong color is evidenced by the caratenoids. The longer ones are red (maple leaves in the fall) or orange (carrots). The shorter ones, for example, butadiene, absorb in the UV region and are therefore colorless.

However, the linear polyenes are more complicated than just described in the one-electron model. There is another excited state at a low energy. This is a state with the same symmetry as the ground state. The transition to this state is thus forbidden in all polarizations. The wave function for this state is a superposition of two singly substituted Slater determinants, corresponding to the HOMO − 1 → LUMO and the HOMO → LUMO + 1 substitutions, respectively. The orbital energy differences are both larger than the HOMO → LUMO substitution, of course, but the correlation effects, due to the interaction between the two substitutions, bring down the energy almost to the lowest excited state.

3.5.5 PPP MODEL

The PPP model is a general, semiempirical CI model for the excited states. It gives equivalent results to the Platt model for systems that are cyclic or almost cyclic, since both methods include states that are singly substituted from the ground state. In the PPP model, one includes a large number of excited states, for example, all states with an energy expectation value above the ground state by at most 3 eV. The CI matrix is automatically constructed and diagonalized.

The electronic ground state is described by a Slater determinant:

$$\Psi(1,2,\ldots,N) = A_N\left[\phi_{1\uparrow}(1)\cdot\phi_{1\downarrow}(2)\cdot\phi_{2\uparrow}(3)\cdot\phi_{2\downarrow}(4)\ldots\phi_{N/2\uparrow}(N-1)\cdot\phi_{N/2\downarrow}(N)\right]. \quad (3.55)$$

Independent variables are $i = (x_i, y_i, z_i, \zeta_i)$. The spin functions are illustrated by arrows.

In the simplest case, the excited state is obtained by replacing one spin orbital in Equation 3.55 with one previously unoccupied spin orbital. The lowest excited electronic singlet state usually has a HOMO → LUMO replacement as the main character.

The wave functions for the ground or excited states are more complicated than a single Slater determinant, as follows from the Platt model. Generally, replacing a spin orbital ϕ_i with a previously unoccupied spin orbital ϕ_a leads to a new configuration, written Φ_i^a. The excited states are obtained by CI between all possible excited states. We may remind ourselves that if we include only singly substituted configurations out of a ground state Slater determinant, the ground state will not be further improved and the excited states will be a linear combination of singly substituted Slater determinants.

Semiempirical methods that also include the σ electrons have been developed and are referred to as INDO and CNDO. NDO stands for Neglect of Differential Overlap. We will not treat these approximations here. C in CNDO stands for Complete and I in INDO stands for Intermediate. Approximations where /S is added at the end mean that the parametrization is relevant for the description of excited states and that CI is used. AM1 (Austin model 1) and PM3 (Parametrized model 3) are more recent models that build on the NDO approximation and are still in use. The AM1 (developed by M. J. S. Dewar) and PM3 (developed by J. J. P. Stewart) methods are useful to get quick results on comparatively large systems. PM3 in particular has been extensively used in this book, mainly for illustration purposes.

A diagram may be used to clarify the situation (Figure 3.15). On the left, we place the orbital energies as calculated using a Hartree–Fock method. The substitution indicated by an arrow is placed along the horizontal scale according to the Hamiltonian expectation value: $\left\langle\Phi_i^a\middle|H\middle|\Phi_i^a\right\rangle$. The final energies after CI are placed on the bottom scale.

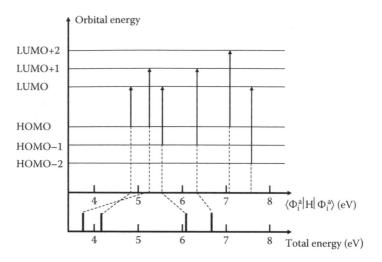

FIGURE 3.15   PPP calculation of electronic spectrum of an aromatic molecule.

With an acceptable accuracy, we can thus write the wave function for an excited state (after CI) as

$$\Psi = \sum_{i,a} C_i^a \Phi_i^a. \tag{3.56}$$

3.5.6   SINGLETS AND TRIPLETS

The ground state wave function is assumed to be a Slater determinant (indicated by bars) and is written as

$$^1\Phi = N!^{-1/2} \left| \prod_{\neq i} (\text{occ}) \phi_i \alpha \phi_i \beta \right|. \tag{3.57}$$

The replacement of ϕ_i by ϕ_a in a Slater determinant can be done either in a spin-up or in a spin-down spin orbital. The orbitals that are the same in the ground state and excited state are denoted as (occ). Two different Slater determinants are obtained and, subsequently, linear combinations may be formed between the two possible Slater determinants. We obtain new electronic states:

$$\Phi_i^a(+) = (2N!)^{-1/2} \left\{ \left| \prod_{\neq i} (\text{occ}) \phi_i \alpha \phi_a \beta \right| + \left| \prod_{\neq i} (\text{occ}) \phi_a \alpha \phi_i \beta \right| \right\} \tag{3.58}$$

$$\Phi_i^a(-) = (2N!)^{-1/2} \left\{ \left| \prod_{\neq i} (\text{occ}) \phi_i \alpha \phi_a \beta \right| - \left| \prod_{\neq i} (\text{occ}) \phi_a \alpha \phi_i \beta \right| \right\}. \tag{3.59}$$

Additionally, there are two excited triplet states with spin flip:

$$^3\Phi_i^a = N!^{-1/2} \left| \prod_{\neq i} (\text{occ}) \phi_i \alpha \phi_i \alpha \right| \tag{3.60}$$

$$^3\Phi_i^a = N!^{-1/2} \left| \prod_{\neq i} (\text{occ}) \phi_i \beta \phi_i \beta \right|. \tag{3.61}$$

In Equations 3.58 and 3.59, we change the order between the last two MOs and obtain

$$^1\Phi_i^a(+) = (2N!)^{-1/2} \left\{ \left| \prod_{\neq i} (\text{occ}) \phi_i \phi_a (\alpha\beta - \beta\alpha) \right| \right\} \tag{3.62}$$

$$^3\Phi_i^a(-) = (2N!)^{-1/2} \left\{ \left| \prod_{\neq i} (\text{occ}) \phi_i \phi_a (\alpha\beta + \beta\alpha) \right| \right\}. \tag{3.63}$$

It is easy to show that Equations 3.60, 3.61, and 3.63 are triplet states and that Equation 3.62 is a singlet state.

The expectation values are not the same for singlet states and triplet states. We calculate the energy expectation values and express them with the help of the orbital energies, ε_i and ε_a, as well as Coulomb integrals. We will use the following notation:

$$\int \phi_i^*(1)\phi_j(1)\frac{1}{r_{12}}\phi_k^*(2)\phi_l(2) = (ij|kl), \tag{3.64}$$

$$J_{ij} = (ii|jj), \tag{3.65}$$

$$K_{ij} = (ij|ij), \tag{3.66}$$

known as the Mulliken notation. The energy expectation values are given without proof:

$$\left\langle {}^1\Phi_i^a \left| H \right| {}^1\Phi_i^a \right\rangle - \left\langle {}^1\Phi_0 \left| H \right| {}^1\Phi_0 \right\rangle = \varepsilon_a - \varepsilon_i - J_{ia} + 2K_{ia}, \tag{3.67}$$

$$\left\langle {}^3\Phi_i^a \left| H \right| {}^3\Phi_i^a \right\rangle - \left\langle {}^1\Phi_0 \left| H \right| {}^1\Phi_0 \right\rangle = \varepsilon_a - \varepsilon_i - J_{ia}, \qquad (3.68)$$

where J_{ia} is a Coulomb repulsion integral ($\approx 2-5$ eV) between the charge distributions of the orbitals ϕ_i and ϕ_a, while K_{ia} is an exchange integral. J_{ia} and K_{ia} are both positive, but K_{ia} is considerably smaller than J_{ia}, usually $0.1-0.2$ eV. Since $K_{ia} > 0$, Equation 3.67 shows that the diagonal matrix elements for the triplet states are lower than the corresponding singlet diagonal matrix element. The final state energies after CI are also lower for the triplet states than for the singlet states. This is consistent with experimental results and is in accordance with Hund's rule: same spin gives lower energy than different spin. In most cases, there are a few triplet states below the first excited singlet state.

4 Nuclear Motion

4.1 INTRODUCTION

The time-independent Schrödinger equation (SE) for a molecular system derives from Hamiltonian classical dynamics and includes atomic nuclei as well as electrons. Eigenfunctions are therefore functions of both electronic and nuclear coordinates. Very often, however, the nuclear and electronic variables can be separated. The motion of the heavy particles *may* be treated using classical mechanics. Particularly at high temperatures, the Heisenberg uncertainty relation $\Delta p \cdot \Delta x \geq \hbar/2$ is easy to satisfy for atomic nuclei, which have a particle mass at least 1836 times the electron mass. The immediate problem for us is to obtain a time-independent SE including not only the electrons but also the nuclei and subsequently solve the separation problem.

4.2 SEPARATION OF ELECTRONIC AND NUCLEAR COORDINATES

The full Hamiltonian of a system in atomic units can be written as

$$H = \sum_{I=1}^{K}\left(-\frac{1}{2M_I}\Delta_I\right) + \sum_{i=1}^{N}\left(-\frac{1}{2}\Delta_i\right) + \sum_{I=1}^{K}\sum_{i=1}^{N}\left(-\frac{Z_I}{|\vec{R}_I - \vec{r}_i|}\right) + \sum_{I<J}^{K}\frac{Z_I Z_J}{|\vec{R}_I - \vec{R}_J|} + \sum_{i<j}^{N}\frac{1}{r_{ij}}.$$

(4.1)

Equation 4.1 contains, in order, the kinetic energy of the K nuclei, the kinetic energy of the N electrons, the Coulomb attraction between nuclei and electrons, the Coulomb repulsion between the nuclei, and the Coulomb repulsion between the electrons. Since the mass of nucleus I (M_I) is at least 1836 times the mass of the electron (m = 1), and the momenta are on an average the same, the kinetic energy of the nuclei can be neglected in the first approximation. If this is done, an equation for the electrons remains, valid for fixed positions of the nuclei. The task to obtain an equation for the nuclei, quantum or classical, was undertaken by the German physicist Max Born and his American assistant Robert Oppenheimer in 1927.

4.2.1 BORN–OPPENHEIMER APPROXIMATION

The *first part* of the Born–Oppenheimer (BO) approximation uses the assumption that the nuclei are fixed. The eigenfunctions, then, are purely electronic, but depend implicitly on the nuclear positions $Q = (\vec{R}_1, \vec{R}_2, ..., \vec{R}_K)$. After the kinetic energy of the nuclei has been neglected, the time-independent SE for the electrons is

$$H_e \Psi_e (1,2,...,N;Q) = E_e (Q) \Psi_e (1,2,...,N;Q). \tag{4.2}$$

Here, H_e is the Hamiltonian in Equation 4.1 with the first term neglected; 1, 2, ..., N stand for the electronic coordinates: $1 = (x_1, y_1, z_1, \zeta_1)$, etc.; and $E_e(Q)$ forms a connected surface as a function of Q, called the *potential energy surface* (PES). By fitting with known mathematical functions, PES may be written with an explicit nuclear coordinate dependence. We will soon show that the PES, that is, $E_e(Q)$, is a potential surface for the motion of the nuclei.

Next, we include the kinetic energy of the nuclei, and write H of Equation 4.1 in the following way:

$$H = H_e - \frac{\hbar^2}{2} \sum_{I=1}^{K} \frac{1}{M_I} \Delta_I, \tag{4.3}$$

where Δ_I is the Laplace operator for nucleus I with mass M_I. Since the repulsion between the nuclei is constant according to the first part of the BO approximation, the electron and nuclear variables may be separated and the wave function for electrons and nuclei written as

$$\Psi(1,2,...,N;Q) = \Psi_e (1,2,...,N;Q) \chi(Q), \tag{4.4}$$

where $\chi(Q)$ is an explicit function of the nuclear positions Q.

4.2.2 NUCLEI MOVE ON PES

Ψ_e contains the electronic coordinates explicitly and the nuclear coordinates implicitly. Ψ is an eigenfunction of the total Hamiltonian H and thus the total SE is

$$H\Psi = E\Psi = E\Psi_e \chi. \tag{4.5}$$

If we use Equation 4.2: $H_e \Psi_e = E_e \Psi_e$, Equation 4.4 may be written as

$$
\begin{aligned}
H\Psi &= \left(-\frac{\hbar^2}{2} \sum_{I=1}^{K} \frac{1}{M_I} \Delta_I + H_e \right) \Psi_e \chi(Q) \\
&= \left(-\frac{\hbar^2}{2} \sum_{I=1}^{K} \frac{1}{M_I} \Delta_I \Psi_e \right) \chi(Q) + \chi(Q) H_e \Psi_e + \left(-\frac{\hbar^2}{2} \sum_{I=1}^{K} \frac{1}{M_I} \Delta_I \chi(Q) \right) \Psi_e \qquad (4.6) \\
&\approx \left(-\frac{\hbar^2}{2} \sum_{I=1}^{K} \frac{1}{M_I} \Delta_I \chi(Q) + E_e (Q) \chi(Q) \right) \Psi_e.
\end{aligned}
$$

Since we may assume that Ψ_e is not very much changed for close values of Q, the first term on the second row, where the derivatives of Ψ_e with respect to the nuclear coordinates are taken ($\Delta_I \Psi_e$), may be neglected. Subsequently, we just divide through by Ψ_e in Equation 4.6. If $\approx$ is replaced by $=$, we obtain

$$\left(-\frac{\hbar^2}{2} \sum_{I=1}^{K} \frac{1}{M_I} \Delta_I \chi(Q) + E_e(Q)\chi(Q) \right) = E\chi(Q). \tag{4.7}$$

This is the general equation for the nuclei obtained by Born and Oppenheimer. For a single ($K = 1$) nuclear coordinate R (for example, diatomic molecules), where R is the distance between the atoms, we have

$$-\frac{\hbar^2}{2M} \Delta\chi(R) + E_e(R)\chi(R) = E\chi(R) \quad \left(M = \text{effective mass} \right). \tag{4.8}$$

Quantization is obtained by using boundary conditions where the wave function exponentially tends to zero as R tends to infinity. E is an energy eigenvalue. Equations 4.7 and 4.8 show that the nuclei move with $E_e(Q)$ as potential energy and with χ as the eigenfunction. This may be referred to as the *second part* of the BO approximation.

The vibrational motion of the nuclei thus results in stationary states with quantized vibrational energy levels via the time-independent SE and its boundary conditions. Transitions between the energy levels give rise to a spectrum in the infrared (IR) region.

The stationary solutions of time-independent SE may be used to form a time-dependent wave packet (Chapter 7). The motion of the wave packet for $t > 0$ may be approximated quite well with classical motion, according to Newton's equations. The potential energy for the classical motion is the same as for the wave packet. A good example of a nearly classical motion is an oscillating side group in a protein. The higher energy levels are densely distributed, rotational levels. At a higher temperature, energy is easily available to form a quantum mechanical wave packet whose motion resembles the classical rotational motion.

If the derivatives of Ψ_e with respect to Q cannot be neglected in Equation 4.6, they have to be included as a correction. This is the *third part* of the BO approximation. Physically, this means that there is probability for jump between PES.

4.2.3 CALCULATION OF PES

If the potential energy for the nuclei is given by $E_e(Q)$, the force on the nucleus is given by

$$\vec{F}(\vec{R}_I) = -\left(\frac{\partial E_e}{\partial X_I}, \frac{\partial E_e}{\partial Y_I}, \frac{\partial E_e}{\partial Z_I} \right), \tag{4.9}$$

in other words, the gradient of the energy surface. In the equilibrium point, this gradient is equal to zero.

The equilibrium geometry of a molecule is determined by the PES with the lowest energy. PES is a complicated function of the coordinates of a number of nuclei. Usually, PES are approximated by known functions in such a way that bond distances, bond angles, and force constants are close to the calculated ones.

To find an energy minimum as fast as possible in computational optimizations, an iterative method is used. SE is first solved for an approximate nuclear geometry. Subsequently, the first and second derivatives of the total energy with respect to the different nuclear coordinates are calculated. The geometry is updated with the help of the derivatives. New energies and derivatives are calculated iteratively until the geometry has converged and the derivatives are equal to zero. In most cases, the geometry optimization requires 100–1000 macro-steps, and in each step, electronic self-consistent field calculations have to be performed, including perhaps 100 iterations, before a self-consistent solution can been obtained.

The second derivatives form a matrix called the Hessian matrix (after Ludwig Hesse):

$$\begin{pmatrix} \dfrac{\partial^2}{\partial Q_1^2} & \dfrac{\partial^2}{\partial Q_1 \partial Q_2} & \cdots & \dfrac{\partial^2}{\partial Q_1 \partial Q_K} \\ \dfrac{\partial^2}{\partial Q_2 \partial Q_1} & \dfrac{\partial^2}{\partial Q_2^2} & \cdots & \dfrac{\partial^2}{\partial Q_2 \partial Q_K} \\ \vdots & \vdots & \cdots & \vdots \\ \dfrac{\partial^2}{\partial Q_K \partial Q_1} & \cdots & \cdots & \dfrac{\partial^2}{\partial Q_K^2} \end{pmatrix}. \tag{4.10}$$

In a local or global minimum, the Hessian is positive definite. If the Hessian has both positive and negative eigenvalues, then there is a saddle point in PES.

A calculation may concern a rotation around a bond. The simplest way to calculate a rotational barrier is to keep the geometry of the substituent constant, turn the double bond in small steps, calculate new ground state energies, and finally connect them to a continuous curve. To obtain an improved result, the geometry of the substituent should be reoptimized for each rotation angle. This is indeed the definition of a potential surface: everything has to be optimized.

4.2.4 ISOTOPE EFFECTS AND ISOTOPE SEPARATION

It is thus the motion of the electrons that generates the potential surface for the motion of the nuclei. The mass of the nuclei does not appear in H_e and hence PES must be independent of isotope mass. The most common chemical properties depend only on the PES, and are therefore conserved under isotope substitution. In fact, the basis for nuclear chemical analysis is that the isotopes, including the radioactive ones, behave chemically in exactly the same way as the natural ones. One may carry out reactions that are specific for a certain element and check in which fraction radioactivity remains (radioactivity may appear in the precipitation).

The vibrational energy levels depend on the nuclear mass through the term M_I of Equation 4.7. A vibrational or rotational spectrum is thus considerably changed when the isotope is changed, but in a predictable way.

4.2.5 Hellman–Feynman Theorem

The repulsive force on a nucleus I due to the presence of another nucleus J is defined by taking the derivative of the energy with respect to position. We take the derivative of the internuclear term in the Hamiltonian Equation 4.1 and obtain

$$\frac{\partial}{\partial X_I} \frac{1}{\left|\vec{R}_I - \vec{R}_J\right|} = -\frac{X_I - X_J}{\left|\vec{R}_I - \vec{R}_J\right|^3}. \tag{4.11}$$

Similarly, we have to take the derivative of the negative of the nuclear attraction:

$$\frac{\partial}{\partial X_I} \frac{1}{\left|\vec{R}_I - \vec{r}_i\right|} = \frac{\partial}{\partial X_I}\left[\left(X_I - x_i\right)^2 + \left(Y_I - y_i\right)^2 + \left(Z_I - z_i\right)^2\right]^{-1/2} = -\frac{X_I - x_i}{\left|\vec{R}_I - \vec{r}_i\right|^3}. \tag{4.12}$$

The force acted on particle I $(-\partial V/\partial X_I)$ is called $\vec{G}(\vec{R}_I)$ and derives its contributions from all other $K - 1$ nuclei and electrons:

$$\vec{G}_{nuc}\left(\vec{R}_I\right) = \sum_{J \neq I}^{K-1} \frac{Z_I Z_J}{\left|\vec{R}_I - \vec{R}_J\right|^3}\left(\vec{R}_I - \vec{R}_J\right). \tag{4.13}$$

In the continuous negative charge distribution ρ of the electrons, we obtain

$$\vec{G}_{el}\left(\vec{R}_I\right) = -\int \frac{\rho(\vec{r})Z_I}{\left|\vec{R}_I - \vec{r}\right|^3}\left(\vec{R}_I - \vec{r}\right)d\tau, \tag{4.14}$$

where the integration region is over all space and spin and the volume element $d\tau = d\tau_1 d\tau_2 \ldots d\tau_N$, with $d\tau_1 = dx_1 dy_1 dz_1 d\xi_1$, etc. We assume that the charge distribution arises from N electrons. Since

$$\rho(\vec{r}) = \int \Psi^* \Psi d\xi_1 d\tau_2 d\tau_3 \ldots d\tau_N, \tag{4.15}$$

(for the first electron, only spin integration) the normalization condition is

$$\int \rho(\vec{r}) dx dy dz = \int \Psi^* \Psi d\tau_1 d\tau_2 d\tau_3 \ldots d\tau_N = 1. \tag{4.16}$$

We may write Equation 4.13 as

$$\vec{G}_{nuc}\left(\vec{R}_I\right) = \left\langle \Psi \left| \sum_{J \neq I}^{K-1} \frac{Z_I Z_J}{\left|\vec{R}_I - \vec{R}_J\right|^3} \left(\vec{R}_I - \vec{R}_J\right) \right| \Psi \right\rangle, \tag{4.17}$$

since Ψ is normalized. The total force on nucleus I consists of repulsion between the other nuclei and attraction between the electrons Equation 4.14, and may thus be written:

$$\vec{G}\left(\vec{R}_I\right) = \vec{G}_{nuc} + \vec{G}_{el}. \tag{4.18}$$

The Hellman–Feynman theorem states that the force $\vec{G}(\vec{R}_I)$ is equal to the force $\vec{F}(\vec{R}_I)$, defined by Equation 4.9, provided that Ψ is the ground state solution of the time-independent SE (with appropriate boundary conditions).

Proof:

It follows, using Equation 4.3 for H_e, that

$$G_{X_I}\left(\vec{R}_I\right) = \left\langle \Psi \left| \frac{\partial H_e}{\partial X_I} \right| \Psi \right\rangle, \tag{4.19}$$

and similarly for Y and Z.

To prove the Hellman–Feynman theorem, we must show that

$$\left\langle \Psi \left| \frac{\partial H_e}{\partial X_I} \right| \Psi \right\rangle = \frac{\partial}{\partial X_I} \left\langle \Psi \left| H_e \right| \Psi \right\rangle \tag{4.20}$$

We have

$$\frac{\partial}{\partial X_I} \left\langle \Psi | H_e | \Psi \right\rangle = \left\langle \partial \Psi / \partial X_I | H_e | \Psi \right\rangle + \left\langle \Psi \left| \frac{\partial H_e}{\partial X_I} \right| \Psi \right\rangle + \left\langle \Psi | H_e | \partial \Psi / \partial X_I \right\rangle. \tag{4.21}$$

Since H_e is a self-adjoint operator, we obtain

$$\left\langle \Psi | H_e | \partial \Psi / \partial X_I \right\rangle = \left\langle H_e \Psi | \partial \Psi / \partial X_I \right\rangle. \tag{4.22}$$

Using $H_e \Psi = E_e \Psi$ and $H_e^* \Psi^* = E_e \Psi^*$, we obtain from Equations 4.21 and 4.22:

$$\frac{\partial}{\partial X_I} \left\langle \Psi | H_e | \Psi \right\rangle = \left\langle \Psi \left| \frac{\partial H_e}{\partial X_I} \right| \Psi \right\rangle + 2E_e \frac{\partial}{\partial X_I} \left\langle \Psi | \Psi \right\rangle = \left\langle \Psi \left| \frac{\partial H_e}{\partial X_I} \right| \Psi \right\rangle, \tag{4.23}$$

since $\left\langle \Psi | \Psi \right\rangle = 1$. The force defined in the way of classical electrostatics is thus the same as the force obtained from the PES.

It is easy to show that the derivative may be taken in a general direction λ. The force in direction λ may be defined as

$$\frac{\partial}{\partial \lambda} = \sum_{g=1}^{K} \left(\frac{\partial X_I}{\partial \lambda} \frac{\partial}{\partial X_I} + \frac{\partial Y_I}{\partial \lambda} \frac{\partial}{\partial Y_I} + \frac{\partial Z_I}{\partial \lambda} \frac{\partial}{\partial Z_I} \right). \tag{4.24}$$

With Equation 4.19 proven, the Hellman–Feynman theorem may thus be summarized as

$$\left\langle \Psi \left| \frac{\partial}{\partial \lambda} H_e \right| \Psi \right\rangle = \frac{\partial}{\partial \lambda} \left\langle \Psi | H_e | \Psi \right\rangle. \tag{4.25}$$

In the proof, we made use of the fact that Ψ satisfies the electronic time-independent SE. It may be shown that the theorem also holds in Hartree–Fock and density functional theory (DFT), which means that the force definitions are consistent, though not necessarily correct, in the two methods.

Using the Hellman–Feynman theorem, one may explain why an orbital with an increased probability density between two atoms is a bonding molecular orbital (MO). If the orbitals of two atoms start to overlap, and there are only electrons to occupy the bonding orbitals, there is an initial increase of the density between the atoms, and therefore a force on both (positively charged) atomic nuclei to approach the region where the electronic density has increased, and thus each other. It is important that the electrons satisfy the time-independent SE for all nuclear positions. The net increase of charge between the atoms, due to overlap of electronic wave functions, is compensated by a decrease in the distant regions. There are secondary charge motions within the atomic shells. The net force on the atomic nuclei is an attraction until the repulsive forces start to build up at shorter distance. When the equilibrium point has been reached, the forces on the nuclei vanish.

4.2.6 Car–Parinello Approach

The Car–Parinello method is an alternative to Born–Oppenheimer. Optimization of geometry and calculation of wave function occur simultaneously. Forces on the nuclei are calculated from the charges as in the Hellman–Feynman approximation. The nuclei are moving classically and their kinetic energy is decreased until the equilibrium for that particular electronic density is found. The electronic charge density is recalculated and a new equilibrium position found.

If DFT methods are used to obtain the electronic density, very large systems can be treated accurately with this method.

4.3 CLASSICAL MOLECULAR DYNAMICS

Electronic motion can be described accurately only by quantum mechanics. Nuclear motion, on the other hand, involves much heavier particles, and thus a classical treatment is usually sufficient particularly for high temperatures. In the classical case too, the PES $E_e(Q)$ forms the potential for the nuclei in their dynamics. The problem with vibrational motion is that the quantum effects are particularly important for the lowest energy levels and cannot be avoided at a low temperature. In the case of bond length vibrations, a classical treatment is unsuitable and the energy levels are strongly quantized with energy differences of more than one-tenth of an electron volt.

If, on the other hand, whole molecules are moving subject to forces much weaker than the bond forces, a classical treatment is relevant. Biological systems are such systems and most often of a size that quantum calculations are impossible. For smaller nuclei, tunneling is of importance, for example, in biological proton transfer, and then a classical treatment is not feasible.

4.3.1 CLASSICAL HARMONIC OSCILLATOR

Close to an equilibrium point, X_0, PES is parabolic. Let us simplify by treating the one-dimensional case. The harmonic approximation to PES is a parabola:

$$E_e(X) = V(X) = \frac{1}{2}k(X - X_0)^2, \qquad (4.26)$$

where k is the force constant. We will first use classical mechanics. Newton's equations may be written as

$$F_X = \frac{d^2X}{dt^2}M; \quad \vec{F} = \vec{a}M, \qquad (4.27)$$

where $\vec{F}$ is the force acting on a body with mass M and a is its acceleration due to this force. F_X may be obtained from PES, $V(X)$, by derivation:

$$F_X = -\frac{dV}{dX} = -k(X - X_0). \qquad (4.28)$$

Newton's equation (4.27) may now be written as a differential equation:

$$\frac{d^2X}{dt^2} + \frac{k}{M}(X - X_0) = 0. \qquad (4.29)$$

Solutions of this equation are the sine and cosine functions, for example,

$$X(t) - X_0 = A\cos\sqrt{\frac{k}{M}}t. \qquad (4.30)$$

Boundary conditions are necessary to determine the final solution of the problem. From Equation 4.30, we see that $X_{max} = X_0 + A$ and $X_{min} = X_0 - A$. Therefore, we choose A equal to the maximum deviation from X_0. $X(t) = X_0$

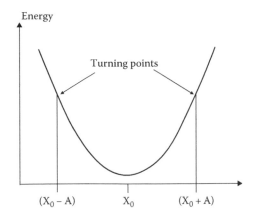

FIGURE 4.1 The classical harmonic oscillator. X_0 is the equilibrium bond length.

occurs for $t = \sqrt{M/k}\,(2n + 1)\pi/2$, where n is an integer number. The largest and smallest values of X, occuring for $t = 2n\pi\sqrt{M/k}$ and $t = \sqrt{M/k}\,(2n + 1)\pi$, are called the *turning points* (Figure 4.1). At the turning points, the kinetic energy is equal to zero, and the potential energy is at a maximum and exactly equal to the maximum kinetic energy for $X = X_0$. The turning points cannot be passed unless more energy is provided for the pendulum.

The vibrational motion of the nuclei in a molecule is approximately described by Equation 4.30, the equation for the classical harmonic oscillator. The bond length is oscillating between $X_0 + A$ and $X_0 - A$. The restoring force is proportional to the deviation from the equilibrium, according to Equation 4.28. The energy of the oscillator may take arbitrary values. In reality, the classical description is only valid as a limit case, but this limit case is the basis for classical molecular dynamics.

The greatest differences between the classical oscillator and the quantum oscillator, to be described in the next section, are quantization and probability distribution. The classical oscillator has a higher probability distribution close to the turning points, while the ground state of the quantum oscillator has the highest probability amplitude at the minimum energy. Since the eigenvalues of the quantum oscillator agree perfectly with the measured values, the quantum oscillator contains the truth, of course. The comparison is quite unjust, however, since the classical description should be compared with the time-dependent quantum description. If a wave packet is formed, it behaves in almost the same way as the classical solution, with almost the same probability distribution (maximum at the turning points). One may show that the acceleration is the same as in the classical case. The wave packet is extended into the forbidden region, which is a typical result of the Heisenberg uncertainty relation. The main difference between quantum and classical is thus that only quantized energies are allowed in the quantum oscillator.

The classical picture is a limit picture that becomes increasingly correct when there is higher energy compared with the energy splittings. Larger particles, such as proteins, move very close as classical particles, while the side groups often show quantization.

4.3.2 ANHARMONIC MOTION

PES for a diatomic molecule is harmonic only close to the equilibrium. A better approximation is a *Morse* function, which may be expressed as

$$E(R) = D_e \left[1 - \exp\left(-A\left(R - R_e\right)\right) \right]^2 - D_e. \tag{4.31}$$

The function is constructed in such a way that it has a minimum equal to $-D_e$ at $R = R_e$, where R_e is the equilibrium interatomic distance. A is a constant to be determined and D_e is the *dissociation energy*. When $R \to \infty$, the exponential function and $E(R)$ tend to zero.

D_e may be measured with the help of spectroscopic methods. In principle, all vibrational transitions are measured. The sum determines the *spectroscopic dissociation energy*, D_0. D_0 and D_e differ by the zero-point energy:

$$D_0 = D_e - \hbar \sqrt{\frac{k}{M}} \cdot \frac{1}{2}. \tag{4.32}$$

The force constant is determined by Taylor expanding $E(R)$ around $R = R_e$. Comparing with Equation 4.26, we obtain

$$A = \sqrt{\frac{k}{2D_e}}. \tag{4.33}$$

The Morse curve correctly predicts a higher density of vibrational states just before dissociation. The harmonic potential incorrectly gives the equidistant vibrational energies.

In principle, we should not allow any approximation in the nuclear potential $E_e(Q)$, but use the calculated one in a numerical expansion, point by point. The disadvantage with the latter approach is that no simple and useful equations are obtained.

4.3.3 SMALL MOLECULAR OSCILLATIONS

Each atomic nucleus in a molecule can move in three directions. If there are K nuclei, we say that there are 3K degrees of freedom. The whole molecule moves translationally as an inert body. As a free molecule, it rotates. Hence, only 3K − 6 degrees of freedom remain for the internal coordinates. If the molecule is linear, however, the rolling motion of the molecule around its axis does not change the geometry, so only 3K−5 degrees of freedom remain. In a diatomic molecule, there are $3 \times 2 - 5 = 1$ degree of freedom, the interatomic distance (R). In a bent triatomic molecule, there are $3 \cdot 3 - 6 = 3$ independent variables (water), and for a linear triatomic molecule, four (CO_2). The internal coordinates are two for bond length, and one for bond angle (H_2O) or two for bond angle (CO_2) if bending in two planes is possible.

We assume that there are L nuclear coordinates (degrees of freedom). We include all L = 3K coordinates in the derivation. The translational and rotational coordinates for the whole molecule may be removed later. We will consider small displacements from the energy minimum (equilibrium point):

$$Q_0 = \left(Q_1^0, Q_2^0, \ldots, Q_L^0 \right), \tag{4.34}$$

on $E_e(Q)$. It is convenient to define generalized coordinates by multiplying the small displacements for each nucleus by the square root of the nuclear mass:

$$q_I = \left(Q_I - Q_I^0 \right) \cdot \sqrt{M_I}. \tag{4.35}$$

The potential $V(Q) = E_e(Q)$ is expanded in a Taylor series around the equilibrium positions:

$$V = V_0 + \sum_{I=1}^{L} \left(\frac{\partial V}{\partial q_I} \right)_0 q_I + \frac{1}{2} \sum_{I,J=1}^{L} \left(\frac{\partial^2 V}{\partial q_I \partial q_J} \right)_0 q_I q_J + \cdots. \tag{4.36}$$

The constant V_0 may be set equal to zero. In a point of equilibrium, the first derivatives are equal to zero, while those higher than the second-order variations are neglected (this is the meaning of the term "small molecular vibrations").

$$V = \frac{1}{2} \sum_{I,J=1}^{L} \left(\frac{\partial^2 V}{\partial q_I \partial q_J} \right)_0 q_I q_J = \frac{1}{2} \sum_{I,J-1}^{L} k_{IJ} q_I q_J, \tag{4.37}$$

where k_{IJ} are force constants.

The expression for kinetic energy in classical mechanics may be written as second derivatives of the position coordinates $\{q_i\}$ with respect to time:

$$T = \frac{1}{2} \sum_{I,J=1}^{L} T_{IJ} \frac{\partial q_I}{\partial t} \frac{\partial q_J}{\partial t}. \tag{4.38}$$

We thus obtain the Lagrangian Л:

$$Л = \frac{1}{2} \sum_{I,J=1}^{L} \left(T_{IJ} \frac{\partial q_I}{\partial t} \frac{\partial q_J}{\partial t} - k_{IJ} q_I q_J \right). \tag{4.39}$$

The classical (Newton) equations of motion are the N equations:

$$\sum_{J=1}^{L} \left(T_{IJ} \frac{\partial^2 q_J}{\partial t^2} - k_{IJ} q_J \right) = 0. \tag{4.40}$$

The solution of these equations is sketched next.

4.3.4   Eigenvalue Equation

In general, all coordinates $\{q_j\}$ are involved in all equations. The standard way to solve Equation 4.40 is to use physical intuition rather than mathematical rigor and do the substitution:

$$q_i = Ca_i \sin\left(-\omega t + \delta_i\right), \qquad (4.41)$$

where C is a general normalization factor and a_i an amplitude factor; δ_i depends on the boundary conditions. Inserting Equation 4.41 into Equation 4.40 leads to a sum of cosine terms. A solution of Equation 4.40 requires the coefficients to be equal to zero:

$$\sum_{j=1}^{N}\left(k_{ij} - \omega^2 T_{ij}\right)a_j = 0. \qquad (4.42)$$

This is an eigenvalue problem of standard form. It has a nontrivial solution only if the determinant is equal to zero:

$$\begin{vmatrix} k_{11} - \omega^2 T_{11} & k_{12} - \omega^2 T_{12} & \cdots \\ k_{21} - \omega^2 T_{21} & k_{22} - \omega^2 T_{22} & \cdots \\ \vdots & \vdots & \cdots \end{vmatrix} = 0. \qquad (4.43)$$

This is a polynomial in ω^2, whose zeroes are the eigen-frequencies of the system. Each solution ω_k^2 of Equation 4.43 has to be inserted into Equation 4.42 to obtain the eigenvectors $(a_{1k}, a_{2k}, \ldots, a_{Nk})$, where k is the index for the eigenfunctions, referred to as *normal modes*. The normal mode has the eigenvalue ω_k.

   If the translational and rotational coordinates have not been removed from the beginning, there are a corresponding number of eigenvalues of Equation 4.42 equal to zero.

4.3.5   Molecular Dynamics Simulation

We may conclude from Section 4.3.1 that treatment of a chemical system using classical mechanics, based on potential calculated by quantum mechanics, should give quite accurate nuclear dynamics, except possibly for protons. The forces are calculated from the PES and included in Newton's equations, which are solved in an iterative way. This model is referred to as *molecular dynamics simulation*.

   The potential function of a particle corresponds to $E_e(Q)$ in Equation 4.7, which we write as $V(\vec{R})$. It consists of binding forces and forces from nonbonded nuclei groups and other nuclei. What is included in the molecular dynamics simulation depends on how detailed we want to be. Usually, it is sufficient to include CH bonds as a united atom and neglect the motion of the hydrogen atoms. In proteins, we may choose to ignore the bond distances and include only the torsion angles since the latter are more important for protein folding. The same holds true for certain other polymers.

Potential functions may be calculated or approximated on the basis of experimental data or taken from calculated results. The bonding potential may be approximated using the harmonic approximation of Equation 4.26:

$$V(R) = \frac{1}{2}k(R - R_e)^2.$$ (4.44)

Anharmonicities may be ignored at normal temperature. R_e is the equilibrium distance between the atoms i and j. Equation 4.44 is part of the potential sum for the bonded atoms.

The potential from bond angles and torsion angles is written in an analogous way:

$$V(\theta) = \frac{1}{2}k_\theta(\theta - \theta_e)^2.$$ (4.45)

Nonbonding forces between atoms or molecules may be included as electrostatic forces, provided a charge can be defined for the molecule or atom. If the atoms are neutral, there may still be attraction due to induced forces. All these forces are referred to as van der Waals forces and are reasonably well accounted for in simulations by the *Lennard–Jones potential*:

$$V(R) = 4a\left[\left(\frac{R_0}{R}\right)^{12} - \left(\frac{R_0}{R}\right)^6\right],$$ (4.46)

where R is the distance between the molecules. $V(r) = 0$ for $R = R_0$. It is easy to show that the minimum energy occurs for $R = 2^{1/6}R_0$. The depth of this minimum is $-a$.

All contributions concerning a given atom are summed to a potential V. The force is then calculated as

$$F_X = -\frac{\partial V}{\partial X} = -\frac{\partial V}{\partial R}\frac{\partial R}{\partial X} + \cdots.$$ (4.47)

$\partial V/\partial R$ may be obtained from Equation 4.44.

Classical equations of motion are used (Newton's equations). Velocity and acceleration are defined by

$$v_x = \frac{dX}{dt} \Rightarrow \frac{dv_x}{dt} = \frac{d^2X}{dt^2}.$$ (4.48)

Equation 4.27 gives

$$F_X = \frac{dv_X}{dt}M \Rightarrow dv_X = \frac{F_X}{M}dt.$$ (4.49)

The time step has to be chosen sufficiently small in Equation 4.49 (usually $dt = 10^{-15}$ sec). The change in velocity, dv_X, is calculated. Velocity and position for $t > 0$ are integrated from the initial values by using Equation 4.48 and $dX = vdt$. The case of three dimensions is not much difficult. The force is a vector:

$$\vec{F} = \left(\frac{\partial V}{\partial x}, \frac{\partial V}{\partial y}, \frac{\partial V}{\partial z} \right), \tag{4.50}$$

and so are position, velocity, and acceleration. Newton's equation reads

$$\vec{F} = \frac{d^2\vec{R}}{dt^2} M, \tag{4.51}$$

and from this equation, $\vec{v}$ and $\vec{R}$ are integrated as in the one-dimensional case.

Velocity and position must have reasonable starting values. If the velocity is too large, the temperature is too high, and ways must be found to "cool down" the system.

Molecular dynamics gives a very detailed simulation of the dynamics in a chemical system. It is usually necessary to make a number of integrations from given initial values. Unfortunately, the total time span that can be reached is usually less than 1 msec. However, in most cases, only uninteresting things take place during large simulation times. For example, the waiting time to pass a barrier is "lost time." It may be advisable to give the system a well-directed kick to make something happen.

In most cases, molecular dynamics is less useful at a lower temperature. Low energy quantum states are involved, where classical mechanics is less successful. At low temperature, the nuclear tunneling phenomenon is more important, relative to the classical motion. Tunneling is not easily taken into account in classical simulations.

4.4 QUANTIZATION OF VIBRATIONS

The SE for vibrational motion is given in Equations 4.7 and 4.8. The kinetic energy for the nuclei is essentially obtained by replacing the classical velocity by the quantum mechanical one:

$$\frac{\partial Q}{\partial t} \rightarrow -i\hbar \frac{\partial}{\partial Q}. \tag{4.52}$$

4.4.1 HARMONIC OSCILLATOR

We solve the time-independent SE for a harmonic oscillator. The vibrational wave function is denoted as Ψ. In one dimension, Equation 4.7 may be written as

$$-\frac{\hbar^2}{2M} \frac{d^2\Psi}{dX^2} + \frac{kX^2}{2} \Psi = E\Psi. \tag{4.53}$$

In a diatomic molecule, the equilibrium distance is R_e and the masses of the nuclei are M_1 and M_2. Here, $X = R - R_e$ and M is the reduced mass:

$$M = \frac{M_1 M_2}{M_1 + M_2}. \tag{4.54}$$

Equation 4.53 is the time-independent SE for the vibrational motion. The boundary conditions are $\Psi \to 0$ exponentially when $X \to \pm\infty$. A possible solution of this type is

$$\Psi(X) = \exp(-aX^2). \tag{4.55}$$

We insert this expression into Equation 4.53 and obtain

$$-\frac{\hbar^2}{2M}\left(4a^2 X^2 - 2a\right)\exp(-aX^2) + \frac{1}{2}kX^2 \exp(-aX^2) = E\exp(-aX^2). \tag{4.56}$$

After canceling $\exp(-aX^2)$, the following equation remains:

$$\left(-\frac{\hbar^2}{M}2a^2 + \frac{1}{2}k\right)X^2 + \frac{\hbar^2}{M}a - E = 0. \tag{4.57}$$

This polynomial in X can only be equal to zero for all X if the coefficient before X^2 is $= 0$ and the terms without X also are equal to zero. This implies that

$$-\frac{\hbar^2}{M}2a^2 + \frac{1}{2}k = 0 \Rightarrow a = \frac{\sqrt{kM}}{2\hbar} \quad \text{and} \quad -\frac{\hbar^2}{M}a = E. \tag{4.58}$$

Thus,

$$\Psi = \exp\left(-\frac{\sqrt{kM}}{2\hbar}X^2\right) \tag{4.59}$$

is a possible solution of Equation 4.53, if

$$E = \frac{\hbar^2}{M}a = \hbar\sqrt{\frac{k}{M}}\cdot\frac{1}{2}. \tag{4.60}$$

Another possible solution of Equation 4.53 is

$$\Psi = P(X)\exp(-aX^2), \tag{4.61}$$

where $P(X)$ is a polynomial in X (Figure 4.2). The energy levels are

$$E_v = \hbar\sqrt{\frac{k}{M}}\left(v + \frac{1}{2}\right) \quad v = 0, 1, 2, \ldots, \tag{4.62}$$

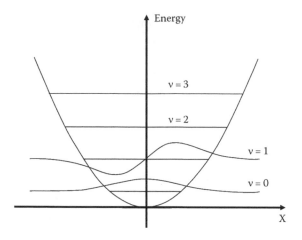

FIGURE 4.2 Energy levels for the harmonic oscillator and the two first eigenfunctions.

Typical for the energy levels of a harmonic oscillator is that the distance between them is a constant $= \hbar\sqrt{k/M}$. In reality, PES is anharmonic. The distance between the energy levels then decreases as we go to higher energies and approach the dissociation limit. Also note that the lowest energy, the zero-point energy, is larger than zero.

The eigenfunctions belonging to Equation 4.53 may be written as

$$\Psi_v(X) = NH_v(\xi)\exp\left[-\frac{\sqrt{Mk}}{2\hbar}X^2\right], \text{ where } \xi = \left(\frac{kM}{\hbar^2}\right)^{1/4}X, \qquad (4.63)$$

where N is a normalization constant. The Hermite polynomial H_v is defined by

$$H_v(\xi) = (-1)^v \exp(\xi^2)\left(\frac{d}{d\xi}\right)^v \exp(-\xi^2). \qquad (4.64)$$

The first Hermite polynomials are given by

$$H_0(\xi) = 1$$

$$H_1(\xi) = 2\xi$$

$$H_2(\xi) = 4\xi^2 - 2$$

$$H_3(\xi) = 8\xi^3 - 12\xi \qquad (4.65)$$

$$H_4(\xi) = 16\xi^4 - 48\xi^2 + 12$$

$$H_5(\xi) = 32\xi^5 - 160\xi^3 + 120\xi.$$

We see that H_0, H_2, and H_4 are even functions of ξ, while H_1, H_3, and H_5 are odd functions of ξ. The ground state ($v = 0$) forms a Gaussian function around the origin. For higher values of v, probability is moved toward the classical turning points. Since the eigenfunctions are alternating between odd and even, it is possible to superpose two adjacent eigenfunctions to form a wave packet that oscillates between the turning points. This quantum motion tends to the classical pendulum motion for large v.

4.4.2 SMALL VIBRATIONS

In the molecule, we have to solve SE for the normal modes, which are the same as in the classical case. The SE can be written as

$$-\frac{1}{2}\frac{\partial^2 \Psi_i}{\partial Q_i^2} + \frac{1}{2}\omega_i^2 Q_i^2 \Psi_i = E\Psi_i. \tag{4.66}$$

The energy levels are given by

$$E_i = \left(v + \frac{1}{2}\right)\hbar\omega_i; \quad v = 1, 2, \ldots. \tag{4.67}$$

The wave function is given by the Hermite polynomial Equations 4.63 through 4.65. The expressions for the solutions are the same, but the frequencies and normal modes are different.

Classically, we may set the energy of an oscillator at any value. The frequency for a normal mode is not affected by the energy. Quantum mechanically, only certain energies are possible according to Equation 4.67. However, the energy differences between different states are equal. For this reason, the wave packet moves with the classical frequency. As we will see in Chapter 7, any energy can be given to the wave packet, just as in the classical case. For example, the first and second states can be mixed in any proportions, but if the coefficient for the second state is very small, the wave packet will hardly move. Just a small disturbance is introduced.

As an example, the normal modes for a water molecule are given in Figure 4.3. The wave number for the stretching modes is approximately 3200 cm^{-1} and for the bending mode, 1600 cm^{-1}. The corresponding classical vibration mode can be used to represent the normal mode (the arrows in Figure 4.3).

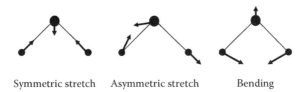

Symmetric stretch   Asymmetric stretch   Bending

FIGURE 4.3   Normal modes for the water molecule H_2O.

4.5 VIBRATIONAL SPECTRA

Vibrational spectra are the fingerprints of a molecule. Equation 4.60 shows that the energy of a vibrational transition is proportional to $\sqrt{(k/M)}$. Strong bonds with a large force constant therefore have high energy. Large M (heavy vibrating atoms) results in low energy and wave number in the spectrum. In most cases, vibrational transitions do not represent a single, pure mode of vibration, since many primitive vibrations may be involved in a quantum mode, as in the symmetric stretch in Figure 4.3 (the bond angle changes slightly).

An IR spectrum for quite ordinary air (in a laboratory with many students) is shown in Figure 4.4. The measured wave numbers agree well with the calculated ones, using Hartree–Fock with a 6-31G** basis set. The main components of air, the diatomic molecules N_2 and O_2, have no dipole moment and therefore no absorption. The broad absorption at 1600 cm^{-1} is due to the bending vibration of the H_2O molecule. The absorption at approximately 700 cm^{-1} corresponds to the bending vibration of the CO_2 molecule and the absorption at 2400 cm^{-1} to the asymmetric stretch of the same molecule. The symmetric and asymmetric stretches for water are outside the diagram. The CO_2 symmetric stretch is at about the same energy as the H_2O bend ($\approx$1600 cm^{-1}), but there is no absorption since CO_2 is a linear molecule (Figure 4.4).

The rotational broadening is considerably smaller for carbon dioxide than for water. Apparently, this depends on the moment of inertia, which is much larger for the oxygen atoms in the carbon dioxide molecule than for the hydrogen atoms in the water molecule.

4.5.1 INTENSITY IN INFRARED SPECTRA

The fundamental equations for interaction between an electromagnetic (EM) field and a molecular system will now be given. The most important quantity to evaluate is the transition moment. If the electronic states are the same before and after

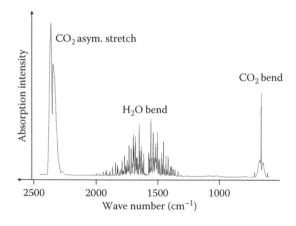

FIGURE 4.4 IR spectrum for humid air. The bending vibrations of CO_2 and H_2O and the asymmetric stretch vibration of CO_2 are seen.

the transition, the transition dipole moment for transition between the vibrational levels v and u in the same electronic state is

$$\vec{M} = \int_0^\infty \Psi_v^*(R - R_e) e\mu(R) \Psi_u(R - R_e) dR, \tag{4.68}$$

where R_e is the equilibrium distance, Ψ_v and Ψ_u are wave functions for the two vibrational states, and $\mu(Q)$ is the dipole moment of the molecule at deviation Q from the equilibrium distance. The derivation of intensity for vibrational transitions is quite complicated for molecules in general. The derivation below is valid for diatomic molecules and is given to illustrate the principles.

The intensity of the absorption line depends on the oscillator strength, which in turn depends on the transition moment squared. The electronic part of the transition moment for x-polarized light and transition $\phi_i \rightarrow \phi_a$ may be written as

$$\mu_x = \int \phi_i^* \cdot ex \cdot \phi_a dx dy dz. \tag{4.69}$$

For y- or z-polarized light, x is replaced by y or z, respectively, in Equation 4.69. A complete derivation will be given later.

The dipole moment of the molecule at distance R is $e\mu(R)$. We may use the Taylor expansion for $\mu(R)$ around the equilibrium position $(Q = R - R_e)$:

$$\mu(R) = \mu(R_e) + \left(\frac{d\mu}{dR}\right)_{R=R_e}(R - R_e) + \frac{1}{2}\left(\frac{d^2\mu}{dR^2}\right)_{R=R_e}(R - R_e)^2 + \cdots. \tag{4.70}$$

We neglect the $(R - R_e)^2$ term since the contribution to the integral in Equation 4.68 is important only for $R \approx R_e$, where $|R - R_e|$ is small. The constant term does not contribute to the integral of Equation 4.68, since with $\mu = \mu(R_0)$, we have an overlap integral between eigenfunctions with different eigenvalues, which is always $= 0$. Only the linear term remains now. If harmonic potentials are used, it is quite easy to show that the transition moment $\vec{M}$ in Equation 4.68 is significantly different from zero only if

$$v - u = \pm 1. \tag{4.71}$$

This is thus the *selection rule* for an IR transition. If $v - u \neq \pm 1$, $M = 0$, and the transition is forbidden.

If $d\mu/dR = 0$ at $R = R_e$, then M vanishes even if $v - u \neq \pm 1$. This is the situation for all homonuclear diatomic molecules, since $\mu = 0$ for all R. In the IR spectrum, there is thus intensity only if the charge distribution of the molecule varies as a function of the atomic distance R. The only molecules of natural air that absorb IR radiation are the greenhouse gases H_2O, CO_2, CH_4, and various nitrogen oxides (NO_x), but neither O_2 nor N_2.

4.5.2 IR FREQUENCY DEPENDS ON TYPE OF BOND

As we have seen in the previous sections, a molecule with many atoms has a mode for each degree of freedom in molecular vibrations. A mode is seldom "clean" in the sense that it is formed by only a single interatomic vibration, as is the case for diatomic molecules. For large molecules, a mode may correspond to a complicated nuclear motion pattern involving, for example, five covalent bonds, as in hexatriene.

In the vibronic (vibrational + electronic) ground state, the molecule is in its lowest electronic state, while the vibrational modes have the lowest quantum numbers: $(v_1, v_2, \ldots, v_{3N-6}) = (0, 0, \ldots, 0)$. The quantum numbers $v_1, v_2, \ldots, v_{3N-6}$ are ordered according to vibration quantum. The first excited vibronic state is thus $(1, 0, 0, \ldots, 0)$. Next state is either $(0, 1, 0, \ldots, 0)$ or $(2, 0, 0, \ldots, 0)$. Due to the selection rules $(\Delta v_i = \pm 1)$, only the former can be reached from the ground state and is visible in the spectrum.

In organic molecules, the CC stretch frequency is usually between 1000 and 1800 cm^{-1} (Figure 4.5). The lowest CC stretch vibrations are at 1057 cm^{-1} for ethane and at 1488 cm^{-1} for ethylene. In both these molecules, the hydrogen atoms follow the carbon atoms in their motion. There is another CC stretch at 1564 cm^{-1} for ethane, and at 1851 cm^{-1} for ethylene, where the hydrogen atoms do not move.

These values are obtain in Hartree–Fock calculations. The experimental value for the lowest ethane stretch frequency is 995 cm^{-1}.

The CH frequencies are situated just above 3000 cm^{-1}. The OH frequencies of water are considerably higher, at 4148 and 4264 cm^{-1} (in the calculation of Figure 4.5). Bending vibrations of different types in organic molecules are located between 1000 and 1600 cm^{-1} (Figure 4.5). They interact to some extent with the CC stretch modes. The lowest frequencies occur for the motion of side groups or, more generally, when a large part moves as a whole against the rest of the molecule. These modes are called librational modes.

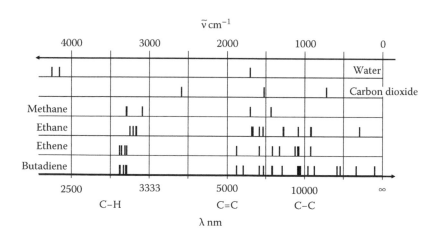

FIGURE 4.5 Vibrational spectra for some small molecules. Hartree–Fock calculation of PES (*ab initio*, 6-31G).

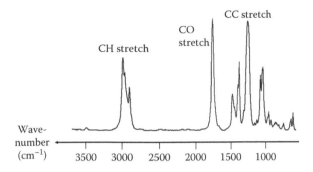

FIGURE 4.6 Raman spectrum (Stokes) for organic molecule. Origin is at the Rayleigh scattering.

In some cases, it is possible to see the transitions to higher vibrational states (overtones). In the harmonic approximation, these transitions are forbidden. Hence, the appearance of overtones is a sign that the vibration is anharmonic.

4.5.3 RAMAN SPECTRA

Another vibration spectroscopy is Raman spectroscopy. Here, the probe is radiated with monochromatic visible light, which is spread by the molecules of the probe. The scattered light has another wave number, which depends on the fact that the dispersing molecule has jumped between rotational and vibrational levels. The difference in wave number between the absorbed light and the scattered light is called the *Raman shift* (Figure 4.6). If the final state is vibrationally excited, then the emitted light is shifted to a lower frequency. This shift is called a *Stokes shift*. If a quantum of vibrational energy instead is added to the energy of the emitted light, the shift is called an *anti-Stokes shift*.

Raman spectra are very useful not just because the selection rules are different from the IR selection rules. The intensity of the peak depends on how much the geometry is changed in the electronically excited state.

4.5.4 ROTATION SPECTRA

A pure rotation spectrum may be observed using microwaves (radar waves or, more correctly, EM waves in the wavelength range 1 mm $< \lambda <$ 30 cm). Rotational spectra can be seen in the gas phase for molecules with a dipole moment and for molecular groups rotating around a bond in larger molecules, for example, in proteins or transition metal complexes.

In a pure rotational spectrum, the upper and lower states of the transition belong to the same vibrational level. If the vibrational state is different in the lower and upper states, then there is a rotational broadening of the vibrational level in the IR region. The selection rule for the rotational transition is the same as in a pure rotational spectrum.

A diatomic molecule that rotates with a constant interatomic distance has the same time-independent SE for the nuclei as the angular part of an electron in the hydrogen atom. Rotation SE is thus the following equation:

$$-\frac{\hbar^2}{2I}\left(\frac{1}{\sin\theta}\frac{\partial}{\partial\theta}\left(\sin\theta\frac{\partial}{\partial\theta}\right)+\frac{1}{\sin^2\theta}\frac{\partial^2}{\partial\varphi^2}\right)Y(\theta,\varphi)=EY(\theta,\varphi).\qquad(4.72)$$

M is the reduced mass of the molecule and $I = MR^2$ is the moment of inertia. The boundary conditions are simply that the wave function is unique when the angles change. Two quantum numbers appear, called J ($\leftrightarrow$l) and M ($\leftrightarrow$m$_l$). MR^2 for a rotating molecule are much larger than mr^2 for electrons, since M is much larger than m while the interatomic distance is roughly the same as the average distance between the electron and the nucleus in the hydrogen atom. The much larger mass thus leads to a very small distance between the quantized energy levels. For diatomic molecules, the energies are

$$E_J = \frac{\hbar^2}{2I}J(J+1) = hcBJ(J+1), \quad J = 0, 1, 2, \ldots . \qquad(4.73)$$

$\hbar^2/2I$ has dimension energy. The rotation constant B is given in wave numbers (reciprocal centimeter). For a given J, there are $2J + 1$ functions with $J \geq M \geq -J$. The degeneracy is thus equal to $2J + 1$. B usually varies between 0.01 and 10 cm^{-1}. At a high temperature, there are thus many molecules on levels with high values of J (Figure 4.7). The relative number is given by Boltzmann's lag as

$$N_{J,M}/N_{0,0} = \exp\left[-J(J+1)\cdot\frac{hcB}{k_BT}\right], \qquad(4.74)$$

where T is the absolute temperature and k_B is Boltzmann's constant. If $N_{0,0}$ is the number of molecules on the lowest level (J = 0), $N_{J,M}$ is the number of molecules on level (J,M). Since the degeneracy is $2J + 1$, the total relative number of molecules, N_J, on level J is (Figure 4.7):

$$N_J/N_0 = (2J+1)\cdot N_{J,M}/N_0 = (2J+1)\cdot\exp\left[-J(J+1)\cdot\frac{hcB}{k_BT}\right], \qquad(4.75)$$

where $N_0 = N_{0,0}$, since the lowest energy level is nondegenerate. The constant hc/k_B = 1.4388 cm · K.

The selection rule for the transition between the diatomic rotational levels in an EM field (microwaves or radio waves) is simple:

$$\Delta J = 0, \pm 2, \Delta v = \pm 1 \text{ for Raman and } \Delta J = \pm 1 \, \Delta v = \pm 1 \text{ for IR.} \qquad(4.76)$$

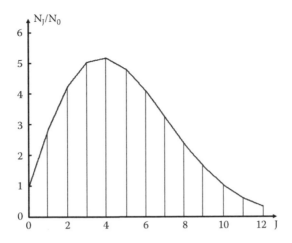

FIGURE 4.7 The number of molecules in an excited state J, relative to the number of molecules in the ground state (J = 0) in carbon monoxide (CO) at 100 K (Equation 4.75).

Thus, Raman and IR spectra have different selection rules. Raman lines can be discovered where there is no IR absorption, and vice versa.

The energy of the transition is (IR)

$$\Delta E_J = hcB \cdot (J+1)(J+1+1) - hcB \cdot J(J+1) = hcB(J+1) \cdot 2. \qquad (4.77)$$

If the transition moment is nonzero, the transition is said to be *allowed*. For rotations, the absorption occurs at the wave numbers $\tilde{v} = 2B(J + 1)$; J = 0, 1, 2, … (Figure 4.8).

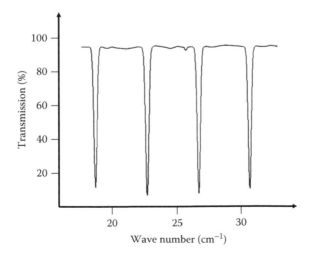

FIGURE 4.8 Typical rotational spectrum for a diatomic molecule (CO).

For allowed transitions, the transition moment is independent of J for each transition (not proven). The intensity is therefore proportional to the number of molecules on a certain level. In the spectrum appears one line for each value of J (Figure 4.8). The first rotational level is at $\tilde{v} = 2B$ and the next one at twice the wave number: $\tilde{v} = 4B$. The distance between the following absorption lines is constant, $\Delta\tilde{v} = 2B$.

4.6 VIBRATIONS IN ELECTRONIC SPECTRA

The electronic states in a molecule form PES in a multidimensional space. This has important consequences if a molecule is excited. The time for excitation is very short, considerably shorter than the time for a vibration cycle. The excitation may thus be considered as "vertical" in the sense that the nuclei may be regarded as fixed in space during the excitation. This is the *Franck–Condon condition*. The excitation ends on a PES, but usually not on the local minimum of a PES. In fact, the excited states of molecules almost always have different equilibrium geometries compared with the ground state. A local excitation lifts one electron from a bonding to an antibonding orbital. Thus, there is less bonding in the excited state and therefore the excited states usually have longer bonds than the ground state. The trivial fact that energy surfaces depend on bond length is important when we want to understand the structural dynamics taking place at excitation.

The structural change at excitation and deexcitation also determines the intensity in the Raman spectrum.

4.6.1 VIBRATIONAL BROADENING

As will be clear in Chapter 7, the absolute value of the transition dipole moment vector (μ_x, μ_y, μ_z) determines the intensity of a transition. The x-component can be written as

$$\mu_x = e\int \Phi^* (x_1 + x_2 + \cdots + x_N)\Phi_i^a d\tau. \qquad (4.78)$$

Here, $d\tau$ is an integration element for both electron and nuclear coordinates. The nuclear part of the transition moment should be included, but since the electronic states are orthogonal (zero overlap), this contribution will vanish.

The total wave functions for the states involved in the excitation are

$$\Phi = \Psi(1,2,\ldots,N)\Xi_{0u}(1,2,\ldots,M) \rightarrow \Phi_i^a = \Psi_i^a(1,2,\ldots,N)\Xi_{i\rightarrow a,v}(1,2,\ldots,M), \qquad (4.79)$$

where Ξ_{0u} and $\Xi_{i\rightarrow a,v}$ are the vibrational wave functions before and after the electronic excitation $\phi_i \rightarrow \phi_a$, respectively. Ψ_i^a is an electronic wave function where one electron has been moved from orbital ϕ_i to orbital ϕ_a. Rotational wave functions are neglected. In most cases, every excited state has its own equilibrium point and vibrational levels.

Equation 4.78 may be rewritten as

$$\mu_x = \int \phi_i^*(x,y,z)\Xi_{0,u}^* \cdot x \cdot \phi_a(x,y,z)\Xi_{i \to a,v}dxdydzdXdYdZ$$

$$= \int \phi_i^*(x,y,z) \cdot x \cdot \phi_a(x,y,z)dxdydz \cdot \int \Xi_{0,u}^*\Xi_{i \to a,v}dXdYdZ \qquad (4.80)$$

$$= \mu_{ex}\langle \Xi_{0u}|\Xi_{i \to a,v}\rangle.$$

The transition moment thus depends on vibrational overlap. Since u is usually the lowest vibrational state (u = 0), it may be written as $\langle \Xi_{00} \Xi_{i \to a,v}\rangle$. The square of the latter is called the Franck–Condon integral. The width of the spectrum depends on how many vibrational levels in the ground state have overlapped with the vibrational levels in the excited state. If the excitation takes place in the gas phase and there are no obvious broadening mechanisms, the spectrum will be resolved in sharp vibration levels. Equation 4.80 is the quantum mechanical expression for the Franck–Condon principle. If the overlap is calculated in Equation 4.80, one obtains the highest intensity for the almost vertical transitions.

One may also conclude from Equation 4.80 that many final vibrational states are populated after the transition. This is, in fact, the main reason for the dynamics that takes place after excitation. The excited state as well as the ground state may involve many atoms. The excited states may be charge transfer states or exciton states, as will be discussed in Chapter 12.

4.6.2 FRANCK–CONDON FACTORS

We may now extend the derivation of the oscillator strength given in Section 3.5.3 to include the nuclear coordinates. In the simplest case, the states involved are the products of electronic and vibration wave functions. The electronic states may be different in a single MO, ϕ_i in the ground state and ϕ_a in the excited state. Assume a diatomic molecule where the vibrational state is changed from $\Psi(Q)$ to $\Xi(Q)$, where Q is the interatomic distance and Q_0 and Q_e are equilibrium points for the ground and excited states, respectively. The vibration wave functions may be written as

$$\Psi(Q) = \left(\frac{\alpha}{\pi}\right)^{1/4} \exp\left[-\alpha(Q-Q_0)^2/2\right] \qquad (4.81)$$

and

$$\Xi(Q) = \left(\frac{\alpha}{\pi}\right)^{1/4} \exp\left[-\alpha(Q-Q_e)^2/2\right]. \qquad (4.82)$$

The overlap between these two wave functions is

$$S_{0,0} = \int_{-\infty}^{\infty} \Xi \Psi dQ = \left(\frac{\alpha}{\pi}\right)^{1/2} \int_{-\infty}^{\infty} \exp\left[-\alpha(Q-Q_e)^2/2\right]\exp\left[-\alpha(Q-Q_0)^2/2\right]\Psi dQ. \quad (4.83)$$

This and similar integrals are calculated in Appendix 6. With $a = Q_0 - Q_e$, we obtain

$$S_{0,0} = \exp\left(-\alpha a^2/4\right). \quad (4.84)$$

In the electronically excited state, all higher vibrational states are of importance and are given in Equations 4.63 and 4.64. To obtain the vibration overlap to the excited state, we have to use recursion formulas for Gaussian integrals (Appendix 6). The result is comparatively simple to obtain using these equations and Equations 4.63 and 4.64 in computer programs. The first one is

$$S_{0,1} = \left(\frac{\alpha}{\pi}\right)^{1/2} a \exp\left(-\alpha a^2/4\right). \quad (4.85)$$

4.7   PES CROSSING

The two PES frequently cross each other. Energy surfaces for excited states run parallel only if the equilibrium geometry and force constants are the same. Another example is ionic states and neutral (covalent, homopolar) states. The ionic $Na^+ + Cl^-$ state is the ground state at the equilibrium geometry of the NaCl diatomic molecule. The neutral $Na + Cl$ state is the ground state at a large interatomic separation. The electronic wave functions of the two states are very different, since the Na 3s orbital is occupied only in the neutral state. If the Na and Cl atoms approach each other in a collision, the energy of the $Na^+ + Cl^-$ state and the $Na + Cl$ state will be equal at a certain interatomic distance. Here, the states interact and there is an avoided crossing. The PES do not intersect. There is a certain probability, which can be calculated in time-dependent quantum mechanics, that the $Na + Cl$ system will exchange an electron and form $Na^+ + Cl^-$.

4.7.1   AVOIDED CROSSING

If we want to describe the curve-crossing problem in physical terms, we may think of a configuration interaction calculation where the two wave functions that represent the ionic and neutral states are included. If the ground state wave function is examined for decreasing values of the interatomic separation, we find, for NaCl, that $Na + Cl$ gradually changes to $Na^+ + Cl^-$ in the ground state.

Far from a possible crossing, the two wave functions, Φ_1 and Φ_2, are both eigenfunctions of the electronic Hamiltonian H:

$$H\Phi_1 = H_{11}\Phi_1 \text{ and } H\Phi_2 = H_{22}\Phi_2. \quad (4.86)$$

Matrix elements are defined as

$$H_{ij} = \int \Phi_i^* H \Phi_j dx, \quad \text{for } i = 1,2; \ j = 1,2; \ (H_{21} = H_{12}).$$ (4.87)

H_{11} and H_{22} are said to represent *diabatic* states, but the diabatic states are true quantum mechanical states only if the interaction matrix element H_{12} is equal to zero. Near the possible PES crossing ($H_{11} = H_{22}$), both wave functions have to be included in a configuration interaction. The eigenvalue problem may be written as

$$\begin{pmatrix} H_{11} & H_{12} \\ H_{21} & H_{22} \end{pmatrix} \begin{pmatrix} C_1 \\ C_2 \end{pmatrix} = E \begin{pmatrix} C_1 \\ C_2 \end{pmatrix}.$$ (4.88)

Equation 4.88 has a nontrivial solution for

$$\begin{vmatrix} H_{11} - E & H_{12} \\ H_{12} & H_{22} - E \end{vmatrix} = 0.$$ (4.89)

The solution of Equation 4.89 is given by (Figure 4.9)

$$E_{\pm} = \frac{H_{11} + H_{22}}{2} \pm \sqrt{\left(\frac{H_{11} - H_{22}}{2} \right)^2 + H_{12}^2}.$$ (4.90)

If $H_{12} = 0$, the two solutions are $E_+ = E_1$ and $E_- = E_2$ with an abrupt change of wave function at the crossing. If $H_{12} \neq 0$, E_+ and E_- do not cross and the two wave functions gradually change character. There is a gap, $\Delta = 2H_{12}$, between a lower surface and an upper surface due to interaction between the two original PES states.

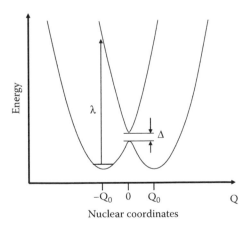

FIGURE 4.9 Curve-crossing problem in the case of equal energies at the equilibrium points $-Q_0$ and Q_0.

In a multidimensional space, "conical intersection" is probably a more used notation than "avoided crossing." We will not dig deeper into the multidimensional problem here, however, but refer to the paper by Yarkony.

Each PES is connected to a wave function with particular properties. Due to the
gap, it is possible to go continuously from an up-going to a down-going PES and thus
remain on the lower surface. Therefore, passing the avoided crossing on the same
PES means that the character of the system changes, smoothly or quickly, depending on the width of the gap. In electron transfer, the passing of the avoided crossing
corresponds to a full electron that moves from one molecule to the next. If there is
large interaction between the two diabatic PES, there is a wide energy gap between
the upper and lower PES. In that case the reality of the diabatic states as the origin
of two PES becomes somewhat confused.

4.7.2 Vibration Spectrum in Double Minimum PES

In a number of systems, two geometries are possible for a molecule, even without
the involvement of any electron. There are systems with inversion center, where
two equivalent states are present and the energy is exactly the same after inversion. Depending on the size of the activation barrier, this type of molecule may
have a spectrum different from the parabolic approximation due to the "interaction"
between two parabolic PES.

Another example is proton transfer. The proton has a possibility to be attached
to two different heavy atoms. This situation appears in a number of other cases, for
example, at proton transfer between symmetrically placed "heavy" atoms (usually
O, N, or F). The proton may be situated in one minimum and may transfer to the
equivalent one.

If there is no energy gain, what then drives the protons to leave a minimum and
move to the equivalent one? The basic reason is that the system is not in a pure quantum state. If it is not, it will move according to time-dependent quantum mechanics,
oscillating between the two almost equivalent sites (see Chapter 7).

An example of an activation barrier is provided by the ammonia umbrella motion
(Figure 4.10). The motion of the nitrogen atom through the plane of the three hydrogens is related to a comparatively low activation barrier.

There is no simple way to calculate the vibrational levels in a double minimum
potential. The two lowest levels are approximately linear combinations of the zero
levels of the separate diabatic parabolas. The lowest state is symmetric and the

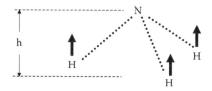

FIGURE 4.10   Umbrella motion vibration in the ammonia molecule.

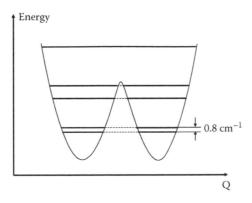

FIGURE 4.11 Double minimum potential with vibrational levels.

higher one antisymmetric. The energy splitting, Δ, between the states is very small if the barrier is high (Figure 4.11).

The higher levels are less affected by the double minimum form of the potential, and are thus at equal distance. The energy difference is determined by the force constant (and the reduced mass).

As we will see later, the transfer of a particle between the energy minima is treated in time-dependent quantum mechanics. The particle is represented by a wave packet. Suppose this wave packet is localized in one of the two minima. For time $t > 0$, the particle will oscillate between the minima. The frequency is high if the energy splitting Δ is large.

As the barrier tends to large values, the gap between the two lowest vibrational states tends to zero. The same applies to the oscillation frequency. In fact, the effect on the particle is so small that it can be neglected. This is the case in very weak hydrogen bonds. The probability for proton transfer is very low.

If the mixture of vibrational and electronic states is not of simple product type, as is prescribed in the BO approximation, the neglect of terms in deriving Equation 4.7 from Equation 4.6 can no longer be carried out. It is necessary to include these terms as corrections to the BO approximation.

5 Statistical Mechanics

5.1 INTRODUCTION

In 1662, Robert Boyle discovered that the volume (V) of a gas in a closed container is inversely proportional to its pressure (P), as long as the temperature (T) is constant. Much later (1802), when temperature could be accurately measured, Joseph-Louis Gay-Lussac was able to show that the constant in Boyle's law (PV = constant) is proportional to the temperature. Sometime later, Benoit Clapeyron wrote the general gas law:

$$PV = nRT, \tag{5.1}$$

where n is the number of moles and R is the gas constant. In 1859, Rudolf Clausius and James Clerk Maxwell developed the kinetic theory of gases. The latter derived a distribution law for the kinetic energy of gas molecules. Ludwig Boltzmann developed the subject of statistical distributions. Scientists like Max Planck and Willard Gibbs also contributed a lot to the development of statistical physics in the latter part of the nineteenth century. A milestone was reached when, in 1875, Boltzmann defined statistical entropy.

Sections 5.2–5.4 contain a simplified derivation of statistical thermodynamics. There are three fundamental laws. Energy exists in various forms, for example, as heat energy (q) or mechanical work energy (w). The principle of conservation of energy states that the total energy U = q + w is constant, but q and w can be transferred into each other.

Entropy is a state function that depends on the system. Change in the content of entropy in a reversible process may be measured as $\delta S = \delta q/T$, where δq is the heat added or removed, and T is the absolute temperature. In a spontaneous (irreversible) process, δS is positive. For example, if two isolated bodies of different temperature are brought together, the temperature will be the same after some time. In this spontaneous process, $\delta S > 0$, which is referred to as the *Clausius inequality*. This is the second law of thermodynamics, which states that the entropy of an isolated system tends to increase until a new equilibrium is established. The second law dates back to Sadi Carnot and was fully developed during the nineteenth century. In statistical terms, a spontaneous change occurs because the resulting state is statistically more likely than the original state. It is said to carry a higher *statistical weight*.

The third law of thermodynamics states that entropy is measured as zero at absolute temperature equal to zero. This law follows automatically in statistical mechanics.

The last section of this chapter treats nonequilibrium statistical mechanics and includes transport processes and molecular dynamics simulations. This section is important for the following chapters, not least for calculating the rate of chemical reactions, including electron transfer reactions.

5.2 PARTITION FUNCTION AND THERMODYNAMIC PROPERTIES

Since this is not a textbook for thermodynamics, only a short derivation will be given, with the aim to obtain quick and useful results. Toward the end of the chapter, the problems in the derivation will be pointed out and an ensemble theory is mentioned. Here, we use simplified versions of the *microcanonical ensemble*.

5.2.1 BOLTZMANN DISTRIBUTION

If heat is supplied, a set of molecules will occupy increasingly higher energy levels. A number of different occupations are possible for a given energy. Every unique occupation is called a configuration. We assume that the particles are distinguishable. Every configuration with the same energy has the same chance to exist but since some configurations occur more often than others, only the most common ones turn out to be of any interest. An example of a quite common configuration is given in Figure 5.1.

The total number of selections from N objects is N!. Since it does not matter in what order the molecules are chosen to a certain energy level, we have to divide the total number by the factorial of the number of molecules on that level (the occupation number n_i). What remains is the *weight* of the configuration. The weight of the configuration is thus

$$W = \frac{N!}{n_1!n_2!n_3!\dots}. \tag{5.2}$$

Two relations have to be satisfied. The total number of particles is constant:

$$N = \sum_i n_i. \tag{5.3}$$

The total energy is constant and may be written as

$$E = \sum_i n_i \varepsilon_i. \tag{5.4}$$

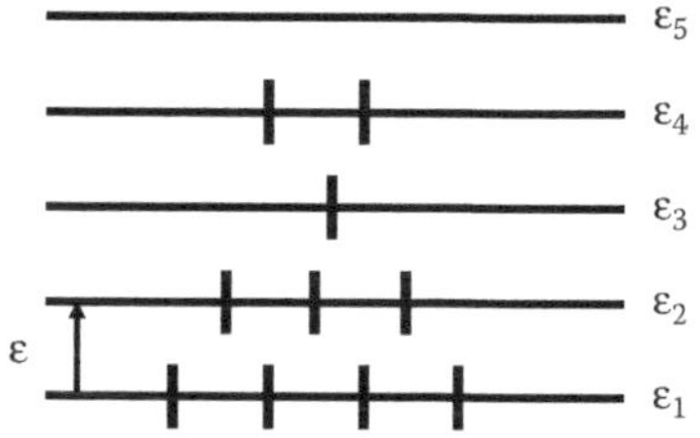

FIGURE 5.1 Energy levels with the indicated occupation numbers ($n_1 = 4$, $n_2 = 3$, $n_3 = 1$, $n_4 = 2$, $n_5 = 0$). According to Equation 5.2, this configuration has the weight $10!/(4! \cdot 3! \cdot 1! \cdot 2!) = 12,600$.

To find the configuration with the largest weight, $\ln W$ is maximized instead of W, since this proves to be simpler. Maximum of W must be a maximum also of $\ln W$, since the function $\ln x$ is monotonously increasing. At this maximum, any change in occupation number δn_i must give $\delta(\ln W) = 0$. Since Equations 5.3 and 5.4 must be satisfied at the same time, we have to introduce Lagrangean multipliers, α and β. αN and $-\beta E$ are added to what is varied, that is, $\ln W$. The differential of the new function is equal to zero:

$$d\left(\ln W + \alpha N - \beta E\right) = 0, \tag{5.5}$$

which implies

$$\sum_i \frac{\partial \ln W}{\partial n_i} dn_i + \alpha \sum_i dn_i - \beta \sum_i \varepsilon_i dn_i = 0. \tag{5.6}$$

According to the Lagrange theorem, the coefficients before dn_i have to be equal to zero for all dn_i, that is,

$$\frac{\partial \ln W}{\partial n_i} + \alpha - \beta \varepsilon_i = 0. \tag{5.7}$$

It is quite simple to calculate $\ln W$ from Equation 5.2:

$$\ln W = \ln N! - \sum_j \ln n_j!. \tag{5.8}$$

According to Stirling's equation, $\ln K! \approx K \ln K$; thus, from Equation 5.2, we obtain

$$\ln W = N \ln N - \sum_j n_j \ln n_j. \tag{5.9}$$

Because of Equation 5.3, $\partial N/\partial n_i = 1$; therefore, Equations 5.7 through 5.9 yield

$$\ln N + N \frac{1}{N} - \ln n_i - n_i \frac{1}{n_i} + \alpha - \beta \varepsilon_i = -\ln \frac{n_i}{N} + \alpha - \beta \varepsilon_i = 0. \tag{5.10}$$

Exponentiation gives

$$\exp\left(-\ln \frac{n_i}{N} + \alpha - \beta \varepsilon_i\right) = 1 \Rightarrow \frac{n_i}{N} = \exp(\alpha)\exp(-\beta \varepsilon_i). \tag{5.11}$$

5.2.2 Partition Function

Since $\Sigma n_i = N$, by summing over i in Equation 5.11, we obtain

$$\sum_i \frac{n_i}{N} = 1 = \sum_i \exp(\alpha)\exp(-\beta\varepsilon_i) \Rightarrow \exp\alpha = 1 / \sum_i \exp(-\beta\varepsilon_i) = 1/Z. \quad (5.12)$$

Z is the *partition function*:

$$Z = \sum_i \exp(-\beta\varepsilon_i). \quad (5.13)$$

Index i runs over all energy levels i. For this reason, we may rewrite Equation 5.11 to

$$P_i = \frac{n_i}{N} = \frac{\exp(-\beta\varepsilon_i)}{\sum_j \exp(-\beta\varepsilon_j)}. \quad (5.14)$$

Equation 5.14 is the theorem of Boltzmann. By comparison with ordinary thermodynamics, it turns out that the multiplier β is the inverse of the temperature, $\beta = 1/k_B T$, where k_B is the Boltzmann constant. The *Boltzmann distribution* gives the fraction of particles in a state i with energy E_i as

$$P_i = \frac{\exp(-E_i/k_B T)}{Z} = \frac{\exp(-E_i/k_B T)}{\sum_j \exp(-E_j/k_B T)}. \quad (5.15)$$

5.2.3 Internal Energy

The internal energy, U, may be calculated from the partition function as

$$U = \frac{N}{Z} \sum_{i=0}^{\infty} E_i \exp(-E_i/k_B T), \quad (5.16)$$

where N is the total number of molecules. We have to multiply by N if E_i is the energy of a single molecule.

If we take the derivative with respect to temperature T of the natural logarithm of Z in Equation 5.13, we obtain

$$N\left(\frac{\partial \ln Z(T)}{\partial T}\right) = N\frac{\partial}{\partial T} \ln\left(\sum_{i=0}^{\infty} \exp(-E_i/k_B T)\right)$$

$$= \frac{N}{Z} \sum_{i=0}^{\infty} \exp(-E_i/k_B T)\frac{E_i}{k_B T^2} = \frac{U}{k_B T^2}. \quad (5.17)$$

Thus, the internal energy may be expressed as

$$U = Nk_BT^2\left(\frac{\partial \ln Z}{\partial T}\right),\tag{5.18}$$

where U is expressed by "system properties" and is not dependent on the way we arrived at the present state of the system. Such a function is called a *state function*. The internal energy of the system may have been acquired either by heat transfer or work on the system. The heat added, δq, and the work applied, δw, are not state functions, but their sum is

$$dU = \delta q + \delta w.\tag{5.19}$$

Equation 5.19 is called the first law of thermodynamics.

In a chemical reaction, the system is transformed into another system with other energy levels. If $\delta w = 0$, as is often the case, vibrational and rotational energies have to compensate for the change of electronic energy levels in Equation 5.16. If the internal energy U remains constant, heating or cooling of the system will take place during the chemical reaction.

Tabulated values of internal energy are given in the *standard state* of a chemical compound, which is the most stable form of the element at 1 bar of pressure and the specified temperature, usually 298.15 K or 25°C. Its symbol is ΔH_f^0.

5.2.4 ENTROPY

Boltzmann defined entropy as $S = k_B \ln W$. From Equation 5.9, we obtain

$$\ln W = N \ln N - \sum_i n_i \ln n_i = -\sum_i n_i \ln \frac{n_i}{N} = -N \sum_i P_i \ln P_i,\tag{5.20}$$

where P_i is the fraction of states at energy E_i. From Equation 5.14,

$$P_i = \frac{\exp(-E_i/k_BT)}{Z}.\tag{5.21}$$

Hence,

$$\ln P_i = -E_i/k_BT - \ln Z.\tag{5.22}$$

Using Equation 5.16, the entropy, $S = k_B \ln W$, may now be written as

$$S = Nk_B \ln Z + U/T.\tag{5.23}$$

At constant temperature T,

$$dS = dU/T + Nk_Bd(\ln Z).\tag{5.24}$$

From Equation 5.24, we may derive relations that show that the entropy S defined in statistical mechanics has the same properties as the entropy of thermodynamics. If two bodies with different temperature are put together, and if dU = 0, we obtain

$$dS = Nk_B d(\ln Z) > 0. \qquad (5.25)$$

This corresponds to a spontaneous process. As long as the bodies are separated, the total weight function is $W = W_1 W_2$ and $S = k_B \ln W = k_B \ln W_1 + k_B \ln W_2 = S_1 + S_2$ (addition theorem for entropy), but as soon as they come in contact, W grows enormously because of the many possible new configurations.

Under other circumstances, work is produced and entropy increased by the first term in Equation 5.23. All this agrees with the phenomenological definition and guarantees that S defined by Equation 5.23 is the correct concept of entropy.

Furthermore, Equation 5.23 shows that entropy is a state function that can be defined on the basis of the properties of the system. S does not depend on the way a state was achieved, however. Like internal energy, entropy is a state function. This is in fact the second law of thermodynamics. The same final state with entropy S can be achieved either by spontaneous change or by performing work.

The entropy at a given temperature and pressure of a gas is a sum of many contributions: (1) the heating of the solid to the melting point; (2) the latent heat of fusion of the solid; (3) the heating of the liquid to the boiling point; (4) the latent heat of vaporization; and (5) heating of the gas to the temperature relevant for the experiment. Large changes in entropy are associated with phase transitions and chemical reactions.

5.2.5 HELMHOLTZ FREE ENERGY

Processes that occur according to Equation 5.25 are, in principle, reversible. According to Equation 5.19, $\delta q = dU - \delta w$. We first assume that $\delta w = 0$, meaning that no pressure–volume work is carried out. Equation 5.24 may be written as

$$dU - TdS < 0 \Rightarrow d(U - TS) < 0 \ (T \text{ const.}), \qquad (5.26)$$

and this is a condition for a spontaneous isothermal process. For a reversible isothermal process, we have equality sign in Equation 5.26.

Equation 5.26 suggests that a new state function may be defined as

$$A = U - TS. \qquad (5.27)$$

dA < 0 is thus the condition for a spontaneous process at constant T and V.

With the help of Equations 5.23 and 5.27 A may be expressed using the partition function Z in a very simple way:

$$A = -Nk_B T \ln Z. \qquad (5.28)$$

Clearly, A is a state function, referred to as *Helmholtz free energy*.

5.2.6 PRESSURE

Pressure is caused by momentum transfer when the molecules collide with the wall. The higher the temperature, the faster the molecules move, so higher temperature leads to increased pressure in a certain amount of gas.

Since, according to Equation 5.19 $dU = \delta q + \delta w$, the internal energy is changed only if heat is added or work is applied on the system in a reversible process. The first term is $\delta q = TdS$. If there is no spontaneous change, we have

$$dS = \frac{\delta q}{T}. \tag{5.29}$$

The second term is the work term $\delta w = dU - \delta q = d(U - TS) = dA$. We assume that the work is pressure–volume work, equal to PdV, and define pressure, using Equation 5.28, as

$$P = -\left(\frac{\partial A}{\partial V}\right)_T = k_B N T \left(\frac{\partial \ln Z}{\partial V}\right)_T. \tag{5.30}$$

5.2.7 ENTHALPY

The internal energy of a system is changed if we apply work to it. In a chemical reaction, for example,

$$2H_2 + O_2 \rightarrow 2H_2O, \tag{5.31}$$

two gases form a fluid (at laboratory temperature). During the reaction, the volume of the reacting gas molecules is drastically decreased to almost zero. The work allowed to be applied by the gas is equal to $\Delta U = -P\Delta V$ (ΔV is negative and ΔU is positive). We need a new thermodynamic function that is not dependent on the work carried out by pressure and therefore define the *enthalpy* H as

$$H = U + PV = N k_B T^2 \frac{\partial \ln Z}{\partial T} + PV. \tag{5.32}$$

PV is obviously a state function, so H is a state function. If we assume that the pressure is invariant, as it is in most applications, the change in PV is equal to $P\Delta V$. In the chemical reaction (5.31), the energy has to be compensated for the term $-P\Delta V$, arising because of decreased volume. ΔH is thus a much better measure of the energy change during the reaction than ΔU. Only "chemical energy" of the reaction is counted in ΔH, since ΔH is independent of the change of volume. H is therefore the most adequate quantity to list in tables.

The question arises as to which experimental quantity the energy eigenvalue in a variational calculation of a molecule is going to be compared with. The answer is the enthalpy, since the presence of the PV term leads to elimination of the kinetic energy of the gas molecules. The kinetic energy of the molecules as particles is not involved in the variational calculation, of course.

Enthalpy is also useful in experimental work, since most reactions are carried out at constant pressure. The *standard enthalpy of formation*, $\Delta_f H^0$ (J/mol), of a compound is the change of enthalpy that accompanies the formation of 1 mol of a substance in its standard state from its constituent elements in their standard states.

5.2.8 GIBBS' FREE ENERGY

If the process occurs at a constant pressure, but $\delta w \neq 0$, one may add PV to A, as in the case of enthalpy, and thereby obtain Gibbs free energy, which is also a state function. We obtain

$$G = A + PV = U - TS + PV = H - TS. \tag{5.33}$$

Using Equations 5.30 and 5.32 we may express the free energy as

$$G = -Nk_B T \ln Z + Nk_B TV \left(\frac{\partial \ln Z}{\partial T} \right). \tag{5.34}$$

The thermodynamic functions U, S, A, and G may thus be written with the help of the partition function. From this, it follows that they are state functions and that variations of them, dU, dS, dA, and dG, are exact differentials.

G is *Gibbs' free energy*. Since chemical reactions often occur at a constant pressure, G is perhaps the most important state function in chemistry. If H is constant and there is no heat exchange with the environment, a spontaneous reaction at constant temperature ($\Delta S < 0$) means that

$$\Delta G = -T\Delta S < 0 \quad (T, P \text{ constant}). \tag{5.35}$$

If the product side of a chemical reaction has a lower value of G than the reactant side ($\Delta G < 0$), the chemical reaction occurs spontaneously. Equation 5.33 implies

$$\Delta G = \Delta H - T\Delta S \quad (T, P \text{ constant}). \tag{5.36}$$

Consider a process where ammonium chloride (NH_4Cl) (salmiac) is formed from ammonia and hydrochloric acid:

$$NH_3(g) + HCl(g) \rightarrow NH_4Cl(s). \tag{5.37}$$

ΔH for this reaction at 298.15 K at normal pressure is -176.2 kJ/mol, while ΔS is -0.285 kJ/mol $\cdot$ K. At constant temperature $\Delta G = \Delta H - T\Delta S$, we obtain

$\Delta G = -91.2$ kJ/mol. The reaction is thus spontaneous since the decrease in enthalpy is larger than the increase in $-T\Delta S$. The reaction starts. The internal energy is also constant and hence there will be an increase in vibrational and rotational energies to compensate for the decrease in H. This will be noted as a heating of the probe. Due to the increased temperature, the term $-S\Delta T$ should be included in Equation 5.36.

Another example is the solvation of salmiac in water. This process is also spontaneous ($\Delta G < 0$). In this case, ΔH is positive, but the increase in entropy ($\Delta S > 0$) is very large, so the reaction starts spontaneously. Heat must now be taken from the system to compensate for the increase in H. This may be noticed as a cooling of the system.

From the previous discussion, it follows that the equilibrium condition is

$$dG = 0 \quad (T, P \text{ constant}). \tag{5.38}$$

5.2.9 MAXWELL RELATIONS

For reversible processes where the pressure–volume work is the only form of work, the change in T, P, or V is related. We start out from

$$dU = \delta q + \delta w = TdS - PdV, \tag{5.39}$$

which follows from the definitions of U and V. This yields, with $H = U + PV$,

$$dH = dU + VdP + PdV = TdS + VdP. \tag{5.40}$$

From $A = U - TS$ and Equation 5.39 follows that

$$dA = dU - SdT - TdS = -SdT - PdV, \tag{5.41}$$

and, finally, from $G = H - TS$ and Equation 5.40:

$$dG = dH - TdS - SdT = -SdT + VdP. \tag{5.42}$$

From Equation 5.42 we obtain the Maxwell relations:

$$\left(\frac{\partial G}{\partial T}\right)_P = -S, \tag{5.43}$$

$$\left(\frac{\partial G}{\partial P}\right)_T = V. \tag{5.44}$$

Other relations of the same type may be obtained from Equations 5.39 through 5.42.

5.3 INTERNAL ENERGY AND HEAT CAPACITY IN GAS PHASE

We will now use the statistical definitions of the previous section to derive the expression for internal energy and heat capacity in the gas phase. We will begin with gases, since they form the simplest case. However, we will leave out the contribution from rotational levels since this contribution is of little interest in the remaining parts of the book.

5.3.1 TRANSLATIONAL CONTRIBUTION

To derive the partition function for an ideal gas as defined in Equation 5.13 we should know the discrete spectrum, at least at a low energy. Imagine the molecules to be moving freely in a cube with the side equal to L. In this three-dimensional box model, the potential is equal to zero inside the cube and infinitely large outside the cube. Since $\Psi = 0$ outside the cube, all wave functions tend to zero at the edge of the cube. Inside the cube, the Schrödinger equation (SE) is

$$-\frac{\hbar^2}{2M}\left(\frac{\partial^2\Psi}{\partial x^2}+\frac{\partial^2\Psi}{\partial y^2}+\frac{\partial^2\Psi}{\partial z^2}\right)=E\Psi, \tag{5.45}$$

where M is the molecular mass. Equation 5.45 allows separation of the variables:

$$\Psi(x,y,z)=\psi(x)\psi(y)\psi(z); \quad E=E_x+E_y+E_z. \tag{5.46}$$

The equation for ψ may be written as

$$-\frac{\hbar^2}{2M}\frac{\partial^2\psi}{\partial x^2}=E\psi. \tag{5.47}$$

The origin is in one corner of the cube and the coordinate axes along the edges. It is easily seen that sin(kx) is a solution of Equation 5.47. This function also satisfies the boundary condition at x = 0. To satisfy the boundary condition at x = L, we must have

$$k_x L = n_x\pi, \quad n=1,2,\ldots, \tag{5.48}$$

which mean that k (k_x,k_y,k_z) can only assume discrete values.

E is determined by putting $\psi=\sin(kx)$ in Equation 5.47:

$$E_n=\frac{\hbar^2}{2M}k^2=\frac{\hbar^2}{2M}\left(\frac{n\pi}{L}\right)^2. \tag{5.49}$$

Since we prefer to put zero at the lowest energy level, Equation 5.49 is written in simplified form as

$$\varepsilon_n=\frac{h^2}{8ML^2}\left(n^2-1\right)=a\left(n^2-1\right). \tag{5.50}$$

This suggests the following partition function:

$$Z_x = \sum_{n=1}^{\infty} \exp\left[-a\left(n^2-1\right)/k_BT\right].$$ (5.51)

Since the particles are molecules with $M \gg 1000m_e$, and since L is of macroscopic dimensions, the number a, which determines the distance between the energy levels, will be extremely small compared with that in the case of electrons. If the temperature is not too small, the sum may be replaced by an integral:

$$\int_0^{\infty} \exp\left(-n^2a/k_BT\right)dn = \frac{1}{\sqrt{a/k_BT}} \int_0^{\infty} \exp\left(-x^2\right)dx = \frac{1}{2}\sqrt{\frac{k_BT\pi}{a}}.$$ (5.52)

The parameter a of Equation 5.50 is inserted, and we obtain

$$Z_x = \left(\frac{2M\pi k_BT}{h^2}\right)^{1/2} L.$$ (5.53)

If we return to the three-dimensional Equation 5.45, the total energy is $E = E_x + E_y + E_z$. The partition function is a product of the partition functions for x, y, and z. Consequently, the partition function for the motion of the molecule in three dimensions is

$$Z = \sum_{n_x,n_y,n_z} \exp\left[-\left(\varepsilon_{nx} + \varepsilon_{ny} + \varepsilon_{nz}\right)/k_BT\right] = Z_xZ_yZ_z$$

$$= \left(\frac{2\pi Mk_BT}{h^2}\right)^{3/2} L^3 = \left(\frac{2\pi Mk_BT}{h^2}\right)^{3/2} V.$$ (5.54)

Note that the partition function for translations is proportional to the volume. Strictly speaking, we have shown Equation 5.54 only for a rectangular box. Since we know that the thermodynamic properties of the gases do not depend on the form of the container, we may assume that the same expressions would be obtained for other geometrical shapes.

The contribution to the internal energy from the translational motion, E, of an ideal gas is obtained by inserting Equation 5.54 into Equation 5.16:

$$U = k_BT^2\left(\frac{\partial \ln Z}{\partial T}\right)$$

$$E = k_BT^2\left(\frac{\partial \ln Q}{\partial T}\right)_V = N_Ak_BT^2 \frac{\partial}{\partial T}\ln\left[\left(\frac{2\pi k_BTM}{h^2}\right)^{3/2} V\right]_V = \frac{3}{2}RT.$$ (5.55)

We put $N = N_A$ equal to Avogadro's number for 1 mol, and substitute $N_A k_B$ with the general gas constant R.

Notice that the internal energy for a given gas is independent of pressure (but not during a chemical reaction, of course).

The translational contribution to the molar heat capacity (at constant volume, V) is then

$$C_V = \left(\frac{\partial E}{\partial T}\right)_V = \frac{3}{2}R. \tag{5.56}$$

This value is the same as obtained in classical mechanics. The reason is that we have substituted the sum in Equation 5.52 by an integral, thereby missing the zero limit value as $T \to 0$. In fact, if T is sufficiently small, we obtain $C_V \to 0$ from the sum formula. This has been experimentally verified at extremely low temperatures (Bose–Einstein condensation).

5.3.2 Internal Energy and Heat Capacity due to Vibrations

Usually, the vibrational motion can be approximated by a harmonic oscillator for the lowest energy states. Since we prefer to count the energies from the lowest possible energy level, we number the energy levels as follows:

$$\varepsilon_j = (j-1) \cdot hc \cdot \tilde{v}; \quad j = 1, 2, \ldots, \tag{5.57}$$

where $\tilde{v}$ is the vibration frequency. Thus, the partition function is given by

$$Z = \sum_{j=1}^{\infty} \exp\left[-(j-1)hc\tilde{v}/k_B T\right] = \frac{1}{1-\gamma}, \tag{5.58}$$

where

$$\gamma = \exp\left(-\frac{\theta}{T}\right) \text{ and } \theta = hc\tilde{v}/k_B. \tag{5.59}$$

In the last step of Equation 5.58, we have used the equation for the sum of an infinite geometric series. The *characteristic vibration temperature* is θ. As is seen in Equation 5.59 Z is negligibly larger than 1 until T begins to approach θ. For $T > \theta$, there is exponential growth.

We need

$$\frac{d\gamma}{dT} = \exp\left(-\frac{\theta}{T}\right)(-\theta)\left(-\frac{1}{T^2}\right) = \frac{\gamma\theta}{T^2} \tag{5.60}$$

and obtain

$$\frac{1}{Z}\frac{dZ}{dT} = (1-\gamma)\frac{d(1/1-\gamma)}{d\gamma}\frac{d\gamma}{dT} = \frac{\gamma}{1-\gamma}\cdot\frac{\theta}{T^2}. \tag{5.61}$$

Thus, the vibrational contribution to the internal energy per mole:

$$E = RT^2\frac{\gamma}{1-\gamma}\cdot\frac{\theta}{T^2} = R\cdot\frac{\gamma}{1-\gamma}\cdot\theta. \tag{5.62}$$

The vibrational contribution to the molar heat capacity (at constant volume) is $C_V = dE/dT$, hence,

$$C_V = R\theta\frac{d(\gamma/1-\gamma)}{d\gamma}\frac{d\gamma}{dT} = R\theta\frac{1-\gamma+\gamma}{(1-\gamma)^2}\frac{d\gamma}{dT} = R\frac{\theta^2}{T^2}\frac{\gamma}{(1-\gamma)^2} \; J/mol\cdot K. \tag{5.63}$$

We see that if $T \ll \theta$,

$$\frac{\gamma}{1-\gamma} \to \frac{\exp(-\theta/T)}{1-\gamma} = \frac{0}{1} = 0, \tag{5.64}$$

while, if $T \to \infty$ (Taylor expansion: $\gamma = \exp(-\theta/T) = 1 - \theta/T + \cdots$),

$$\frac{1}{1-\gamma} \to \frac{T}{\theta}. \tag{5.65}$$

For $T \to \infty$, we have

$$E \to RT \text{ and } C_V \to R. \tag{5.66}$$

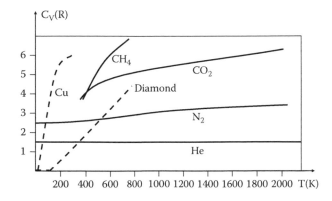

FIGURE 5.2 Heat capacity of an atom (He), some molecules (N_2, CO_2, CH_4), diamond insulator, and copper metal as a function of temperature.

We may now summarize the results for heat capacity (Figure 5.2). Helium is an atomic gas and therefore contributes (=3/2 R) to the heat capacity from the kinetic energy of the atoms. The nitrogen molecule shares this contribution since it is a gas. One may derive that the rotational energy of a diatomic molecule is =R. The excitation energy is so small that this value is already attained at the boiling point of nitrogen. Therefore, the heat capacity for 100 K is 5/2 R. The heat capacity for nitrogen increases in the region of 400–1200 K. This is due to the vibrational levels, which start to receive energy by thermal excitation. In CH_4 and CO_2, there are more vibrations and therefore the heat capacity increases more than for N_2. In copper and diamond, the heat capacity arises mainly from the vibrations (phonons).

5.4   CHEMICAL REACTIONS

In a chemical reaction, the system, consisting of different kinds of particles, spontaneously changes its composition. Some particles unite to form a new kind of particle and others may split. A chemical reaction is *exothermic* if heat is given off and *endothermic* if heat is absorbed. The reaction may need thermal energy to get started (activation barrier), but this topic will be discussed in Chapter 8. Here, we will discuss the thermodynamic properties.

5.4.1   CHEMICAL POTENTIAL

We consider a system with different components (i) with N_i particles of each kind or $n_i = N_i/N_A$ moles. G depends on the number of particles of each kind. The change dG in G may be expressed with the help of derivatives and differentials. We also include the dependence of temperature T and pressure P:

$$dG = \left(\frac{\partial G}{\partial T}\right)_{P,n_i} dT + \left(\frac{\partial G}{\partial P}\right)_{T,n_i} dP + \left(\frac{\partial G}{\partial n_1}\right)_{P,T,n_2...} dn_1 + \left(\frac{\partial G}{\partial n_2}\right)_{P,T,n_1,n_3...} dn_2. \quad (5.67)$$

A derivative with respect to n_i is called a *partial molar free energy*, but is better known as *chemical potential* (μ_i):

$$\mu_i = \left(\frac{\partial G}{\partial n_i}\right)_{P,T,n_j,j\neq i}. \quad (5.68)$$

The *chemical potential* is the driving force for the change in the number of molecules in a chemical process. The molecules of kind i will decrease in number spontaneously as we approach equilibrium. If G decreases when the number of molecules of kind i is decreased, μ_i is positive. As is the case with temperature and pressure, and heat capacity, the chemical potential is an *intensive* property, which is independent of the size of the system. *Extensive* properties or quantity properties, on the other hand, are volume, mass, internal energy, and free energy, which depend on the amount of molecules, that is, the size of the system.

If we change the amount, we may thus write the differential of G as

$$dG = -SdT + VdP + \sum \mu_i dn_i.$$

(5.69)

If T and P are constant (dT = dP = 0), we may integrate Equation 5.60 to macroscopic amounts if the proportions (=composition) are not changed. μ_i is constant and we obtain

$$G = \sum \mu_i n_i.$$

(5.70)

5.4.2 GIBBS–DUHEM EQUATION

If the composition is changed, μ_i in Equation 5.70 will vary. From this equation follows

$$dG = \sum \mu_i dn_i + \sum d\mu_i n_i.$$

(5.71)

If T and P also change, Equations 5.69 and 5.71 are both valid only if

$$\sum d\mu_i n_i = -SdT + VdP.$$

(5.72)

Equation 5.72 is called the *Gibbs–Duhem equation* and expresses how the chemical potentials for the different components depend on each other at constant temperature and pressure. If the amount of one component, n, is changed, we obtain from Equation 5.72

$$d\mu = -\frac{\partial S}{\partial n} dT + \frac{\partial V}{\partial n} dP.$$

(5.73)

The derivatives are called *partial molar entropy* and *partial molar volume*, respectively. The partial molar volume does not have to be the same as the molar volume. In a concentration range when alcohol is diluted with water to form normal vodka, the volume of vodka is smaller than the volume of the two components together before mixing. Measuring the amount by mass before measuring by volume is therefore preferable.

5.4.3 GIBBS–HELMHOLTZ EQUATION

Gibbs energies have been measured and listed for a number of chemical compounds. The temperature dependence of the Gibbs energy may be expressed in different ways. Since, according to Equation 5.36 G = H − TS and according to Equation 5.43, $(\partial G/\partial T)_P = -S$, we have

$$G = H + T \left(\frac{\partial G}{\partial T} \right)_P .$$

(5.74)

This type of equation is unpractical if we want to determine G from experimental numbers, since both G and its derivative with respect to T are present at the same time. We may solve the problem with the help of a trick. The following identity is obtained by just taking the derivative of the quotient G/T:

$$\frac{\partial (G/T)}{\partial T} = \frac{T(\partial G/\partial T) - G}{T^2} .$$

(5.75)

With the help of Equation 5.74, we obtain

$$\left(\frac{\partial (G/T)}{\partial T} \right)_P = -\frac{H}{T^2} \text{ or } \left(\frac{\partial (\Delta G/T)}{\partial T} \right)_P = -\frac{\Delta H}{T^2} .$$

(5.76)

This is the *Gibbs–Helmholtz equation*. Since Equation 5.76 is linear in G and H, we may replace them by ΔG and ΔH, respectively, for a process, for example, a chemical reaction. If ΔH for the reaction may be determined with the help of a calorimeter, ΔG may also be determined.

5.4.4 GIBBS ENERGY FOR IDEAL GAS

According to Equation 5.44, from the Maxwell relations we obtain

$$\left(\frac{\partial G}{\partial P} \right)_T = V.$$

(5.77)

We further assume that we are dealing with an ideal gas and apply the ideal gas law:

$$V = \frac{nRT}{P} .$$

(5.78)

If the pressure changes for constant temperature, ΔG changes as follows:

$$\Delta G = G(P_1) - G(P_2) = \int_{P_1}^{P_2} \left(\frac{\partial G}{\partial P} \right)_T dP = \int_{P_1}^{P_2} \frac{nRT}{P} dP = nRT \ln \frac{P_2}{P_1} .$$

(5.79)

In other words, free energy is related to pressure by a logarithmic function. If there are several components, we have to use partial pressure. Furthermore, partial pressure is proportional to the concentration of the substance in the fluid phase. Equation 5.79 may therefore be used for the fluid phase if we replace the pressure with concentration in Equation 5.79.

5.4.5 Law of Mass Action

Chemical equilibrium is achieved between reactants and products when the amounts of the participants in the reaction remain constant. Forward and backward reactions still occur simultaneously. We write a chemical reaction in the following general way:

$$|v_A| A + |v_B| B + \cdots \rightleftarrows v_P P + v_Q Q + \cdots. \tag{5.80}$$

For practical reasons, we have chosen the stoichiometric coefficients for the reactants as negative ($v_A, v_B, \ldots, < 0$). The amounts of the substances change linearly during the reaction according to

$$n_i = n_i(0) + v_i \xi, \tag{5.81}$$

where $n_i(0)$ is the number of moles at the start of the reaction and n_i is the current number of moles. We may thus express all dn_i with the help of $d\xi$. Equation 5.69 may be written as

$$dG = -SdT + VdP + d\xi \sum_i \mu_i v_i. \tag{5.82}$$

At equilibrium and constant T and P, the following equation holds:

$$\frac{\partial G}{\partial \xi} = 0 \Rightarrow \sum_i \mu_i v_i - 0. \tag{5.83}$$

If we use Equation 5.79 we have for each reactant and product

$$\mu_i = \mu_i^\ominus + RT \ln C_i, \tag{5.84}$$

where C_i is the concentration of component i. From Equations 5.83 and 5.84 follows

$$\sum_i \mu_i v_i = \sum_i v_i \mu_i^\ominus + RT \sum_i \ln\left(C_i^{v_i}\right) = 0. \tag{5.85}$$

Taking the exponent, we obtain

$$\prod_i C_i^{v_i} = K. \tag{5.86}$$

Equation 5.86 is called the *law of mass action*. K is called the thermodynamic equilibrium constant.

5.4.6   CONNECTION BETWEEN $G^{\ominus}$ AND K

If we call

$$\sum_i v_i \mu_i^{\ominus} = \Delta G^{\ominus}, \tag{5.87}$$

the equilibrium condition of Equation 5.85 may be written as

$$\Delta G^{\ominus} = -RT \ln K. \tag{5.88}$$

Equations of this type serve the bookkeeping and transfer of thermodynamic data between different measurements.

5.4.7   TEMPERATURE DEPENDENCE OF EQUILIBRIUM CONSTANT

If we apply the Gibbs–Helmholtz Equation 5.76 on $\Delta G^{\ominus}$, we obtain

$$\left(\frac{\partial \ln K}{\partial T}\right)_P = \frac{\Delta H^{\ominus}}{RT^2}. \tag{5.89}$$

This is *van't Hoff's equation*. For example, in the Haber–Bosch process

$$N_2 + 3H_2 \rightarrow 2NH_3 \quad \Delta H^{\ominus} = -46.11\,\text{kJ/mol}. \tag{5.90}$$

If heat is generated in the reaction (exothermal reaction), K is decreased when T is increased. If, on the other hand, heat is added to the reaction (endothermal reaction, $\Delta H^{\ominus} > 0$), the equilibrium constant is increased when the temperature is increased and the amount of the reaction product is increased.

If pressure is increased, the reaction is shifted to the right since fewer moles (2) are formed than are reacting (4). This is summarized in the useful thumb-rule called *le Chatelier's principle*, which can be stated as follows:

> If the outer conditions are changed in a chemical reaction, the equilibrium is shifted in such a way that the change is counteracted.

Since $\Delta G^{\ominus} = \Delta H^{\ominus} - T\Delta S^{\ominus}$ (Equation 5.36), we have, according to Equation 5.88,

$$\ln K = \frac{\Delta S^{\ominus}}{R} - \frac{\Delta H^{\ominus}}{RT}. \tag{5.91}$$

If K is measured as a function of the temperature T and if ln K is given as a function of $1/T$, we may, according to Equation 5.91, obtain $-\Delta H^{\ominus}/R$ as the slope and $\Delta S^{\ominus}/R$ as the limit value as $1/T \rightarrow 0$. If Equation 5.89 is used in Equation 5.91 we get

$$\Delta S^{\ominus} = R \left[T \left(\frac{\partial \ln K}{\partial T} \right)_P + \ln K \right]. \tag{5.92}$$

$\Delta H^{\ominus}$ and $\Delta S^{\ominus}$ are almost independent of T.

5.5 EQUILIBRIUM STATISTICAL MECHANICS USING ENSEMBLES

So far in this chapter we have tried to give a simplified derivation of the most useful equations of statistical thermodynamics. As usual, when shortcuts are made, questions and problems appear later in the text. For example, we have assumed that there is no interaction between particles and no interaction between the various degrees of freedom. If there are strong interactions among a number of particles, for example, molecular bonds, statistical methods are less useful at least at low temperatures. The same applies even to the electrons of molecules. Except when the ground state is almost degenerate, excitation energies are too large to be of any importance in statistical mechanics at normal temperatures.

On the other hand, valence electrons in a metal may be treated as free electrons moving in a box with infinite walls. Having the same sign, electrons repel each other with very strong forces, but these forces may be thought of as compensated by a uniform positive background. Since there are a large number of electrons, it is practical to use a statistical treatment in dealing with phenomena such as conductivity and heat transfer. The border between occupied and unoccupied levels is called the *Fermi level*.

However, a statistical treatment of the electrons in a metal using the methods that we have used so far would certainly fail. All electrons would be in the lowest state when T = 0, but if this is the case, the fundamental Pauli principle stating that electrons are fermions and form an antisymmetric wave function would not be satisfied. As we have seen, the Pauli principle implies that each spin orbital can be occupied by only one electron.

5.5.1 Phase Space

The derivation of statistical mechanics on the basis of quantum mechanics was done in a simplified version above. What could be simpler than just counting the number of possibilities and taking the logarithm to get manageable numbers for the entropy? The important historical question, then, is how Boltzmann derived statistical mechanics the first time, long before quantum mechanics was known. In fact, the original Boltzmann derivation is sometimes much simpler to apply to some problems than the quantum mechanical one; therefore, we need to understand how statistical mechanics is derived on the basis of classical mechanics.

A hint may be obtained from the electron in a box problem. The kinetic energy ($p^2/2m$) is given in Equation 1.17. The variation δp in momentum is between $p = hn/L$ and $-hn/L$. The variation in distance δx is L and hence $\delta p \cdot \delta x = hn$. The space consisting of a p-axis and an x-axis is the *phase space* in one dimension. In the case of particle in a box, the system has width L on the x-axis. The width along

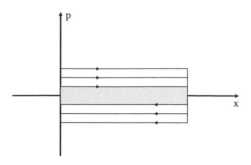

FIGURE 5.3   Volume of phase space.

the p-axis depends on the number of particles we fill up with and their spin. The integration in the shaded area in Figure 5.3 goes from $p = -hn/L$ to $p = hn/L$ on the p-axis, and from $x = 0$ to $x = L$ on the x-axis. One state of each spin is permitted in this area. As the integration region gets larger, the number of possible states increases (Figure 5.3).

In classical mechanics, the coordinates $\{q_i\}$ and the momenta $\{p_i\}$ for each particle i define a point in phase space, that is, a possibility for the system to exist. q_i and p_i are functions of time. It is, of course, impossible to simulate the motion of a great number of particles and we are probably not interested in the fate of each single particle. Instead of calculating the time average for a single system, we calculate the average of a number of systems (called an *ensemble*) at a given time. We believe that the same result will be obtained, but we will not prove it here (this belongs to the *theory of ergodicity*).

In a completely isolated system, we may assume that the total energy is within a certain interval dE. This *microcanonical ensemble* corresponds quite closely to the treatment that we have already carried out in Sections 5.2–5.4.

We found that the distances between the energy levels become extremely small when the width of the box or the mass of the particle has macroscopic dimensions. We also commented on the fact that a wave packet, for example, one describing a macroscopic body, obeys Newton's equations and has the same group velocity as the body. In other words, quantum mechanics has an easily reached classical limit when everything behaves as we are used to in the macroscopic world.

5.5.2   PROBLEMS IN THE EARLIER DERIVATION

In Section 5.2.1, we made the assumption that particles are distinguishable. We also assumed that every single particle is free to occupy any one-particle state. At the same time, we also wanted to apply the theory on electrons, where this is impossible. Electrons are indistinguishable and they are fermions, and therefore only a single electron can occupy each spin orbital. The statistics we derived for distinguishable particles is referred to as *Maxwell–Boltzmann statistics*. In the case of electrons, this statistics is only valid for very high temperatures when the likelihood to have more than a single electron in each spin orbital is vanishing anyway (and when the metal would be vaporized).

The microcanonical ensemble is realistic only if the condition of equal energy is replaced by a condition that the total energy has to be in a certain interval dE. This is the only way to do the derivation in the classical case. On the other hand, dE may be chosen arbitrarily small, with still correct end results.

In any case, statistical methods work better if there is a variation in the total energy. This is the main reason to use ensembles other than the microcanonical ensemble. An ensemble consists of copies of the same system, called members of the ensemble. In the *canonical ensemble*, the volume is the same and all members are in contact with the same heat bath at a given temperature T. However, the total energy of all the members of the ensemble has to be the same. In this way, it is not necessary to assume that the members have almost the same energy as is done in the microcanonical ensemble.

The canonical ensemble prescribes a constant number of particles, N. Alternatively, the number of particles may also be allowed to vary. This is done in the *grand canonical ensemble*. The new condition is that the total number of particles in all ensembles has to be constant. The purpose of ensembles is to resemble a natural system as much as possible, for example, a piece of metal or a certain volume.

5.5.3 CANONICAL ENSEMBLE

If the molecules interact, we should in principle solve the time-independent SE for the whole system and consider it as a single molecule. We would obtain a great number of excited states, but no general law for the statistical probability of the various excited states can be realized. In one way or another, we must allow the total energy to vary. For that purpose, we consider a great number of subsystems (N), all with the same number of particles and the same volume (V) but with varying energy: $E_1, E_2, \ldots, E_N$. The sum of the energies for all subsystems is assumed to be constant:

$$E_1 + E_2 + \cdots + E_N = E. \tag{5.93}$$

All subsystems are considered to be in the same heat bath, and the temperature (T) is therefore a constant. The collection of all N subsystems, with constant N but with varying E, are considered to represent all aspects of the original system. This collection is called the *canonical ensemble*.

When the Boltzmann distribution was derived in Section 5.2, we assumed a constant number of particles (N) and a given volume (V). We also assumed that the energy was constant. In the canonical ensemble, we imagine that the subsystems (members of the canonical ensemble) may exchange energy but not particles with each other. The ensemble that we are supposed to use is the one where the assumptions agree with the experimental conditions. Since the total energy is constant in the microcanonical ensemble, the definition of such a simple concept as temperature has to be done indirectly.

The different subsystems have the same energy levels since the external conditions are identical. The total number of particles is constant. We assume that there are n_1 subsystems with energy E_1, n_2 subsystems with energy E_2, etc. The set of numbers $\{n_i\}$ is called n. Evidently, we have

$$E(n) = \sum_i n_i E_i. \tag{5.94}$$

We use the same theory as in Section 5.2 to derive the canonical partition function. The relationship between the entities follows from Table 5.1.

The mathematical problems are identical and the canonical partition function may be calculated in the same way as the microcanonical was calculated previously (N and E constant, and N and E constant, respectively). For the canonical partition functions, we thus obtain the partition function:

$$Z = \sum_i \exp\left[-\beta E_i\right]. \tag{5.95}$$

This is, in fact, formally the same result as was obtained for independent particles with the microcanonical ensemble, since the energy steps are the same.

5.5.4 GRAND CANONICAL ENSEMBLE

The thermodynamic system with volume V, whose properties we are going to calculate, is in contact with a heat bath as earlier, but we do not require that the number of particles is constant in each subsystem. On the other hand, the total number of particles must be a constant. Every subsystem with N particles is in a quantum mechanical state with energy $E_j(N,V)$. If the number of particles (N) in the subsystem is changed, the subsystem will have other eigenstates, thus defining a dependence of N. The ensemble of all possible subsystems with varying N and E is called the *grand canonical ensemble*. We let $n_j(N)$ be the number of subsystems that contain N molecules in the state j. The set of all n_j is $n = \{n_j(N)\}$. We imagine that the components of the system have constant chemical potentials $\mu_1, \mu_2, \dots$. The total number of subsystems is N. The total energy is E and the total number of particles, N_t. The weight of a certain configuration of this type, $\Omega_t(n)$, may be calculated in the same way as W in Section 5.2.1:

$$\Omega(n) = \frac{\left[\sum_{j,N} n_j(N)\right]!}{\prod_{j,N} n_j(N)!}. \tag{5.96}$$

TABLE 5.1

Comparison of Derivations of the Partition Function

Earlier Derivation	Canonical Ensemble
N = total number of particles	N = total number of subsystems
n_i = number of particles with energy ε_i	n_i = number of subsystems with energy E_i
$N = \sum_i n_i$	$N = \sum_i n_i$
Total energy: $E = \sum_i n_i \varepsilon_i$	Total energy: $E = \sum_i n_i E_i$

The auxiliary conditions may be summarized as

$$\sum_{j,N} n_j(N) = N, \tag{5.97}$$

$$\sum_{j,N} n_j(N)E_j(N,V) = E, \tag{5.98}$$

$$\sum_{j,N} n_j(N)N = N_t. \tag{5.99}$$

We need three Lagrangean multipliers, α, β, and γ, to satisfy the three conditions at maximization of $\ln \Omega(n)$. The result is analogous to Equations 5.5 through 5.15. The number of systems in state j with the number of particles N is

$$P_j(N) = n_j^*(N)/N = e^{-\alpha}e^{-\beta E_j(N,V)}e^{-\gamma N}, \tag{5.100}$$

where P_j is the fraction and $n_j^*(N)$ is the most common distribution. From Equations 5.96 and 5.100 follows:

$$1 = \sum_{j,N} P_j(N) = e^{-\alpha} \sum_{j,N} e^{-\beta E_j(N,V)}e^{-\gamma N} \Rightarrow e^{\alpha} = \sum_{j,N} e^{-\beta E_j(N,V)}e^{-\gamma N} = \Xi. \tag{5.101}$$

Ξ is the grand canonical partition function.

Using of Equations 5.100 and 5.101, we may generalize the equations that we had derived earlier for the microcanonical ensemble. The average energy, for example, is

$$\bar{E} = E/N = \sum_{j,N} P_j(N)E_j(N,V). \tag{5.102}$$

The average number of particles in a subsystem is

$$N_t/N = \sum_{j,N} P_j(N)N. \tag{5.103}$$

Comparison with Equation 5.30 gives the average pressure (P):

$$P = -\sum_{j,N} P_j(N)\left[\frac{\partial E_j(N,V)}{\partial V}\right]_N. \tag{5.104}$$

Equation 5.100 may be written as

$$P_j(N) = e^{-\beta E_j(N,V)} e^{-\gamma N} / \Xi. \qquad (5.105)$$

If the logarithm is taken, we obtain

$$E_j(N,V) = -\frac{1}{\beta}\left(\gamma N + \ln P_j(N) + \ln \Xi\right). \qquad (5.106)$$

The inner energy is equal to the average energy of a subsystem. According to Equation 5.83 we obtain

$$U = \bar{\bar{E}} = \sum_{j,N} P_j(N) E_j(N,V). \qquad (5.107)$$

For the differential dU, this gives

$$dU = \sum_{j,N} E_j(N,V) dP_j(N) + \sum_{j,N} P_j(N) dE_j(N,V). \qquad (5.108)$$

If Equation 5.106 is inserted and the differential in E substituted with a differential in V, we obtain

$$dU = -\frac{1}{\beta}\sum_{j,N}\left[\gamma N + \ln P_j(N) + \ln \Xi\right]dP_j(N) + \sum_{j,N} P_j(N)\frac{\partial E_j(N,V)}{\partial V}dV. \qquad (5.109)$$

The first term is an expression for how the inner energy U is changed when occupation N on level j is changed. The total change in the number of particles is

$$dN = \sum_{j,N} N dP_j(N). \qquad (5.110)$$

Furthermore,

$$\sum_{j,N} \ln P_j(N) dP_j(N) = d\left(\sum_{j,N} P_j \ln P_j(N)\right) \quad \text{since} \quad \sum_{j,N} dP_j(N) = 0. \qquad (5.111)$$

According to Equation 5.104, the second part of the right member of Equation 5.109 may be identified with PdV. We may also rewrite Equation 5.109 in the following way:

$$-\frac{1}{\beta}d\left(\sum_{j,N} P_j \ln P_j(N)\right) = dU + PdV + \frac{\gamma}{\beta}dN. \qquad (5.112)$$

This may be compared with Equation 5.11 at constant temperature and we thus obtain

$$TdS = dU + PdV - \mu dN, \tag{5.113}$$

where we have identified

$$-\frac{1}{\beta} d \left(\sum_{j,N} P_j \ln P_j(N) \right) \leftrightarrow TdS \tag{5.114}$$

and

$$-\frac{\gamma}{\beta} \leftrightarrow \mu. \tag{5.115}$$

Equations 5.100 and 5.101 may be rewritten as $(k = k_B)$

$$P_j(N,V,T,\mu) = \frac{e^{(-E_j+N\mu)/kT}}{\Xi}, \tag{5.116}$$

where

$$\Xi = \sum_{j,N} e^{(-E_j+N\mu)/kT} = \sum_N e^{N\mu/kT} \sum_j e^{-E_j/kT} = \sum_N Q(N,V,T) e^{N\mu/kT}$$

$$= Q(0,V,T) + Q(1,V,T)\lambda + Q(2,V,T)\lambda^2 + \cdots \tag{5.117}$$

and

$$\lambda = e^{\mu/kT}. \tag{5.118}$$

The grand canonical partition function, Ξ, may thus be expressed with the help of the canonical partition function, Q, and the chemical potential, μ.

The thermodynamic state functions may now be derived.

5.5.5 FERMI–DIRAC AND BOSE–EINSTEIN STATISTICS

If we are dealing with nondistinguishable particles, the number of quantum mechanical states is much lower than for distinguishable particles. For example, the lithium atom has three spin orbitals $(\chi_{1s\alpha}, \chi_{1s\beta}, \chi_{2s\alpha})$. One many-electron wave function is $\Psi_1(1,2,3) = \chi_{1s\alpha}(1) \cdot \chi_{1s\beta}(2) \cdot \chi_{2s\alpha}(3)$; another is $\Psi_2(1,2,3) = \chi_{1s\alpha}(2) \cdot \chi_{1s\beta}(1) \cdot \chi_{2s\alpha}(3)$. Since electrons satisfy Fermi–Dirac (FD) statistics, however, there is only one possible wave function, the Slater determinant:

$$\Psi(1,2,...,N) = \frac{1}{\sqrt{N!}} \sum_P (-1)^P P\left[\phi_1(1)\phi_2(2)...\phi_N(N)\right].$$ (5.119)

There is a single quantum mechanical state (in the case of the Li atom), not $W = 3! = 6$, as had been the case if the electrons had been distinguishable, or $W = 3!/2! = 3$ if the states had been considered spatial orbitals capable of accepting two electrons, according to Equation 5.2.

Proceeding similarly if the wave function is the same (does not change sign) on interchange of two particle coordinates, we may write the wave functions that correspond to Equation 5.119:

$$\Psi(1,2,...,N) = \frac{1}{\sqrt{N!}} \sum_P P\left[\phi_1(1)\phi_2(2)...\phi_N(N)\right].$$ (5.120)

Let us assume that we have three energy levels, a twofold degenerate lower one (for example, with two different spin) with energy $E_1 = E_2 = 0$, and a higher one with energy $E_2 = \Delta E$. The number of possible states, Ω_{FD}, Ω_B, and Ω_{BE}, if we first consider the particles as distinguishable in the Boltzmann case, are given in Table 5.2.

5.5.5.1 FD Statistics

The energy levels for single electrons (orbital energies) are called ε_i. We imagine that n_i subsystems in the canonical ensemble have energy ε_i. We apply Equation 5.117 for the canonical partition function, where the summation over N is taken over all possible sets with occupation numbers $\{n_i\}$:

The number of electrons may be anything from 0 to enormously large, while n_i can only be 1 or 0:

$$\Xi = \sum_{\{n_i\}} e^{\left(\mu \sum_i n_i - \sum_i n_i \varepsilon_i\right)/kT} = \sum_{\{n_i\}} e^{\sum_i (\mu - \varepsilon_i) n_i/kT} = \sum_{\{n_i\}} \prod_i e^{(\mu - \varepsilon_i) n_i/kT}.$$ (5.121)

TABLE 5.2

Number of Possible States for Different Statistics (Fermi–Dirac [Ω_{FD}], Boltzmann [Ω_B], Bose–Einstein [Ω_{BE}])

Energy	Ω_{FD}	Ω_B	Ω_{BE}
$3\Delta E$	0	1	1
$2\Delta E$	0	3	1
ΔE	1	3	2
0	0	1	2
	1	8	6

It is then possible to write Equation 5.121 in the following way in the case of FD statistics:

$$\Xi = \sum_{\{n_i\}} \prod_i e^{(\mu-\varepsilon_i)n_i/kT} = \prod_i \sum_{n_i=0,1} e^{(\mu-\varepsilon_i)n_i/kT} = \prod_i \xi_i, \qquad (5.122)$$

where

$$\xi_i(V,T) = 1 + e^{-\varepsilon_i(V)/kT}\lambda, \text{ where } \lambda = e^{\mu/kT}. \qquad (5.123)$$

The average number of electrons may then be written as

$$\bar{n}_j = \frac{n_j\Xi}{\Xi} = \frac{n_j \sum_{\{n_i\}} e^{\sum_i(\mu-\varepsilon_i)n_i/kT}}{\sum_{\{n_i\}} e^{\sum_i(\mu-\varepsilon_i)n_i/kT}} = \frac{n_j\xi_j \prod_{i\neq j}\xi_i}{\prod_i \xi_i} = \frac{0\cdot1 + 1\cdot e^{-\varepsilon_j(V)/kT}\lambda}{1 + e^{-\varepsilon_j(V)/kT}\lambda}, \qquad (5.124)$$

$$\bar{n}_j = \frac{1}{e^{(\varepsilon_j-\mu)/kT} + 1}. \qquad (5.125)$$

This has been illustrated in Figure 5.4. At 0 K, the energy levels are filled to the Fermi level. When the temperature increases, the upper electrons will be lifted to formerly unoccupied orbitals.

5.5.5.2 BE Statistics

In the same way as in Equation 5.121 the following equation holds:

$$\Xi = \sum_{\{n_i\}} \prod_i e^{(\mu-\varepsilon_i)n_i/kT} = \prod_i \sum_{n_i=1,\infty} e^{(\mu-\varepsilon_i)n_i/kT} = \prod_i \xi_i. \qquad (5.126)$$

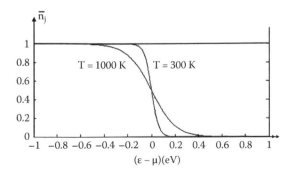

FIGURE 5.4 The FD distribution at different temperatures (T).

But in this case with

$$\xi_j(V,T) = 1 + e^{-\varepsilon_j(V)/kT}\lambda + e^{-2\varepsilon_j(V)/kT}\lambda^2 + \cdots = \frac{1}{1 - e^{-\varepsilon_j/kT}\lambda}, \qquad (5.127)$$

where

$$\lambda = e^{\mu/kT}. \qquad (5.128)$$

The average number may now be written as

$$\bar{n}_j = \frac{1}{e^{(\varepsilon_j - \mu)/kT} - 1}. \qquad (5.129)$$

We notice that

$$\frac{1}{e^{(\varepsilon_j - \mu)/kT} + 1} \leq e^{-(\varepsilon_j - \mu)/kT} \leq \frac{1}{e^{(\varepsilon_j - \mu)/kT} - 1}. \qquad (5.130)$$

That is,

$$\bar{n}_j(FD) \leq \bar{n}_j(B) \leq \bar{n}_j(BE). \qquad (5.131)$$

When $T \to 0$, $n_j \to 0$ in BE statistics if $\varepsilon_j > \mu$. If $(\varepsilon_j - \mu)/kT \to 0$ from above, which happens at a low temperature, $n_j \to \infty$. This means that all states end up in the lowest state when the temperature is decreased to zero.

In FD statistics, n_j varies between 1 and 0 as $(\varepsilon_j - \mu)/kT$ varies between $-\infty$ and $+\infty$. The electrons in a metal will occupy only the levels that are below the Fermi level at $T \to 0$. If $\varepsilon > \mu$ and $(\varepsilon - \mu) > kT$ and increasing, the different distributions will begin to agree.

5.6   NONEQUILIBRIUM STATISTICAL MECHANICS

In many systems found in nature, there is a continuous flux of matter and energy so that the system cannot reach equilibrium. Equilibrium statistical mechanics says nothing about the rate of a process. Chemical reaction rates will be discussed in Chapter 8. The nonequilibrium processes to be discussed here are transport processes like diffusion, heat transfer, or conductivity, where the statistics are expressed as time-evolving probability distributions. Transport processes are due to random motion of molecules and are therefore called *stochastic*. The equations are partial differential equations describing the time evolution of a probability function rather than properties of equilibrium.

5.6.1   MAXWELL VELOCITY DISTRIBUTION

How fast do molecules move around in the gas phase? The velocity distribution, f(v), depends on temperature, of course. If the molecular mass is M, the kinetic energy is

$E = Mv^2/2$. The number of molecules with velocity in a range dv, df(v), is proportional to the Boltzmann factor multiplied by the volume of a shell with thickness dv ($k = k_B$):

$$df(v) \sim \exp\left(-\frac{Mv^2}{2kT}\right) \cdot 4\pi v^2 dv. \tag{5.132}$$

If we normalize to unity, we obtain

$$f(v) = 4\pi \left(\frac{M}{2\pi kT}\right)^{3/2} v^2 \exp\left(-\frac{Mv^2}{2kT}\right). \tag{5.133}$$

This expression is known as *Maxwell's velocity distribution*.

In this equation, we multiply Avogadro's number, N_A, in the numerator and denominator to obtain the molecular mass $M = 1000 \cdot m \cdot N_A$ u (the unit u corresponds to gram/mole) in the numerator and the gas constant $R = N_A k$ in the denominator (thus we use here the notation M for the molecular mass and m (kilogram) for the mass of the molecule). $\bar{v}$ is obtained from Equation 5.133:

$$\bar{v} = 4\pi \left(\frac{M \cdot 10^{-3}}{2\pi RT}\right)^{3/2} \int_0^\infty v^3 \exp\left(-\frac{M \cdot 10^{-3} \cdot v^2}{2RT}\right) dv = \left(\frac{8RT}{\pi M \cdot 10^{-3}}\right)^{1/2}. \tag{5.134}$$

Since $R = 8.31$ J/mol $\cdot$ K, the mean velocity of ordinary small molecules such as N_2 and O_2 is approximately $(1000 \cdot T)^{1/2}$ m/sec. At room temperature $T = 300$ K, the average velocity is ≈ 550 m/sec. The sound velocity (340 m/sec) is directly related to this, since sound is a shock wave of molecules. The dependence of sound velocity on temperature and mass is, in fact, the same as that in Equation 5.134.

The mean velocity of a gas is thus proportional to the square root of the temperature and inversely proportional to the square root of the molecular weight. Contrary to the case with chemical properties, here we are dealing with a property that depends on the molecular weight.

Mean velocities may be defined in other ways. For example, we may determine the "most probable velocity" by checking which velocity in the Maxwell distribution corresponds to the maximum velocity (set the derivative of the Maxwell distribution equal to zero). This velocity is lower than the mean velocity. On the other hand, if the velocity is calculated from the mean value of the kinetic energy, one obtains a slightly higher velocity. It is interesting to note that the mean velocity depends on the mass of the molecule, contrary to the chemical properties in general. Diffusion may be used for isotope separation.

5.6.2 KINETIC THEORY OF GASES

We also need an expression for how far a molecule moves on an average in a gas before it collides with another molecule. For simplicity, we imagine that all the molecules are spherical and that the distance between the molecules when they touch

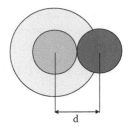

FIGURE 5.5 The two smaller bodies represent colliding molecules with an infinitely hard surface. The collision cross section, πd^2, also includes the larger surface with twice the diameter.

each other is d. The molecules are assumed to have a hard surface, unable to penetrate each other. πd^2 is the size of the cross-sectional area that another molecule cannot reach during the collision (Figure 5.5).

πd^2 is called the *collision cross section*. If the mean velocity when a molecule approaches another molecule is $\bar{v}_{rel}$, there will be per second sweep of a volume where it collides with a number of other molecules. The number of collisions must be the number density N/V multiplied by the volume of the forbidden area:

$$z = \bar{v}_{rel}\pi d^2 \frac{N}{V}. \tag{5.135}$$

$\bar{v}_{rel}$ is directly related to the average velocity of the molecules $\bar{v}$. $\bar{v}$ is determined from the velocity distribution:

$$\bar{v} = \int_0^\infty v f(v) dv, \tag{5.136}$$

where $f(v)$ is the Maxwell's velocity distribution given in Equation 5.134. Next, we will need the relative mean velocity, $\bar{v}_{rel}$, by which two molecules approach each other. This calculation is a bit complicated, but the final expression is simple:

$$\bar{v}_{rel} = \left(\frac{8RT}{\pi\mu \cdot 10^{-3}} \right)^{1/2}, \tag{5.137}$$

where μ, the effective molecular mass, simply replaces the molecular mass in Equation 5.134. This holds for two molecules in the same volume of gas with different mass. If we are dealing with the same kind of molecules, $\mu = M/2$. Molecules that approach each other, apparently do it with an average velocity $\bar{v}_{rel}$, that is, $\sqrt{2}$ times larger than the ordinary mean velocity. The molecules cross each other's path at different angles. If the crossing is perpendicular, we would obtain the factor $\sqrt{2}$, but this factor thus holds on an average.

We may thus calculate the number of collisions per second and obtain

$$z = \bar{v}_{rel}\pi d^2 \frac{N}{V} = \frac{\bar{v}_{rel}\pi d^2 P}{kT}. \qquad (5.138)$$

Here, we have also substituted volume V with pressure P and temperature T according to the ideal gas law.

The *mean free path* is now simple to calculate:

$$\lambda = \frac{\bar{v}}{z} = \frac{kT}{\pi d^2 P\sqrt{2}} = \frac{kV}{\pi d^2 N\sqrt{2}} = \frac{1}{\pi d^2 \rho\sqrt{2}}. \qquad (5.139)$$

5.6.3 MOLECULAR DYNAMICS: THE ENTROPY PROBLEM

In principle, classical molecular dynamics involves motion on the potential energy surface (PES). We assume that the temperature is sufficiently high to avoid problems with motion involving the lowest vibrational levels, where quantum mechanics tends to complicate the problem. We also assume that the motion takes place on the ground state energy surface. There is no surface-crossing problem of the type that we discussed in Section 4.7.

We also assume that there is some way to calculate the temperature. The boundary conditions must be such that they truly represent the statistical situation at the start of the simulation. If the system is isolated, no energy change will take place. A solvation problem may be simulated by assuming a grain of rock salt (NaCl) surrounded by water. Actually, this solvation problem is "clean" enough to discuss whether the temperature increase or decrease will be faithfully represented. According to Equation 5.36 the free energy change may be written as

$$\Delta G = \Delta H - T\Delta S \quad (\text{T constant}). \qquad (5.140)$$

The enthalpy change ΔH is included in the PES, obtained, for example, from the SE. In molecular dynamics, we solve the Newton equations for the great number of particles with PES as potential function.

A more difficult problem is to account for the entropy change, ΔS, in a simulated process. We must be very observant about the simulation method to certify that entropy is correctly included. Looking through tables of formation constants, we find many cases where ΔH is negative and ΔS is positive. This simply means that ΔG is negative so that solvation or another process proceeds. Since the total internal energy is constant, there will be an increase in the average momentum of the particles; in other words, heat will be given off.

If, on the other hand, ΔH is positive, solvation may proceed anyway, since the increase in S may be so large that $\Delta G < 0$. In this case, there will be a slowing down of the particles. The problem is to calculate this slowing down in molecular dynamics.

To have a macrosystem to compare with, we may imagine a male rolling skater. If he skates down the hill, his speed will, of course, increase at the bottom of the ramp and then decrease again on the other side. If the speed is enough to continue above the edge, he may end up in a flat landscape where there are so many possibilities to "get lost" that he will very unlikely find his way back to the down-going ramp. The entropy increase prevents him from reaching the bottom of the ramp. We conclude that entropy should be included in a molecular dynamics calculation, but this problem is far from trivial and depends on how reasonable the boundary conditions are.

A molecular dynamics problem of this sort is simulation of ligand escape. For example, one may be interested in calculating the rate of O_2 escaping into hemoglobin, the oxygen carrier in blood. Unfortunately, we cannot discuss this problem here.

5.6.4 DIFFUSION

Diffusion may be considered as random motion. If we know where the particles are located at different times, as well as their velocity, we may follow their motion by solving Newton's equations and use molecular dynamics. A statistical mean value is obtained for varied boundary conditions and the results are sampled.

Historically, "transport processes" originate from "fluid mechanics." In fluid mechanics, one neglects the existence of atoms and writes differential equations for the continuum. The basic equations were derived and solved already in the nineteenth century. Transport processes have become very important today. One wants to simulate the airflow around rockets, airplanes, and cars with the help of computers. We cannot treat this subject in any detail here, of course.

The laws of Fick and Stokes are good examples on transport processes in a continuum. Adolf Fick (1829–1901), who derived the law of diffusion in 1855, was in fact a physiologist from Kassel in Germany. The Irishman George Stokes (1819–1903) was a pure mathematician. For us it is important to find out how the diffusion constant (D) and the viscosity coefficient (η) are related to entities in thermodynamics and kinetic gas theory.

In Figure 5.6, J is the flux through a surface perpendicular to the flow direction, expressed in moles per meter squared. The concentration is expressed in moles per meter cubed. The change of concentration per step δx, $\delta C/\delta x$, has to be expressed in moles per quartic meter. If δx tends to zero, $\delta C/\delta x$ converges to the derivative dC/dx of the concentration with respect to distance from the source of the diffusion.

The driving force for the flow is the change of concentration per unit of length, dC/dx. Thus, J will be proportional to dC/dx. This is the *first law of Fick*:

$$J = -D\frac{\partial C}{\partial x}, \qquad (5.141)$$

where D is the *diffusion constant*.

Let two parallel and equally large surfaces be perpendicular to the direction of the diffusion current at a distance δx from each other (Figure 5.7). Let the number of molecules (N) in the region between the surfaces (Figure 5.7) be changed by δN because of the current:

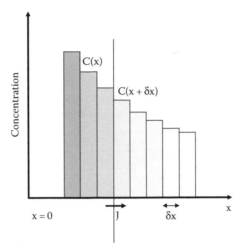

FIGURE 5.6 The concentration of a diffusing substance as a function of the distance from the source ($x = 0$).

$$\delta N = \left(J_{in}A - J_{out}A\right)\delta t. \tag{5.142}$$

If Equation 5.143 is divided by $A\delta x$, we obtain

$$\frac{\delta N}{A\delta x} = \frac{\left(J_{in} - J_{out}\right)\delta t}{\delta x} = -\frac{\delta J \delta t}{\delta x}. \tag{5.143}$$

Since $\delta N/A\delta x$ is equal to the change of concentration δC in the region between the surfaces, we thus obtain

$$\delta C = -\frac{\delta J \delta t}{\delta x}. \tag{5.144}$$

Letting δx tend to zero and using the first law of Fick, we obtain

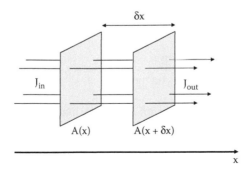

FIGURE 5.7 Diffusion current through surface A.

$$\frac{\partial C}{\partial t} = D\frac{\partial^2 C}{\partial x^2}. \qquad (5.145)$$

This is the *second law of Fick*, referred to as the *diffusion equation*.

Using the kinetic theory of gases, it is possible to derive an equation based on the mean free path (λ) and the mean time it takes to move the mean free path. The mean free speed is then (λ/τ). The final equation may be written as

$$P(x,t) = \left(\frac{2\tau}{\pi t}\right)\exp\left(-\frac{x^2\tau}{2\lambda^2 t}\right). \qquad (5.146)$$

This equation is called the *Einstein–Smoluchowski equation*. The function $P(x,t)$ satisfies the diffusion equation with

$$D = \frac{\lambda^2}{2\tau}. \qquad (5.147)$$

This equation explains the so-called *Brownian motion*. Under magnification, one may distinguish small particles (5 μm or smaller) suspended in water, moving in a random "drunken sailor" motion. The phenomenon was discovered at the beginning of the nineteenth century by Robert Brown, a Scottish botanist. Brown pioneered the use of the microscope. When he discovered "Brownian motion," he used pollen grains suspended in water and correctly concluded that the motion was caused by the surrounding molecules.

In 1905, Einstein explained Brownian motion in terms of mathematical equations. He derived the following equation:

$$\frac{\overline{(\Delta x)}^2}{2\Delta t} = \frac{RT}{N_A f}. \qquad (5.148)$$

The average move during time Δt is measured; f is the frictional coefficient and N_0 is Avogadro's number. If Δx is identified with λ and Δt with t, just comparing with the Einstein–Smoluchowski equation 5.146 simply provides the diffusion constant D.

Equation 5.149 provides a possibility to obtain an approximate value of Avogadro's number:

$$\frac{\lambda^2}{2t} = \frac{RT}{N_A 6\pi\eta a} \Rightarrow N_A = \frac{t}{\lambda^2}\frac{RT}{3\pi\eta a}. \qquad (5.149)$$

This equation was used by Perrin in 1908 when he made the first accurate determination of Avogadro's number. It is interesting that this successful use of a differential equation for molecular motion was considered as a final proof that atoms and molecules exist! In fact, until then, this had been doubted by a few physicists. This is explainable if we recall that ordinary thermodynamics does not assume anything about the existence of atoms and molecules.

6 Ions in Crystals and in Solution

6.1 INTRODUCTION

Electropositive atoms, such as alkali atoms, tend to donate the weakly bound electrons of their valence shell to electronegative atoms, such as halogens, which tend to accept electrons in the empty valence orbitals, thereby completing the shell for both ions. Ordinary rock salt (NaCl) thus consists of positive ions (Na^+) and negative ions (Cl^-). In NaCl, one orbital is a linear combination of 3p orbitals of the liganding Cl^- ions, all pointing inward toward the sodium ion. This orbital reminds of the 3s orbital in Na, but is considerably smaller, thus allowing the Na^+ and Cl^- ions to come closer and increase the ionic binding energy.

We will first treat ionic crystals formed by positive ions from the first (alkali) and second (alkaline earth) group in the periodic table, together with negative ions from the sixth (chalcogen) group and seventh (halogen) group in the periodic table.

To ionic crystals also belong the transition metal compounds, where the negative ions come from group 6 or 7, and the transition metal is from periods 4–6, where the d subshells are being filled. Transition metal ions are often involved in electron transfer reactions.

6.2 IONS IN AQUEOUS SOLUTION

An alkali atom and a halogen atom at a large interatomic distance have tagged a lowest electronic state in which the atoms are neutral. If the atoms come closer, the lowest state smoothly changes to an ionic ground state. For example, Na + F become Na^+ + F^- when closer than 8.3 Å. The stability of the ionic form is high in the crystal or aqueous solution, where every ion is surrounded by several ions of opposite charge or by water molecules with a high dipole moment. Water solutions of ionic solids are called *electrolytes*, since neutral atoms are dissolved as ions. The electrons are removed from the valence shell of metal atoms and added to the valence shell of halogens (F, Cl, Br, and I) and chalcogens (O, S, Se, and Te). In a dilute solution, the positive metal ion is surrounded by a shell of water molecules, all with a negative oxygen lone pair oriented toward the positive metal ion. This *electrolytic dissociation* in water is easy to understand with our present knowledge of electronic structure and the periodic table, but it was not immediately accepted when first proposed by the Swedish physical chemist Svante Arrhenius (1859–1927) in his thesis with the title "Theory for electrolytic dissociation." Since then, the term *ion* is used for a charged atom, *cation* for a positive ion, and *anion* for a negative ion.

6.2.1 SOLVENT STRUCTURE AROUND IONS

Ionic forces are large at a small distance between two ions, with the charges $\pm Z \, |e|$. The attractiive energy E(R) at a distance R between two ions with charges Z_1 and Z_2 is given by Coulomb's law (atomic units):

$$E(R) = \frac{Z_1 Z_2}{R}. \tag{6.1}$$

The interatomic distance R is of the order of 3 Å or 6B. The repulsion energy at this distance if the charge is 1 (for example, Li^+, Na^+, or K^+) is thus 1/6 H in vacuum and, since 1 H is equal to 27.2 eV, the attractive potential energy is about 5 eV or 500 kJ/ mol. This is an enormously large number. In the next section, we will calculate the enthalpy gain, the Born energy, when ions are placed in polarizable solvents.

The spatial distribution of water molecules around a positive ion may be studied using molecular dynamics calculations. One obtains a radial distribution function g(R) and the integrated number of ligands n(R) ("the running integration number"). If an alkali ion is placed in water, it is surrounded by H_2O molecules with the oxygen pointing toward the sodium ion. The distribution function g(R) starts to grow at R = 2.0 Å (for Na^+). There is a sharp maximum at 2.35 Å followed by a minimum at 3.2 Å. For larger R, there is another quite strong maximum at 4.4 Å. At the minimum, the running integration number is n(3.2) = 6.5 water molecules. The first hydration shell at radius 2.35 Å thus contains 6.5 water molecules (a time-averaged number) and a second shell at 4.37 Å.

Some data are shown in Table 6.1. Of course the larger the metal ion, the larger the radius of the hydration shell. An important corollary is that the attraction of a water layer around K^+ (as measured by the free energy of hydration $\Delta G^{\ominus}$) is considerably smaller than a similar layer around a smaller alkali ion. Thus, potassium ions throw off their accompanying water molecules much more easily than sodium ions, if this is necessary to enter a channel of a cell membrane in living matter. Potassium penetrates the cell wall more easily in an ionic channel, since it may pass the channel without a load of water molecules around it.

TABLE 6.1
Some Experimental Data Taken from the Simulation of I. Benjamin

Ion	R_{max} (Å)	$\Delta G^{\ominus}$ (kcal/mol)
Li^+	2.08	126.4
Na^+	2.36	101.3
K^+	2.8	84.1

Note: $\Delta G^{\ominus}$ is the free energy of hydration.

6.2.2 BORN EQUATION

The polarization energy may be calculated with the help of a Born model, if we know the dielectric constant ε. A charged sphere with radius R gives rise to a field F(r) outside the sphere:

$$F = \frac{-Z}{4\pi\varepsilon_0\varepsilon r^2}(J/C) = \frac{-Z}{\varepsilon r^2}(H),$$ (6.2)

where r is the distance to the nucleus and ε is the dielectric constant of the medium. The polarization energy is given by

$$U = \frac{1}{2}\int_R^\infty FDdv = \frac{2\pi}{\varepsilon}\int_R^\infty D^2 4\pi r^2 dr(H),$$ (6.3)

where the electric displacement D is given by

$$D(r) = \frac{Z}{4\pi r^2}.$$ (6.4)

If D(r) is inserted into Equation 6.3, we obtain

$$U = \frac{1}{2\varepsilon}\int_R^\infty \frac{Z^2}{4\pi r^4} 4\pi r^2 dr = -\frac{Z^2}{2\varepsilon R}.$$ (6.5)

The difference in field energy between a sphere in vacuum (ε = 1) and a sphere in a dielectric medium with dielectric constant index ε is then

$$U_{med} - U_{vac} = \left(1 - \frac{1}{\varepsilon}\right)\frac{Z^2}{2R}.$$ (6.6)

Equations 6.5 and 6.6 give the polarization energy according to Born. It is the energy gained by an ion in a polar or polarizable solvent. The magnitude is easily estimated. For water, $\varepsilon = 81$. If $Z = |e|$ and $R = 4$ B, the polarization energy is roughly 1/8 a.u. ≈ 3.4 eV ≈ 78 kcal/mol, a very large number compared to thermal energies (kT = 1 eV if T = 11616 K). In nonpolar solvents, $\varepsilon = n^2$, where n is refractive index. It should be remembered that the Born energy is an enthalpy and not directly comparable to the numbers in Table 6.1.

Large polarization in the Born model depends on the fact that the solvent molecules are ordered in the field. A positive ion in water solvent is surrounded by H_2O molecules, which turn the negatively charged oxygen toward the ion. Consequently, the orbital energy for a positive ion is increased to less negative values (lower ionization energy). The great variation in orbital energy for differently charged ions in vacuum is then much smaller in water solution. Since reduction and oxidation generally take place in solution, this means that reduction and oxidation potentials vary much less than electron affinity and ionization energy.

6.3   CRYSTALS

Solids form crystals because the energy gain by ordering is large. Normally, order remains after heating to the melting point. In other cases, the energy gain by ordering is quite small, and the system is ordered only at low temperatures. It is also possible that different phases of crystal structure exist in different temperature ranges. Tin is a good example. Above 13°C, the stable structure is tetragonal and forms a silver-like metal, called β tin or white tin. This structure is stable to the melting point. Below 13°C, the stable structure is a gray powder called α tin or gray tin, with less density than β tin. On a closer look, the grains form a diamond crystal structure. Thus, below 13°C, β tin goes over to the less precious α tin form, although the process is very slow above −40°C.

6.3.1   CRYSTAL DIRECTIONS AND PLANES: UNIT CELL AND RECIPROCAL SPACE

The *crystal planes* pass through an infinite number of equilibrium positions of the atoms. Planes that are parallel to each other form a *family* of crystal planes (Figure 6.1). For such a family, it holds that all crystal planes are at the same distance from the next plane. Call this distance d. All lattice points will belong to one of the crystal planes of a given family.

A two dimensional crystal lattice (Figure 6.1) may be used to illustrate the main principles for characterization of a crystal. The crystal planes are defined by choosing an origin in an arbitrary lattice point and require that the plane passes infinitely through many lattice points. *Crystal directions* are straight lines passing infinitely through many points. The crystal axes are chosen as three of the crystal directions,

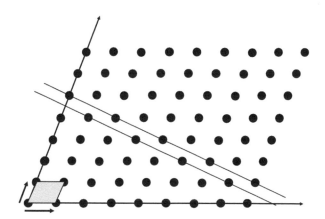

FIGURE 6.1   Two-dimensional crystal as an illustration. Black circles are lattice points. Crystal directions (arrows) define the unit cell (gray). Two members of the (2,1) family of crystal planes are marked out (in two dimension as lines).

of course not in the same plane. There are many ways to choose the crystal axes, but a useful choice is to connect one lattice point (for example, the left-most one in Figure 6.1) and then go to the closest lattice points, where the atom or molecule is the same. In three dimensions, we obtain three axes that are not in the same plane. These three axes define a *unit cell*.

In a lattice of the type shown in Figure 6.1 (where a two-dimensional lattice is shown), each point consists of an atom, an atomic ion, a molecule, or a molecular ion. It is possible to move one step in all crystal directions and get back the (infinite) crystal (translational symmetry). It is required from the unit cell that (1) translational symmetry is still valid, (2) all unit cells together fill the volume of the crystal, and (3) different unit cells do not overlap. Despite these requirements, a unit cell is not unique, since the crystal axes are not unique, but can be defined in more than one way. The lengths of the unit cell axes are called the *lattice constants*.

Let (c_1, c_2, c_3) be a unit vector, orthogonal to a crystal plane. If (x,y,z) is a point in the plane, its scalar product with (c_1, c_2, c_3) has to be zero if the plane passes through the origin. If the plane does not pass through the origin but belongs to the same family as the first plane, we must have

$$c_1 x + c_2 y + c_3 z = nd, \tag{6.7}$$

where n is the number of the plane counted from the plane through the origin. d is the distance between the planes.

In a family of crystal planes, the planes are at the same distance from each other. All lattice points will belong to one of the planes of a given family. The least distance between two planes in a family is d.

Since the plane with $n = 1$ has to pass through the given lattice points $(a_1, 0, 0)$, $(0, a_2, 0)$, or $(0, 0, a_3)$ the latter must satisfy Equation 6.7 and we obtain

$$c_1 = \frac{d}{a_1}; \quad c_2 = \frac{d}{a_2}; \quad c_3 = \frac{d}{a_3}. \tag{6.8}$$

Thus, we find that the equations for the planes in a family may be written as

$$\frac{x}{a_1} + \frac{y}{a_2} + \frac{z}{a_3} = n, \quad n = 0, \pm 1, \pm 2, \dots. \tag{6.9}$$

The vector

$$\left(\frac{1}{a_1}, \frac{1}{a_2}, \frac{1}{a_3} \right) = \left(\frac{h}{a}, \frac{k}{a}, \frac{l}{a} \right) \tag{6.10}$$

thus defines the family of planes. The family shown two dimensionally in Figure 6.2 has the vector (2,1).

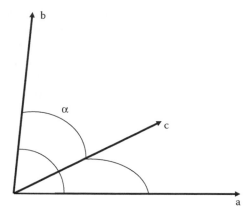

FIGURE 6.2   Definition of the intervector angles.

The distance d between the planes may be obtained. In the cubic case,

$$d = \frac{1}{\sqrt{a_1^{-2} + a_2^{-2} + a_3^{-2}}} = \frac{a}{\sqrt{h^2 + k^2 + l^2}} \tag{6.11}$$

The following vectors define the *reciprocal lattice*:

$$\vec{b}_1 = 2\pi \frac{\vec{a}_2 \times \vec{a}_3}{\vec{a}_1 \cdot (\vec{a}_2 \times \vec{a}_3)}$$

$$\vec{b}_2 = 2\pi \frac{\vec{a}_3 \times \vec{a}_1}{\vec{a}_2 \cdot (\vec{a}_3 \times \vec{a}_1)} \tag{6.12}$$

$$\vec{b}_3 = 2\pi \frac{\vec{a}_1 \times \vec{a}_2}{\vec{a}_3 \cdot (\vec{a}_1 \times \vec{a}_2)}.$$

The scalar–vector product in the denominator defines a number, in fact the volume of the parallelepiped spanned by the three vectors $\vec{a}_1$, $\vec{a}_2$, and $\vec{a}_3$.

Miller indices (h, k, l in Equation 6.10) are a way of denoting a family of crystal planes and are defined by the intersection of the crystal planes with the crystal axes. The crystal planes, shown in the twodimensional case in Figure 6.2, may be denoted as (9,4.5) and (10,5). Usually, we divide by 4.5 and 5, respectively, and obtain (2,1) in both cases; that is the two lines drawn belong to the same family of planes (lines). The notation of a family may thus be obtained from either Miller indices or the reciprocal space.

The intervector angles (α,β,γ) are defined as in Figure 6.2. Each family of planes thus corresponds to a point in the reciprocal lattice. A factor 2π is sometimes introduced to make it easier to write the equation for a plane wave.

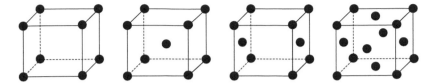

FIGURE 6.3 Different types of centering of the unit cell. In order from the left: primitive (P); body centered (bc) (I); side centered (sc) in the c direction (C); and face centered (fc) (F).

A diffraction method was originally developed by the English scientists W. H. Bragg and W. L. Bragg (father and son) in order to determine crystal structures and from that, molecular distances.

6.3.2 CRYSTAL SYSTEMS

There are four different kinds of centering of the unit cells. Figure 6.3 shows from left to right: primitive (no centering), body centered, side centered, and face centered.

In three dimensions, there are seven crystal systems (Table 6.2). The crystal systems are further divided according to centerings (Figure 6.3) into 14 Bravais lattices: cubic (3), tetragonal (2), orthorhombic (4), hexagonal (1), trigonal (1), monoclinic (2), and triclinic (1). Auguste Bravais was a French mathematician.

Crystal structure refers to a crystal system and centering. For example, CsCl is a primitive cubic, and NaCl face centered cubic (fcc) (Figure 6.4).

The description of a crystal is not unique. The choice used may differ from one author to the next. Usually, cations and anions are looked on separately. In Figure 6.4, the Cl^- ions have been considered as the main actors. They form a simple cubic sublattice for CsCl and an fcc lattice for NaCl. The reader is advised to consult various textbooks on the subject.

TABLE 6.2
Crystal Systems

Crystal System	Unit Cell	Symmetry	Bravais Lattices
Cubic	$A = b = c; \alpha = \beta = \gamma = 90°$	3 tetrade; 4 triade; 8 diade axes	P,F,I
Tetragonal	$A = b \neq c; \alpha = \beta = \gamma = 90°$	1 tetrade; 8 diade axes	P,I
Orthorhombic	$A \neq b \neq c; \alpha = \beta = \gamma = 90°$	3 diades (or mirror planes)	P,F,I,C
Hexagonal	$A = b \neq c; \alpha = \beta = 90°; \gamma = 120°$	1 hexade	P
Trigonal	$A = b \neq c; \alpha = \beta = 90°; \gamma = 120°$	1 triade	P
Monoclinic	$A \neq b \neq c; \alpha = \beta = 90°; \gamma = 90°$	1 diade (or mirror planes)	P,C
Triclinic	$A \neq b \neq c; \alpha \neq \beta \neq \gamma$	None	P

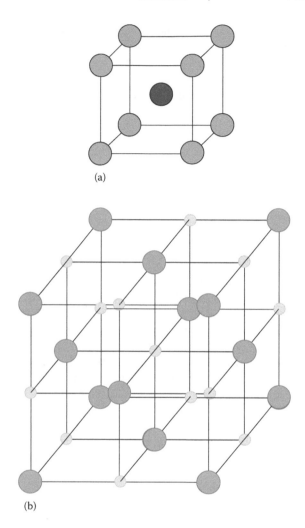

FIGURE 6.4   (a) CsCl structure (primitive cubic) and (b) NaCl structure. The larger spheres are Cl$^-$, forming an fcc cubic structure.

6.3.3 Lattice Enthalpy

The lattice enthalpy of an ionic crystal is the enthalpy lowering when 1 mol of crystal is formed from its ions. It consists particularly of the attraction between ions of different signs of charge, according to Equation 6.1. The standard formation enthalpy of crystal is the energy lowering from the pure phases, usually the metal phase for the metal ion and the halogen gas phase at standard conditions.

If a very electronegative atom is bonded with a very electropositive one, in a condensed medium, one electron will be transferred to the electronegative one, which forms a negative ion. One example is LiF formed from the ions Li$^+$ and F$^-$. Each

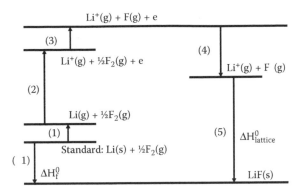

FIGURE 6.5 Born–Haber cycle.

Li$^+$ ion is coordinated to six F$^-$ ions, and every F$^-$ ion coordinated to six Li$^+$ ions. The crystal is an insulator, but there is a strong absorption of infrared radiation. No electronic transitions are expected at a low energy, so the low energy absorption in the infrared must be due to vibrations.

Since the lattice enthalpy for ionic crystals cannot be determined experimentally in any simple way, it has to be determined from other experimental numbers. Below we describe a model for calculating the reaction enthalpy, ΔH_f^0, for the formation of the LiF ionic crystal, referred to as the *Born–Haber cycle* developed by Max Born and Fritz Haber (Figure 6.5).

We want to know the enthalpy for formation of the LiF lattice from its components of Li$^+$ and F$^-$ in the gas phase. This quantity cannot be directly measured, of course. Instead, we have to obtain the lattice formation enthalpy of LiF (step 5 in Figure 6.5) by summing a number of measurable quantities. Step (−1) is the formation enthalpy of LiF when it is obtained from lithium metal and fluorine gas in the standard state. The numbers we are using here are listed in tables, in textbooks, or on the Internet. Experimentally, (−1) is obtained by reacting a weighed amount of sodium metal in a surplus of chlorine gas in a calorimeter.

Step (1) is the vaporization energy of Li. Step (2) is the charge separation step from Li gas to Li$^+$ and free electrons, which may be calculated from ionization energies. Step (3) is the dissociation energy of F$_2$, which may be obtained from vibration spectra or from accurate quantum chemical calculations. Step (4) is the electron affinity of the fluorine atom.

The experimental lattice enthalpy obtained from the Born–Haber cycle (step 5) may be compared to a theoretical estimation. The major component of the lattice enthalpy is the Coulomb attraction (U) between the ions, given in Equation 6.1. As the ions approach each other, the electronic shells will start to overlap and repulsive forces become important.

$$U_{Rep} = \frac{B}{R^n},\qquad\qquad (6.13)$$

TABLE 6.3

Negative of Calculated Lattice Enthalpies for Some Salts

Compound	Latt. Enth. (kJ/mol)	Compound	Latt. Enth. (kJ/mol)	Compound	Latt. Enth. (kJ/mol)
LiF	1021	$LiNO_3$	848	CaF_2	2630
LiCl	841	$NaNO_3$	756	MgO	3930
LiBr	808	AgF	915	CaO	3480
NaF	908	AgCl	860	SrO	3220
NaCl	774	AgBr	845	Ar	8
NaBr	745	AgI	835	Cl_2	31

where B is a parameter, which may be obtained by various means. The sum of U and U_{Rep} is the potential energy E_p, also called the Born–Landé Equation (1918). Alternatively, E_p may be obtained theoretically and referred to as the Madelung correction:

$$E_p = -\frac{N_A M Z^+ Z^- e^2}{4\pi\varepsilon_0 R}\left(1 - \frac{1}{n}\right). \tag{6.14}$$

In Equation 6.14, N_A is Avogadro's constant; M is the Madelung constant, which may be calculated from the lattice structure; Z^+ is the number of electronic charges on the positive ion and Z^- is the number of electronic charges on the negative ion, $e = 1.6022 \times 10^{-19}$ C; ε_0 is the permittivity of vacuum $= 8.8541 \times 10^{-12}$ C²/Jm; R is the interatomic distance; and n is called the Born exponent. It may be determined from the compressibility of the solid.

The lattice enthalpy (Table 6.3) of an ionic solid is a measure of the strength of the bonds in that compound. The decrease of the lattice energy from LiF to LiBr or from NaF to NaBr depends mainly on the increased bond distance. The large values for MgO to SrO depend on the higher charge of the ions.

6.4   CRYSTAL FIELD THEORY FOR TRANSITION METAL IONS

Most of the transition metal oxides and salts may be treated as ionic crystals. The valence electrons of s symmetry (4s, 5s, or 6s) are stripped off. The metal forms an ion that is considerably smaller than the atom. The inner $(n - 1)d$ electrons (3d, 4d, or 5d) are degenerate in the central field approximation of the atom. In the next step, multiplets are formed (Section 2.4). In the third step, the transition metal ions interact with the crystal field, which is dominated by repulsion from the neighboring negative ions. This leads to a splitting of the 3d orbital energies. The theory describing the splitting of the electronic states in a crystal field is due to the American physicists Hans Bethe and John van Vleck and is called *crystal field theory* (CFT).

6.4.1 TRANSITION METAL OXIDES AND SALTS

A short review of transition metal systems that are suitable to treat with CFT will be given below. The concentration will be on the first row of transition metal atoms (Sc, Ti, V, Cr, Mn, Fe, Co, Ni, Cu), where the number of valence electrons vary from one to nine 3d electrons. It is meaningful to also include d^0 systems, where the d orbitals can be reached by excitation of electrons from the ligand valence orbitals (ligand to metal [LM] transition). To this group belongs in the first row (with O^{2-} as ligands): Sc^{3+} (scandate), Ti^{4+} (titanate), V^{5+} (vanadate), Cr^{6+} (chromate), and Mn^{7+} (permanganate). The ending -ate indicates the presence of at least one other atom in the compound. The same oxygen ligands appear in the oxides: Sc_2O_3 (scandium oxide), TiO_2 (titanium dioxide), V_2O_5 (vanadium pentoxide), etc. The oxidation states are Sc(III), Ti(IV), V(V), Cr(VI), and Mn(VII).

Scandates and titanates are colorless, since the lowest energy absorption is in the UV region. Vanadates are white-yellow, chromates orange-yellow, and permanganate violet. The intensity is high since transition is allowed.

The d shell is filled from Sc^{2+} ($3d^1$) to Zn^{2+} ($3d^{10}$). Zinc oxide (ZnO) is colorless, but the others are at least weakly colored. The color is due to symmetry-forbidden transitions within the d shell, which become slightly allowed by vibrational coupling. If the complex is tetrahedral and missing a symmetry center, the color tends to be stronger since mixing of components in the states involved occurs, which allow the transition. Zn^{2+} has a filled 3d shell and thus no transitions are allowed within the 3d shell.

Typical for transition metal ions is that an inner shell (3d) is being filled. The following examples of ground state configurations may be mentioned: Sc^{2+} [Ar]$3d^1$, Ti^{2+} [Ar]$3d^2$, Ti^{3+} [Ar]$3d^1$, V^{2+} [Ar]$3d^3$, Cr^{3+} [Ar]$3d^3$, Mn^{2+} [Ar]$3d^5$, Fc^{2+} [Ar]$3d^6$, Fe^{3+} [Ar]$3d^5$, Co^{2+} [Ar]$3d^7$, Co^{3+} [Ar]$3d^6$, Ni^{2+} [Ar]$3d^8$, and Cu^{2+} [Ar]$3d^9$. Notice that in CFT the assumption is made that only atomic 3d (4d, 5d) orbitals are treated. This limits the applicability of the CFT method.

6.4.2 ENERGY LEVELS

CFT may be divided into weak field theory (WF-CFT) and strong field theory (SF-CFT). WF-CFT is a state interaction model of the type just mentioned. For example, the Ni^{2+} ion with eight 3d electrons has a ^3F ground state, consistent with Hund's rule. The crystal field is introduced as a perturbation and the ^3F is split into three states.

In strong fields, the orbitals of Ni^{2+} are first split. For example, in an octahedral field, created for example by negative ligands in the points $(\pm R,0,0)$, $(0,\pm R,0)$, $(0,0,\pm R)$, the d orbitals of the type $x^2 - y^2$ and $2z^2 - x^2 - y^2$, which are directed toward the negative charges, have a higher energy than the orbitals xy, yz, and zx, which are directed in between the ligands.

In principle, WF-CFT does not deal with single electrons moving in orbitals. Nevertheless, as we move from the weak field limit to stronger fields, the total energy levels must go over smoothly to the strong field limit. This is the principle for the energy diagrams introduced by Tanabe and Sugano. Tanabe–Sugano diagrams are available for most ligand geometries.

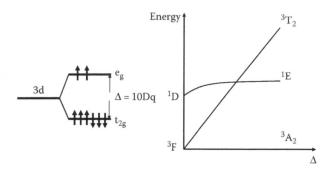

FIGURE 6.6 CFT orbital energy diagram (left) and total energy diagram (right) for the octahedral field in the case of a d^8 state. Δ is the CFT gap corresponding to the energy splitting between t_{2g} and e_g.

In an octahedral complex, there are six ligands on the x-, y-, and z-axes. The energies of the d orbitals pointing toward the ligands are higher by Δ (Figure 6.6), a parameter that is commonly fitted to an experimental spectrum (10Dq is an old notation for Δ). A diluted aqueous solution of $NiCl_2$ contains $Ni(H_2O)_6^{2+}$ complex ions and a solution of $CuSO_4$ contains $Cu(H_2O)_4^{2+}$ ions. The water molecules turn the negative oxygen atom toward the positively charged metal ion. The spectrum of the NiO crystal is similar to that of $Ni(H_2O)_6^{2+}$. The Ni atoms in NiO are also surrounded by six oxygen atoms in an octahedral geometry. These oxygen atoms can be considered as O^{2-} ions in both the NiO crystal and in a dissolved $Ni(H_2O)_6^{2+}$ complex in water.

In Ni^{2+} and Cu^{3+}, eight out of ten electrons are present and used to occupy the 3d orbitals and form the atomic multiplet structure. The ground state is a spin triplet state for Ni^{2+} (^3F). In the crystal field, a triplet state is still the ground state (^3A) (Figure 6.6). This state is the one with the lowest spin orbitals occupied (Figure 6.6). Arising also from ^3F is a state ($^3T_2$) where one electron has been excited from $t_{2g}\downarrow$ to $e_g\downarrow$. As Δ increases, this state will separate from the ground state. The transition is one of the ligand field transitions.

Reality is more complicated than this, however. By the same reasoning, CuO would be expected to be greenish-blue like $Cu(H_2O)_4^{2+}$ in water, but is black. This is due to other types of excitations that are not included in the CFT. The black color of CuO may be due to intermetal coupling superposed on the CFT spectrum. Another possibility, also not within the CFT, is that the excitation is from the ligand levels of the 3d levels on the same metal complex. Some of these transitions, like the intermetal transitions, would have a high intensity.

6.4.3 HIGH SPIN AND LOW SPIN

In the SF-CFT case, the Δ splitting is large in comparison to the Hund's rule splitting. This happens when the charge on the transition metal ion is larger than or equal to +3 (for example, Co^{3+}). A high charge pulls the ligands closer to the metal ion, whereby Δ increases. The orbital ordering is determined by Δ and Hund's rule is overshadowed. The lowest d levels in the strong crystal field are always occupied.

For example, in the cobalt(III) ion (d^6), $Co(OH)_6^{3-}$, the metal charge is $+3$, and therefore is a strong field case. The t_{2g} orbital is much lower than e_g and hence fully occupied. An equal number of orbitals of each spin are occupied and the total spin state is a singlet state (low spin). In the strong field case, multiplet stucture is treated as a perturbation after the ligand field splitting has been taken into account (for example, using a DFT calculation).

In the weak field case, Hund's rule holds, and the state with the highest spin is the ground state (high spin). In the case of Mn^{2+} (d^5), Δ is small for the octahedral $Mn(OH_2)_6^{2+}$ complex, and therefore the five d electrons have the same spin direction ($2S + 1 = 6$; thus a spin sextet).

Most metal ions with $+2$ or lower charge form weak field metal complexes where Hund's rule wins and the maximum possible total spin is obtained in the ground state. The so-called Mott insulators (MnO, FeO, CoO, NiO, and CuO) are typical examples. However, Δ depends on both metals and ligands. Ammonia (NH_3) ligands have a 25% larger Δ than water, and Cl^- and Br^- have a 25% smaller Δ than water. The tendency for a strong field is higher in the 4d and 5d series than in the 3d series.

4f and 5f electrons may also be included in a special kind of CFT; however, in this case, the weak field case is dominating. The spectrum consists essentially of the atomic multiplets.

6.4.4 Problems with Crystal Field Theory

The main problem with CFT is that the 3d electrons are living their own life independent of the electrons of the ligands. More correctly, they are allowed to feel the fixed field of electrons and nuclei, but they are not allowed to mix with the ligand electrons and form molecular orbitals (MOs) and bonds.

This is not the only problem with CFT, however. We also made the assumption that all the electrons of a ligand ion, for example, Cl^-, are located at the ligand nucleus. In reality, they are spread out around this nucleus in atomic orbitals. It is not difficult to imagine that all ligand electrons together form an almost spherically symmetric density, which does not contribute to the important nonspherical parts of the field that cause the splitting among the d electrons. In other words: are the Cl^- ions looked on as negative ions by the metal ion? If the electron clouds are smeared out, the metal ion should see the ligand nuclei, which, however, are positive.

This is, in fact, the case. If the spatial distribution of the ligand electrons is taken into account, the nuclei become the main contributor to the field, and the ordering of the metal electrons should be opposite to what we know is correct from the experiments. CFT survives just because the $\Delta = 10Dq$ parameter can be assigned any value that fits the experimental results. In other words, it is not the electrostatic field that causes the crystal field splitting, but something else.

Thus, CFT does not agree with the physics of the system. The conclusion is that 3d orbitals cannot be singled out. They take part in the ordinary formation of MOs. The crystal field splitting is due to the fact that the MOs becoming occupied in a metal complex are antibonding MOs. CFT has to be replaced by ligand field theory (LFT).

6.5 LIGAND FIELD THEORY

LFT is simply MO theory for the positive transition metal ion and the closest neutral or negatively charged ligands. LFT is designed to treat mainly the 3d, 4d, and 5d electrons, mixing with the ligand levels. It is recognized that d orbitals are at a higher orbital energy than the orbitals of the ligands. The interactions between the metal atoms are ignored in the first approximation. LFT is, in a way, a theory that is opposite to the intermetal band formation idea, which seems to be dominating solid state physics. The latter theory is not wrong, of course, but should be thought of as a limit case for ionic transition metal oxides and salts. CuO and some other oxides, particularly many sulfides and selenides, are not fully describable as ionic systems, but are metallic in the sense that intermetallic MOs are formed.

6.5.1 EXTENSION TO LFT

CFT and LFT have a historical background from the end of the 1920s and therefore we should not be surprised if the exact definitions of the two models vary between different textbooks. CFT and LFT together form the background knowledge for all discussions on the electron structure of transition metal systems. They serve to interpret experimental spectra as well as the results of quantum mechanical *ab initio* calculations.

Transition metal ions in aqueous solution are surrounded by water molecules, halide, or other ions from the solution. LFT is the simplified MO theory that involves the metal d orbitals and ligand valence orbitals (for example, 2p for an oxygen ion, Figure 6.4). The 3d splitting is caused primarily by MO formation between a metal 3d orbital (χ_M) and a ligand orbital (χ_L). The metal orbitals that are directed toward the ligands have a greater overlap with the ligand orbitals than those that are directed in between the ligands. Therefore, there is a larger energy splitting between bonding and antibonding orbitals if the metal orbitals are pointing toward, rather than in between, the ligands.

One exception is when the ligands are diatomic molecules. In that case, the unoccupied orbitals of, say N_2 or CO, are also taking part in the interaction. Here, we are going to restrict ourselves to the case with ligands that are "atomic." For example, the H_2O ligand does not have any unoccupied MO that can be used as acceptor MO, and therefore H_2O behaves as an atomic ion (O^{2-}) in LFT.

The octahedral geometry in the spin triplet Ni^{2+} complex is a consequence of the occupation by eight electrons. The most antibonding MO (e_g) is twofold degenerate and is occupied by two electrons with equal spins according to Hund's rule, forming together a triplet state. The resulting charge density may be obtained by writing down the expressions for the two degenerate MOs, denoted $3d(z^2)$ and $3d(x^2 - y^2)$, and singly occupied (z^2 is short-hand notation for $2z^2 - x^2 - y^2$). The 3d part of the MOs may be written as (Equation 2.12)

$$3d\left(z^2\right) = R_{32}\left(r\right)\frac{1}{2}\sqrt{\frac{5}{4\pi}}\frac{2z^2 - x^2 - y^2}{r^2}, \tag{6.15}$$

$$3d\left(x^2 - y^2\right) = R_{32}\left(r\right)\frac{1}{2}\sqrt{\frac{15}{4\pi}}\frac{x^2 - y^2}{r^2}. \tag{6.16}$$

The sum of the squares add up to a charge density ρ:

$$\rho\left(x,y,z\right) = R_{32}^2\left(r\right)\frac{5}{16\pi^2 r^4}\left(x^4 + y^4 + z^4 - x^2 y^2 - y^2 z^2 - z^2 x^2\right). \tag{6.17}$$

We see that the charge density ρ arising from both of these singly occupied orbitals attains a maximum along the coordinate axes x, y, and z.

The general picture of the interaction in a transition metal complex is given in Figure 6.7. The antibonding ligand field levels (LF) correspond directly to the crystal field levels. The bonding levels below are mainly ligand levels, for example, 2p levels of oxygen. A number of the latter are nonbonding orbitals. In an octahedral complex, they are ungerade orbitals. Since the metal 3d orbitals are gerade, there is no mixing. On the other hand, there are allowed transitions from nonbonding orbitals to metal orbitals.

In square planar geometry, for example, $Cu(H_2O)_4^{2+}$, the b_{1g} MO is the most anti-bonding level. Its 3d component is $x^2 - y^2$ (Equation 6.16). In a Cu^{2+} complex, b_{1g} is occupied by a single electron. In a Cu^{3+} complex, it is empty. Its energy is much higher than that of the other ligand levels (corresponding to a high field case). Hund's rule is overshadowed, and so the Cu^{3+} complex is diamagnetic.

Cr^{3+} complexes are d^3 complexes with t_{2g}^* orbitals occupied by the maximum number of three electrons of the same spin. The ground state is a quartet state $(2S + 1 = 4)$. The transition $t_{2g}^* \rightarrow e_g^*$ is in the visible region and gives a strong green color to the system. Cr^{3+} is a strong field case, but the spin is maximum due to the degeneracy.

Co^{3+} complexes are also of the strong field type. All the six 3d electrons enter into the three possible t_{2g} orbitals, which point in between the ligands. The ground state is a spin singlet and the system is diamagnetic.

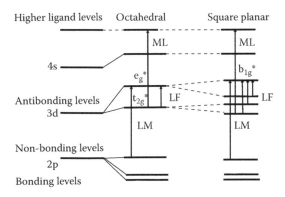

FIGURE 6.7 MOs in a transition metal complex with a single metal atom (LFT model). LF denotes ligand field transitions.

6.5.2 LOCALIZED OR DELOCALIZED EXCITATIONS

If crystals are formed from aqueous solutions, they in most cases, have the same color as the solution. The ligand field effects are mainly due to the closest ligands and therefore the spectral properties do not change very much. One may ask whether there is anything that prevents us from using LFT also in crystals, where there are not isolated metal complexes always. Generally, this is possible if there is negligible *coupling* between the metal ions. In cases when the metal ions are well separated, the coupling is small and the spectrum in the crystal is the same as in a solution where the ligands of the metal ions are the same. This implies two important conclusions: (1) We have to be careful when spectra are calculated. Delocalized methods where the orbitals are extended over the whole crystal cannot be used. (2) If the spectrum is different in complex and in crystal, there may be strong interactions between the metal ions.

The transition from localized behavior to delocalized behavior is an important one. As an example, nickel oxide (NiO) has about the same color as Ni^{2+} complexes in aqueous solution. Cu^{2+} complexes in aqueous solution are blue, but tend to be green if the ligands are halide ions or violet if the ligands are ammonia molecules. Copper(II) oxide (CuO), on the other hand, is black. Very likely, the wide absorption in the visible region is from an allowed transition that covers the weak ligand field transitions. It cannot be an LM transition since there is no reason to believe that LM transition are in the visible region just in the case of CuO.

Photoconduction experiments show that we are dealing with a metal–metal (MM) transition. As we will see in later chapters, MM transition energies are strongly dependent on the distance between the metal ions. For small distances, they go down into the visible region for some metal ions.

6.5.3 FIRST-ORDER JAHN–TELLER EFFECT

The $3d^8$ configuration of Cu^{3+} forms strong field complexes. In an octahedral complex, one would still expect two electrons with the same spin, since e_g orbitals are degenerate. This high spin case has to compete with a low spin case that has another occupation of the orbitals, however. If e_g orbitals are degenerate, we may decide to occupy one of the orbitals with two electrons and leave the other one empty. It turns out that the lowest energy is obtained if the orbital with metal character according to Equation 6.15 (the $3d(z^2)$ orbital) is doubly occupied.

As always, the bond distances depend on occupation. Thus, if we occupy antibonding $3d(z^2)$, the bond lengths along the z-axis will increase. The $3d(z^2)$ is now less antibonding. In the iterative determination of a consistent geometry, we will obtain convergence for a metal complex where the four ligands in the plane are tightly bonded, but the axial ligands are at a much longer distance from the copper ion. On the other hand, if we occupy the $3d(x^2 - y^2)$ orbital with two electrons, we will end up with a complex that is compressed along the z-axis. Both complexes are in spin singlet states.

The third possible occupation of the orbitals is to use the condition that the final state is a triplet state. The $3d(z^2)$ and $3d(x^2 - y^2)$ spin orbitals, say the α spin orbitals, are now democratically occupied with one electron each. The geometry remains octahedral.

We cannot say *a priori* which of the three cases has the lowest energy. It happens to be the axially elongated complex in most cases. The important thing is that three different states are possible. The degenerate situation that existed for the original octahedral complex has ended up in three possible states with different geometry and *without degeneracy*.

This is an example where the so-called *Jahn–Teller theorem* comes into play. Hermann Jahn, a British physicist of German descent, and the perhaps more well known and outspoken Hungarian physicist Eduard Teller, proved that degeneracies cannot exist. All possible symmetries distort into a lower symmetry where the degeneracies have disappeared. This is the "first-order Jahn–Teller theorem" (FOJT).

There is also a "second-order Jahn–Teller theorem" (SOJT). In this case, there is no original degeneracy, but there are lower excited states that have the same effect as the degenerate states. In some cases, there is distortion of the geometry. A familiar case is the molecule cyclo-octatetraene (C_8H_8), which is not planar like benzene (C_6H_6). There is another SOJT theorem proven by the French theoretical chemist Lionel Salem. A more general proof has been given by the American inorganic chemist Ralph G. Pearson.

6.5.4   L → M CHARGE DONATION

Mixing of metal and ligand orbitals accounts for covalency. Covalency is often incorrectly referred to as "hybridization." The term hybridization should be reserved for mixing of orbitals on the same atom with different ℓ quantum numbers. In LFT, a bonding and an antibonding MO with the same symmetry indices are formed, as already mentioned. This linear combination may be written as:

$$\text{Bonding: } \phi_1 = \cos\theta\,\chi_L + \sin\theta\,\chi_M$$
$$\text{Antibonding: } \phi_2 = -\sin\theta\,\chi_L + \cos\theta\,\chi_M. \tag{6.18}$$

This form of MO assumes that the atomic orbitals χ_L and χ_M are orthonormalized, but we will not discuss this here. In Equation 6.18, trigonometric coefficients ($\cos\theta$ and $\sin\theta$) are used to conserve orthonormalization. CFT corresponds to the case $\theta = 0$, where only the 3d (4d, 5d) orbitals are treated. Notice that if both ϕ_1 and ϕ_2 are fully occupied, there is no charge or spin transfer contribution because $\cos^2\theta + \sin^2\theta = 1$. Equation 6.18 has been verified in several calculations without restrictions on form. θ may be determined using hyperfine structure, spectroscopy, or calculations, provided, of course, that the orbital pair in Equation 6.18 is not fully occupied.

As Equation 6.17 and Figure 6.8 show, charge flow between ligands and metal ion is allowed in LFT. If only the lower, bonding orbital ϕ_1 is occupied, the amount of 3d character is $\sin^2\theta$, arising from the interaction of one of the 3d orbitals with the ligand orbital with the same symmetry label. This is referred to as ligand to metal (L → M) charge donation. In the normal case, the ligand energy levels are at a much lower energy than the metal orbitals and therefore the mixing charge coefficient $\sin^2\theta$ is small (a few percent). The "effective number of 3d electrons" is larger in LFT than in CFT due to L → M charge donation.

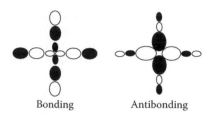

Bonding   Antibonding

FIGURE 6.8   The $\phi_1 = b_{1g}$ (bonding) and $\phi_2 = b_{1g}{}^*$ (antibonding) pair of MOs with the symmetry $x^2 - y^2$. The latter MO is unoccupied in the Cu^{3+} $3d^8$-singlet ground state.

L → M charge donation is large for Cu^{2+} and Cu^{3+} complexes and this has unfortunately led to confusion in the assignment of electronic states and valence states to metal sites. The majority of chemists prefer to use the valence state label used in the CFT model ($3d^8$), of course, aware of the fact that $3d^8$ is a formal notation, suggesting nothing about the charge distribution. In part of the physics literature, one instead counts the L → M charge donation to the metal charge. In the cuprates, where the L → M charge donation is close to a full electron, one considers the $3d^8$ complex as a $3d^9$ complex. This is misleading since it would imply that there is an unpaired spin.

Detailed calculations have been carried out on two complexes of $KCuO_2$: CuO_4^{5-} and $Cu_3O_8^{7-}$ using the CASSCF model. In the smaller system, there are 25 inactive orbitals. The active space contained essentially the ligand valence and metal 3d levels. The number of configurations was 45 for the smallest active space and 825 for the largest. In the larger system, the active space had to be reduced as much as possible. This choice generated 4950 configurations. The calculated L → M charge donation is consistent with experimental data. The "number of 3d electrons" on Cu^{3+} is, as expected, close to 9. Other properties than charge (for example, spin) are better understood if we refer Cu^{3+} as a $3d^8$ system. The notation should be used to avoid confusion, and therefore $3d^8$ is the preferred notation.

7 Time-Dependent Quantum Mechanics

7.1 INTRODUCTION

The time-dependent Schrödinger equation (SE) or the *wave equation* may be written as

$$i\eta \frac{\partial}{\partial t} \Psi(x,t) = H\Psi(x,t), \tag{7.1}$$

where x contains, in principle, all the space and spin coordinates of all particles of the system. Using Equations 1.12 through 1.14 it is easy to check that a wave packet satisfies the wave equation.

Time dependence is important for understanding electron transfer (ET) and excitation energy transfer reactions. Fast ET reactions happen on the time scale of femtoseconds (1 femtosec = 10^{-15} sec). In reactions with activation barriers, the whole range is covered, from picoseconds (1 picosec = 10^{-12} sec) to seconds. In this chapter, we will study how time dependence enters into quantum mechanics and its relationship to classical mechanics.

7.2 WAVE EQUATION

Time dependence is necessary for the simple reason that quantum mechanics must be able to predict processes that evolve in time, as is the case with classical mechanics. The boundary condition is the time-independent wave function at the beginning of the time counting (t = 0). We will find that, in some cases, time-dependent quantum mechanics deviates significantly from classical mechanics, for example, in the case of degeneracy.

7.2.1 TIME-INDEPENDENT ENERGY LEVELS AND COEFFICIENTS

Let us think of x as including only electrons, or even a single electron on the x-axis. Such a drastic approximation is justified by the fact that in most processes only one or two electrons are *active*. The majority of electrons remain in their orbitals. We may think of nuclei as moving classically in the first approximation. If H has no explicit time dependence, its eigenfunctions, $\psi_n(x)$, are time independent and satisfy

$$H\psi_n = E_n\psi_n. \tag{7.2}$$

In other words, ψ_n is a solution of the time-independent SE. A time-dependent wave function that satisfies the wave equation, may be written as

$$\Psi_n(x,t) = \exp(-iE_n t/\hbar)\psi_n(x). \tag{7.3}$$

An arbitrary time-dependent wave function may be expanded in the set of functions given by Equation 7.3:

$$\Psi(x,t) = \sum_n C_n \exp(-iE_n t/\hbar)\psi_n(x), \tag{7.4}$$

with coefficients $\{C_n\}$ that are independent of time. $|C_n|^2$ gives the probability for the system to be in a state n with energy E_n. If $C_k = 1$ and $C_n = 0$ for $n \neq k$, the system is in a stationary state (k). In Equation 7.4, Ψ satisfies the wave Equation 7.1 since

$$H\Psi = H\sum_n C_n \exp(-iE_n t/\hbar)\psi_n = \sum_n C_n \exp(-iE_n t/\hbar)H\psi_n$$

$$= \sum_n E_n C_n \exp(-iE_n t/\hbar)\psi_n = \sum_n C_n i\hbar \frac{\partial}{\partial t}\exp(-iE_n t/\hbar)\psi_n = i\hbar\frac{\partial\Psi}{\partial t}. \tag{7.5}$$

Knowing the wave function at time $t = 0$ means that the coefficients C_n are known. From Equation 7.5, we know the wave function at all times $t > 0$.

The energy expectation value over the time-dependent wave function Ψ is a constant, since it follows from Equation 7.4 that

$$E = \langle\Psi|H|\Psi\rangle = \sum_n |C_n|^2 E_n = \text{const.} \tag{7.6}$$

In time-dependent quantum mechanics, any energy can be chosen. If we include only one term in Equation 7.4, the system remains in this eigenstate and nothing happens. If the boundary conditions force us to include another state, no matter how small its expansion coefficient, there will forever be a probability that the system is found in this other state.

As an example, we consider a case where two fixed protons, A and B, attract a single electron. The electrons are described by the 1s orbitals of the protons, χ_a and χ_b, respectively.

We first assume that two states are available with equal energies. The eigenvalues of the time-independent problem are determined from the following secular equation:

$$\begin{vmatrix} H-E & H_{12} \\ H_{12} & H-E \end{vmatrix} = 0. \tag{7.7}$$

($H = H_{11} = H_{22}$). The solution is as follows ($H_{12} < 0$):

$$
\begin{aligned}
E_+ &= H + H_{12} \qquad \phi_+ = \left(\chi_a + \chi_b\right)/\sqrt{2} \\
E_- &= H - H_{12} \qquad \phi_- = \left(\chi_a - \chi_b\right)/\sqrt{2},
\end{aligned}
\tag{7.8}
$$

where we assume that χ_a and χ_b are non-overlapping.

The interaction matrix element H_{12} varies with the distance between the two protons and is large in absolute value when the latter are close to each other. The time-dependent wave function in Equation 7.4 may be written as

$$
\Psi(x,t) = C_+ \frac{\chi_a + \chi_b}{\sqrt{2}} \exp\left(-iE_+ t/\hbar\right) + C_- \frac{\chi_a - \chi_b}{\sqrt{2}} \exp\left(-iE_- t/\hbar\right),
\tag{7.9}
$$

which is an equation for a moving "wave packet." C_+ and C_- are determined by the boundary conditions for $t = 0$. If we require that the electron is on atom A at time $t = 0$, we have to choose $C_+ = C_- = 1/\sqrt{2}$. A state satisfying the boundary condition at $t = 0$ is called a *prepared state*. The prepared state is usually not a stationary state. With $C_+ = C_- = 1/\sqrt{2}$, Equation 7.9 may be written

$$
\Psi = \exp\left(-iHt/\hbar\right)\left[\chi_a \cos\left(H_{12}t/\hbar\right) - \chi_b i \sin\left(H_{12}t/\hbar\right)\right].
\tag{7.10}
$$

The electronic density (the probability to find the electron) $\Psi^*\Psi$ may be written

$$
\rho(r,t) = \Psi^*\Psi = \chi_a^2(r)\cos^2\left(H_{12}t/\hbar\right) + \chi_b^2(r)\sin^2\left(H_{12}t/\hbar\right).
\tag{7.11}
$$

At time $t = 0$, we have $\rho = \chi_a^2$ in accordance with the boundary conditions (Figure 7.1). For $t = \hbar\pi/2H_{12}$, we have $\rho = \chi_b^2$ and for $t = \hbar\pi/H_{12}$, $\rho = \chi_a^2$ again. Apparently, the electronic density oscillates between the two protons with the frequency $H_{12}/\hbar\pi = (E_+ - E_-)/h$. This frequency is the same as the frequency for the electromagnetic (EM) radiation that can excite the system from E_+ to E_- (Bohr's frequency condition). When the protons are close to each other and $|H_{12}|$ is large, the

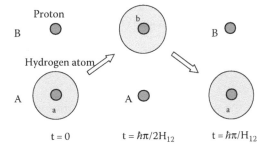

FIGURE 7.1   Electron oscillating between two fixed protons, A and B.

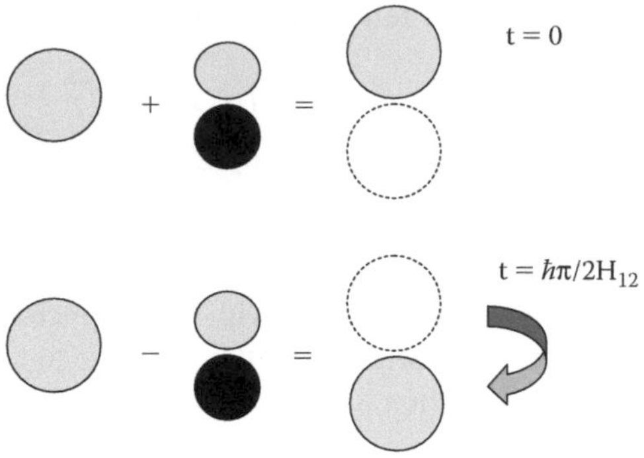

FIGURE 7.2 Electron oscillating in an atom.

oscillations are fast. When the distance between the protons increases, the oscillations eventually stop completely, since $|H_{12}|$ tends to zero.

There are many interesting problems of this simple type. The same mathematical machinery may be used if we want to know how electrons move in an atom (Figure 7.2). Suppose we know that an electron is on one side of the nucleus in the shell with principal quantum number n = 2 at time t = 0. Then we may describe this situation by superposing one 2s and one 2p orbital at t = 0, in such a way that the wave function is cancelled by interference of the 2s and 2p orbital on the other side. For t > 0, the electron will oscillate back and forth within the n = 2 shell.

For highly excited states, we let an electron in the form of a wave packet be accepted in a rather well-defined region far from the nucleus. For t = 0, we need to superpose atomic orbitals with all kinds of n, , and m quantum numbers. For t > 0, an electron will move almost classically if the principal quantum number is high, at least for some time. The limit motion is planetary motion according to Newton's equations.

7.2.2 Time-Dependent Energy Levels

We will now study the more useful case when the nuclei are moving by classical mechanics with a Born–Oppenheimer potential energy surface (PES). The points passed by the nuclei are called the *trajectory* and denoted Q(t). Q(t) is the spatial coordinates for the heavy particles (nuclei) in the system:

$$Q(t) = \left[X_1(t), Y_1(t), Z_1(t), X_2(t), \ldots, Z_N(t) \right]. \tag{7.12}$$

We have to solve the wave equation for the electrons as the nuclei move along this trajectory. A general solution, given by the Russian physicist E. E. Nikitin, is sketched below.

The wave equation for an electronic Born–Oppenheimer state n is written:

$$i\hbar \frac{\partial}{\partial t} \Psi_n(x,Q,t) = H(Q)\Psi_n(x,Q,t) = E_n(Q)\Psi_n(x,Q,t). \qquad (7.13)$$

H(Q) is the electronic Hamiltonian in Q along the trajectory. We try to satisfy Equation 7.13 by expressing the time dependence in the following way:

$$\Psi_n(x,Q,t) = \psi_n(x,Q)\exp\left[-\frac{i}{\hbar}\int_0^t E_n(Q)d\tau\right]. \qquad (7.14)$$

In analogy with Equation 7.4, the time-dependent wave function may be expanded as

$$\Psi(x,Q,t) = \sum_n C_n(t)\psi_n(x,Q)\exp\left[-\frac{i}{\hbar}\int_0^t E_n(Q)d\tau\right]. \qquad (7.15)$$

This wave function has to satisfy the wave equation. If the Hamiltonian H works on Ψ, we obtain

$$H\Psi(x,Q,t) = \sum_n C_n H\psi_n(x,Q)\exp\left[-\frac{i}{\hbar}\int_0^t E_n(Q)d\tau\right]$$

$$= \sum_n C_n E_n \psi_n(x,Q)\exp\left[-\frac{i}{\hbar}\int_0^t E_n(Q)d\tau\right]. \qquad (7.16)$$

If the time derivative operates on Ψ, we obtain

$$i\hbar \frac{\partial}{\partial t}\Psi(x,Q,t) = \sum_n i\hbar \frac{\partial C_n}{\partial t}\psi_n(x,Q)\exp\left[-\frac{i}{\hbar}\int_0^t E_n(Q)d\tau\right]$$

$$+ i\hbar \sum_n C_n \frac{\partial \psi_n}{\partial t}\exp\left[-\frac{i}{\hbar}\int_0^t E_n(Q)d\tau\right] \qquad (7.17)$$

$$+ \sum_n C_n \psi_n E_n \exp\left[-\frac{i}{\hbar}\int_0^t E_n(Q)d\tau\right].$$

The wave equation requires Equation 7.16 = Equation 7.17, hence

$$\sum_n \frac{\partial C_n}{\partial t} \psi_n(x,Q) \exp\left[-\frac{i}{\hbar}\int_0^t E_n(Q)d\tau\right] = -\sum_n C_n \frac{\partial \psi_n}{\partial t} \exp\left[-\frac{i}{\hbar}\int_0^t E_n(Q)d\tau\right]. \quad (7.18)$$

Multiplying with ψ_k^* and integrating, we obtain the following set of equations:

$$\frac{\partial C_k(t)}{\partial t} = -\sum_n C_n(t)\left\langle \psi_k \left| \frac{\partial \psi_n}{\partial t} \right\rangle \exp\left[-\frac{i}{\hbar}\int_0^t \left(E_n(Q) - E_k(Q)\right)d\tau\right]. \quad (7.19)$$

This differential equation may be solved exactly in simple cases or be integrated numerically.

In this equation, a number of interesting time behaviors are hidden, particularly regarding curve crossings, or generally in more than one dimension, a PES crossing.

First, we may consider how Equation 7.19 behaves if the nuclei are fixed. The eigenfunctions are the same for all times. In other words, they are orthogonal and noninteracting. The overlap between one eigenfunction and the derivative of another one in Equation 7.19 is equal to zero, since $\partial \Psi_n/\partial t$ is equal to zero. C_k is then constant and we are back to Equation 7.9 with constant coefficients.

Another example is a curve-crossing problem. Generally, this is a crossing between many-dimensional PES, but very often a single coordinate along a certain pathway may be singled out. We simply assume one-dimensional motion. The potential energy curves are shown in Figure 7.3.

We assume that at $t = 0$, the system moves on the lower energy curve in electronic ground state Ψ_1. For $t = 0$, we have $C_1 = 1$ and $C_2 = 0$. The upper state corresponds to Ψ_2, thus for this state $C_2 = 1$ and $C_\mu = 0$. Equation 7.19 corresponds to

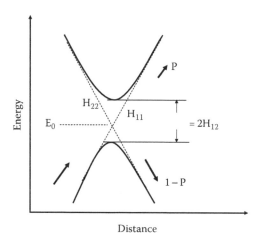

FIGURE 7.3   Curve-crossing problem.

$$\frac{\partial C_1(t)}{\partial t} = -C_1(t)\left\langle \psi_1 \left| \frac{\partial \psi_1}{\partial t} \right\rangle - C_2(t)\left\langle \psi_1 \left| \frac{\partial \psi_2}{\partial t} \right\rangle \exp\left[-\frac{i}{\hbar}\int_0^t (E_2(Q) - E_1(Q))d\tau \right]$$

$$\frac{\partial C_2(t)}{\partial t} = -C_1(t)\left\langle \psi_2 \left| \frac{\partial \psi_1}{\partial t} \right\rangle \exp\left[-\frac{i}{\hbar}\int_0^t (E_1(Q) - E_2(Q))d\tau \right] - C_2(t)\left\langle \psi_2 \left| \frac{\partial \psi_2}{\partial t} \right\rangle.$$

$$(7.20)$$

Since the change in ψ_μ is orthogonal to ψ_μ, the diagonal terms disappear in Equation 7.20.

The time integration is done with small, finite increments, δt. In each step, we solve the time-independent SE to obtain ψ_1 and ψ_2. As long as the system is far from the avoided crossing, ψ_1 and ψ_2 remain almost constant, and hence, because of the derivatives, there is no change in C_1 and C_2 either. During the passage of the avoided crossing region, state 1 gradually obtains the character of state 2 at $t = 0$. After the passage, the character of the two states has swapped compared to the situation for $t = 0$. In the beginning, it is the term with C_1 in the second equation that is most important. Particularly, if E_1 and E_2 are very different in the avoided crossing region (a wide gap), there is a large contribution.

Furthermore, if the systems are moving slowly through the avoided crossing region, the integral gets a larger value than if the system moves fast. We may replace the derivative with respect to t with a derivative with respect to Q:

$$\frac{\partial}{\partial t} = \frac{\partial}{\partial Q}\frac{\partial Q}{\partial t} = \frac{\partial}{\partial Q}v, \qquad (7.21)$$

and rewrite Equation 7.20 as

$$\frac{\partial C_1}{\partial Q} = -C_2\left\langle \psi_1 \left| \frac{\partial \psi_2}{\partial Q} \right\rangle \exp\left[-\frac{i}{v\hbar}\int_0^Q (E_2 - E_1)dQ' \right]$$

$$\frac{\partial C_2}{\partial Q} = -C_1\left\langle \psi_2 \left| \frac{\partial \psi_1}{\partial Q} \right\rangle \exp\left[-\frac{i}{v\hbar}\int_0^Q (E_1 - E_2)dQ' \right].$$

$$(7.22)$$

It is now seen more clearly that if $E_2 - E_1$ is small in the avoided crossing region, or if the velocity v is large, the increase of coefficient C_2 is small. The system tends to continue through the avoided crossing region with a very slight change in the wave function.

7.2.3 Electron Transfer Dynamics

We will make a very simple application to an ET system. We assume that there are four protons, H_1, H_2, H_3, and H_4, and three electrons in the system, organized as in Figure 7.4. The perpendicular distance R_0 is quite large compared to the distances

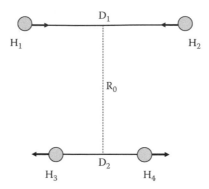

FIGURE 7.4 $H_2H_2^+$ ET system.

between the upper two and lower two protons. Hence, geometry optimization leads to one H_2^+ molecular ion ($D_1 \approx 1.4$ Å), as in Figure 7.4.

From this equilibrium geometry, at $t = 0$, the nuclei move as follows. The interatomic distance of the upper pair of protons (D_1) is decreased to the equilibrium distance for H_2. The interatomic distance of the lower pair (D_2) is increased to the equilibrium distance for H_2^+.

The original state at $t = 0$ is called the *precursor state*. The final state when the motion is completed with $D_1 = 0.7$ Å and $D_2 = 1.4$ Å is called the *successor state*. We integrate Equation 7.22 and read off the electronic density at the H_1 and H_3 atoms (which is the same as the density at the H_2 and H_4 atoms, respectively).

At the successor state geometry, the lowest energy of the system is obviously when the upper system has two electrons and the lower system a single electron.

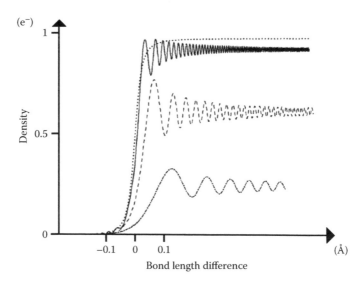

FIGURE 7.5 Probability density at H_1 as a function of D_2; for different velocities v and $R = 7$ B.

Figure 7.5 shows the quantum mechanical probability density at H_1 during the indicated nuclear motion. This assumes that R_0 is not too large. If R_0 is increased, less electron charge will move across.

The velocity is also important. The upper curve is obtained for a slow motion. The lower curve is obtained for a considerably faster motion.

The example illustrates how ET is induced by the nuclear motions. Electrons are coupled to the structure of molecules by the fact that bond lengths and bond angles are slightly different (sometimes very different) depending on the number of electrons in the molecule. Thus, if we bind together a molecule with its ion, the two parts of the composite molecule are slightly different. If the electron moves from one part to the other, all bond angles and bond lengths will be swapped.

7.2.4 LANDAU–ZENER APPROXIMATION

The Russian physicist L. Landau solved the problem of avoided crossing of Figure 7.3. The important result is the average values of the probability before and after the crossing. The PES are defined by

$$H_{11} = E_0 + \frac{dH_{11}}{dt}t \quad H_{22} = E_0 + \frac{dH_{22}}{dt}t \quad \frac{d(H_{11} - H_{22})}{dt} > 0 \qquad (7.23)$$

(dashed lines in Figure 7.3). At $t = 0$, we have $H_{11} = H_{22}$. There is almost always some interaction between two degenerate states, and this interaction splits the degeneracy. This can be simulated by a simple eigenvalue problem with the secular equation:

$$\begin{vmatrix} H_{11} - E & H_{12} \\ H_{12} & H_{22} - E \end{vmatrix} = 0, \qquad (7.24)$$

where the avoided crossing is $\Delta = 2H_{12}$. The solution is the full-drawn curves in Figure 7.3. Landau solved the wave equation for the case when the nuclei are moving by classical mechanics and under the simplifying assumptions that H_{11} and H_{22} are straight lines that cross each other (Equation 7.23). We will not give the full solution here. If the absolute value of the interaction matrix element $|H_{12}|$ is reasonably large and the system slowly passes the avoided crossing, one might suspect that the system should follow the lower PES. However, if the speed increases or $|H_{12}|$ decreases, one expects that the avoided crossing should be passed in a way that the states "do not know about each other." This is also the case. The result is as follows (the *Landau–Zener approximation*):

$$P = \exp\left(-\frac{2\pi \cdot H_{12}^2}{\hbar \left|\dfrac{d}{dt}(H_{11} - H_{22})\right|}\right). \qquad (7.25)$$

The probability to end up on the lower surface increases if $|H_{12}|^2 > \hbar d|H_{11} - H_{22}|/dt$.

7.3　TIME DEPENDENCE AS A PERTURBATION

7.3.1　TIME-DEPENDENT PERTURBATION THEORY

We introduce a set of time-independent state wave functions $\{\psi_n^0\}$, which are eigenfunctions of a time-independent Hamiltonian:

$$H_0\psi_n^0 = E_n^0\psi_n^0. \tag{7.26}$$

In the absence of time-dependent perturbations, Equation 7.5 holds for a solution of the time-dependent SE. If time dependence is introduced as a perturbation V, the time-independent state wave functions $\{\psi_n^0\}$ may still be used, but the coefficients $C_n(t)$ will be time dependent. $\Psi(t)$ satisfies the time-dependent SE:

$$i\hbar\frac{\partial\Psi}{\partial t} = \left(H_0 + V\right)\Psi(t) \Rightarrow i\hbar\sum_n \frac{dC_n}{dt}\exp\left(-iE_n^0 t/\hbar\right)\psi_n^0 + \sum_n C_n E_n^0 \exp\left(-iE_n^0 t/\hbar\right)\psi_n^0$$

$$= H_0\sum_n C_n \exp\left(-iE_n^0 t/\hbar\right)\psi_n^0 + V\sum_n C_n \exp\left(-iE_n^0 t/\hbar\right)\psi_n^0$$

$$\Rightarrow i\hbar\sum_n \frac{dC_n}{dt}\exp\left(-iE_n^0 t/\hbar\right)\psi_n^0 = V\sum_n C_n \exp\left(-iE_n^0 t/\hbar\right)V\psi_n^0. \tag{7.27}$$

Multiplying by ψ_n^{0*} and integrating, we obtain

$$i\hbar\frac{dC_m}{dt}\exp\left(-iE_m^0 t/\hbar\right) = \sum_n C_n \exp\left(-iE_n^0 t/\hbar\right)\left\langle\psi_m^0\left|V\right|\psi_n^0\right\rangle$$

$$\Rightarrow i\hbar\frac{dC_m}{dt} = \sum_n C_n \exp\left[-i\left(E_n^0 - E_m^0\right)t/\hbar\right]\left\langle\psi_m^0\left|V\right|\psi_n^0\right\rangle. \tag{7.28}$$

Equation 7.28 is a system of simultaneous linear homogeneous differential equations called the equation of motion for C_m. Note that the summation includes n = m. In matrix form, Equation 7.4 may be written as

$$i\hbar\frac{d}{dt}\begin{pmatrix} C_1 \\ C_2 \\ \vdots \\ \vdots \\ C_n \end{pmatrix} = \begin{pmatrix} V_{11} & V_{12}\exp(i\omega_{12}) & V_{13}\exp(i\omega_{13}) & \cdots & V_{1k}\exp(i\omega_{1k}) \\ V_{21}\exp(i\omega_{21}) & V_{22} & V_{23}\exp(i\omega_{23}) & \cdots & V_{2k}\exp(i\omega_{2k}) \\ \vdots & \vdots & \vdots & \cdots & \vdots \\ \vdots & \vdots & \vdots & \cdots & \vdots \\ V_{k1}\exp(i\omega_{k1}) & V_{k2}\exp(i\omega_{k2}) & V_{k3}\exp(i\omega_{k3}) & \cdots & V_{kk} \end{pmatrix}\begin{pmatrix} C_1 \\ C_2 \\ \vdots \\ \vdots \\ \vdots \end{pmatrix}, \tag{7.29}$$

where

$$\omega_{mn} = \frac{E_m^0 - E_n^0}{\hbar}. \tag{7.30}$$

One may now derive expressions for the coefficients $\{C_m\}$ for different situations. Assume, for instance, that the system is originally ($t = -\infty$) in a stationary state n of the time-independent Hamiltonian, so that

$$C_n(-\infty) = 1 \quad C_m(-\infty) = 0 \quad \text{for } m \neq n. \tag{7.31}$$

The derivatives at time $t = -\infty$ may be obtained by inserting Equation 7.30 into the right member of Equation 7.28:

$$i\hbar \frac{dC_m(t)}{dt} = \exp(i\omega_{mn}t)\langle \psi_m^0 | V | \psi_n^0 \rangle \quad \text{for } m \neq n. \tag{7.32}$$

Provided that the coefficients C_m remain small for $m \neq n$, we may obtain the expression for these coefficients by integrating Equation 7.32.

$$C_m(t) = -\frac{i}{\hbar} \int_{-\infty}^t \exp(i\omega_{mn}\tau)\langle \psi_m^0 | V | \psi_n^0 \rangle d\tau \quad \text{for } m \neq n \tag{7.33}$$

If the perturbation dies out after some time, the upper integration limit may be replaced by ∞. Equation 7.33 gives the probability amplitude for transition from state n to state m.

7.3.2 Decay Rates: Fermi Golden Rule

If the perturbation V does not depend on time, Equation 7.33 can be easily integrated. If the initial time is chosen as $t_0 = 0$, we obtain

$$C_m(t) = -\frac{i}{\hbar} \int_0^t \exp\left[\frac{i}{\hbar}\left(E_m^0 - E_k^0\right)\tau\right]\langle \psi_m^0 | V | \psi_n^0 \rangle d\tau = \frac{\langle \psi_m^0 | V | \psi_k^0 \rangle}{\hbar\omega_{mk}}\left[1 - \exp(i\omega_{mk}t)\right],$$

$$\tag{7.34}$$

where

$$\omega_{mk} = \frac{E_m^0 - E_k^0}{\hbar}. \tag{7.35}$$

This implies

$$|C_m(t)|^2 = 2 \frac{|\langle m|V|k\rangle|^2}{\hbar^2} \frac{(1 - \cos\omega_{mk}t)}{\omega_{mk}^2} \quad (k = \psi_k^0). \tag{7.36}$$

Equation 7.36 gives the probability to find the system in the state m if it was in the state k for t = 0.

Let us now assume that there is a continuum of energy levels of density $\rho(E)$. Such a continuum may arise due to the presence of solvent vibrational levels or in the gas phase due to rotations and kinetic energy. As an example, we may consider decay from a state in the helium atom where both electrons are in the 2s orbital. This $(2s)^2$ state may exist for some time despite the fact that its energy is higher than the energy of the He^+ ground state. This means that there are state energies of the type (1s ∞s) where one electron is free with a continuum of energy levels. We do not need to know the exact expression for the perturbation, but assume that $\langle m |V| k\rangle$ is a constant. The continuum of (1s ∞s) states interact with the $(2s)^2$ state. The density of states curve will remain more or less the same, but the amount of 2s character in the states will have a maximum at the energy of the $(2s)^2$ state. Alternatively, one may say that the $(2s)^2$ state acquires a certain width. This is the time-independent picture.

For the time-dependent picture, Equation 7.36 serves quite well. The probability for a certain continuum state, $|C_m(t)|^2$, increases with time. This means that the probability of a state of type (1s ∞s) also increases with time. The latter state is a so-called Auger state. Thus, the $(2s)^2$ state decays by moving one electron to 1s and ejecting the other. The total energy of the states before and after is the same.

The largest contribution to the decay is for small values of ω_{mk}. For a given time t in Equation 7.36, the function $(1 - \cos\omega_{mk}t)/\omega_{mk}^2$ goes to t^2 for $\omega_{mk} \to 0$. We thus obtain

$$\lim_{\omega_{mk}\to 0} |C_m(t)|^2 = \frac{\langle m|V|k\rangle}{\hbar^2} t^2 \quad \text{for } m \neq k. \tag{7.37}$$

For a large time, there is an oscillatory behavior that leaves positive and negative contributions even for small ω_{mk}. Let us assume a density of states $\rho(E)$, which is constant in a region around E_k^0. The rate of decrease of the total probability of the state Ψ_k^0, which was equal to unity at time t = 0, may be obtained by integrating over the range of continuous energy levels $(E = \hbar\omega)$:

$$\int_{-\infty}^{\infty} |C_m(t)|^2 \rho(E)dE = 2 \frac{\langle m|V|k\rangle^2}{\hbar} \rho(E_k^0) \int_{-\infty}^{\infty} \frac{1 - \cos\omega t}{\omega^2} d\omega. \tag{7.38}$$

The *decay rate* W is the derivative of this function with respect to time:

$$W = \frac{d}{dt} 2 \frac{\langle m|V|k \rangle^2}{\hbar} \rho(E_k^0) \int_{-\infty}^{\infty} \frac{1-\cos\omega t}{\omega^2} d\omega = 2 \frac{\langle m|V|k \rangle^2}{\hbar} \rho(E_k^0) \int_{-\infty}^{\infty} \frac{\sin\omega t}{\omega} d\omega$$

$$(7.39)$$

$$= 2\pi \frac{\langle m|V|k \rangle^2}{\hbar} \rho(E_k^0).$$

Equation 7.39 is *Fermi's golden rule*.

Since W is constant, the decay law is exponential: If P_k is the probability to remain in the original state, then

$$P_k = |C_k|^2; \quad P_k = 1 \text{ for } t = 0; \quad dP_k = WP_k dt. \tag{7.40}$$

From this equation follows that

$$P_k = \exp(-Wt) = \exp(-t/\tau), \tag{7.41}$$

where τ is equal to the lifetime.

We must keep in mind that the decay law is derived under restrictive conditions. One is that there is a continuum of states in the vicinity of the one that is occupied by the system (state k). It is required that the states other than k, to which we have a continuous decay, do not decay back to state k. In the case of Auger or fluorescence decay, this last requirement is fulfilled. In the case of electron resonance between two possible locations, this is not satisfied.

To calculate the coefficients C_m of the "drain" states, we write Equation 7.29 as

$$i\hbar \frac{dC_m}{dt} = \langle m|V|k \rangle C_k \exp(i\omega_{mk}t) \quad \text{for } m \neq k. \tag{7.42}$$

Compared to Equation 7.1, we permit C_k to decrease below unity, but the change in C_m is still considered to depend on the decrease of C_k only. C_m will not be affected by return from the drain states. Equation 7.42 is integrated to

$$C_m(t) = -\frac{i}{\hbar} \langle m|V|k \rangle \int_0^t C_k(t') \exp(i\omega_{mk}t') dt' \quad \text{for } m \neq k. \tag{7.43}$$

For t = 0, we have $C_k = 1$ and, as suggested by Equations 7.39 and 7.41, we may try the substitution:

$$C_k(t) = \exp\left(-\frac{W}{2}t\right). \tag{7.44}$$

Equation 7.44 is inserted into Equation 7.41. We obtain the coefficients:

$$C_m(t) = -\frac{i}{\hbar}\langle m|V|k\rangle \int_0^t \exp\left[-\frac{i}{\hbar}\left(E_k^0 - E_m^0 + i\hbar\frac{W}{2}\right)t'\right]dt'$$

$$= \langle m|V|k\rangle \frac{1 - \exp\left(-\frac{W}{2}t\right)\cdot\exp\left[-\frac{i}{\hbar}\left(E_k^0 - E_m^0\right)t\right]}{E_m^0 - E_k^0 + i\hbar\frac{W}{2}}. \tag{7.45}$$

It follows that the probability that the system decays into state m is

$$|C_m(t)|^2 = \langle m|V|k\rangle^2 \frac{1 - 2\exp\left(-\frac{W}{2}t\right)\cos\left[-\frac{i}{\hbar}\left(E_k^0 - E_m^0\right)t\right] + \exp\left(-\frac{W}{\hbar}t\right)}{\left(E_m^0 - E_k^0\right)^2 + \frac{(\hbar W)^2}{4}}. \tag{7.46}$$

With $\Gamma = \hbar W$ and $V_{mk} = \langle m|V|k\rangle$, we obtain the probability amplitude, C_m, for the appearance of a state with energy E_m. For $t \to \infty$, Equation 7.46 tends to

$$I = |C_m(\infty)|^2 = \frac{|V_{km}|^2}{\left(E_m^0 - E_k^0\right)^2 + \frac{\Gamma^2}{4}}. \tag{7.47}$$

This is a Lorentzian function that has a maximum at $E_m^0 = E_k^0$ and decreases in a bell-shaped manner. The maximum value is

$$I_{max} = \frac{4|V_{mk}|^2}{\Gamma^2}. \tag{7.48}$$

Half the maximum value is obtained for $E_m^0 = E_k^0 \pm \Gamma/2$. The width of the bell at half maximum is thus equal to Γ.

    The theory presented here may be applied to the absorption process. At $t = 0$, there is an excitation ("a photon is absorbed") to the unstable state k. However, not only does this sharp state appear, but also an infinite number of states distributed on the energy scale, as shown in Equation 7.47. The intensity is proportional to I. The faster the decay rate, the larger the width Γ of absorption or emission lines. In various spectroscopies, for example, NMR, the line widths may be used to probe the interactions with the nuclei.

The decay rate assumed in Equation 7.41 corresponds to a half life of $\tau = (\ln 2)/W$. We thus have the relation between the energy width Γ and τ:

$$\Delta E \cdot \Delta t = \Gamma \tau = \ln 2 \, \frac{\Gamma}{W} = \ln 2 \cdot \hbar. \tag{7.49}$$

The broadening we have calculated is close to the smallest possible broadening of an emission or absorption line consistent with the time–energy uncertainty principle and is called the *natural line width*.

8 Chemical Kinetics

8.1 INTRODUCTION

We will first deal with the rate of chemical reaction and how it is measured. The rate of even a simple reaction depends on the rate of several elementary reaction steps and this will be discussed in Section 8.3. In Section 8.4, we will treat the special case of reactions where the amount or concentration of the reactants is maintained at a constant level. This applies to a great number of reactions in nature. Photochemistry is another case where the amount of substance and the radiation density is constant.

8.2 RATE OF CHEMICAL REACTIONS

Rates of chemical reactions differ by many orders of magnitude. Even explosions may appear as slow compared to the primary absorption steps of several photochemical reactions. On the other hand, there are reactions that are much slower than explosions, for example, fermentation reactions. Slowest are probably geochemical reactions. When the earth went from fluid to solid, the reactions still taking place in the crust became so slow that the thermodynamic equilibrium was of no importance, since it will almost certainly be achieved. In the present section, the *rate* of a chemical reaction will be defined and ways to measure it will be mentioned.

8.2.1 REVERSIBLE AND IRREVERSIBLE REACTIONS

If a reaction at equilibrium has proceeded so far that the reactants can no longer be found in the reaction product mixture, it may be considered as finished for all practical purposes. The reaction is said to be *irreversible*. This happens also if the reaction products are dispersed during the reaction, as is the case with explosions. A *reversible* reaction does not proceed completely to the side of the reaction products, but to some thermodynamic equilibrium. This does not mean that the reaction is over, only that the number of molecules that react is the same as the number that backreact.

The simplest of all reactions is irreversible unimolecular decay. The reaction is

$$A \rightarrow B. \tag{8.1}$$

In kinetics, one measures in some way the concentration of A ([A]) at different times. It is reasonable that the change of concentration of A that takes place per time unit is proportional to [A]. Hence, the following equation must hold:

$$\frac{d[A]}{dt} = -k[A]. \tag{8.2}$$

This differential equation has the solution $[A] = [A]_0 \exp(-kt)$, where $[A]_0$ is the original concentration at $t = 0$, which is easy to check by insertion. The rate constant k obviously has the dimension "amount per time unit" (moles per second).

At a chemical reaction, the system moves from a local free energy minimum of the reactants to another, lower free energy minimum of the products (Figure 8.1). In a slightly more advanced picture, there is a valley on the left, a pass at the top of the barrier in the middle of the figure, and another valley on the right. In a photochemical reaction, the dynamics takes place on an excited potential energy surface (PES).

The *reaction path* connects the two minima. In a simple elementary reaction, the system follows an energy valley to the saddle point and then another valley to the product minimum (Figure 8.1).

8.2.2 ACTIVATION ENERGY

Kinetic energy is needed to get across the thermal barrier and this energy is called the *activation energy*. The latter is taken from the thermal motion in the reactant mixture or from electromagnetic (EM) radiation (see Chapter 13).

Explosion of gunpowder and dynamite takes place along an energy surface similar to that in Figure 8.1. Despite the fact that the energy of the products is considerably lower than that of the reactants, no reaction takes place. The reaction may be started using high pressure or heating. In the latter case, sufficient amounts of energy are added to enable a number of molecules to get across the barrier. During the reaction, enough energy is produced to continue the reaction until the reactants are finished, which perhaps takes 0.001–0.1 sec at an explosion.

In which way does the reaction rate depend on temperature? In general, most of the molecules do not have enough kinetic energy to cross the barrier. As a crude

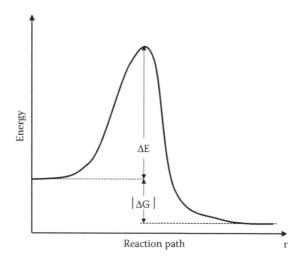

FIGURE 8.1 PES for an elementary reaction. Free energy ΔG and activation energy ΔE are indicated.

approximation, one may say that the rate of reaction is proportional to the number of molecules that have succeeded in crossing the barrier since they have the energy $E > E_0$. We should at least integrate the Boltzmann distribution from E_0 and upward as $E \to \infty$. Since the number decreases exponentially when the energy increases, it suffices to keep proportionality only to the number of molecules only at the top of the barrier. Boltzmann's law gives this number as

$$k = C \exp\left(-\frac{E_0}{k_B T}\right). \tag{8.3}$$

If k is measured and ln k plotted as a function of $1/T$, a straight line will be obtained with the slope $-E_0/k_B$ and y intercept C for $T \to \infty$. In 1889, the Swedish physical chemist Svante Arrhenius wrote down Equation 8.3. E_0 is called the activation barrier or the activation energy.

8.2.3 ELEMENTARY REACTIONS

By a mechanism for a chemical reaction, one means essentially, how the reaction is composed in terms of elementary reactions. Modern chemistry deals with the rate of elementary reaction steps. A total reaction such as

$$A_2 + B_2 \to 2AB, \tag{8.4}$$

may be composed of a number of elementary reaction steps (Table 8.1).

The reaction rate for formation of A is

$$\frac{d[A]}{dt} = 2k_2[A_2] - 2k_{-2}[A]^2 + k_4[B][A_2]$$
$$- k_{-4}[AB][A] - k_5[A][B_2] + k_{-5}[AB][B]. \tag{8.5}$$

Similar differential equations may be written for $d[B_2]/dt$ and $d[AB]/dt$. If all rate constants are known, theoretically or experimentally, it is a simple task to calculate all concentrations at all times by simply integrating equations of type Equation 8.5

TABLE 8.1
Possible Elementary reactions and Rate constants in the Total Reaction (8.4)

Elementary Reaction	Forward Reaction	Backward Reaction
$A_2 + B_2 \leftrightarrows 2AB$	k_1	k_{-1}
$A_2 \leftrightarrows 2A$	k_2	k_{-2}
$B_2 \leftrightarrows 2B$	k_3	k_{-3}
$B + A_2 \leftrightarrows AB + A$	k_4	k_{-4}
$A + B_2 \leftrightarrows AB + B$	k_5	k_{-5}

from boundary conditions at $t = 0$ with sufficiently small time steps, dt. Of course, the elementary rate constants are not known from the beginning, but have been obtained after much experimental work, or are still unknown. Furthermore we want to express the whole reaction with the help of mathematical equations rather than using computers where the insight is in programs that are hidden from the users. It is necessary to carry out experiments under different boundary conditions, neglect some very slow reactions, use equilibrium conditions, irreversibility, etc., to obtain the final mechanism of the reaction. Finally, these reactions may be compared to those that are calculated from molecular theory, in order to obtain a detailed understanding of how the reaction proceeds.

The rate constant for the chemical reaction

$$v_A A + v_B B + \cdots \rightarrow v_P P + v_Q Q + \cdots \tag{8.6}$$

may be expressed with the help of $-d[A]/dt$, $-d[B]/dt$, $d[P]/dt$, or $d[Q]/dt$. If the reaction (8.6) is a simple reaction step, it is easy to realize that the following equation holds:

$$-\frac{1}{v_A}\frac{d[A]}{dt} = -\frac{1}{v_B}\frac{d[B]}{dt} = \frac{1}{v_P}\frac{d[P]}{dt} = \frac{1}{v_Q}\frac{d[Q]}{dt}. \tag{8.7}$$

It is thus necessary to specify the concept reaction rate, for example, whether we mean the rate of consumption for A or B, or the rate of formation for P or Q.

For the total reaction, which consists of many elementary reaction steps, the rate usually cannot be expressed with the help of stoichiometric coefficients, as in Equation 8.7. Normally, however, the rate of reaction for irreversible reactions is proportional to the powers of the concentration of one or many of the reacting components:

$$-\frac{d[A]}{dt} = k[A]^{\alpha_A}[B]^{\alpha_B}[C]^{\alpha_C}. \tag{8.8}$$

$\alpha_A + \alpha_B + \alpha_C$ is called the *order* of the reaction.

It is common for some of the products to increase the reaction rate. Such substances are called *catalysts* (or *inhibitors* if the rate is slowed down), and are usually not part of the total reaction, since their concentration is constant. If the product formed speeds up the reaction itself, we are talking about an *autocatalyst*. When the reaction is slowed down, we talk about *autoinhibition*.

Equations that express a disappearance or formation rate with the help of rate constants and concentrations of reactants and products are called rate equations. As examples, we choose two reactions where one would expect the same rate dependence on concentration, but where there is a great difference:

$$H_2 + I_2 \leftrightarrow 2HI \qquad \frac{1}{2}\frac{d[HI]}{dt} = k[H_2][I_2] - k'[HI]^2, \tag{8.9}$$

$$H_2 + Br_2 \leftrightarrow 2HBr \quad \frac{1}{2}\frac{d[HBr]}{dt} = \frac{k[H_2][Br_2]^{1/2}}{1+k'[HBr]/[Br_2]}. \quad (8.10)$$

The reaction of Equation 8.9 is simple since [HI] changes as the difference in rate between the forward reaction and back reaction. This ought to be the usual case for a reversible reaction where $k \neq k'$. There is no reason to assume the existence of elementary reactions in this case. When the change in [HI] is equal to zero, the right member is also equal to zero, and equilibrium has been achieved with the equilibrium constant $K = k/k'$ (see below). Even though the total reaction appears to have ceased, equal amounts of products are formed as the reactants are consumed in the back reaction.

The reaction in Equation 8.10 is much more complex, however; so complicated that it once required decades of experimental and theoretical work to nail down the mechanism. Please notice that Equation 8.10 has no direct connection with the stoichiometric components. Still it depends, of course, on the concentrations of the reactants. There are reactions that do not depend on the concentrations (zeroth-order reactions; see below).

8.2.4 RATE MEASUREMENTS

Clearly, we need to find ways to measure the concentration of reactants and products at different times to be able to correctly determine the rate constant and make conclusions on the mechanism. This may be done in different ways:

1. Terminate the reaction by precipitating one or more of the reactants and measure how much has already been consumed or formed among the products. For natural reasons, this can only be done for slow reactions.
2. Monitor *pressure changes*: The number of molecules in the gas phase may be different among reactants and products, for example, in the reaction $2N_2O_5 \rightarrow 4NO_2 + O_2$, where, of course, the pressure increases at a constant volume as the reaction proceeds.
3. *Spectrophotometry*: In the reaction: $H_2 + Br_2 \rightarrow 2HBr$, only Br_2 is colored (has absorption in the visible range). The intensity is measured at different times and this can be related to the concentration of HBr.
4. *Conductivity*: $(CH_3)_3CCl$ is an insulator, but if it is dissolved in water, we have the reaction: $(CH_3)_3CCl(aq) + 2H_2O(l) \rightarrow (CH_3)_3COH(aq) + H_3O^+ + Cl^-$, where ions are formed. Conductivity grows with ion concentration.
5. *Flash photolysis*: The principles are the same as in spectrophotometry (3), but here complicated laser systems and electronics are used. A laser pulse is divided in two parts. One is delayed a given time, and it becomes possible to measure what has happened in the part that passed the probe in the meantime when the two parts unite (see Chapter 12). With this technique, it is possible to measure times that are shorter than a picosecond (10^{-12} sec).
6. Other methods are mass spectrometry, gas chromotography, nuclear magnetic resonance (NMR), and electron paramagnetic (spin) resonance (EPR, ESR).

8.3 INTEGRATED RATE EQUATIONS

8.3.1 IRREVERSIBLE REACTIONS OF FIRST ORDER

By irreversibility, we mean, of course, that the back reaction can be neglected. In Equation 8.8, $\alpha_A = 1$ while all other α's are equal to zero. The reaction is thus

$$\frac{d[A]}{dt} = -k[A] \text{ or } \frac{d\left([A]_0 - x\right)}{dt} = -k\left([A]_0 - x\right). \tag{8.11}$$

Here, x stands for the decrease of the concentration of A. This equation may be integrated to

$$[A] = [A]_0 \exp(-kt) \text{ or } \left([A]_0 - x\right) = [A]_0 \exp(-kt). \tag{8.12}$$

This is realized if we rewrite Equation 8.11:

$$\frac{d\left([A]_0 - x\right)}{[A]_0 - x} = kdt \Rightarrow \ln\left([A]_0 - x\right) - \ln[A]_0 = -kt. \tag{8.13}$$

Thus, k may be obtained by measuring the concentration of reactants in time intervals and plotting the logarithm as a function of time. A linear plot is obtained where k is the slope.

The time it takes for the concentration to decrease to half of the original concentration is called the *half life time* and is denoted $t_{1/2}$. By substituting $[A] = [A]_0/2$ in Equation 8.13, we find that

$$t_{1/2} = \ln 2/k. \tag{8.14}$$

More frequently the *average life time*, t_{av}, defined by Equation 8.15, is used:

$$t_{av} = \int_0^\infty te^{-kt}\,dt \bigg/ \int_0^\infty e^{-kt}dt = 1/k \tag{8.15}$$

The average life time is simply the reciprocal of the rate constant.

8.3.2 IRREVERSIBLE REACTIONS OF SECOND ORDER

We first consider the case

$$-\frac{d[A]}{dt} = k[A]^2. \tag{8.16}$$

The rate constant k has the dimension $M^{-1}s^{-1}$.

Equation 8.16 is rewritten as

$$-\frac{d[A]}{[A]^2} = kdt, \qquad (8.17)$$

which is integrated to

$$[A]^{-1} - [A]_0^{-1} = kt, \qquad (8.18)$$

or, if we want to use the consumed concentration of A, called x,

$$\left([A]_0 - x\right)^{-1} - [A]_0^{-1} = kt. \qquad (8.19)$$

The half-time is, in this case of an irreversible reaction of second order,

$$\left([A]_0/2\right)^{-1} - [A]_0^{-1} = kt_{1/2} \Rightarrow t_{1/2} = 1/k[A]_0. \qquad (8.20)$$

We now assume instead a second-order reaction of the type $v_A A + v_B B \rightarrow$, where

$$-\frac{d[A]}{dt} = k[A][B]. \qquad (8.21)$$

In order to integrate Equation 8.21, we have to find a connection between the concentrations [A] and [B] to integrate the differential Equation 8.21 with the help of stoichiometry. Evidently, if $[A] = [A]_0 - v_A x$ holds at time t, $[B] = [B]_0 - v_B x$ at the same time t. We may then write Equation 8.21 as

$$-\frac{d[A]}{dt} = -\frac{d\left([A]_0 - v_A x\right)}{dt} = v_A \frac{dx}{dt} = k\left([A]_0 - v_A x\right)\left([B]_0 - v_B x\right). \qquad (8.22)$$

We must now distinguish two cases. Assume first that A and B are mixed in stoichiometric ratios: $[A]_0/v_A = [B]_0/v_B$. We may then rewrite Equation 8.22 as

$$\frac{dx}{dt} = v_B k \left(\frac{[A]_0}{v_A} - x\right)^2, \qquad (8.23)$$

which may be integrated to

$$\int_0^x \frac{v_A^2 dx'}{\left([A]_0 - v_A x'\right)^2} = \int_0^t v_B k dt' \Rightarrow \frac{1}{[A]_0 - v_A x} - \frac{1}{[A]_0} = \frac{v_B}{v_A} kt$$

$$\qquad (8.24)$$

$$\Rightarrow \frac{1}{[A]} - \frac{1}{[A]_0} = \frac{v_B}{v_A} kt; \quad \text{if } \frac{[A]_0}{v_A} = \frac{[B]_0}{v_B}.$$

If we do not start with stoichiometric ratios, the integration is slightly more complicated. We first write Equation 8.22 as

$$v_B k dt = \left[\frac{1}{a-x} - \frac{1}{b-x} \right] \frac{dx}{b-a}; \quad \text{where } a = \frac{[A]_0}{v_A} \text{ and } b = \frac{[B]_0}{v_B}. \quad (8.25)$$

This may be integrated to obtain

$$\frac{kt}{v_A} = \frac{v_B kt}{v_A v_B} = \frac{1}{v_A v_B (b-a)} \left[-\ln|a-x| + \ln|b-x| \right]_0^x. \quad (8.26)$$

Thus,

$$\frac{kt}{v_A} = -\frac{1}{[B]_0 v_A - [A]_0 v_B} \ln \left| \frac{[B]_0 ([A]_0 - v_A x)}{[A]_0 ([B]_0 - v_B x)} \right|; \quad \text{if } \frac{[A]_0}{v_A} \neq \frac{[B]_0}{v_B}. \quad (8.27)$$

8.3.3 IRREVERSIBLE REACTIONS OF ZEROTH ORDER

A reaction is said to be of zeroth order if the rate is independent of the concentrations of all reactants:

$$-\frac{d[A]}{dt} = k. \quad (8.28)$$

Integration yields

$$[A]_0 - [A] = kt. \quad (8.29)$$

This simply means that the reactants are consumed with constant rate.

8.3.4 UNIMOLECULAR REVERSIBLE REACTION

If the reversible reaction

$$A \leftrightarrow B \quad (8.30)$$

is of first order, the rate of consumption of A becomes

$$-\frac{d[A]}{dt} = k_+ [A] - k_- [B]. \quad (8.31)$$

It is practical to use: $x = [A]_0 - [A] = [B] - [B]_0$ and then write Equation 8.31 as

$$-\frac{d[A]}{dt} = -\frac{d([A]_0 - x)}{dt} = \frac{dx}{dt}$$

$$= k_+ ([A]_0 - x) - k_- ([B]_0 + x) = (k_+ [A]_0 - k_- [B]_0) - (k_+ + k_-)x.$$

(8.32)

Integration of Equation 8.32 between times 0 and t yields

$$t = \int_0^x \frac{dx'}{k_+ [A]_0 - k_- [B]_0 - (k_+ + k_-)x'}$$

$$\Rightarrow \ln \frac{(k_+ [A]_0 - k_- [B]_0)}{(k_+ [A]_0 - k_- [B]_0) - (k_+ + k_-)x} = (k_+ + k_-)t.$$

(8.33)

$$\left(\text{Use: } \int \frac{1}{a + bx} dx = \frac{1}{b} \ln(a + bx). \right)$$

At equilibrium, $dx/dt = 0$ and $x = x_e$. From Equation 8.33 follows

$$(k_+ [A]_0 - k_- [B]_0) - (k_+ + k_-)x_e = 0.$$

(8.34)

If Equation 8.33 is inserted in Equation 8.32, we obtain

$$\ln \frac{x_e}{x_e - x} = (k_+ + k_-)t \Leftrightarrow x = x_e (1 - \exp[-(k_+ + k_-)t]).$$

(8.35)

If Equation 8.35 is rewritten as $x_e - x = x_e \cdot \exp[-(k_+ + k_-)t]$, it follows that a unimolecular reaction approaches equilibrium as a first-order reaction with the rate constant $k = k_+ + k_-$. If we determine x for the reaction (8.30) at a number of times and furthermore measure x_e at equilibrium, we may determine $k_+ + k_-$. The separate rate constants k_+ and k may be obtained subsequently by using the stoichiometric equilibrium constant and Equation 8.31:

$$K = \frac{[B]_e}{[A]_e} = \frac{[B]_0 + x_e}{[A]_0 - x_e} = \frac{k_+}{k_-}.$$

(8.36)

According to Equation 8.36, the stoichiometric equilibrium constant for a reversible, unimolecular reaction is given by the quotient between the rate constants for the forward reaction and the back reaction. The results are general.

8.4   CONSECUTIVE REACTIONS

8.4.1   RATE DERIVATION

Suppose that a substance A reacts and forms a product B, which in turn continues to react and finally forms a stable end product C. The reaction consists of two consecutive (irreversible) elementary reactions:

$$A \xrightarrow{k_1} B \xrightarrow{k_2} C. \tag{8.37}$$

This type of reaction occurs at radioactive decay. An example from chemistry is the decay of acetone at high temperatures:

$$CH_3COCH_3 \rightarrow CH_2CO + CH_4$$
$$2CH_2CO \rightarrow 2CO + C_2H_4. \tag{8.38}$$

For the elementary reactions given by Equation 8.37, we may formulate the rate equations as

$$\frac{d[A]}{dt} = -k_1[A], \tag{8.39}$$

$$\frac{d[B]}{dt} = k_1[A] - k_2[B], \tag{8.40}$$

$$\frac{d[C]}{dt} = k_2[B]. \tag{8.41}$$

We assume that only A is present at the start of the reaction at $t = 0$, with concentration $[A]_0$. Equation 8.39 integrates readily to

$$[A] = [A]_0 \exp(-k_1 t). \tag{8.42}$$

Equation 8.40 is rewritten as

$$\frac{d[B]}{dt} + k_2[B] = k_1[A]_0 \exp(-k_1 t). \tag{8.43}$$

This equation is multiplied by $\exp(k_2 t)$:

$$\frac{d[B]}{dt} \exp(k_2 t) + k_2 \exp(k_2 t) \cdot [B] = k_1[A]_0 \exp((k_2 - k_1)t). \tag{8.44}$$

Since the left member is the derivative of $[B] \cdot \exp(k_2 t)$, Equation 8.44 may be readily integrated:

$$[B]\exp(k_2 t) = [A]_0 \frac{k_1}{k_2 - k_1}\left[\exp(k_2 - k_1)t - 1\right]. \tag{8.45}$$

The integration constant was chosen in order to satisfy $[B] = 0$ for $t = 0$. Equation 8.45 is finally multiplied by $\exp(-k_2 t)$ to obtain

$$[B] = [A]_0 \frac{k_1}{k_1 - k_2}\left[\exp(-k_2 t) - \exp(-k_1 t)\right]. \tag{8.46}$$

To obtain [C], we use the boundary condition that [B] and [C] are equal to zero at $t = 0$ and hence, at all times:

$$[A] + [B] + [C] = [A]_0. \tag{8.47}$$

From Equations 8.42, 8.45, and 8.47, it follows that

$$[C] = [A]_0 \left[1 + \frac{k_2 \exp(-k_1 t) - k_1 \exp(-k_2 t)}{k_1 - k_2}\right]. \tag{8.48}$$

The solution is schematically given in Figure 8.2 in a typical case.

If $[A]_0$ is obtained from Equation 8.42 and inserted in Equation 8.46, we obtain

$$\frac{[B]}{[A]} = \frac{k_1}{k_2 - k_1}\left[1 - \exp\left[-(k_2 - k_1)t\right]\right]. \tag{8.49}$$

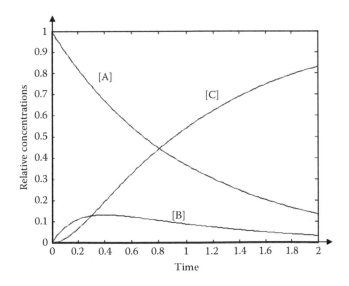

FIGURE 8.2 Concentrations as a function of time for consecutive reactions with $k_1/k_2 = 0.2$.

From this equation, it is clear that if $k_2 \gg k_1$, the exponential term is negligible. Then it also holds that $[B] \approx k_1[A]/k_2$ and $d[B]/dt \approx (k_1/k_2)d[A]/dt \approx 0$. Furthermore, if $k_1 = k_2$ and the exponent is expanded in a Taylor series, we find that $k_2 - k_1$ may be eliminated. It follows that $[B] = [A]k_1t$.

8.4.2 STEADY STATE ASSUMPTION

The result of Section 8.4.1 may be generalized. If the rate constant for the initial step is small compared to the rate of the subsequent steps, the concentration of the intermediate products is similar to the one found for B above. A *steady state* will appear in the concentrations of the intermediary products. The time derivatives are so small that they may be assumed equal to zero. This is a useful approximation in order to solve the complicated kinetics of many photochemical and biological reactions.

9 Proton Transfer

9.1 INTRODUCTION

Protons bind with an ordinary chemical bond particularly to the atoms N, O, and F and form molecules such as NH_3, H_2O, and HF. The bound protons may additionally bind to N, O, and F in neighboring molecules, but with a weaker bond, called a *hydrogen bond* (Figure 9.1). Hydrogen bonds are responsible for the remarkable molecular ordering in *water* and the structure of several polymers and are the driving force for the base pairing in DNA.

Reactions such as $NH_4^+ + H_2O \rightarrow NH_3 + H_3O^+$ and $NH_3 + H_2O \rightarrow NH_4^+ + OH^-$ are referred to as *proton transfer* (PT) reactions. In a PT reaction, the proton is moved between heavier atoms (Figure 9.1). Protons and PT reactions are behind acid–base equilibria and are referred to as *protolysis*. Protolysis and *rates* of PT reactions are concepts of great importance in life processes, particularly in the transport of energy within the cell membrane.

9.1.1 Acid–Base Concept of Brønsted

In 1923, a theory for acids and bases was developed by two Danish chemists, Johannes Brønsted and Niels Bjerrum, and independently by the English chemist Martin Lowry. The chemical equilibrium between acids and bases is given by

$$\text{Acid}\left(\text{HA}\right) = \text{donor of protons: } HA \rightleftarrows H^+ + A^-, \tag{9.1}$$

$$\text{Base}\left(\text{B}\right) = \text{acceptor of protons: } B + H^+ \rightleftarrows BH^+. \tag{9.2}$$

Thus, HA and BH^+ are acids and B and A^- are bases. HCl is a strong acid in aqueous solution and NH_3 is a base. Protons cannot exist in a free state in a water solution, but are bonded to the water molecules. Thus, in an aqueous solution, Equation 9.1 has to include a water molecule

$$HA + H_2O \rightleftarrows A^- + H_3O^+ \text{ or } HA + OH^- \rightleftarrows A^- + H_2O. \tag{9.3}$$

Similarly, Equation 9.2 should be written as

$$B + H_3O^+ \rightleftarrows BH^+ + H_2O \text{ or } B + H_2O \rightleftarrows BH^+ + OH^-. \tag{9.4}$$

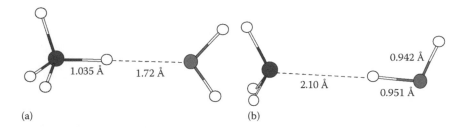

FIGURE 9.1 Hydrogen bond (dashed line) between NH_4^+ and H_2O (a) and between NH_3 and H_2O (b), as obtained in an *ab initio* calculation. In (a), all remaining NH bonds are 1.009 Å.

"Strong acid" (HA) and "strong base" (B) mean that the equilibria in Equations 9.3 and 9.4, respectively, are shifted far to the right. A^- is thus a weak base and BH^+ is a weak acid. In an aqueous solution, HCl is transformed into a hydronium ion (H_3O^+) and a chloride ion (Cl^-). HCl is a strong acid, while Cl^- is a weak base. In the same way, NH_3 is a strong base, while NH_4^+ is a weak acid. H_2O can act as an acid as well as a base, forming OH^- (hydroxyl ion) and H_3O^+, respectively. The equilibrium

$$2H_2O \rightleftarrows H_3O^+ + OH^- \tag{9.5}$$

has to be included in the protolysis in aqueous solutions.

9.1.2 Acid–Base Equilibrium in Water

The acid–base equilibrium in water may be written in the following way:

$$HA(aq) + H_2O(l) \rightleftarrows H_3O^+(aq) + A^-(aq). \tag{9.6}$$

Here, l stands for "liquid." Please notice that we have to add (aq) to indicate that the molecules are dissolved in an aqueous solution. Thus, a calculation of the reaction barriers in Equation 9.6 cannot be carried out in vacuum, but must include a great number of water molecules. The net reaction is a PT reaction, where the proton on HA has been transferred to a molecule of water and bonded with the help of the lone pairs of electrons of the oxygen atom in a hydrogen bond. Strong bases have easily available lone pairs of electrons (for example, NH_3).

In neutral water, $[H_3O^+] = [OH^-] \approx 10^{-7}$ mol/l. However, the equilibrium constant for the PT reaction (9.6) is best written with the help of *activities*. Unfortunately, it would go too far to discuss the very useful activity concept here. We refer to standard textbooks of physical chemistry. In the first approximation, activity may be replaced by the concentration [X] of the compound X. Therefore, we write the equilibrium constant as

$$K_a = \frac{[H_3O^+][A^-]}{[HA][H_2O]}. \tag{9.7}$$

In the acidity constant K_a for an acid HA in water, the concentration of water can be considered to be constant and can be calculated to be 55.56 mol/l according to (1000 g/l/18g/l). When this value is included into the acidity constant K_a, it takes the form

$$K_a = \frac{[H_3O^+] \cdot [A^-]}{[HA]}. \qquad (9.8)$$

This is usually expressed by the negative of the logarithm of K_a:

$$pK_a = -\log K_a \quad (\log = {}^{10}\log). \qquad (9.9)$$

Some typical values of acidity constants are given in Table 9.1. For example, pK_a for the ammonium ion $= -\log(5.6 \times 10^{-10}) = 9.25$. The larger the value of pK_a, the weaker is the acid.

Acidity constants are measures of acidity. Strong acids have acidity constants larger than unity. The strongest acid is hydroiodic acid. The concentration of HI is extremely small in an aqueous solution.

For water, the *autoprotolysis constant* is used:

$$K_w = [H_3O^+] \cdot [OH^-] \text{ and } pK_w = -\log K_w. \qquad (9.10)$$

At 25°C, $K_w = \approx 10^{-14}$. The concentration of H_2O in pure water is 55.6 mol/l, K_a for water is 1.8×10^{-16}, hence $pK_a = -15.75$.

Since the molar concentrations of the hydroxyl anion OH^- and the hydronium ion H_3O^+ have to be equally large in pure water, we have $[H_3O^+] = 10^{-7}$ (at room temperature, 298 K). The following notation is commonly used:

TABLE 9.1
Examples of Acidity Constants

Acid	Symbol	K_a
Hydroiodic acid	HI	10^{11}
Hydrochloric acid	HCl	10^7
Sulfuric acid	H_2SO_4	10^2
Sulfonic acid	$-SO_3H$	0.2
Hydrofluoric acid	HF	7.2×10^{-4}
Acetic acid	CH_3COOH	1.8×10^{-5}
Carbonic acid	H_2CO_3	4.3×10^{-7}
Ammonium ion	NH_4^+	5.6×10^{-10}
Phenol	C_6H_5OH	10^{-10}

$$pH = -\log[H_3O^+],\qquad(9.11)$$

$pH = 7$ in pure water. Please notice that a *more* acid solution has a *lower* pH value.

In a 1 M water solution of a strong acid, for example, HCl (hydrochloric acid), the concentration of hydronium ions and chloride ions is very close to 1 M, since the dissociation of the HCl molecule into H_3O^+ and Cl^- ions is almost complete. This means that the pH = 1 in this solution.

As an example, we let a water solution contain 1 M sodium acetate and add carboxylic acid in an amount that corresponds to 0.001 M H_3O^+ in the solution. We have the reaction: $Ac^- + H_3O^+ \rightleftarrows HAc + H_2O$ with $K_a = 1.8 \times 10^{-5}$ and the equilibrium relation (x = [HAc]; 1 − x = [Ac⁻]):

$$(1-x)(10^{-3} - x)/x = 1.8 \times 10^{-5}.\qquad(9.12)$$

If 1 − x is approximated by 1, we obtain: $x \approx 10^{-3}$. Thus, $[H_3O^+]$ will be close to zero, while $[HAc] = x = 10^{-3}$. This means that the solution is almost neutral, despite the fact that we have added acid corresponding to pH = 3 in pure water. The much higher pH value is due to the presence of sodium acetate. The latter compound is said to be a *buffer*. The solution of NaAc in water acts as a buffer ("bumper" in American English) that absorbs the acid shock.

9.1.3 PROTON AFFINITY

Proton affinity is the negative of the free energy gain if a proton is added to a base (A⁻ or B) in vacuum:

$$A^- + H^+ \rightarrow HA \quad \left(\Delta G^0 = -E_{pa}\right)$$
$$B + H^+ \rightarrow BH^+ \quad \left(\Delta G^0 = -E_{pa}\right).\qquad(9.13)$$

As far as known, proton affinities are always positive in vacuum (Table 9.2). Neutral and negative bases cannot be directly compared, due to the difference in charges.

The hydrogen negative ion and CH_3^- have very high proton affinities. The products formed (H_2 and CH_4) are very stable compounds and very weak acids (they are normally not referred to as acids at all). The negative bases with the lowest values of proton affinity in vacuum usually correspond to the strongest acids.

Proton affinity enthalpies in vacuum may be obtained by calculating the energy of A⁻ and AH separately (or B and BH⁺) and subtracting. It is not surprising that negatively charged ions have a higher proton affinity than neutral molecules in general.

TABLE 9.2

Proton Affinities in Vacuum for Some Neutral and Negative Bases

Neutral Base (B)	E_p (kJ/mol)	Neutral Base (B)	E_p (kJ/mol)	Negative Base (A^-)	E_p (kJ/mol)
He	178	CO_2	548	Cl^-	1395
O_2	422	CH_4	552	COO^-	1458
HF	490	HCl	564	F^-	1554
N_2	495	C_2H_6	601	H^-	1675
NH_3	880	CS_2	632	CH_3^-	1743
X_e	496	H_2O	712		

Proton affinity and pK_a measure the acidity of a molecule. Proton affinity depends only on the molecule itself, while pK_a values depend on the surrounding molecules. pK_a values may be defined for any solvent, and reveal how important the solvent is in practice. Although pK_a values are not unique for the molecule, they are more useful than proton affinities. This depends also on the fact that pK_a is related to the free energy, which depends on both enthalpy and entropy. The free energy difference may be obtained via the acidity constant, if we use the definition of Equation 9.7 together with Equation 5.88.

9.1.4 HYDRATION

Proton affinity differences are more meaningful and may be measured more easily than proton affinities themselves. When we deal with pK_a values, we are actually comparing the proton affinity differences between water and a base A^- (or between water and a base B). It is understood that the comparison is done in an aqueous solution.

Calculation of proton affinity in a solvent by quantum chemical methods is a very difficult problem, if at all solvable. Suppose two calculations are done, one where the proton is placed near A, and one where it is placed near one of the water molecules (Figure 9.2). The difference between the two ground state energies should be taken. In both calculations, the geometry should be optimized. But geometry optimization can only lead to the lowest enthalpy situation. The geometry of the other state is not well-defined.

9.2 HYDROGEN BONDING

The discovery of the hydrogen bond was not a historical event like, for example, the discovery of x-rays. Rather, it was gradually realized that hydrogen binds weakly to electronegative atoms other than those it is bonded with in normal chemical bonds (Latimer and Rodebush 1920). Often, quite strong intermolecular bonds are formed. The PT in bond has become a very important subject, since the processes of life rely on coupled electron and proton transfer.

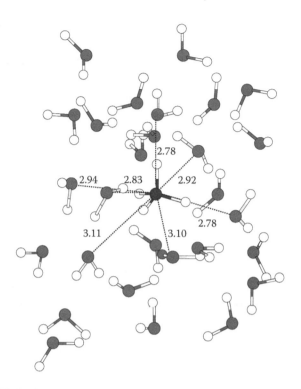

FIGURE 9.2 Hydration around the ammonium ion NH$_4^+$. The water molecules of the first solvent shell are connected to the nitrogen atom by dashed lines.

9.2.1 TYPICAL HYDROGEN BONDS

A proton between two "heavy atoms" (O, N, or F) usually forms one hydrogen bond and one ordinary bond. In rare cases, the hydrogen atom binds equally strongly to two electronegative atoms, with identical bond lengths.

Hydrogen bonds are important in condensed forms of water and, as is clear from the earlier sections of this chapter, are intimately connected to the concepts of acidity and basicity. Hydrogen bonds hold peptides together in proteins. The strands of DNA, carrying genetic information, are partly held together by hydrogen bonds, which stabilize the unique structure and make replication possible. Many other polymers, for example, carbohydrates like starch and cellulose, are also connected by hydrogen bonds. Examples of hydrogen bonds are shown in Figure 9.3.

Some left–right symmetric hydrogen bonds, for example, in [F–H–F]$^-$, possess a single equilibrium point. The bonding power is then distributed equally on the left and right. The strength of the bond is in between an ordinary covalent bond and a hydrogen bond.

Hydrogen bonding leads to high boiling points for NH$_3$, H$_2$O, and HF, if the comparison is made to the heavier analogs PH$_3$–AsH$_3$–SbH$_3$, H$_2$S–H$_2$Se–H$_2$Te, or to HCl–HBr–HI, respectively. For example, H$_2$S is a gas, while H$_2$O is a fluid at room temperature. The reason why atoms from the third and higher periods are engaged

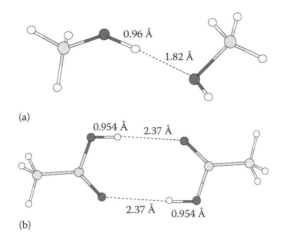

(a)

(b)

FIGURE 9.3 Typical hydrogen bonds: (a) methyl alcohol dimer and (b) acetic acid dimer.

less strongly in hydrogen bonds than corresponding atoms from the second period is quite obscure.

9.2.2 HYDROGEN BONDS IN PROTEINS

Proteins are polymers formed from peptides in condensation reactions. Condensation of two amino acids means that the OH group of carboxyl of one peptide and a hydrogen of the amino group of the other expel a water molecule and form a peptide bond (Figure 9.4). Further condensation reactions lead to the protein. The *sequence* of peptides is referred to as the *primary structure* of the protein. The 20 amino acids composing natural proteins are given in Table 9.3.

Natural amino acids are all α-amino acids, meaning that there is only one carbon atom (C_α) between the carboxyl group and the amino group of the peptide (Figure 9.4). C_α is part of the peptide chain and possesses a characteristic side group. The oligomer has $-NH_3^+$ at one end (N-terminus) and $-COO^-$ at the other end (C-terminus).

Hydrogen bonds form four units between the CO group and the NH group further along the peptide chain (Figure 9.5). These hydrogen bonds determine the *secondary structure* of the protein (α-helix or β-sheet). Figure 9.5 shows an α-helix.

The pK_a value of the N-terminus is ≈7.8 and the pK_a value of the C-terminus ≈3.7. The pK_a values of the side groups, given in Table 9.3, vary between 3.9 (Asp) and 12.0 (Cys).

Hydrogen bonds also form between side groups. Figure 9.6 shows a chain of hydrogen bonds (dotted) between side groups that are part of the structure of the D1 protein of photosystem II. At the Mn complex, water is oxidized to oxygen molecules. For each oxygen molecule formed, four electrons are given away and sucked into the so-called reaction center (not shown) via the tyrosine side group on the left. Each electron is localized for some time on this tyr 161 side group (the number refers to the number in the sequence of peptides). Very likely, its motion is correlated with PTs along the hydrogen bonding chain. Notice that the side groups involved belong

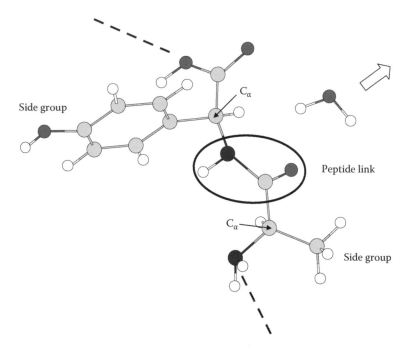

FIGURE 9.4 **(See color insert)** Formation of a dipeptide from two amino acids, tyrosine and alanine.

to the peptides that are many units away from each other in the peptide chain. The distance between the oxygen and nitrogen atoms along the hydrogen bond chain is between 2.6 and 2.8 Å.

9.2.3 STRENGTH OF A HYDROGEN BOND

The enthalpy gain when the hydrogen bond is formed from single molecules is defined as the strength of the hydrogen bond. Usually, the strength is between 5 and 30 kJ/mol. This is weaker than covalent or ionic bonds, but stronger than other distant interactions, such as van der Waals forces and London forces. In water, the boiling point is much higher than the boiling point of ammonia and hydrogen sulfide, and this is due to the strong hydrogen bonds between the water molecules.

In living systems, hydrogen bonds are very important. In DNA, there are strong bonds between the bases, which help keep the strands together. Strong hydrogen bonds determine the secondary structure of proteins, sugar, starch, and cellulose.

9.2.4 POTENTIAL ENERGY SURFACE

The potential energy surface (PES) is obtained in the Born–Oppenheimer approximations. We have to include at least two "heavy atoms" (N, O, F in discussions of

TABLE 9.3
Twenty Common Amino Acids in Natural Proteins

Peptide	ab. 1	ab. 2	Chemical Formula	Acidity Constants (pK$_a$)
Alanine	Ala	A	$NH_2CH(CH_3)COOH$	$pK_{COOH} = 2.35$; $pK_{NH_2} = 9.87$
Arginine	Arg	R	$NH_2CH(sc)COOH$ $sc = CH_2CH_2CH_2NHC(NH)$ NH_2	$pK_{COOH} = 2.0$; $pK_{NH_2} = 9.0$; $pK_{side\ chain} = 12.1$
Asparagine	Asn	N	$NH_2CH[CH_2CO(NH_2)]COOH$	$pK_{COOH} = 2.1$; $pK_{NH_2} = 8.84$
Aspartic acid	Asp	D	$NH_2CH(CH_2COOH)COOH$	$pK_{COOH} = 2.10$; $pK_{NH_2} = 9.82$; $pK_{side\ chain} = 3.86$
Cysteine	Cys	C	$NH_2CH(CH_2SH)COOH$	$pK_{COOH} = 1.71$; $pK_{NH_2} = 10.78$; $pK_{side\ chain} = 8.27$
Glutamic acid	Glu	E	$NH_2CH(CH_2CH_2COOH)$ $COOH$	$pK_{COOH} = 2.16$; $pK_{NH_2} = 9.96$; $pK_{side\ chain} = 4.32$
Glutamine	Gln	Q	$NH_2CH(CH_2CH_2CONH_2)$ $COOH$	$pK_{COOH} = 2.17$; $pK_{NH_2} = 9.13$
Glycine	Gly	G	NH_2CH_2COOH	$pK_{COOH} = 2.35$; $pK_{NH_2} = 9.78$
Histidine	His	H	$NH_2CH(CH_2C_3N_2H_3)COOH$	$pK_{COOH} = 1.82$; $pK_{NH_2} = 9.17$; $pK_{side\ chain} = 6.00$
Isoleucine	Ile	I	$NH_2CH[CH(CH_3)CH_2CH_3]$ $COOH$	$pK_{COOH} = 2.32$; $pK_{NH_2} = 9.76$
Leucine	Leu	L	$NH_2CH[CH_2C(CH_3)_2]COOH$	$pK_{COOH} = 2.33$; $pK_{NH_2} = 9.74$
Lysine	Lys	K	$NH_2CH[CH_2CH_2CH_2CH_2NH_2]$ $COOH$	$pK_{COOH} = 2.18$; $pK_{NH_2} = 9.12$; $pK_{side\ chain} = 10.53$
Methionine	Met	M	$NH_2CH(CH_2CH_2SCH_3)COOH$	$pK_{COOH} = 2.17$; $pK_{NH_2} = 9.27$
Phenylalanine	Phe	F	$NH_2CH(CH_2C_6H_5)COOH$	$pK_{COOH} = 1.83$; $pK_{NH_2} = 9.13$
Proline	Pro	P	$NHCH(CH_2CH_2CH_2)COOH$	$pK_{COOH} = 1.95$; $pK_{NH_2} = 10.64$
Serine	Ser	S	$NH_2CH(CH_2OH)COOH$	$pK_{COOH} = 2.21$; $pK_{NH_2} = 9.15$
Threonine	Thr	T	$NH_2CH[CH(OH)CH_3]COOH$	$pK_{COOH} = 2.17$; $pK_{NH_2} = 9.00$
Tryptophan	Trp	W	$NH_2CH(CH_2\ indole)COOH$	$pK_{COOH} = 2.38$; $pK_{NH_2} = 9.44$
Tyrosine	Tyr	Y	$NH_2CH(CH_2\ phenol)COOH$	$pK_{COOH} = 2.20$; $pK_{NH_2} = 9.11$; $pK_{side\ chain} = 10.07$
Valine	Val	V	$NH_2CH[CH(CH_3)CH_3]COOH$	$pK_{COOH} = 2.32$; $pK_{NH_2} = 9.62$

hydrogen bonds and PT) and the molecules to which these atoms belong. There will be one absolute minimum and one (or several) local minima. To be able to calculate the relative energies of the minima in a protein, a very large part of the protein has to be included. It is desirable to know local acidity constants from which the free energy PES can be calculated.

Calculation of the activation energy may be done by carrying fixed nuclei calculations of the total energy for a number of proton positions along a pathway between two minima. However, a much more accurate result is obtained if the positions of the other atoms in the system are optimized in each step.

Generally, the attraction a proton feels to the closest heavy atom is quite independent of the presence of the other heavy atom. As the proton is moved away from

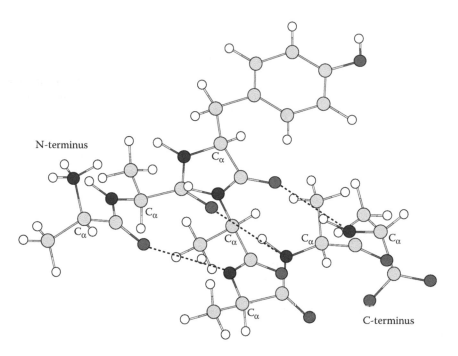

FIGURE 9.5 (See color insert) α-Helix with peptides in order, from N-terminus, Ala-Ala-Tyr-Ala-Ala-Ala-Ala.

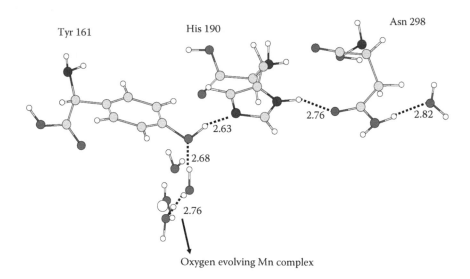

FIGURE 9.6 Hydrogen bond chain between side groups in a protein (PSII-D1). Distances are in angstrom. Dashed lines are hydrogen bonds between 2.6 and 3.0 Å.

the equilibrium geometry, the PES is thus parabolic with roughly the same force constant as in the free acid. When the proton is some distance away from the heavy atom, the potential is smoothed out and, as it approaches the acceptor atom, again becomes parabolic.

The presence of two minima always leads to a modification of the harmonic vibrational structure. If the two minima are equivalent, the lowest level will have one symmetric and one antisymmetric proton wave function. The larger the barrier, the closer the two lowest vibrational levels, until they become almost degenerate. The wave function of the proton is then a superposition of the symmetric state and the antisymmetric state, localized in either of the two minima.

Lippincott and Schröder have used curve fitting to construct PES for several hydrogen bonds. Biczó, Ladik, and Gergely have calculated vibrational levels. These results demonstrate very well the picture that often applies to asymmetric hydrogen bonds. The proton usually has preference of one of the states. The hydrogen stretch frequency of 3600–4000 cm^{-1} corresponds to a vibrational energy splitting of 10 kcal/mol. Since kT at room temperature is about 210 cm^{-1}, the Boltzmann factor is about exp(-18) for excitation to the first vibration level. This means that vibrational activation of a stretching mode for the bond between the heavy atom and proton appears as an unlikely reason for PT. The barriers for PT are in most cases considerably lower than the first vibrational level. PT between two heavy atoms is indeed a quantum problem.

9.2.5 COORDINATE SYSTEM

Imidazole (Figure 9.7) forms a planar, conjugated ring system with three carbon atoms and two nitrogen atoms. In a conjugated ring system, nitrogen behaves more or less as a carbon atom, except that it is capable of submitting two electrons to the π-system. In imidazole, however, one of the nitrogens is missing its hydrogen; therefore, this nitrogen itself must provide two electrons to form a lone pair instead of the single electron necessary for a bond the hydrogen atom (Figure 9.7). The two nitrogen members of the ring thus provide three electrons to the π-system. The three carbon atoms provide one each. There are six π-electrons, and this gives the molecule the same stability as benzene.

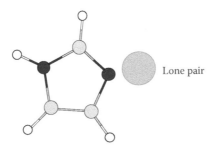

Lone pair

FIGURE 9.7 Imidazole molecule with lone pair orbital.

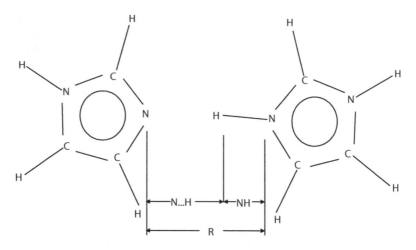

FIGURE 9.8   Symmetric proton transfer between imidazole and protonated imidazole.

The pK$_a$ value is 14.5 for imidazol, making it a very weak acid. The protonated imidazole is shown to the right of Figure 9.8. The base imidazol has a pK$_a$ value of 7. Molecules that can act as both acid and base are called *amphoteric*. In Figure 9.8, the proton is situated between two nitrogen atoms belonging to different imidazole molecules.

We assume for simplicity that the hydrogen bonds are symmetric and linear. First coordinates and distances have to be defined (Figure 9.9). Let the distance between the heavy atoms be R. The equilibrium distance between the proton and the heavy atom it is bonded to is R$_0$ in the absence of the hydrogen bond. The equilibrium coordinate for the proton without interaction is

$$Q_0 = R/2 - R_0. \qquad (9.14)$$

R is the length of the hydrogen bond, assuming that the bond is linear and, of course, that R is so small that it is meaningful to talk about a hydrogen bond

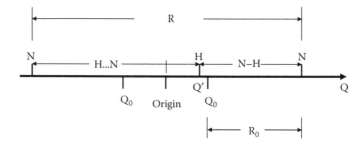

FIGURE 9.9   Coordinates for proton transfer in a hydrogen bond between two nitrogen atoms. Q′ and Q$_0$ are coordinates of the proton, in the presence and absence of the other heavy atom, respectively.

$(R < 3.5$ Å$)$. A diabatic PES is simply the energy surface of the proton around a heavy atom. There is one diabatic surface for each heavy atom, thus at least two diabatic PES. When the distance R between the heavy atoms is decreased, the interaction via the proton becomes stronger. The coordinate of the proton is $Q' < Q_0$, where Q' is defined by

$$Q' = R/2 - X - H. \qquad (9.15)$$

In Figure 9.7, X (=N) is the right N atom. X–H is calculated in Equation 9.15. The distance (X–H) between the proton and the heavy atom (X) it is bonded to by hydrogen bonds is thus increased, since $Q_0 > Q'$, while the distance H...N is decreased.

9.2.6 Parabolic Model (Marcus Model for Proton Transfer)

When the proton is moved away from the heavy atom it is bonded to, it moves away from a parabolic energy minimum ($Q = -Q_0$ in Figure 9.10). In reality, more than one distance will be involved, but we neglect the others for simplicity. At the same time, it approaches a new equilibrium position ($Q = Q_0$ in Figure 9.10). Therefore, the total energy will reach a maximum and start to decrease again after the proton has been moved a certain distance.

The PES for the proton is a parabolic curve where the force constant k is consistent with the vibrational frequencies of a typical hydrogen atom covalent bond. The vibrational frequencies are rather large since the low mass of the proton becomes the effective mass of the vibration and since a covalent bond to hydrogen is quite strong:

$$H_{11} = 1/2 \, k \left(Q + Q_0 \right)^2 \qquad (9.16)$$

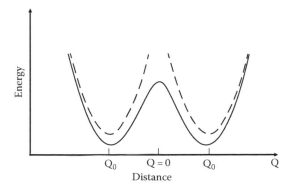

FIGURE 9.10 Symmetric double minimum potential. Diabatic curves are dashed.

If Equation 9.16 holds, the actual PES would continue as a parabola, making PT impossible. However, as mentioned above, as the proton approaches the other heavy atom, the other diabatic PES becomes more important:

$$H_{22} = 1/2\,k\left(Q - Q_0\right)^2. \tag{9.17}$$

We know from experience that there is no crossing between H_{11} and H_{22}. The crossing is avoided due to interaction between the two parabolas. The interaction matrix element may have to be obtained by curve fitting to experimental values. We simply call this matrix element H_{12}. According to the theory given in Chapter 4, $2\,|H_{12}|$ is equal to the avoided crossing in the case when the minima are equally deep (symmetric PT). The secular equation may thus be written as

$$\begin{vmatrix} H_{11} - E & H_{12} \\ H_{12} & H_{22} - E \end{vmatrix} = 0. \tag{9.18}$$

Equation (9.18) gives a solution that agrees well with actual PES. The eigenvalues to Equation 9.18 are given by

$$E_{\pm} = \frac{H_{11} + H_{22}}{2} \pm \sqrt{\left(\frac{H_{11} - H_{22}}{2}\right)^2 + H_{12}^2} = \frac{k}{2}\left(Q^2 + Q_0^2\right) \pm \sqrt{k^2 Q_0^2 Q^2 + H_{12}^2}. \tag{9.19}$$

The function E_- is shown in Figure 9.10.

It is suitable to define the *reorganization energy* λ as the value of H_{11} at $Q = Q_0$ (or alternatively, the value of H_{22} for $Q = -Q_0$):

$$\lambda = H_{11}\left(Q_0\right) = 2kQ_0^2 \tag{9.20}$$

To obtain the minima, we set the derivative of E_- with respect to $Q = 0$. If $kQ_0^2 > |H_{12}|$, there are two minima:

$$Q = \pm Q_0\left(1 - \frac{H_{12}^2}{k^2 Q_0^4}\right)^{1/2} = \pm Q_0\left(1 - \left(\frac{2H_{12}}{\lambda}\right)^2\right)^{1/2}. \tag{9.21}$$

Otherwise, there is only a single minimum: $Q = 0$. If $|H_{12}|$ is relatively small, the new minima are slightly different from the above-mentioned diabatic equilibria: $\pm Q_0$. If there are two minima, there will be a barrier between the new minima, which can easily be shown to have the size

$$E_a = \frac{\left(2|H_{12}| - \lambda\right)^2}{4\lambda}. \tag{9.22}$$

If $H_{12} = 0$, we obtain $E_a = \lambda/4$.

We write Equation 9.21 as

$$Q' = \pm Q_0 \left(1 - \frac{\Delta^2}{\lambda^2}\right)^{1/2}.$$ (9.23)

This shows that the distance between the heavy atoms will also change. In crystals, where it is common with symmetric PT, this decrease may be shown using neutron spectroscopy (similar to x-ray crystallography, but the protons are detectable due to their magnetic nuclei).

If the distance between the heavy atoms is sufficiently small in the molecular ion $H_5O_2^+$ (O ... O < 2.5 Å), there is a single energy minimum with O−H = 1/2 O ... O. This is thus the case when the proton is placed at equal distances from the heavy atoms. In this case, the activation energy for PT disappears.

9.3 PROTON TRANSFER

PT in hydrogen bonds is of great importance. Living systems are characterized by PT pathways, along which protons can move. These are protein enzymes with more than 50 peptides in the peptide chain. The pathways are formed by parts of the peptide chains and possibly water molecules that are part of the structure, where the free energy for PT is small compared with kT. The activation energy also has to be small. As is clear from the parabolic model, the activation energy is small if Q_0 is small. According to Equation 9.15 Q_0 is the difference between R/2 and the bond distance between the proton and the heavy atom it is bonded to (in the absence of the other heavy atom). So, if R is small, the activation energy for PT is small (Equations 9.20 and 9.22).

9.3.1 RATES OF PT REACTIONS

NMR spectroscopy may be used, provided the rate corresponds to the characteristic rate of NMR. IR and Raman spectroscopy are other possibilities in other time regions.

The calculation of rates for PT reactions proceeds as for other chemical reactions. The activation barrier for the proton is determined in the Born–Oppenheimer approximation. Ideally, the rate satisfies the Arrhenius equation. In the case of PT, the tunneling contribution to the rate is large.

PT is arguably the most fundamental and common of chemical reactions, since it involves a single particle. Particularly important is proton diffusion in hydrogen-bonded systems. The anomalously high rate of proton diffusion in water relative to cations of similar ionic radius (that is, K^+) has led to speculation concerning the existence of different mechanisms for the diffusion of protons versus other cations. The presence of a hydrogen bond network in water led Grotthuss to suggest a proton hopping mechanism in which the proton jumps from one water molecule to the next.

The PT process is then one of chemical rather than hydrodynamic diffusion. Other ions cannot take advantage of this network and must diffuse via Stokes' law. The Grotthuss mechanism is currently the most commonly accepted theory describing PT in bulk water.

The mechanism of PT in bulk water is a topic of great interest. One of the recent studies used a two-state empirical valence bond model (Lobaugh and Voth 1996) to determine the ground state adiabatic surface for H_5O_2 solvated in water. Tuckerman et al. (1995) carried out a thorough study of the solvation of hydronium (H_3O^+) in bulk water using the Car–Parrinello (CP) molecular dynamics (MD) technique and identified a possible mechanism for PT. The degree of coordination of the first solvation shell of hydronium was found to be the key factor determining the transfer of the proton.

9.3.2 Proton Tunneling

Since the proton is a relatively light particle, tunneling becomes important, particularly at a low temperature. It is clear from the above that the activation barrier is very high if the distance between the heavy atoms is large. The impetus for passing the barrier is obtained from other motions, such as the motion of side groups in a protein. One very important mechanism for PT should be very low frequency vibrations of motion of large parts of the systems toward each other. In that case, the distance for PT oscillates and possibly becomes small enough for PT during parts of the oscillation cycle.

If a free particle hits a barrier, it is reflected, according to classical mechanics. In quantum mechanics, on the other hand, the solution of the time-independent Schrödinger equation (SE) in the barrier is a decreasing exponential in absolute value. The reason is that the wave function must have a continuous derivative. The wave function in the barrier does not become equal to zero. At the right boundary, it has to have a continuous derivative into the region with zero potential. It has a lower amplitude than to the left of the barrier.

In the symmetric case, the proton is described by the wave functions:

$$\Psi_1 = \left(\Psi_A + \Psi_B\right)/\sqrt{2} \quad \text{and} \quad \Psi_2 = \left(\Psi_A - \Psi_B\right)/\sqrt{2}. \qquad (9.24)$$

In each of the eigenstates, the probability is thus equally large that the system is described by Ψ_A or Ψ_B. The situation when the proton is hydrogen-bonded to A (t = 0) is described as a superposition between Ψ_1 and Ψ_2. In time-dependent quantum mechanics, the proton will oscillate between A and B.

9.3.3 Grotthuss Effect

A symmetric hydrogen bond is found in $H_2O–H–OH_2^+$. A symmetric hydrogen bond, as defined here, is one where the proton equally likely sits on either of the heavy atoms. However, this does not mean that there is no activation barrier. In the case of $H_2O–H–OH_2^+$, it has been found in the calculations that the activation barrier is in fact absent.

In pure water too, the activation barrier for PT is shallow or even absent. PT reactions in water have proven to be very fast and this is referred to as the Grotthuss effect. Theodore von Grotthuss (1785–1822) was a German-Lithuanian baron who contributed to both electrolysis and photochemistry. At this time, neither protons nor the electrodissociation theory was known, of course (it was not even known that the formula for water is H_2O). Still, Grotthuss developed a primitive theory that H^+ and OH^- ions are involved in conductivity. This depends on the fact that the water molecules form chains of hydrogen bonds. Protons move stepwise and are almost activationless. Covalent bonds become hydrogen bonds as the charge is passed from

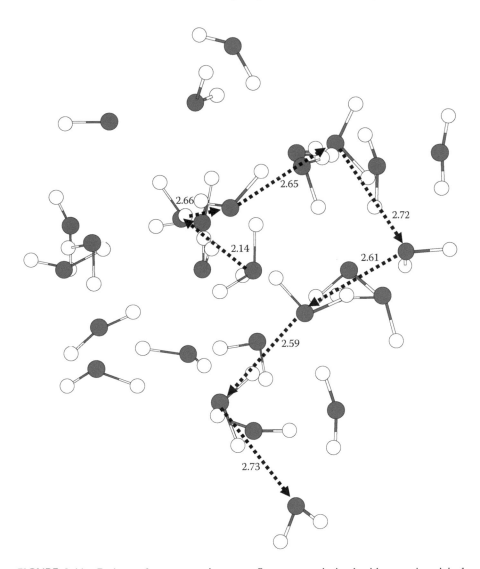

FIGURE 9.11 Pathway for a proton in water. Structure optimized with a semiempirical method.

one water molecule to the next. The next bond transfer takes place on "the other side" of the water molecule, now a hydronium ion. Protons are thus consecutively moved through the hydrogen-bonding network. The activation energy (2–3 kcal/mol) is probably due to heavy ion activation.

What is correct in the Grotthuss mechanism is the stepwise transfer mechanism. If the mechanism had been one of ordinary diffusion of a given (classical) proton through water, there would have been all kinds of barriers. The important thing is that the proton that transfers is not the "same" proton. In Figure 9.11, a cluster of water molecules is geometry optimized. The molecule at the center is a hydronium molecule. The distance to one of the other molecules is very short (2.14 Å) and almost certainly barrierless. From this molecule, there is a chain of hydrogen bonds suitable for successive PT.

The figure is obtained at $t = 0$. It is not necessary that the next proton in turn takes the pathway indicated. After the proton has moved one step, there will be a decrease of the bonds around it, making it possible for another proton to transfer using a small distance and a very small barrier. The difference between transfer in water and in the protein is that the structure is flexible in water, so that after some small motion of a water molecule, it is always possible to make another barrierless transfer to a new water molecule.

This gradient is composed of both the pH gradient and the electrical gradient. The pH gradient is a result of the H^+ ion concentration difference. Together, the electrochemical gradient of protons is both a concentration and charge difference and is often called the proton motive force (PMF). In mitochondria, the PMF is almost entirely made up of the electrical component; but in chloroplasts, the PMF is made up mostly of the pH gradient. In either case, the PMF needs to be about 50 kJ/mol for the ATP synthase to be able to make ATP.

10 Electron Transfer Reactions

10.1 INTRODUCTION

In this chapter, we will study the fundamentals of electron transfer (ET) reactions. In ET reactions, the only moving particle *appears* to be an electron. No bonds are broken or formed. Spectral changes are noticeable after transfer of an electron, and this gives a possibility to study the kinetics of very fast ET reactions by time-resolved spectroscopy. Still, small structural changes do occur at ET, and these changes are the main reason why ET reactions need activation energy. In this chapter and in Chapter 11, we primarily have thermal activation in mind. In Chapter 13, we will discuss photoinduced ET (PIET) reactions.

Homogeneous ET reactions take place in a homogeneous medium from a *donor* (D) to an *acceptor* (A). Electrons are exchanged between molecules or ions, for example, one electron may transfer from Fe^{2+} to Co^{2+}. As always, a *metal complex* (metal + ligands) is the smallest unity for our attention, not just the metal ions. *Heterogeneous* ET reactions take place between a solid material and ions in solution. One example is the well-known Daniell electrochemical cell, where Cu^{2+} ions in a solution receive electrons from a copper electrode and are deposited there as copper metal.

10.2 HOMOGENEOUS ET REACTIONS

Electron exchange between identical metal (complex) ions began to attract the interest of scientists in the 1940s when radioactive isotopes became available for marking. For example, Fe^{2+}/Fe^{3+} electron exchange can be monitored if the Fe nucleus in $FeCl_3$ is radioactive. Chemically, reactions and products are the same and hence $\Delta G^{\ominus} = 0$ in this *symmetric ET reaction*. Still, the reaction rates vary over more than 10 orders of magnitude for different metal ions and ligands. This was explained when it became clear that the reorganization energy, due to small changes in bond lengths and solvent, is the important factor for the rate. A number of phenomena, for example, mobility of electrons in organic crystals, polymers, and molecular wires have been rationalized in terms of the *Marcus model*.

10.2.1 INNER AND OUTER SPHERE ET REACTIONS

In a *redox* reaction, the oxidation states of the participating reactants are changed during the reaction, for example, $0 \rightarrow 1$ for H and $0 \rightarrow -2$ for O in

$$2H_2 + O_2 \rightarrow 2H_2O \tag{10.1}$$

In this case, the electrons of H_2 pass only *formally* from H to O when the water molecule is formed. It has been *agreed* that H and O have valence 0 in the H_2 and O_2 molecules, and that H has valence +1 and O the valence –2 in the water molecule (as in most of its compounds). Systematic use of oxidation numbers in chemistry serves a single, but very useful purpose: keeping track of the electrons. The oxidation state of an atom in a molecule says very little about how the electronic density is distributed within the molecule. Only by solving the time-independent Schrödinger equation (SE), we may calculate the charge density exactly, but there is no useful and unique way to tell to which atom the charge "belongs."

In gas phase, the free proton has no electron density around it, of course. When it binds to O^{2-}, it buries itself in the electronic charge of this ion, belonging to the O 2s and O 2p orbitals. The density close to the hydrogen nucleus is about the same as at a hydrogen atom in, say, CH_4. The electron density can be considered as belonging either to H 1s or to the atomic orbitals of a heavier atom. Fortunately, the assignment of electrons to individual atoms within a single molecule is unnecessary and uninteresting in ET reactions.

Actual transfer of electrons occurs in reaction (10.2), where the electrons in the initial reaction step are passed into an antibonding MO of the MnO_4 ion:

$$C_2H_5OH + 2MnO_4^- \rightarrow CH_3CHO + 2MnO_4^{2-} + 2H^+. \qquad (10.2)$$

The two permanganate ions accept electrons from the alcohol molecule. During the reaction, the color changes from the beautiful violet of permanganate to the dull green of manganate, while the dubious smell of schnapps is replaced by the wonderful odor of acetaldehyde. The reaction (10.2) is not a typical ET reaction, however, since bonds are broken and new molecules formed.

In the ET reactions of interest here, the molecules or metal complexes involved conserve their structure, up to small, but important, changes of bond lengths, possibly bond angles, and solvent structure. These ET reactions are most common in special groups of compounds, such as transition metal ions and aromatic systems, where the number of electrons can easily be changed without losing the structural identity. The acceptor A is reduced (accepts an electron) and the donor D is oxidized (loses an electron). A metal complex such as $Fe(H_2O)_6^{3+}$ may receive an electron and become $Fe(H_2O)_6^{2+}$. The metal–ligand (ML) bond lengths tend to increase by about 0.1 Å as a result. At the same time, the solvent changes its average structure around the complex. We will see below that the greater the change of geometric parameters, the slower is the ET reaction.

In 1953, Henry Taube demonstrated the *inner sphere mechanism*, where two metal ions bind to the same ligand during the ET process. The total reaction is

$$Co(NH_3)_5 Cl^{2+} + Cr(H_2O)_6^{2+} \rightarrow Co(NH_3)_5 H_2O^{2+} + Cr(H_2O)_5 Cl^{2+} \qquad (10.3)$$

Co increases the occupancy of its 3d subshell by one unit and the valence state from +3 to +2, while Cr decreases the occupancy of its 3d shell by one unit and the valence state from +2 to +3. In other words, one electron is transferred from the Cr to the Co complex. It is known from tracer experiments that the Cl⁻ ion in $Cr(H_2O)_5Cl^{2+}$ originates from $Co(NH_3)_5Cl^{2+}$. It is well known in inorganic chemistry that Cr^{3+} and Co^{3+} are both very slow ("substitutionally inert") in exchanging their ligands. This can be related to the large ligand field stabilization energy of these metal ions in the configurations t_{2g}^3 and t_{2g}^6, respectively. Consequently, all ligand exchanges in Equation 10.3 have to take place when the metal ions are in their +2 oxidation states. Cr must attach itself to Cl⁻ before ET, while Co cannot lose Cl⁻ until after ET. ET thus occurs while the metal complexes are connected (Figure 10.1).

Since Cl^{-2} ions are nonexistent in condensed matter, we can be sure that the electron does not travel with the Cl⁻ ion. There is no other place available for the electron, so it apparently leaves the valence shell of cobalt and appears immediately in the valence shell of chromium. The total number of electrons has to be constant.

This type of motion, where the electron is "chemically forbidden" in certain regions of space, resembles the tunneling mechanism in physics, first discovered in radioactive nuclear decays. Since the term "tunneling" has been an accepted name for decades, there is no reason to adopt another name. However, the original tunneling model is a qualitative model. In ordinary quantum chemical calculations, there is no simple way to calculate tunneling barriers for the electron and there is also no reason to do so. Quantitative results can be obtained with the help of molecular orbital (MO) methods. The *electron tunneling* model is based on the overlap between the D and A wave functions and this still holds true in MO models.

Electron tunneling would be very fast and depend negligibly on temperature if only electronic motion is involved. An equally important contributor to the rate is the *nuclear factor*, which depends on how much the molecular structure is changed during ET. The nuclear factor depends on vibrational structure and hence on temperature. In fact, the nuclei find an activation barrier in their motion due to ET. Also, in this case we may talk about tunneling through a barrier, or *nuclear tunneling*.

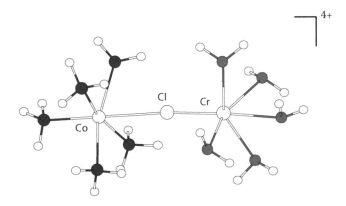

FIGURE 10.1 Intermediate inner sphere complex $(NH_3)_5CoClCr(H_2O)_5^{4+}$ formed in reaction (10.3).

However, a bonded connection between metal ions is by no means necessary. ET reactions often occur directly through the solution between D and A. The reaction in Equation 10.4 in water solution is called an *outer sphere ET reaction* since the ligand shell remains intact:

$$*Fe(H_2O)_6^{3+} + Fe(H_2O)_6^{2+} \rightarrow *Fe(H_2O)_6^{2+} + Fe(H_2O)_6^{3+}. \qquad (10.4)$$

The rate equation is of second order since the rate is proportional to the concentration of both Fe^{2+} and Fe^{3+}. In the case of reaction (10.4), the rate constant is $k = 3\ M^{-1}\ s^{-1}$. If the ligands are bipyridine (bpy) as in

$$*Fe(bpy)_3^{3+} + Fe(bpy)_3^{2+} \rightarrow *Fe(bpy)_3^{2+} + Fe(bpy)_3^{3+}, \qquad (10.5)$$

the rate is several orders of magnitude faster: $k > 10^6\ M^{-1}s^{-1}$. On the other hand, the rate of

$$*Co(NH_3)_6^{3+} + Co(NH_3)_6^{2+} \rightarrow *Co(NH_3)_6^{2+} + Co(NH_3)_6^{3+} \qquad (10.6)$$

is much slower or $k = 10^{-6}\ M^{-1}s^{-1}$.

It is important to be able to explain the difference in rates between Equations 10.2 and 10.4 on the basis of electron and molecular structure. In fact, it depends on how much the ML bond lengths change during oxidation and reduction. The difference between Equations 10.5 and 10.6 depends also on the change of spin from low spin for Co^{3+} to high spin for Co^{2+}.

10.2.2 ELECTRON TRANSFER COUPLED TO PROTON TRANSFER

ET is a well-defined phenomenon sometimes coupled with proton transfer (PT), for example, in living systems. As an example of a symmetric PT reaction, we may choose

$$C_6H_5O^- + C_6H_5OH \rightarrow C_6H_5OH + C_6H_5O^-. \qquad (10.7)$$

No electrons are involved in this reaction. The highest occupied molecular orbit (HOMO) is on phenolate. There is a similar occupied π orbital on phenol with a lower orbital energy (Figure 10.2).

Now suppose that the pair of phenolate–phenol molecules are a step on an ET chain (Figure 10.2). Since phenolate has no proton, the HOMO is less strongly bound and the electrons in it are "activated." One of its electrons continues along the ET chain, and when it has disappeared on the left, the proton in the hydrogen bond gets a driving force for moving over to phenolate in a PT reaction:

$$C_6H_5O + C_6H_5OH \rightarrow C_6H_5OH + C_6H_5O. \qquad (10.8)$$

In this second step, an electron accompanies the proton and fills the HOMO of phenol.

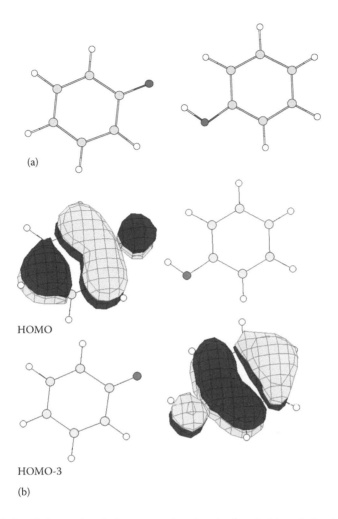

FIGURE 10.2 Hydrogen-bonded system phenolate^{-1}–phenol (a) and the two orbitals involved in ET (b).

Finally, the right molecule accepts an electron from the ET chain on the right. The process depends on the external driving force for the electrons of the chain.

It should be noticed that in the reaction step of Equation 10.8, the electron moves, separated from the proton. We cannot be dealing with hydrogen atom transfer, since the moving electron is in a π orbital, which cannot mix with H 1s. The rate of this reaction depends on both the activation barrier for PT and the activation barrier and coupling for ET. The calculation of rate will be discussed later.

10.3   ELECTROCHEMISTRY

Electrochemistry existed as a scientific field for a full century before the electron was discovered in 1897. In France, Antoine Lavoisier discovered the gas called *oxygen*, which, as the name suggests, generates oxides. In Sweden, Carl Wilhelm Scheele discovered the same gas and found that it makes organic compounds sour or acidulous. In Swedish, oxygen is therefore called "syre" (maker of acid or "syra"). *Oxidation* is the process when electrons are donated, while *reduction* is the process when electrons are accepted.

Electrochemistry is a background field for a modern theory of ET based on quantum mechanics. A short summary will be given here. Electrochemical systems are heterogeneous systems where the action takes place at the interface of a solid metal or semiconductor called the *electrode*, and a fluid, ionic conductor called the *electrolyte*. Electrochemistry started with Galvani, who serendipitously discovered that muscles of an animal may still move shortly after the animal has died, if its muscles come into contact with two different metals at the same time. Later, it was understood that electricity was involved and caused by the two metals rather than by the dead animal. In the year 1800, Volta constructed a pile where the voltage depended on the number of cells and therefore could become high if many cells were piled on top of each other with electrolytes, sucked up in an absorbing material, in between.

10.3.1 ELECTROCHEMICAL CELLS

In an electrochemical cell (voltaic cell, galvanic cell), reduction and oxidation take place in such a way that electric voltage is generated between two electrodes. The *anode* is the electrode where positive ions are solvated in the electrolyte, forming an *anode current*, or negative ions (*anions*) are coming onto the electrode from the electrolyte. At the other electrode, the *cathode*, the current flows in the opposite direction. By adding a counter potential, it is possible to make the anode and cathode currents of equal magnitude but with different signs.

An electrochemical cell may be divided into two half cells. A half cell consists of, for example, a rod of zinc in a solution of Zn^{2+} ions (for example, $ZnSO_4$ in water). The half-cell reaction is

$$Zn(s) \rightleftarrows Zn^{2+} + 2e^-. \tag{10.9}$$

No measurable transport of charge takes place if the half cell is not connected to another half cell. Otherwise, accumulation of charge would generate a counter field. Very quickly, a voltage is created between the metal rods. If the rods are connected, a current will flow between the two electrodes. The voltage can be measured with the help of a voltmeter.

Daniell cell (Figure 10.3) is a cell where one of the electrodes is a rod of zinc immersed in an aqueous solution of $ZnSO_4$ and the other a rod of copper in $CuSO_4$ solution. The two solutions are connected with a "salt bridge" consisting of some conducting aqueous solution. The voltage depends on the concentration of the electrolytes. In the Daniell cell, the voltage is 1.1 V if the electrolytes are 1 M.

If a piece of iron or zinc is dropped in a water solution of $CuSO_4$, the metal piece is dissolved and copper metal is produced. Cu has a greater "desire" to remain a metal than Fe and Zn. In the Daniell cell, the free energy is used to do

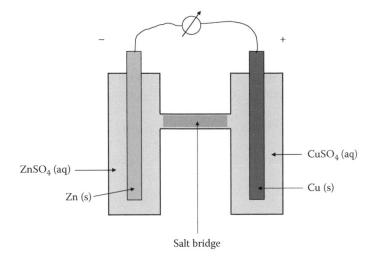

FIGURE 10.3 Daniell cell. The voltage is measured by a voltmeter and the current with an amperemeter.

useful work with the help of the electric current in the wire between the two electrodes. In the solution, there is a transport of positive ions from the Zn electrode to the Cu electrode and Cu^{2+} ions are successively replaced by Zn^{2+} ions. The voltage difference is called *electromotive force* (emf).

The negative pole is written to the left and the positive pole to the right. The electrodes are separated from the solutions with a vertical bar. Aggregation state, pressure, and concentration are put in parentheses. Thus, a Daniell cell is described as follows:

$$-Zn(s)\,|\,ZnSO_4\,(aq;c_1M)\,|\,CuSO_4\,(aq;c_2M)\,|\,Cu(s)+ \qquad (10.10)$$

meaning that metallic zinc is the negative pole and is surrounded by c_1 molar aqueous solution of zinc sulfate, while metallic copper is the positive pole and is surrounded by c_2 molar aqueous solution of copper sulfate. The two solutions are in contact in a way that excludes mixing (except due to diffusion).

10.3.2 THERMODYNAMICS OF THE CELL

Since it is simple to keep reversible conditions in an electrochemical experiment, thermodynamic information can easily be obtained. According to the first and second laws of thermodynamics, reversible processes must satisfy

$$\Delta U = q_{rev} + w_{rev} = T\Delta S - P\Delta V + w_{el}, \qquad (10.11)$$

w_{el} is the work that is added to the system by the electric current:

$$w_{el} = \Delta U + P\Delta V - T\Delta S = \Delta G. \qquad (10.12)$$

The amount of electricity that passes through the amperemeter is measured in moles of electrons. A mole corresponds to 1 Faraday (F), where 1 F = 96 485 As/mol is the *Faraday constant*. If the cell potential is E, and n moles are passed through the cell, the work performed by the cell is $E \cdot n \cdot F$ J/mol.

We count the electric work performed on the environment as positive if $\Delta G < 0$. Thus, it follows from Equation 10.12 that

$$\Delta G = -EnF(T,P \text{ constant}). \tag{10.13}$$

Since $(\partial G/\partial T)_p = -S$ according to Equation 5.69, we have from Equation 10.13

$$\Delta S = nF(\partial E/\partial T)_p. \tag{10.14}$$

Since $\Delta G = \Delta H - T\Delta S$ (Equation 5.33), we obtain the following expression for enthalpy change:

$$\Delta H = -nFE + nFT(\partial E/\partial T)_p. \tag{10.15}$$

Using Equations 5.82 and 5.84 we obtain

$$\Delta G = \Delta G^{\ominus} + RT \ln \prod_i c_i^{v_i} \tag{10.16}$$

from which it follows that

$$E = E^{\ominus} - \frac{RT}{nF} \ln \prod_i c_i^{v_i}. \tag{10.17}$$

$E^{\ominus}$ is the cell potential if all the concentrations are equal to 1 M. Equation 10.17 is the fundamental Nernst equation (Walther H. Nernst, German physicist and chemist). The concentrations of the products are in the numerator and the concentrations of the reactants in the denominator in the concentration product. A more accurate treatment is obtained if we use activity instead of concentration, but this will not be further discussed here.

10.3.3 ELECTROCHEMICAL SERIES

Standard potentials are sometimes called *reduction potentials*. The more negative the value of the reduction potential, the more difficult it is to reduce the positive ion to metal. If electrode reactions are ordered according to standard potentials, the *electrochemical series* (Table 10.1) is obtained. First in the series is the reduction reaction that occurs most easily, as in this case, the reduction of Au^+ to gold metal. The precious metals appear high up on the list. At the bottom we find alkali metal ions, which are the most difficult to reduce to metals. At the same time, alkali metals most easily oxidize to metal ions. Zero is set for the reaction where H_2 is split into

TABLE 10.1
Electrochemical Series at 25°C

Electrode Reaction	Reduction Potential ($E^{\ominus}$ V)	Electrode Reaction	Reduction Potential ($E^{\ominus}$ V)
$Au^+ + e^- \rightarrow Au$	+1.692	$Sn^{2+} + 2e^- \rightarrow Sn$	−0.138
$Au^{3+} + 3e^- \rightarrow Au$	+1.498	$Cr^{3+} + 3e^- \rightarrow Cr$	−0.74
$Cl_2(g) + 2e^- \rightarrow 2Cl^-$	+1.358	$Fe^{2+} + 2e^- \rightarrow Fe$	−0.447
$O_2 + 4H^+ + 4e^- \rightarrow 2H_2O$	+1.229	$Zn^{2+} + 2e^- \rightarrow Zn$	−0.762
$Ag^+ + e^- \rightarrow Ag$	+0.800	$Al^{3+} + 3e^- \rightarrow Al$	−1.662
$Fe^{3+} + e^- \rightarrow Fe^{2+}$	+0.771	$Mg^{2+} + 2e^- \rightarrow Mg$	−2.372
$Cu^+ + e^- \rightarrow Cu$	+0.521	$Ca^{2+} + 2e^- \rightarrow Ca$	−2.868
$O_2 + 2H_2O + 4e^- \rightarrow 4OH^-$	+0.40	$Na^+ + e^- \rightarrow Na$	−2.714
$Cu^{2+} + 2e^- \rightarrow Cu$	+0.342	$Rb^+ + e^- \rightarrow Rb$	−2.92
$Cu^{2+} + e^- \rightarrow Cu^+$	+0.16	$Cs^+ + e^- \rightarrow Cs$	−2.92
$2H^+ + 2e^- \rightarrow H_2$	0	$Ba^{2+} + 2e^- \rightarrow Ba$	−2.90
$Fe^{3+} + 3e^- \rightarrow Fe$	−0.037	$K^+ + e^- \rightarrow K$	−2.931
$Pb^{2+} + 2e^- \rightarrow Pb$	−0.126	$Li^+ + e^- \rightarrow Li$	−3.05

two protons, on the surface of platinum metal flushed with hydrogen gas (*standard hydrogen electrode* [SHE]).

The half-reactions may be added or subtracted to obtain a full reaction. If $E^{\ominus} > 0$, $\Delta G^{\ominus} < 0$ and the reaction occurs spontaneously. Equation 10.16 is used to obtain $\Delta G^{\ominus}$. $\Delta G^{\ominus}$ for the sum reaction is obtained. Finally, the value of $E^{\ominus}$ is obtained.

As an example of the use of Table 10.1, we may take copper and copper ions. Cu^{2+} is readily reduced to Cu^+ with a reduction potential of 0.16 V. The reduction of Cu^{2+} to Cu metal has a reduction potential of 0.342 V. The reduction of Cu^+ to Cu metal may be calculated as follows. The reduction potential for $Cu^{2+} + 2e^- \rightarrow Cu$ is 0.342, according to Table 10.1. $\Delta G^{\ominus}$ is thus equal to $-2 \times 0.342 = -0.684$. $E^{\ominus}$ for the reaction $Cu^{2+} + e^- \rightarrow Cu^+$ is equal to 0.16, thus $\Delta G^{\ominus}$ is equal to −0.16. To obtain $\Delta G^{\ominus}$ for the reaction $Cu^+ + e^- \rightarrow Cu$, we subtract the above-mentioned second reaction from the first one and obtain $\Delta G^{\ominus} = -0.68 + 0.16 = -0.52$. $E^{\ominus}$ for the reaction $Cu^+ + e^- \rightarrow Cu$ is thus equal to 0.52.

The values used in Table 10.1 refer to standard potentials measured in water solution under standard conditions. Concentrations change the values according to Equation 10.17. Furthermore, acidity and complex formation are important. In proteins and solids, there may be great differences due to protein structure and crystal structure, respectively.

10.3.4 LATIMER AND FROST DIAGRAMS

Latimer diagrams show successive reduction potentials of half-reactions that connect various oxidation states of an element. Manganese is a good example since it occurs in a number of different oxidation states from 0 (metal) to +7 (permanganate) in aqueous solution: $MnO_4^- \rightarrow MnO_4^{2-}$ (0.564); $MnO_4^{2-} \rightarrow MnO_4^{3-}$ (0.274); $MnO_4^{3-} \rightarrow MnO_2$ (4.27);

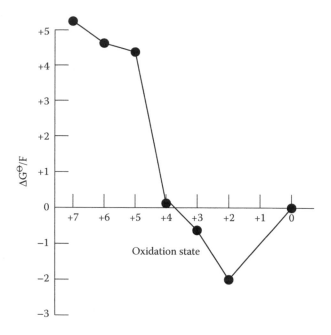

FIGURE 10.4 Frost diagram for the oxidation states of manganese.

$MnO_2 \rightarrow Mn^{3+}$ (0.95); $Mn^{3+} \rightarrow Mn^{2+}$ (1.51); and $Mn^{2+} \rightarrow Mn$ (−1.18) (reduction potentials in V in parenthesis). The oxidized species appear on the left. If we are interested in the reduction potential when permanganate (MnO_4^-) is reduced to Mn^{2+}, all the numbers are added and the sum is divided by the number of electrons, in the present case 5.

Frost diagrams (Figure 10.4) are oxidation state diagrams giving the relative free energy as a function of oxidation state. The diagrams show the properties of the different oxidation states of a species.

10.4 MARCUS PARABOLIC MODEL FOR ET

The simplicity of ET reactions as compared with ordinary chemical reactions is due to the fact that the molecular structure is conserved, disregarding small changes in solvent, bond lengths, and angles. The potential energy surfaces (PES) are often parabolic except at a single (avoided) crossing where the electron leaps from one molecule to the next. If the system is considered a single molecule, the whole ET drama is usually concentrated in two MOs, which form two orthogonal linear combinations of the orbital occupied by the electron before and after the reaction.

In a *symmetric* ET reaction, the driving force for ET is absent; nevertheless, ET takes place. For example, the following ET reaction is symmetric:

$$M_A^+ + M_B \rightarrow M_A + M_B^+ \tag{10.18}$$

between two sites A and B in a crystal. In solution, the kinetics of such a system may be followed using isotope replacement:

$$^*M^+ + M \rightarrow \, ^*M + M^+. \tag{10.19}$$

Notice, however, that the vibrational energy levels are of importance at low temperature, particularly for light elements. In that situation, Equation 10.19 depends on the isotope.

The total system thus consists of two molecules connected in a symmetric configuration. The number of electrons is odd, either one more or one less electron than in the ground state with filled shells. We assume two equivalent sites, M_A and M_B. The two possible states are of the type $M_A^- M_B$ and $M_A M_B^-$. If an electron is missing ("hole state"), the characters are $M_A^+ M_B$ and $M_A M_B^+$. In general, the states are mixed, that is, the electron or hole cannot be said to belong to a single site. In the general case also, M_A and M_B are inequivalent and the total system is asymmetric.

In Section 4.7.1, we found that if the diabatic energy surfaces, H_{11} and H_{22}, interact with the interaction matrix element, H_{12}, we may approximately write the secular equation

$$\begin{vmatrix} H_{11} - E & H_{12} \\ H_{12} & H_{22} - E \end{vmatrix} = 0. \tag{10.20}$$

Equation 10.20 has the solution

$$E_\pm = \frac{H_{11} + H_{22}}{2} \pm \sqrt{\left(\frac{H_{11} - H_{22}}{2} \right)^2 + H_{12}^2}. \tag{10.21}$$

H_{11} and H_{22} are assumed to be degenerate but with different equilibrium points, corresponding to equivalent but different locations of the electron.

10.4.1 Adiabatic and Nonadiabatic Transfer

The energy of the total system depends on the bond lengths and bond angles in both subsystems. For simplicity, we assume that all dependence is in a single coordinate, the ML average bond length. The total energy surface thus depends on R_1 and R_2, the radii of the two subsystems. R_1 and R_2 may vary independently of each other. The minima are obtained for R_1^0 and R_2^0, respectively. The total energy and diabatic energy sur-faces thus have min-ima in (R_1^0, R_2^0) and (R_2^0, R_1^0), each corresponding to different locations of the electron. The total energy diagram is shown in Figure 10.5. If we assume no interaction between the two PES and the same (local) force constant (independent of charge), we obtain two paraboloids centered in the two parabolic minima. In the intersection between the two paraboloids, the PES have to change due to interaction between the two paraboloids of the type seen in Equations 10.20 and 10.21. After the interaction, there is a lower and an upper PES, since there is an avoided crossing between them (Figure 10.5).

For simplicity, we restrict the reaction path to the simplest one possible, along the dashed line in Figure 10.5. The independent variable is called Q. The equilibrium points appear for

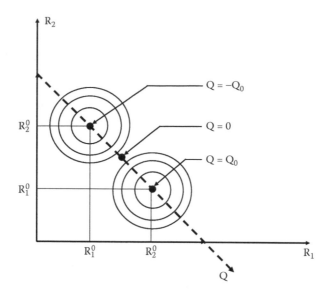

FIGURE 10.5   Coordinate system for the electron transfer system.

$$Q = Q_0 \text{ and } Q = -Q_0. \tag{10.22}$$

Along the Q-axis, we have

$$\delta R_1 = -\delta R_2 \tag{10.23}$$

meaning an increase of the ML distance (δR_1) in one complex and a simultaneous decrease of the ML distance (δR_2) in the other complex. From the figure, we obtain

$$\delta Q / \sqrt{2} = \delta R_1 = -\delta R_2. \tag{10.24}$$

The two parabolic PES may be described by the following equations:

$$H_{11} = \frac{1}{2} k (Q + Q_0)^2$$

$$H_{22} = \frac{1}{2} k (Q - Q_0)^2. \tag{10.25}$$

The interaction matrix element, H_{12}, is chosen to fit the actual energy difference between the PES when $R_1 = R_2$. This energy difference can be calculated in a quantum chemical calculation where $Q = 0$. If the calculated energy difference is Δ, we choose $H_{12} = \Delta/2$. With this substitution and the one of Equation 10.25 the new PES are obtained from Equation 10.21:

$$E_{\pm} = \frac{k}{2} (Q^2 + Q_0^2) \pm \sqrt{k^2 Q_0^2 Q^2 + (\Delta/2)^2}. \tag{10.26}$$

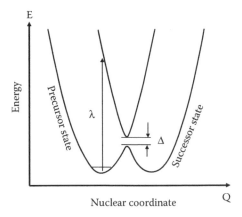

FIGURE 10.6 PES for a symmetric ET system. λ is the reorganization energy; Δ the avoided crossing gap at $Q = 0$.

These solutions, the interacting Marcus parabolas, are given in Figure 10.6.

PES go through an avoided crossing (Section 4.7.1). The electronic wave functions can be calculated in principle. At the avoided crossing, corresponding to a totally symmetric system (the same bond length in the left and right molecules), they are evenly distributed over the two molecules. The ground state wave function is symmetric and the other wave function is antisymmetric, as is the case in chemical bonding.

The "avoidedness," that is, the value of Δ, clearly depends on how closely the two metal complexes are located to each other, whether they have any ligand in common, etc. If we assume that the distance between the two complexes tends to infinity, the energy difference $|\Delta|$ at the avoided crossing clearly tends to zero and we have two parabolic energy surfaces without interaction, called *diabatic* PES.

At the normal distance, there is always some interaction ($\Delta \neq 0$). If $|\Delta|$ is so small that the final PES cannot be easily distinguished from the diabatic ones, we are talking about *nonadiabatic* ET. If, on the other hand, $|\Delta|$ is so large that the lower PES may be regarded as an ordinary ground state of the system at all $Q - Q_0$, we are talking about *adiabatic* ET.

We are usually interested only in the ground state PES (E_-). The location of the two minima may be obtained if we set the derivative of E_- (Equation 10.26) equal to 0. If $kQ_0^2 > |H_{12}|$, there are two minima:

$$Q = \pm Q_0 \left(1 - \frac{H_{12}^2}{k^2 Q_0^4} \right)^{1/2}. \tag{10.27}$$

If $kQ_0^2 < |H_{12}|$, there is only one minimum, for $Q = 0$. Each minimum corresponds to an equilibrium point for the total system.

The difference between the two diabatic PES at the point $Q = -Q_0$ is called the *reorganization energy* and is denoted as λ. We easily find from Equation 10.25 that

$$\lambda = 2kQ_0^2. \tag{10.28}$$

Q_0 is a measure of how much the equilibrium geometries change as the electron moves. λ grows rapidly with Q_0.

A maximum between the minima exists for $\lambda = 2kQ_0^2 > 2|H_{12}| = \Delta$. The activation energy, E_a, can be calculated from Equations 10.26 and 10.27:

$$E_a = \frac{\left(|H_{12}| - kQ_0^2\right)^2}{2kQ_0^2} = \frac{\left(\Delta/2 - \lambda/2\right)^2}{\lambda} = \frac{\lambda}{4} \cdot \left(1 - \frac{\Delta}{\lambda}\right)^2. \tag{10.29}$$

$E_a = \lambda/4$ for $\Delta = 0$ and $E_a = 0$ for $\lambda = \Delta$. Except for short distances between the donor and the acceptor, $|H_{12}| \ll \lambda$. In that case, the activation energy depends only on λ and is equal to $\lambda/4$.

If $|H_{12}| > \lambda/2$, the barrier disappears. One can no longer talk about D being oxidized and A reduced, since the electron belongs to both subsystems. This case is found if the electron can localize itself on a bridge between D and A.

10.4.2 REORGANIZATION ENERGY (λ)

The vertical energy, denoted as λ in Figure 10.6, is the reorganization energy. In other words, λ is the energy difference between the minima of the reactants PES and the products PES at the same abscissa. To be able to calculate the rate constant k from first principles according to the Marcus model, it is necessary to know λ and Δ, and in the case of asymmetric ET reaction, the free energy of the reaction ($\Delta G^{\ominus}$).

λ may be calculated if it is known how much the bond lengths are changed in the subsystem, when an electron is removed from the reduced subsystem or added to the oxidized subsystem (λ_{bond}). λ_{bond} is the *structural reorganization energy*. In addition we must include the *solvent reorganization energy*, λ_{solv}. The latter is a bit tricky since we are dealing with Coulomb forces extending to infinity. Thus, $\lambda_{bond} + \lambda_{solv} = \lambda$.

In the original theory for metal complexes, Marcus considered all reorganization energy as solvent reorganization energy. He was interested in ET between metal ions and considered everything outside the metal ion itself as "solvent." N. Hush improved the theory in 1958 and treated the first layer of ligands as bonded to the metal ion. If the distance between the ligands and the metal ion for different oxidation states and, furthermore, the force constants are known, we may calculate the reorganization of the first shell of ligands more accurately from the parabolic PES. Thereby, the solvent reorganization energy contribution is a smaller part of the total reorganization energy.

We begin with the solvent reorganization energy, λ_{solv}. The average position and orientation of a solvent molecule depend on the local charges. Still, it is fair to hope that λ_{solv} may be taken into account via the dielectric constant by application of the Born approximation. In the case of metal complexes, we may assume that they are spherical. In the case of π-systems, this is not, of course, a reasonable approximation and improved expressions have now been derived.

The dielectric constant depends on phase, temperature, and pressure. For water, the relative dielectric constant is $\varepsilon_r = 80$ at $T = 293$ K ($20°C$). At the freezing point $T = 273$ K ($20°$ lower), we have $\varepsilon_r = 88$, and at $40°C$ ($20°$ higher) we have $\varepsilon_r = 72$. The dependence on temperature is thus linear at these temperatures.

We assume that the two metal complexes are not necessarily identical and let the radii be a_1 and a_2 and the distance between them R, counted from the centers. We assume that the charge of the whole complex is increased by one unit. The Born energy for polarization energy around a charge Z with radius a_1 may be written as

$$U = \frac{Z^2}{2a_1}\left(1 - \frac{1}{\varepsilon}\right) H \text{ or } U = \frac{Z^2}{2 \times 4\pi\varepsilon_0 a_1'}\left(1 - \frac{1}{\varepsilon}\right), \quad J/\text{mol} \qquad (10.30)$$

depending on whether atomic units or SI units are used. In the left expression, we obtain U in Hartree (H), but we must remember to measure a_1 in Bohr (B) (multiply distance in Å by 1.89). In the right expression, we obtain U in joules per mol, but must measure the radius (called a_1') in meters. Henceforth we will use atomic units only.

In Equation 10.30, ε is the relative dielectric constant ($= \varepsilon_r$). There are two contributions to the dielectric constant from the solvent: the polarization due to electronic polarizability when the water molecules are fixed and the polarization caused by the change of geometry and orientation. The first part is an electronic part, which should not be counted since it does not depend on the nuclei and is already (in principle) included in the calculation of the electronic wave function.

The geometry and orientation are the same by definition for a given value of Q. The upper end of the arrow for $Q = -Q_0$ in Figure 10.6 corresponds to the energy when the electron has reached the other metal ion, but none of the solvent coordinates have changed. The contribution to the solvent reorganization is thus obtained by letting the solvent relax around the new position of the electron. This energy is the same as the Born energy for a charged sphere with radius a_1. However, we must subtract the electronic part to obtain a reorganization energy. This contribution (U_1) to the reorganization energy is

$$U_1 = U_{total} - U_{solv} = \frac{1}{2a_1}\left(1 - \frac{1}{\varepsilon}\right) - \frac{1}{2a_1}\left(1 - \frac{1}{n^2}\right) = \frac{1}{2a_1}\left(\frac{1}{n^2} - \frac{1}{\varepsilon}\right). \qquad (10.31)$$

As the system moves from $-Q_0$ to Q_0, the energy of the upper PES is lowered since the solvent geometry readjusts in the move. U_1 is thus a positive contribution to the positive reorganization energy.

There is another contribution to the reorganization energy (U_2) because we move a charge a distance R from one metal ion to the next. The electronic Coulomb energy, which in principle can be calculated if the solvent is included in the calculation, is

$$U' = -\frac{1}{n^2 R}. \qquad (10.32)$$

U' is already included in both adiabatic PES, except a part that depends on solvent molecular geometry. This part is equal to

$$U'' = -\frac{1}{\varepsilon R}. \qquad (10.33)$$

Screening leads to less Coulomb attraction between the electron and the hole, that is, the total energy would have been lower without the screening. When the reorganization energy due to solvent molecular geometry is switched on, the total energy is lowered. The upper curve at $Q = -Q_0$ is too high because the Coulomb energy lowering is not yet switched on. We thus have to add to the upper PES the negative contribution U_2, which is equal to the difference between Equations 10.32 and 10.33:

$$U_2 = U' - U'' = -\frac{1}{R}\left(\frac{1}{n^2} - \frac{1}{\varepsilon}\right). \qquad (10.34)$$

The total solvent reorganization energy is thus $\lambda_{solv} = U_1 + U_2$. It may thus be written

$$\lambda_{solv} = \left(\frac{1}{2a_1} + \frac{1}{2a_2} - \frac{1}{R}\right)\left(\frac{1}{n^2} - \frac{1}{\varepsilon_0}\right). \qquad (10.35)$$

The charge transferred was one atomic unit. If the charge Z is transferred, Equations 10.30 through 10.35 should be multiplied by Z^2. Improvements of Equation 10.35 become necessary if the shape of the molecule is not spherical. Another improvement may be necessary if the distance for ET is short, so that the molecular spheres overlap. In that case, another correction may appear since less than a full charge is transferred. We have used atomic units in Equation 10.35. Electron volts are obtained when multiplication by 27.21. Nonpolar solvents leave no contribution to λ_{solv} since $\varepsilon_0 = n^2$.

The derivation was done by assuming that the molecules are neutral before ET and become charged +1 and −1 a.u. after the transfer. However, most metal complexes are charged. For example, $Fe(H_2O)_6^{2+}$ has the charge +2. The result will be the same as above if we change from +2 to +3 on one site and from +3 to +2 on another site.

The theory for solvent reorganization energy given above may be regarded as a summary of the more comprehensive theory given by R. A. Marcus and others. References may be found in the Marcus paper from 1990.

Structural reorganization energy may be obtained using bond length changes and force constants. A quite good approximation is to sum all distance changes (δr_i) squared, multiplied by a relevant force constant:

$$\lambda_{bond} = \sum_{bond=i} \frac{1}{2} k_i \delta r_i^2. \qquad (10.36)$$

The force constants may be obtained from the IR spectrum.

Another way to obtain reorganization energies is simply to calculate the PES. A quantum chemical calculation without solvent molecules gives only λ_{bond}. The solvent reorganization energies may be obtained using the Poisson–Boltzmann equation.

If all contributions are added, we obtain the total reorganization energy λ:

$$\lambda = \lambda_{bond} + \lambda_{solv}. \tag{10.37}$$

λ is a measure of how much the structure is changed when the electron leaps. If $|\Delta| \ll \lambda$, the activation barrier is linear in λ according to Equation 10.29, since $|\Delta|/\lambda$ may be neglected in this reaction. If the bond lengths are much changed, the reaction barrier for ET will be large, the reaction rate low and also strongly temperature-dependent.

10.4.3 Localized and Delocalized Mixed Valence

It is possible to make molecules or crystals of *mixed valence system* in different chemical ways. If two solutions with different valences are mixed, the chances are high that a mixed crystal is formed rather than two different crystals. It has also been possible to connect metal ions in different valence states with organic molecules. In the latter case, we may talk about a mixed valence molecule.

We may consider the interaction within a mixed valence molecule or between two adjacent metal ions in different valence states in a crystal or even in a solution. Mixed valence systems with two and three valence states are referred to as MV-2 and MV3, respectively. Figure 10.6 shows that in such a system there is absorption with energy ranging between λ and $\Delta = 2H_{12}$.

If $\Delta \geq \lambda$, there is only one minimum (Equation 10.27), and we have to consider the unpaired electron as physically delocalized. The valences are then the same and fractional. In the minimum at $Q = 0$, the electron is equally distributed over the two sites. Even for $Q \neq 0$, the distribution is close to 50/50%.

If $\lambda \gg \Delta$, the system is localized and the metal complexes have different charges. In a solution, there is a wide distribution of possible geometric arrangements. The coupling Δ is an energy difference that depends directly on orbital overlap and therefore may vary between wide limits. As long as we are dealing with an outer sphere complex, we may expect that $\lambda \gg \Delta$. Furthermore, if Δ is small, there are two almost independent parabolas, and the average excitation energy is equal to λ. The intensity of the mixed valence transition may be very weak.

If, on the other hand, $\Delta > \lambda$, there is no barrier and the excitation energy is equal to Δ. The absorption should be narrower and the energy less. The American chemists Carol Creutz and Henry Taube managed to synthesize a number of complexes where the metal ions Ru^{2+} and Ru^{3+} are connected by a bridge.

The dimetallic complex with a pyrazine bridge is shown in Figure 10.7 and is referred to as "the Creutz–Taube ion" in the case when the total charge is +5. Another possible bridge, dipyridine, is shown in Figure 10.8. $(NH_3)_5RuC_4N_2H_4Ru(NH_3)_5^{5+}$, the Creutz–Taube complex, has very distinct spectral properties from the others, immediately suggesting delocalization. λ is approximately equal to 1 eV.

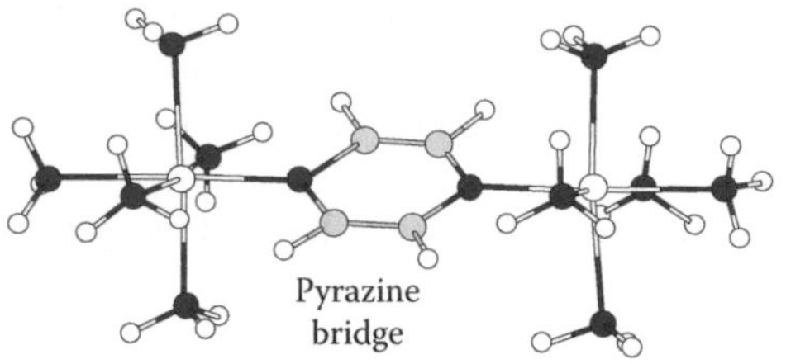

FIGURE 10.7   Creutz–Taube mixed valence system: $(NH_3)_5Ru-pyrazine-Ru(NH_3)_5^{5+}$.

Carol Creutz confirmed (1983) the delocalization by changing the polarity of the solvent. In the localized complexes, the spectrum depends almost linearly on λ. However, this is not the case for the Creutz–Taube complex. Thus, the charges in this bimetallic complex are delocalized, as has been found in a number of experiments. A number of additional experiments to find out if the Creutz–Taube ion is really delocalized have been carried out. Reymers and Hush have discussed these experiments with the conclusion that the Creutz–Taube ion is, in fact, delocalized. This means that the oxidation states of Ru atoms are the same and equal to +2.5.

If the pyrazin bridge is replaced by a adipyridine bridge, we obtain a localized system. In the localized case, each minimum of the lower PES corresponds to different integer valences on the two metal ions. On the upper PES, the valences are swapped compared with the ground state PES. Hence, the transition in the localized case is an *intervalence transition*.

It is important that the spectrum reveals whether a system is localized or delocalized (Figure 10.9). Localization is a physical property that depends on the shape of the PES.

Robin and Day defined three classes of localization, but there are no sharp borders:

Class I: Localized (trapped) valences. For example, in proteins there are often iron or copper ions in different valence states. The same is the case with sparse impurity ions in solids. The spectra are sum spectra for the different valences.

Class II: Localized valences with mixed valence spectra. In this case, ET between metal ions takes place readily. Examples: Molybdenum blue, containing molybdenum in the oxidation states IV, V, and VI: Ru^{2+}/Ru^{3+} complexes of Creutz–Taube

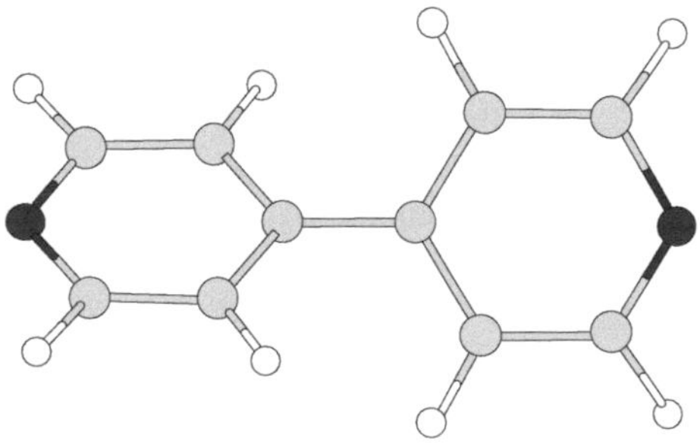

FIGURE 10.8   Dipyridine $(C_{10}N_2H_8)$ used as an alternative bridge in Figure 10.7.

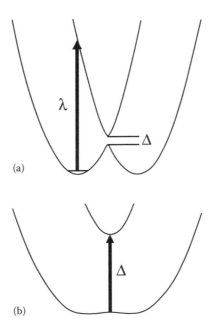

(a)

(b)

FIGURE 10.9 Intervalence transition in localized and delocalized case.

type but with bipyridine or other bridges that do not lead to delocalized valences. Prussian blue is asymmetric. Fe^{2+} is bonded to six carbon atoms of CN^- ions. Fe^{3+} are on alternant sites with the N atoms bonded.

Class III: Delocalized valences. This is a rare case with the Creutz–Taube $Ru^{2+}/$ Ru^{3+} ion as the best example. Mixed valence crystals of Mn^{3+} and Mn^{4+} may become delocalized in certain concentration ranges. The unpaired electron is in an e_g spin orbital. The spin is determined by t_{2g} electrons, and becomes unidirectional (ferromagnetic) due to the mixed valence interaction.

Localization/delocalization is an important concept in a wider perspective. We may think of a molecular wire where alkali has been added and donated to the wire. Conductivity assumes that electrons are not trapped. The condition is that $\Delta > \lambda$ (class III) for each ET step, otherwise the conductivity will be low. The latter condition is therefore the *condition for delocalized electrons*. It is often believed that a repetitive system together with unfilled orbital bands is a satisfactory condition for delocalization (metals). However, even if infinite orbitals may be constructed, they are less useful in a quantum description, if the electrons are trapped.

The above-mentioned Mn^{3+}/Mn^{4+} complex is exemplified by $La_{1-x}Ca_xMnO_3$. In this ionic crystal, we assume as usual that the ions are La^{3+}, Ca^{2+}, and O^{2-}. For $x = 0$, Mn is in the oxidation state +3, a d^4 high spin configuration. One electron occupies a strongly antibonding e_g orbital, which leads to Jahn–Teller effect and trapping. For x closer to unity, Mn goes into the oxidation state +4, a d^3 high spin configuration without Jahn–Teller distortion. In between $x = 0$ and $x = 1$, there is a mixed valence system, which permits electrons to float through delocalized orbitals.

10.4.4   WAVE FUNCTIONS

In an ET system, the number of electrons is odd. Let us assume a symmetric ET system, so that the subsystems D (donor) and A (acceptor) are identical. We further assume that the highest occupied molecular orbitals (HOMO) are filled. Orbitals of the subsystems with the same energy combine to form orbitals of the type $\phi_+ = \phi_1 + \phi_2$ and $\phi_- = \phi_1 - \phi_2$, where ϕ_1 and ϕ_2 are located on D and A, respectively, If ϕ_1 and ϕ_2 are lowest unoccupied MOs (LUMOs), ϕ_+ is the new LUMO, extended over the whole system. Let us occupy ϕ_+ by one electron to obtain an ET system.

Generally, the orbitals may be written as (ϕ_1 and ϕ_2 are assumed orthogonal)

$$\phi_+ = \left(\cos\xi\right)\phi_1 + \left(\sin\xi\right)\phi_2$$
$$\phi_- = -\left(\sin\xi\right)\phi_1 + \left(\cos\xi\right)\phi_2. \tag{10.38}$$

Written in this way, the MOs are automatically orthonormalized.

In the precursor system $\xi = 0°$, $\phi_+ = \phi_1$, and $\phi_- = \phi_2$. The electron is in ϕ_1 on D. Moving along the pathway for ET is equivalent to changing from $\xi = 0°$ to $\xi = 90°$. $\xi = 45°$ corresponds to the symmetric point: $Q = 0$. ϕ_+ gradually changes from the donor orbital ϕ_1 on D to the acceptor orbital ϕ_2 on A. ϕ_- changes from $\phi_- = \phi_2$ to $\phi_- = \phi_1$. At $\xi = 90°$, since ϕ_+ is occupied by one electron and ϕ_- is empty, the final system has one electron at A and none at D, in the LUMO. The orbitals up to and including HOMO are filled.

The coefficients and orbital energies are obtained from the eigenvalue problem:

$$\begin{pmatrix} H_{11} - \varepsilon & H_{12} \\ H_{12} & H_{22} - \varepsilon \end{pmatrix} \begin{pmatrix} \cos\xi \\ \sin\xi \end{pmatrix} = \begin{pmatrix} 0 \\ 0 \end{pmatrix}. \tag{10.39}$$

We multiply the upper equation in Equation 10.39 by $\sin\xi$ and the lower one by $\cos\xi$ and obtain

$$\left(H_{11} - e\right)\cos\xi \, \sin\xi + H_{12} \sin^2\xi = 0$$
$$H_{12} \cos^2\xi + \left(H_{22} - e\right)\sin\xi \, \cos\xi = 0. \tag{10.40}$$

Subtracting these two equations and using $2\sin\xi \cos\xi = 2\sin 2\xi$ and $\sin^2\xi - \cos^2\xi = \cos 2\xi$ gives

$$\text{tg}2\xi = \frac{2H_{12}}{H_{22} - H_{11}}. \tag{10.41}$$

On the other hand, adding the two equations leads to

$$\varepsilon_\pm = \frac{H_{11} + H_{22}}{2} \pm \frac{H_{12}}{\sin 2\xi}. \tag{10.42}$$

The solution with the minus sign is obtained if we use the eigenvector $(\sin \xi, -\cos \xi)$. Thus, ξ may be determined from Equation 10.41 and inserted into Equation 10.42 to get the eigenvalues.

The difference between the eigenvalues is

$$\varepsilon_+ - \varepsilon_- = \frac{\Delta}{\sin 2\xi}. \tag{10.43}$$

Hence, the minimum difference is for $\xi = 45°$ at the avoided crossing, when the eigenvalue difference is equal to Δ.

The eigenvalue problems of Equations 10.20 and 10.39 are identical as mathematical eigenvalue problems, but the physics behind them is different. Equation 10.43 was used for an orbital, while the former eigenvalue problem was used to construct the Marcus PES. By Koopmans' theorem, Δ of the orbital problem happens to be an approximation to the correct Δ at the avoided crossing (see Section 10.6.4). A many-electron calculation of Δ may be performed alternatively, as mentioned above.

If $|H_{12}|$ is small compared with $|H_{11} - H_{22}|$, Equation 10.41 shows that

$$\xi \approx \frac{H_{12}}{|H_{11} - H_{22}|} = \frac{\Delta}{2\lambda}. \tag{10.44}$$

In most cases, $|\Delta|$ decreases roughly exponentially as a function of distance R between the donor and the acceptor:

$$\Delta \approx \Delta_0 \exp(-\alpha R) \tag{10.45}$$

α is usually in the range $0.5 \text{ Å}^{-1} < \alpha < 1.5 \text{ Å}^{-1}$.

10.4.5 INTENSITY OF INTERVALENCE TRANSITION

Using Equation 10.38 the transition dipole moment is

$$\begin{aligned}
\langle \phi_+ | x | \phi_- \rangle &= \langle (\cos \xi)\phi_D + (\sin \xi)\phi_A | x | -(\sin \xi)\phi_D + (\cos \xi)\phi_A \rangle \\
&= -\cos \xi \cdot \sin \xi \langle (\phi_D | x | \phi_D) + \sin \xi \cdot \cos \xi \langle \phi_A | x | \phi_A \rangle.
\end{aligned} \tag{10.46}$$

We assume that the donor and acceptor are at some distance, and hence the integrals that contain both ϕ_D and ϕ_A are small compared with the integrals that contain only one of the terms, as assumed in Equation 10.46. We let the donor and acceptor be located on the x-axis with distance R in between and the origin at the donor D.

The first integral in Equation 10.46 is close to zero, since x has small absolute values in this region with varying sign. In the second integral, x is approximately equal to R and may thus be taken outside the integration. The remaining integral is a normalization integral. Hence, we obtain

$$\mu_x = \langle \phi_+ | x | \phi_- \rangle = \cos \xi \cdot \sin \xi \cdot R. \tag{10.47}$$

μx is thus equal to ξR for small ξ, and equal to $R/2$ for $\xi = 45°$. In the localized case (small ξ), Equations 10.44 through 10.46 may be used to obtain

$$\mu_x = \xi R = \frac{R\Delta}{2\lambda} = \frac{R\Delta_0 e^{-\alpha R}}{2\lambda}. \tag{10.48}$$

This equation tells us that the intensity of the allowed intervalence transition decreases fast with the distance between the two metal ions.

In the delocalized case, the excitations are predominantly in the avoided crossing region. Using Equation 10.47 with $\xi = 45°$, we obtain

$$\mu_x = R/2. \tag{10.49}$$

The intervalence transition is often in the visible region and since the transition is allowed, a mixed valence complex often has a very strong color. A good example is Prussian blue, which can be made by adding Fe^{3+} to $K_4Fe(CN)_6$, where the valence state of Fe is +2. Alternatively, Fe^{2+} may be added to $K_3Fe(CN)_6$, where the valence state of Fe is +3. If compounds with the same valence are mixed, there is no strong color. In the present case, the system is not strictly symmetric since one Fe is connected to C while the other is connected to N in CN^-. The lack of symmetry guarantees a unique product ("Turnbull's blue" is the same as Prussian blue).

10.5   RATE OF ET REACTIONS

In the most common case, the ET reaction is asymmetric with a driving force $\Delta G^\ominus < 0$. If we assume that the reaction starts from the left energy minimum, the right minimum will be lower, since the Marcus parabolas are supposed to be free energy PES (Figure 10.10). The rate is obtained via an Arrhenius equation. It is important to multiply the exponential with an electronic factor that accounts for the probability to pass the avoided crossing region.

10.5.1   GIBBS FREE ENERGY OF ET REACTION

As an example, we may take the reaction

$$Fe^{3+} + Cu^+ \rightarrow Fe^{2+} + Cu^{2+}. \tag{10.50}$$

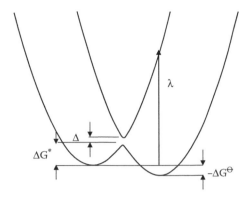

FIGURE 10.10 Marcus model with important parameters: reorganization energy λ, coupling $H_{12} = \Delta/2$, free energy of reaction $\Delta G^{\ominus}$, and activation energy ΔG^*.

If we use Table 10.1, we find that the reaction $Fe^{3+} + e^- \rightarrow Fe^{2+}$ has $\Delta G^{\ominus} = -0.771$ eV, while the reaction $Cu^{2+} + e^- \rightarrow Cu^+$ has $\Delta G^{\ominus} = -0.16$ eV. Subtracting the two equations leads to $\Delta G^{\ominus} = -0.77 + 0.16 = -0.61$, thus the reaction Equation 10.50 is spontaneous.

The Marcus curves are not enthalpy curves, but free energy curves. We may think of them as projected out from a many-dimensional surface. Usually, only one nuclear motion, corresponding to a single vibrational mode, is involved in the promotion of ET. In the case of metal complexes, this motion is a decrease in some ML distances in one metal complex and an increase in some ML distances in the other complex. The corresponding mode is referred to as the *breathing mode*. Motions that do not directly affect the ET problem have to be averaged out and (in principle) included in every point on the Marcus curve. Furthermore, the entropy term in $\Delta G = \Delta H - T\Delta S$ has to be included, although it is unimportant if the gross geometrical structure of the system remains intact during the ET process. If one or more atoms dissociate (for example, I^- in CH_3I^-), the entropy term may become important at elevated temperatures.

10.5.2 ADIABATIC, ASYMMETRIC SYSTEM

ET reactions belong to the few cases when the rate constant k may be calculated on a fundamental level. If the height of the barrier is called ΔG^*, k may be written as

$$k = v_n \cdot \kappa \cdot \exp\left(-\frac{\Delta G^*}{k_B T}\right). \tag{10.51}$$

κ is the *electronic (transmission) factor*, equal to the probability to pass the avoided crossing region on the lower PES. We will first assume that κ is equal to unity. This is called *adiabatic* ET and physically means that the system moves exclusively on the lower PES. After the barrier has been passed, the electronic state corresponds to a state where the electron has moved to the other site.

v_n is the vibrational frequency that corresponds to the nuclear motions, which comprise the promoting nuclear coordinate Q in Figure 10.10. In a metal complex, we are dealing with the breathing modes of ligands around the nucleus, usually located at $\approx 500 \text{ cm}^{-1}$.

We will show next how ΔG^* may be calculated or measured. Usually, it is possible to determine the free energy $\Delta G^{\ominus}$ of the reaction experimentally by measuring the reduction potentials for the acceptor and donor and subtract. If we assume $\Delta S^{\ominus} = 0$, we have ordinary PES based on $\Delta H^{\ominus}$. PES may be assumed to be parabolic. It is also possible to define an effective force constant.

The reorganization energy λ is defined in Figure 10.10. We denote the minimum points on the Q-axis, $-Q_0$ and Q_0 $(Q_0 > 0)$, and write the equations of the parabolas as $k/2 \times (Q + Q_0)^2$ and $k/2 \times (Q - Q_0)^2$. Since $\lambda = 2kQ_0^2$ and k is measurable as an effective force constant, we may determine Q_0 if we know λ and k. The height of the barrier ΔG^*, assuming that $\Delta = 0$, may now be expressed in two ways where Q is the crossing point:

$$\Delta G^* = \frac{k}{2}\left(Q + Q_0\right)^2 = \frac{k}{2}\left(Q - Q_0\right)^2 + \Delta G^{\ominus}. \tag{10.52}$$

If Q is eliminated, we obtain

$$\Delta G^* = \frac{\lambda}{4}\left(1 + \frac{\Delta G^{\ominus}}{\lambda}\right)^2. \tag{10.53}$$

ΔG^* is the activation energy for the ET reaction.

It follows from Equation 10.53 that the activation barrier ΔG^* disappears if $\Delta G^{\ominus} = -\lambda$. If, on the other hand, $\lambda \gg \Delta G^{\ominus}$, ΔG^* depends linearly on λ. We may insert Equation 10.53 into Equation 10.51 and take the logarithm and obtain Figure 10.11. The rate k of the ET reaction is increased with increasing driving force until a maximum is reached for $\Delta G^{\ominus} = -\lambda$. If the driving force is further increased, the rate decreases with increasing driving force (Marcus inverted region) (Figure 10.11).

This behavior appears strange. One is used to seeing the speed of a motor vehicle increase when more energy is applied. Originally, it was quite difficult to find reactions with a driving free energy $|\Delta G^{\ominus}| > \lambda$. Rehm and Weller used PIET to obtain a large $|\Delta G^{\ominus}|$, and were able to find examples of PIET in a solvent where $|\Delta G^{\ominus}| > \lambda$. They found a maximum rate for $|\Delta G^{\ominus}| = \lambda$, but for $|\Delta G^{\ominus}| > \lambda$ the rate remained at the highest value. The rate at least did not continue to increase linearly with increasing $|\Delta G^{\ominus}|$, but it did not follow the expected Marcus inverted behavior.

Later, experimental examples that proved that the inverted region really exists were found (see Miller's experiment in Section 10.5.6). In addition, there was a theory that showed that the Rehm and Weller experiment does not disprove the Marcus inverted region.

10.5.3 ET RATE FOR NONADIABATIC REACTION

Equation 10.51 is a typical expression for a rate constant if it is assumed that the nuclei are moving by classical mechanics with the ground state PES as potential.

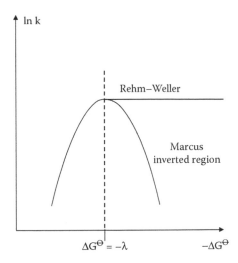

FIGURE 10.11 The natural logarithm of the reaction rate k as a function of driving force $(-\Delta G^\ominus)$ according to the Marcus model.

The derivation here, taken from N. Sutin and B. S. Brunschwig, leads to the simplest approximation for the rate. However, in ET reactions it is unavoidable that we have to consider not only the barrier, but also the avoided crossing between two different PES. In other words, there is a probability that the system may jump to the upper surface. In that case, the system is in a new electronic state, but the character of the wave function does not change, meaning that ET does *not* take place.

The probability for passing the avoided crossing on the ground state PES determines the rate of ET. A probability below unity means that not every attempt to move classically to the left across the barrier is successful. The system bounces back to the starting point in the left minimum. If the donor and acceptor are separated by more than 10 Å, as is usually the case in biological ET, the rate is much reduced. One may say that being at a large distance from each other, the donor and acceptor "do not know about each other." The energy gap at the avoided crossing is extremely small, and this means that the down-going Marcus parabola is ignored most of the times when the system passes the avoided crossing.

10.5.4 ELECTRONIC FACTOR κ

The probability for passing from the lower PES to the upper is expressed in the Landau–Zener approximation. We will use this approximation to calculate κ in Equation 10.51. The Landau–Zener probability P for crossing from the ground state PES to the upper state may be expressed in terms of the slopes of the matrix elements, H_{11} and H_{22}, just outside the crossing region and the coupling $H_{12} = 1/2\Delta$:

$$P = \exp\left[-\frac{\pi\Delta^2}{2\hbar \frac{d}{dt}\left(H_{11} - H_{22}\right)} \right].$$

(10.54)

Note that if Δ tends to zero, P tends to unity.

Since the probability for choosing the down-going PES is $1 - P$, the system "ignores" the down-going PES and no ET takes place, if P tends to unity. However, sooner or later, the system turns back on the up-going PES and again approaches the avoided crossing. The same probabilities are valid (P and $1 - P$). If the system continues on the same PES, all further ET probability is lost. ET only happens if the system switches to the down-going PES at the avoided crossing with probability P. The different possibilities at repeated crossings are given in Figure 10.12.

We conclude that the probability for ET is given by the total probability to switch to the down-going curve. This probability is the electronic factor κ, given by the following series:

$$k = 1 - P + P(1-P)P + P(1-P)^3 P + P(1-P)^5 P + P(1-P)^7 P + \qquad (10.55)$$

Each time the avoided crossing region is passed, the probability is P for crossing and $(1 - P)$ for staying on the same surface. Except for the first $1 - P$, the remaining terms form a geometric series with the quotient $(1 - P)^2$. Hence, we obtain

$$\kappa = 1 - P + \frac{P^2(1-P)}{1-(1-P)^2} = \frac{(1-P)-(1-P)^3 + P^2(1-P)}{1-(1-P)^2} = \frac{2(1-P)}{(2-P)}. \qquad (10.56)$$

We first approximate H_{11} and H_{22} in Equation 10.54 by straight lines in the avoided crossing region:

$$\frac{d}{dt}(H_{11} - H_{22}) = \frac{d}{dQ}(H_{11} - H_{22})\frac{dQ}{dt}. \qquad (10.57)$$

Using Equations 10.25 and 10.29, we obtain

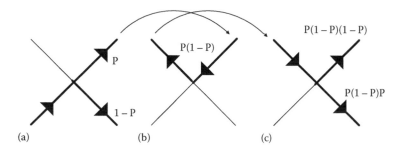

FIGURE 10.12 Probability to pass the barrier coming from the reactant state on the left. In (a) the successor state is reached with probability $(1 - P)$. (b) Going back on the negative Q-axis, the ground state is avoided with probability $P(1 - P)$. This probability enters (c) from the excited state going forward. The successor state is reached with probability $P(1 - P)P$.

$$\left(\frac{d}{dQ}(H_{11} - H_{22})\right)_{Q=0} = 2kQ_0 = 2k\sqrt{\frac{\lambda}{2k}} = \sqrt{2k\lambda}. \tag{10.58}$$

In the simplest case, we assume that the two oscillators (H_{11} and H_{22}) have the same force constant k (it may be shown that an "effective" force constant can be used). K is the force constant for the vibrational motion. The nuclear frequency v_n is

$$v_n = \frac{1}{2\pi}\sqrt{\frac{k}{M}} \Rightarrow \sqrt{k} = \sqrt{M}v_n 2\pi. \tag{10.59}$$

Inserting Equation 10.59 into Equation 10.58 implies

$$\left(\frac{d}{dQ}(H_{11} - H_{22})\right)_{Q=0} = \sqrt{2k\lambda} = 2\pi\sqrt{2\lambda M}v_n, \tag{10.60}$$

where dQ/dt is the system velocity. The Maxwell average is

$$\frac{dQ}{dt} = \left(\frac{2k_B T}{M\pi}\right)^{1/2}, \tag{10.61}$$

where k_B is Boltzmann's constant and T is the absolute temperature. We may now write Equation 10.57:

$$\frac{d}{dt}(H_{11} - H_{22}) = \frac{d}{dQ}(H_{11} - H_{22})\frac{dQ}{dt} = 4\pi v_n\left(\frac{\lambda M k_B T}{M\pi}\right)^{1/2} = 4v_n\left(\pi\lambda k_B T\right)^{1/2}. \tag{10.62}$$

Before this expression is inserted into the expression for the Landau–Zener probability, it is convenient to introduce an *electronic frequency* v_{el} (without particular interpretation), which has the same dimension as the nuclear frequency:

$$v_{el} = \frac{\pi\Delta^2}{2\hbar}\left(\frac{1}{4\pi\lambda k_B T}\right)^{1/2}. \tag{10.63}$$

The Landau–Zener probability Equation 10.54, can now be expressed as

$$P = \exp\left(-\frac{v_{el}}{2v_n}\right). \tag{10.64}$$

The electronic factor κ Equation 10.56, is thus equal to

$$\kappa = \frac{2\left[1 - \exp\left(-v_{el}/2v_n\right)\right]}{2 - \exp\left(-v_{el}/2v_n\right)}. \tag{10.65}$$

We now insert Equation 10.65 into Equation 10.51 and obtain an expression for the ET rate:

$$k = v_n \cdot \frac{2\left[1 - \exp\left(-v_{el}/2v_n\right)\right]}{2 - \exp\left(-v_{el}/2v_n\right)} \cdot \exp\left(-\Delta G^* / k_B T\right). \tag{10.66}$$

10.5.5 ADIABATIC AND NONADIABATIC LIMITS

If the electronic frequency is large compared with the nuclear frequency ($v_{el} \gg v_n$), the exponents in Equation 10.66 can be ignored and we obtain the electronic factor equal to unity ($\kappa = 1$). This is the limit for large Δ. Since the electronic factor is unity, the rate depends on Δ only via the activation barrier. This dependence is expressed for symmetric ET in Equation 10.29, but for asymmetric ET we used the approximation $\Delta = 0$ in Equation 10.53. The derivation of the barrier height is not difficult in the asymmetric case, however, since we just add $-\Delta G^\ominus$ to H_{22} in Equation 10.24 and derive the slightly more complicated expression similar to E_- in Equation 10.26. If Δ is very large so that the activation barrier disappears, the electron (or hole) becomes delocalized over the donor and acceptor in the lowest electronic state. Oscillations between the donor and acceptor are induced if the electron is "dropped" on the donor (meaning that a boundary condition is used that makes the electron localized to the donor at time t = 0). The rate of these oscillations will be independent of nuclear relaxations. Another way is to see that the electron is in an extended MO across the donor and acceptor, with the system preferably considered as a single molecule.

If, on the other hand, the electronic frequency is small compared with the nuclear frequency ($v_{el} \ll v_n$), the exponentials in Equation 10.66 may be Taylor expanded:

$$\kappa = \frac{2\left[1 - \exp\left(-v_{el}/2v_n\right)\right]}{2 - \exp\left(-v_{el}/2v_n\right)} \approx \frac{v_{el}}{v_n}. \tag{10.67}$$

Hence, the following expression for the rate:

$$\kappa = v_n \cdot \frac{v_{el}}{v_n} \cdot \exp\left(-\frac{\Delta G^*}{k_B T}\right) = \frac{\Delta^2}{4\pi\hbar}\left(\frac{\pi^3}{\lambda k_B T}\right)^{1/2} \exp\left(-\frac{\Delta G^*}{k_B T}\right). \tag{10.68}$$

Notice that this rate constant is independent of v_n. The only nuclear dependence is via the activation energy ΔG^*.

10.5.6 MILLER'S EXPERIMENT

The American chemist, John R. Miller, set out to prove the Marcus inverted region by creating a large $|\Delta G^\ominus|$. Using x-rays in pulsed radiolysis, he was able to create trapped electrons in an organic solvent frozen at liquid helium temperatures.

To avoid mobility of the molecules involved in ET, experiments in glass were invented. Some of these reactions are very exothermic, that is, with a large value of $-\Delta G^{\ominus}$ and are in the Marcus inverted region. The solutions were frozen at 77 K. Electron donors were created in the form of trapped electrons by pulsed radiolysis. Miller and Beitz obtained a reduction by a factor of 10^5 for very exothermic reactions well into the Marcus inverted region.

Miller explained the experiment by assuming that there is ET even if the molecules do not touch each other. The rate due to electron tunneling was considered to follow the equation

$$k(R) = k(0)\exp(-R/a),\qquad(10.69)$$

where R is the distance between the acceptor and the trapped electron and a was assumed to be 1 Å. We know that this figure is close to what has been obtained in later calculations. k(R) is the rate at distance R between the molecule and the trapped electron.

The Marcus model was also proven in intramolecular ET by Gerhardt L. Closs and J. R. Miller. They synthesized molecules of the kind seen in Figure 10.13. The spacer is a molecule found in nature (steroid). At the end points, a donor D and an acceptor A are attached. They assumed that the coupling through the cyclohexane spacer is almost the same for the different acceptor molecules.

Miller and Closs used again pulsed radiolysis. They were able to create trapped electrons in an organic solvent frozen at liquid helium temperatures. Which had enough free energy to enter the inverted region. Closs and Miller found a downgoing curve in Figure 10.11 for energy $|\Delta G^{\ominus}| > \lambda$. Thus, it was proven that the Marcus model also works for intramolecular charge transfer.

10.6   ELECTRONIC COUPLING

The concept of electronic coupling appears in a number of situations. In Chapter 6, we discussed magnetic coupling between metal ions, referred to by P. W. Anderson as *superexchange*. As detailed by P. Bertrand, magnetic coupling is irrelevant to the coupling in ET problems. R. A. Marcus used a very crude model for coupling. As we know, coupling is unimportant for the rate in adiabatic ET reactions. J. Halpern and L. E. Orgel, and H. McConnell, attempted to include coupling to treat nonadiabatic ET.

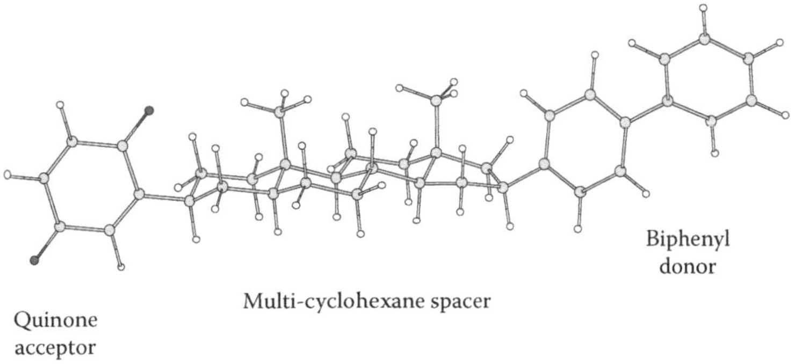

Biphenyl donor

Multi-cyclohexane spacer

Quinone acceptor

FIGURE 10.13   One of the molecules of Closs and Miller.

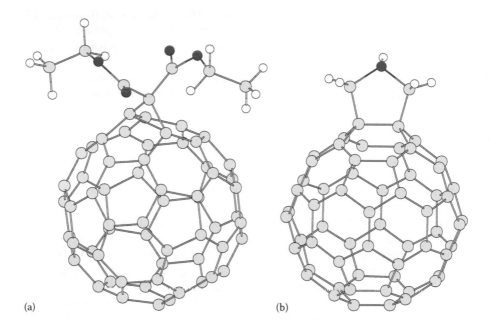

(a)   (b)

FIGURE 3.12 C_{60} molecule with groups covalently bonded to C_{60}.

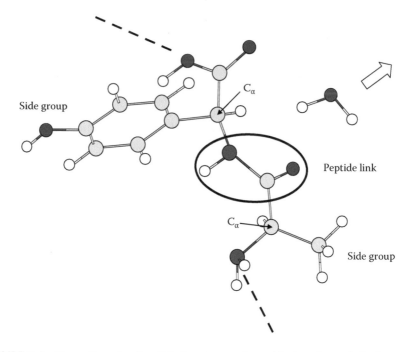

Side group

C_α

Peptide link

C_α

Side group

FIGURE 9.4   Formation of a dipeptide from two amino acids, tyrosine and alanine.

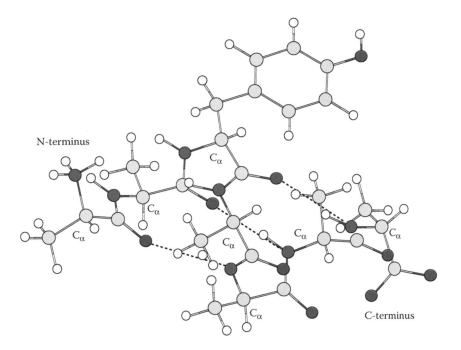

FIGURE 9.5   α-Helix with peptides in order, from N-terminus, Ala-Ala-Tyr-Ala-Ala-Ala-Ala.

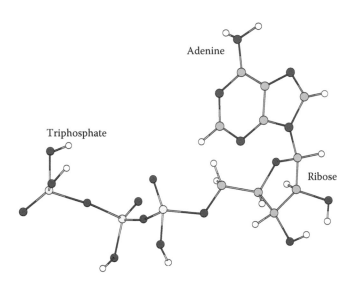

FIGURE 11.2   ATP. The adenine group has four nitrogens (dark blue). The sugar group is a pentose with three oxygens (red). In ATP, there are three phosphorous atoms (yellow) bonded to four oxygen atoms (red); in ADP only two and in AMP only one phosphorous atom.

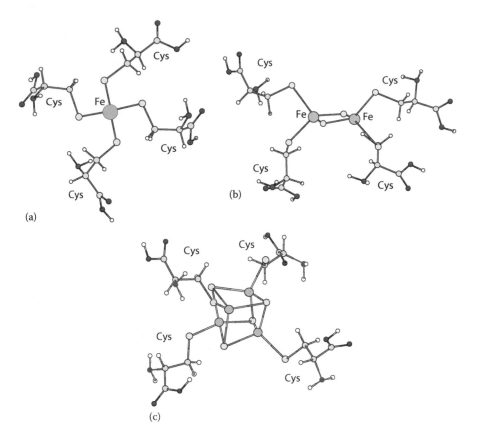

FIGURE 11.11 FeS centers and their connection to cystein (Cys) peptides in the protein, with one (a), two (b), and four (c) iron atoms.

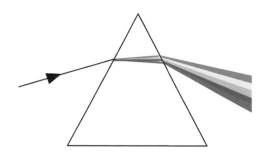

FIGURE 12.2 Refraction: Light ray passing a prism.

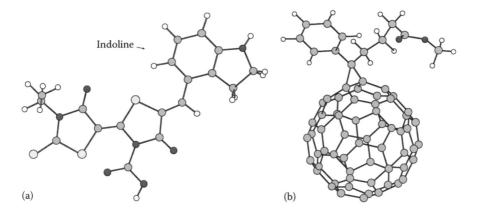

(a)

(b)

FIGURE 13.10 Examples of photosensitizers in Grätzel cells: (a) indoline-like molecule and (b) PCBM.

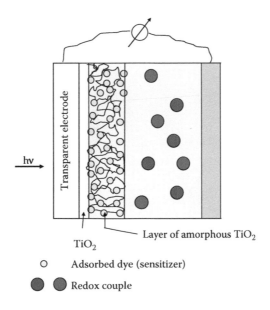

FIGURE 13.12 Grätzel cell.

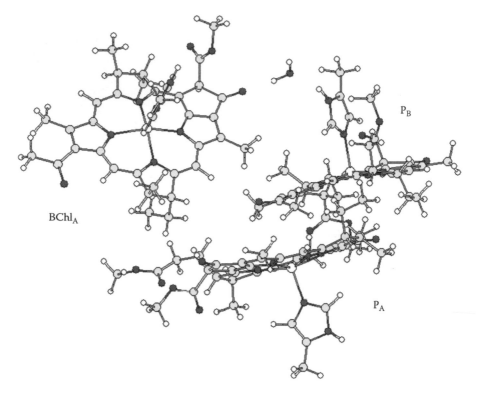

FIGURE 15.9   Detail of the bacterial RC. Notice the water molecule between BChl$_A$ and P$_B$.

(a)   (b)

FIGURE 16.7   Nickel oxide (a) and aqueous solution of NiSO$_4$ (b). The green color is due to ligand field transitions in the NiO$_6$ complex.

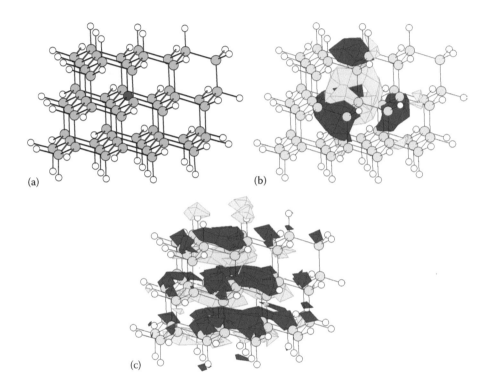

FIGURE 16.10 Piece of diamond saturated by hydrogen atoms and with a nitrogen atom (dark blue) replacing a carbon atom (a, left). In (b) we see the HOMO of a spin polarized calculation (center); the LUMO is seen in the right picture (c). Both the HOMO and the LUMO are considerably delocalized.

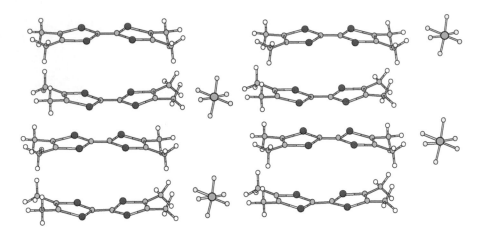

FIGURE 18.10 8(TMTSF)⁺ and 4PF₆⁻ as a part of two stacks.

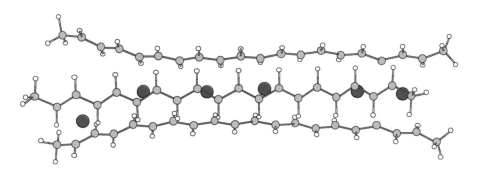

FIGURE 18.13 The result of geometry optimization of three PA chains and six alkali atoms (violet spheres).

FIGURE 18.16 Hole state in oxidized PPY.

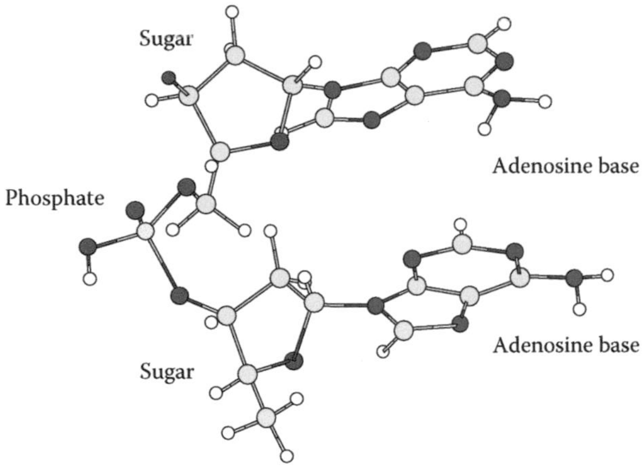

FIGURE 18.23 Part of a DNA strand with two adenosine bases.

Unfortunately, the term "superexchange" was revived as a concept important for ET in the beginning of the 1980s and became mistakenly ascribed to McConnell. In order to avoid mixing between coupling in ET phenomena and magnetic coupling, we will *not* use the term "superexchange" in this book. Needless to say, the electronic coupling in ET problems is generally less well understood, and it is also hard to calculate accurately. Therefore, we devote the present section to the concept of *coupling* in ET problems.

10.6.1 GAMOW MODEL

Since the rate depends on Δ^2, it is important to know the long distance behavior of Δ, for example, in the case of proteins. Before 1980, it was commonly assumed that a saturated σ system prevents ET, while electrons can more or less freely transfer in π-systems. This is incorrect, as we will see. The first indications of a through-bond interaction came when x-ray photoelectron spectroscopy (XPS) was applied on symmetric molecules. Symmetrically placed lone pairs give rise to two lines in the spectrum with equal intensity. Lone pair electrons are often the first to be ionized and the ionization energies therefore correspond to the orbital energies of the HOMO and HOMO − 1, according to Koopmans' theorem. The energy splitting of the ionization energies is thus to be compared with Δ, which was derived in Subsection 10.3.3. The energy splitting of the lone pair electron in XPS shows that there is interaction through the sigma-bonded system between them.

The term "through-bond interaction" as opposed to "through-space interaction" was first used in the case of XPS. Due to the mentioned identification via Koopmans' theorem, this applies to ET as well. The photoelectron splitting can be identified with the "tunneling" mechanism for the D-to-A motion of the electrons. ET is consistent with the qualitative textbook model for tunneling derived by Gamow (Section 1.4.3). Assuming a rectangular potential barrier for an electron of height V, the tunneling rate Ψ^2 may be calculated using

$$\Psi = C\exp\left[-\sqrt{2m(V-E)/27.21}\cdot(R-R_0)\cdot 1.89/\hbar\right] = C\exp\left[-k(R-R_0)\right],$$

where

$$k = \sqrt{2m(V-E)}\frac{1.89}{\hbar}. \tag{10.70}$$

In Equation 10.70, V and E should be expressed in electron volts, and R and R_0 in Ångström. Due to the factors 27.21 and 1.89, a.u. may be used for $\hbar$ and m; thus, $\hbar = 1$ and m = 1. If we use the reasonable value V − E = 2 eV, we obtain k = 0.73 Å^{-1}. However, we are interested in the density, thus the exponential decrease for electrons is given by

$$\Psi^2(R) \sim \exp(-2k(R-R_0)) = \exp(-\beta(R-R_0)), \tag{10.71}$$

from which we obtain $\beta = 1.45$ Å^{-1} for electrons. This order of magnitude is often found experimentally for valence electrons in the case of ET between metal ions in proteins.

Electrons that transfer are loosely bound valence electrons. If we use $\beta = 1$ Å^{-1}, the ET rate decreases by a factor of e ≈ 2.72 per Å. If the distance is 3 Å, the rate decreases by e$^3 \approx 20$. For every 3 Å increase in the distance between D and A, the rate thus decreases by a factor of 20. For a distance of 15 Å, the decrease factor is consequently $20^5 = 3.2 \times 10^6$. Such a low tunneling rate is usually fully acceptable in a biological system. Only the charge separation process requires a million times faster rate. The chlorophyll units exchanging electrons are therefore in close contact with each other.

Unfortunately, the Gamow model is incapable of accounting for chemical structure. Is the function of distance always exponential or are other mechanisms possible?

10.6.2 ORBITAL INTERACTION MODEL

It is easy to show that Δ very likely is increased if there is an orbital contact between two electron exchanging systems. As pointed out in Section 1.4, Δ may be calculated at the top of the activation barrier in a symmetric system. In the case of an asymmetric system, the transition state should occur when the relevant MOs are localized half on the donor and half on the acceptor. In principle, one may find the activated geometry by modifying relevant bond lengths on the donor and/or acceptor ("relevant" means consistent with the physical meaning of the reaction path).

π-systems are usually very good in coupling metal ion d orbitals and this is illustrated in Figure 10.14. In this figure, the orbital interactions are shown in the Creutz–Taube mixed valence system. In the left part, the relevant π orbitals are seen on the left from above the plane of the molecule. In the central part of Figure 10.14, the orbitals are turned 90°.

The most important orbitals of the pyrazine bridge for mediating the interactions between the metal ions are the HOMO and LUMO, shown on the left. The π orbital is occupied by two electrons, and the π^* orbital is empty.

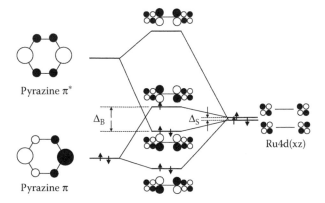

FIGURE 10.14 Illustration of electronic coupling via an aromatic bridge. Δ_S is coupling through space. Δ_B is coupling through bonds.

The D and A MOs of the Ru 4d subshell are shown on the right. Since these orbitals are well separated, direct interaction through space, Δ_S, is very small. Ru^{2+} ion has six occupied electrons in the 4d subshell. It is clear that the six electrons are in t_{2g} orbitals, pointing in between the ligands. Ru^{3+} ion has five 4d electrons. The lowest energy t_{2g} orbitals are filled. The highest t_{2g} orbital is antibonding and singly occupied and is the one that is directed toward the pyrazine bridge and interacts with its π orbitals.

The result of the interaction is seen in the middle of the figure. Bonding and anti-bonding combinations are formed with the ligand orbitals. Ru4d(zx) orbitals include three electrons from the Ru ions. In ordinary ligand field theory, there is only a single occupied orbital on the ligands, which interacts with the metal d orbitals. In the present case, the empty π orbital is also involved and forms a strongly bonded MO with the metal 4d orbitals. Together with the antibonding π – Ru4d(xz) MO, this orbital forms the new HOMO and LUMO of the whole mixed valence system. The orbital energy difference between the HOMO and LUMO (Δ_B) has increased compared with the Ru4d orbitals interacting through space. The reason is that orbital interactions are involved. We cannot be sure of the final ordering of the orbitals without detailed calculations.

If the larger dipyridine molecule is used as a bridge, the symmetric and anti-symmetric bridge orbitals are much closer in energy. The final energy gap is thus much less.

The discussion may be formalized using the partitioning technique of Appendix 4. The transition state may be found by inspection of the LUMO and LUMO + 1 in the electron case and the HOMO and HOMO − 1 in the hole case. According to Equation 10.20 MOs have the character $\phi_1 + \phi_2$ and $\phi_1 − \phi_2$ at the transition state. This form is easily checked out by inspection. For this geometry, the Fock matrix may be written as (see Appendix 4)

$$\overline{H}_{da} = H_{da} - \eta\left(H_{bb} - E\right)^{-1}\vartheta^{\dagger} = H_{da} - \sum_{k}\frac{\eta_{ak}\vartheta_{kd}}{\left(E - E_k\right)}. \tag{10.72}$$

The summation runs over the MOs of the bridge. H_{da} represents the direct interaction between the donor and the acceptor. Usually, this matrix element is negligible. The matrices η_{ak} and ϑ_{kd} represent the interaction between D and A and the bridge. For simplicity, we assume that the contacts are at just two bridge atomic orbitals: i for D and j for A. The matrix elements η_{ak} and ϑ_{kd} are between the donor orbital on D and the bridge at i and between the acceptor orbital on A and the bridge, respectively. We may write

$$\eta_{ak} \to \eta C_{ik} \text{ and } \vartheta_{kd} \to \theta C_{jk}, \tag{10.73}$$

where η and θ are matrix elements. We may now rewrite Equation 10.72 as

$$\overline{H}_{da} = H_{da} - \eta\left(H_{bb} - E\right)^{-1}\vartheta^{\dagger} = H_{da} - \eta\vartheta\sum_{k}\frac{C_{ik}C_{jk}}{\left(E - E_k\right)}. \tag{10.74}$$

The effective interaction between the bridge on the one hand and D and A on the other may thus be thought of as due to contributions from all MOs of the bridge. E is the energy of the electron to be transferred and E_k is the energy of the occupied or unoccupied bridge orbitals. The contribution is thus larger, the closer the HOMO–LUMO gap of the bridge orbital is and the larger the coefficients of the bridge orbital at the contact points j and i to D and A, respectively.

The type of interaction that is of relevance is of the following type (Figure 10.14). The symmetric combination of donor–acceptor MO interacts with a symmetric bridge MO, while the antisymmetric donor–acceptor orbital interacts with an antisymmetric bridge MO. Since the distance between the donor and the acceptor is large, the energy splitting without the bridge is extremely small. The interaction with the bridge orbital increases this energy splitting. There are a great number of these contributions. The final behavior as a function of distance is very complicated, but very often there is an exponential decrease "in average" (see below).

10.6.3 STATE INTERACTION MODEL

We first consider configuration interaction (CI) as described in Section 3.5. In particular, we are interested in the excited states that arise if orbital substitutions are done on the bridge. The quantum mechanical description of the donor and acceptor is not changed. The partitioning technique may be used to treat this case (Appendix 4). What was the donor–acceptor matrix in the orbital case previously, now corresponds to the two states where the electron is on the donor and acceptor, respectively. At a geometry that corresponds to the activated state, the electronic states, separated by Δ, are symmetric and antisymmetric combinations of these states. We diagonalize the Hamiltonian matrix where substitutions are carried out on the bridge. The final equation for the effective interaction matrix between the donor and acceptor looks the same as Equation 10.72:

$$\overline{H}_{da} = H_{da} - \sum_k \frac{\eta_{ak}\vartheta_{kd}}{\left(E - E_k\right)}, \tag{10.75}$$

where E refers to the ground state energy at the activated state; $\{E_k\}$ refer to the excited state energies, except, of course, the lowest excited state corresponding to the upper "Marcus state"; and η_{ak} and ϑ_{kd} are interaction matrices between the "Marcus states" and states corresponding to excited states of the bridge (see "pathway model" below). Also notice that there is not even a hint of exponential decrease in Equation 10.75. Exponential decrease is hidden in η_{ak} and ϑ_{kd}, in orbital overlap terms, etc.

Although the term superexchange should be reserved for magnetic, bridged interactions, it has been used for state interactions of the type expressed in Equation 10.75, particularly if there is only one important excited state of the bridge. The latter interaction is in this connection written as

$$V_{super} = \frac{V_{db}V_{ba}}{\Delta E_{db}}. \tag{10.76}$$

If ΔE_{db} is small and the interaction matrix elements are large, V_{super} will be large. The exponential decrease with distance is totally absent in this model. If the couplings V_{db} and V_{ba} are small, as is usually the case, the excited state of the bridge must have a very low excitation energy to obtain a large V_{super}.

In that case, there may be another possible mechanism for ET. The electron may be accepted in the empty orbital of the bridge for some time, before it continues to the acceptor. The bridge itself is a temporary acceptor. The mechanism is *not* tunneling.

10.6.4 DIRECT CALCULATION OF ELECTRONIC COUPLING

The calculation of Δ is far from trivial. We have to calculate the total energy difference between the ground state of a system with an odd number of electrons and its lowest excited state in the symmetric geometry at $Q = 0$. Normally, the Unrestricted Hartree–Fock (UHF) method is used on odd-electron systems. At the equilibrium points, there is a large lowering of energy if the restriction to same spatial orbital for different spin is relaxed. This is also the case for the symmetric point $Q = 0$, but to a much less extent. In a system with a weak coupling ($\lambda \gg \Delta$), the system gains energy instead by breaking the symmetry and localizing the electron to one of the subsystems. It is impossible to obtain a value of Δ pertaining to a symmetric system.

Two amendments are possible. The first is to use the Restricted Hartree–Fock (RHF) method. In that case, the system remains symmetric, but the final energy is *higher* than the final UHF energy (since the latter method has less number of restrictions). The other possibility is to symmetrize the UHF solution. This can be done by superposing the two solutions when the electron is on either of the two metal ions. Since these two solutions are not orthogonal, it is a very difficult problem, originally solved by the Dutch quantum chemists Ria Broer and Willem Nieuwpoort. In this process, the final energy is *lower* than the UHF energy. The best solution is thus obtained by first using UHF and then CI. RHF tends to give bad approximations in some cases. Nowadays, there are standard programs that include symmetrization (for example, RASSCF in the MOLCAS case).

However, the Hartree–Fock method may be used in a simplified way, which usually gives accurate results. Instead of calculating the total energies, we calculate the ionization energies via Koopmans' theorem. According to the latter theorem, we have in the hole case

$$I_+ = E_{N-1}^+ - E_N \approx -\varepsilon_+ \quad I_- = E_{N-1}^- - E_N \approx -\varepsilon_-, \tag{10.77}$$

$$\Delta = E_{N-1}^+ - E_{N-1}^- \approx \varepsilon_- - \varepsilon_+. \tag{10.78}$$

Notice that we obtain the total energy difference for a system with $N - 1$ electrons, as we want. The orbital energies though belong to a system with filled orbitals (N electrons).

Alternatively, we may use electron affinities:

$$A_+ = E_N^+ - E_{N+1} \approx -\varepsilon_+ \quad A_- = E_N^- - E_{N+1} \approx -\varepsilon_- \qquad (10.79)$$

$$\Rightarrow \Delta = E_{N+1}^+ - E_{N+1}^- \approx \varepsilon_- - \varepsilon_+. \qquad (10.80)$$

The orbital energies belong to the closed shell system where one electron has been removed. We thus add or subtract an electron to restore the original neutral system and subsequently obtain Δ from the orbital energy difference between the symmetric and antisymmetric orbitals that are involved in the ET process. An example is given in Figure 10.15.

The two subsystems exchanging electrons are two ethylene (=ethene) molecules, both connected to a cyclohexane "bridge molecule" by covalent bonds. There are 10 carbon atoms with 4 valence electrons each and 16 hydrogen atoms with 1 electron each, altogether 56 electrons and 28 orbitals are occupied in the ground state. Unsurprisingly, the HOMO and HOMO − 1 are linear combinations of the two π MOs located on the two ethylene groups. ET is possible in the system with one missing electron, that is, with 55 electrons. In the positive molecular ion, the linear combination of the two given MOs can be used for the leaping electron. The calculations behind Figure 10.15 are carried out for the neutral molecule because Koopmans' theorem is used to estimate the energy difference between the two relevant many-electron states with 55 electrons.

The main cyclohexane part of the molecule serves as a bridge. The two π MOs (27 and 28) are split because of interaction through the bridge. In Fig. 10.15, we see

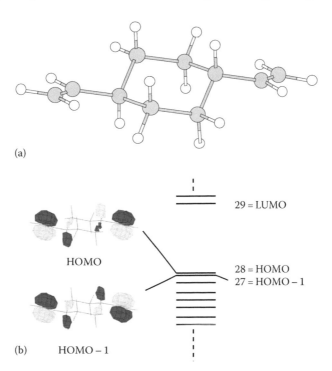

(a)

29 = LUMO

HOMO

28 = HOMO
27 = HOMO − 1

(b)   HOMO − 1

FIGURE 10.15   Diethene-cyclohexane (a) and the two MOs important for ET (b).

that the two MOs are slightly mixed with cyclohexane components. This mixing is visible evidence for coupling through the bridge.

In the neutral molecule, the orbital energies correspond to the symmetric (ε_+) and antisymmetric (ε_-) orbitals, respectively. The HOMO and HOMO − 1 are symmetric or antisymmetric and their energy difference is Δ.

In the case of ET (not hole transfer), we instead use Koopmans' theorem for electron affinities. It is well known that the theorem is less accurate in this case. However, in this particular application, the errors seem to cancel out very well, according to experience. We thus obtain Δ as the orbital difference between the LUMO and LUMO + 1, calculated for the neutral system.

If the ET system is asymmetric, a special procedure has to be used. There is a problem already by the fact that the coupling is not well-defined.

To define the coupling, we take the pragmatic approach that the gap $|\Delta|$ is the quantity that goes into the Landau–Zener approximation; in other words, the gap marked out in Figure 10.10. The gap is thus at the same time the minimum gap between two PES. Koopmans' theorem may be used to obtain the gap, no matter whether this is the minimum gap. This gap is simply equal to $|H_{22} - H_{11}|$, but we cannot know beforehand where the minimum is situated along the Q-axis. On the other hand, we know that when the gap attains the minimum value, the wave functions are equally distributed on the donor D and acceptor A.

The following procedure may be used to obtain the coupling. First, we have to find out which vibrational mode (or nuclear motion in general) is the crucial one for the ET. This mode reveals itself quite easily after some computer experiments. We calculate the system for a trial geometry in between the donor and the acceptor. The relevant MOs for ET are inspected with regard to D and A character. We are interested in the situation with equal distribution on D and A. It is unlikely that this value on the Q-axis is found in the first attempt, and we therefore modify the geometry in such a way that the minimum component is favored. This procedure is repeated with a better choice of trial value of Q.

When the coupling is small, it may be hard to obtain any sizeable mixing between D and A. The point of equal mixing may then be obtained by using Equation 10.44. Calculate tg2ξ by inspection of the wave function and multiply $H_{22} - H_{11}$ by tg2ξ to obtain H_{12}.

If the gap is calculated using total energies, one may obtain the point of equal distribution on D and A in the same way.

Broo and Braga used *ab initio* methods to obtain accurate results for the ET coupling. In our work, we simply calculated the total energy splitting used in the Landau–Zener approximation. Broo showed that an aliphatic chain by no means gives an exponential decrease with distance. On the contrary, it is the stretched out "all-*trans*" form that gives the slowest exponential decrease with distance and hence the fastest ET. If the dihedral angles of the chain are changed from 180°, the rate goes down despite the fact that the distance is decreased. There was later experimental confirmation of that result.

Braga used *ab initio* methods and found cases where the decrease is nonexponential and obtained convincing agreement with the above-mentioned experimental results of Closs and Miller.

10.6.5 PATHWAY MODEL

Additional insight into the mediated ET problem may be obtained if we write the bridge matrix (Section 10.6.2 and Appendix 4) as

$$H_{bb} - E = D + B, \tag{10.81}$$

where D is a diagonal matrix and B is an interaction matrix with zeroes in the diagonal.

We consider the one-electron case, so the interactions are between atomic orbitals. By repeated use of the following matrix identity:

$$\left(D + B\right)^{-1} = D^{-1} - \left(D + B\right)^{-1} BD^{-1} \tag{10.82}$$

we obtain

$$\left(D + B\right)^{-1} = D^{-1} - D^{-1}BD^{-1} + D^{-1}BD^{-1}BD^{-1} - \tag{10.83}$$

In Equation 10.74, the first term is the direct matrix element between the donor and the acceptor. The contributions from the sum may be obtained by rewriting $H_{bb} - E$ as in Equation 10.75 and using expansion (10.83). The first term will be

$$-H_{\mu i}D_{ii}^{-1}H_{iv} = -\sum_i \frac{H_{\mu i}H_{iv}}{E_i - E}. \tag{10.84}$$

This term has interactions from the donor to atomic orbital I on the bridge multiplied by interaction from the latter atomic orbital to the acceptor orbital. The second term in Equation 10.83 gives

$$H_{db}D^{-1}BD^{-1}H_{ba} = \sum_{i,j} \frac{H_{\mu i}H_{ij}H_{jv}}{\left(E_i - E\right)\left(E_j - E\right)}. \tag{10.85}$$

In this term, the interactions from the donor orbital μ via atomic orbitals i and j to the acceptor orbital v are summed. The matrix elements in this summation are likely to be larger than the first mentioned one, since some distances between atomic orbitals are now much smaller.

In the same way, we finally arrive at the following expression for the effective interaction matrix element between the donor and the acceptor:

$$\overline{H}_{da} = H_{\mu v} - \sum_i \frac{H_{\mu i}H_{iv}}{E_i - E} + \sum_{i,j} \frac{H_{\mu i}H_{ij}H_{jv}}{\left(E_i - E\right)\left(E_j - E\right)} - \cdots. \tag{10.86}$$

This equation expresses the donor–acceptor interaction as a sum of contributions from various "pathways" (Figure 10.16). Considering the fact that the matrix

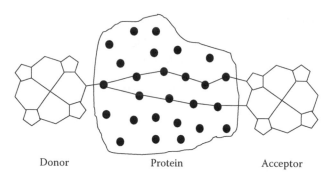

FIGURE 10.16 Pathway model. The example used is ET between two porphyrins. The bridge is a protein and the black dots are the atoms of the protein.

elements all decrease exponentially with the distance between the respective atomic orbitals, it is likely that the maximum contributions occur when only short neighboring interactions are used and the total pathway is as short as possible. With this model, it is reasonable that there is an exponential decrease in total, since each time a new atom has to be inserted between the donor and the acceptor, there has to be a multiplication with an additional factor whose absolute value is much smaller than unity.

A much simplified but still useful pathway model has been developed by Beratan and Onuchic, in particular. Bonds are connected by certain decay factors.

10.6.6 NONEXPONENTIAL DECREASE

A. Haim, S. Isied, H. B. Gray, and many others have obtained experimental results for ET through bridges. In most cases, exponential decrease, as suggested in the Gamow tunneling model and the pathway model, has been obtained. However, when the latter model is improved, there is no definite reason why exponential decrease will be obtained when positive and negative contributions are added up. There is no guarantee that positive and negative contributions will not cancel each other, and such behavior, when Δ has a node, has in fact been obtained for certain series of bridges. We remind the reader that if $\Delta = 0$, there is no ET.

The experimental results of Haim and many others suggest that ET in π-bridges decreases considerably slower than in other types of bridges. We will show theoretically that metal ions with a linear polyene bridge, in the Hückel model, converge to a finite value for the coupling, as the length of the bridge increases. It will be noted, however, that a more advanced Hückel method, where coupling depends on bond length, gives alternating bond lengths and a slight decrease in the electronic factor. Nevertheless, the "proof" will be given since it clearly shows which conditions have to be met to have a nonexponential behavior.

We begin with Equation 10.72. In this equation, we have to insert coefficients and energies that are valid for a π-bridge. For a linear polyene, there are exact Hückel energies and orbitals, given in Equations 3.38 and 3.39. If these expressions are inserted into Equation 10.72 and $H_{ad} = 0$, we obtain

$$\Delta = \frac{2\eta_1\eta_2}{\beta} \sum_{v=1}^{n} \frac{(-1)^{v+1}}{n+1} \frac{\sin^2\left[nv/(n+1)\right]}{\cos\left[nv/(n+1)\right]},\qquad(10.87)$$

where η_1 and η_2 are the couplings at the ends of the bridge and n is the number of carbon atoms in the polyene chain. The largest contributions come from $v = n/2$ and $v = n/2 + 1$, since these orbital energies are closest to the energy of the transferring electron, which is equal to zero. We now let $n \to \infty$ and sum the series:

$$\Delta \approx \frac{4\eta_1\eta_2}{\beta} \frac{2}{\pi} \sum_{v=1}^{\infty} \frac{(-1)^{v+1}}{2v+1} = \frac{2\eta_1\eta_2}{\beta}.\qquad(10.88)$$

We thus obtain a constant value. However, this result depends mainly on the contribution from the HOMO and LUMO of the bridge. We have used an equation for the eigenvalues where the the HOMO–LUMO gap converges to zero. This is not always true. In a few cases, the gap goes to a finite value. Still, it is clear from Equations 10.87 and 10.88 that the π-system offers better conditions for ET than the σ system and this has also been found experimentally.

10.6.7 ELECTRON TRANSFER THROUGH A SOLVENT

An important question is how ET takes place in a solvent between two material bodies with planar surfaces (Figure 10.17). If the solvent is absent, the coupling will go down exponentially since the overlap between the MOs of the bodies goes down exponentially. If we put a solvent in between the bodies, will the coupling increase or will it be hindered by the solvent molecules (grey circles in Figure 10.17)? It is, in fact, unlikely that it will be hindered.

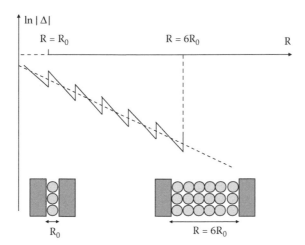

FIGURE 10.17 Exponential decrease of the coupling with distance in a solvent.

In Figure 10.17, the perpendicular distance between the planes is gradually increased until there are six layers of solvent molecules between the bodies. If the distance is increased from zero, there is first a decrease in empty space when no solvent molecules are present. When the distance is large enough for one single layer to be inserted, there will be a jump of $\ln |\Delta|$ (the vertical line for $R = R_0$). If the distance is further increased, there will be a steep decrease since no new layers of solvent molecules can be inserted. When the second layer can be inserted, $\ln |\Delta|$ will increase again. Clearly, the picture is idealized in Figure 10.17, but on an average, the slowly decreasing dotted line will be obtained. In most cases, the exponential decrease of $|\Delta|$ is less steep in the solvent than in open space.

10.7 DISPROPORTIONATION

We will now consider a different type of ET. So far, we have learnt that the oxidation states of two metal ions of the same metal atom have to be different in a single unit to have ET. If the metal ions are in the same oxidation state, ET is possible only if one electron can move from one metal ion to the next. Two new atoms or ions are formed with oxidation states different in two units. Such a reaction is called a *disproportionation* reaction and may be written as

$$M + M \rightarrow M^+ + M^- \left(\Delta G^\ominus \right) \tag{10.89}$$

10.7.1 EXAMPLES OF DISPROPORTIONATION

One example is gold in oxidation state +2. Gold is *not* stable as $M = Au^{2+}$, but disproportionates into Au^{3+} and Au^+. The same is the case for Ag^{2+} and Tl^{2+}, and indeed for all "missing valence states," such as Bi^{4+}, Sb^{4+}, and Pb^{3+}. If the intermediate valence state of M is unstable or missing, Equation 10.89 may have a negative $\Delta G^\ominus$, if M^+ and M^- are stable. If, on the other hand, M is stable, as for example, in Cu^{2+} compounds, $\Delta G^\ominus > 0$. In this case, the reaction (10.89) has a positive $\Delta G^\ominus$ and M^+ and M^- are said to *comproportionate*.

Cu$^+$ disproportionates according to

$$2Cu^+ \rightarrow Cu + Cu^{2+}. \tag{10.90}$$

We know from Table 10.1 that $E^\ominus$ (Cu^+, Cu) is equal to +0.52 eV and that $E^\ominus$ (Cu^{2+}, Cu^+) is equal to 0.16 eV. If the corresponding ET reactions are subtracted, we obtain $E^\ominus$ for the reaction in Equation 10.87: $E^\ominus = 0.52 - 0.16 = 0.36$ eV.

It follows from the Frost diagram (Figure 10.4) that two Mn^{6+} ions have a lower $\Delta G^\ominus$ than one Mn^{7+} and one Mn^{5+}. Two Mn^{4+} ions (in MnO_2) have a lower $\Delta G^\ominus$ than one Mn^{5+} and one Mn^{3+}. In fact, the Frost curve allows disproportionation only if there is a concave connection curve, as for Mn^{5+} to Mn^{3+}. On the other hand, it should be kept in mind that Frost diagrams apply mainly to ET at electrodes. If other cases are considered, for example, a solid, the reduction potentials may not be exactly the same. Nevertheless, the Frost diagrams are helpful as a guide.

Disproportionation may also occur for organic molecules. C_{60}^{-3}, for example, disproportionates into C_{60}^{-4} and C_{60}^{-2}.

10.7.2 DAY–HUSH DISPROPORTIONATION MODEL

The existence of a multitude of valence states for a given element has been evident for 200 years. This fact formed the background for the establishment of the periodic table. Missing valence states are commonplace and not considered as surprising. Why do we need a theory? One reason is that part of the electronic spectra can be explained only as metal to other metal transitions, and then three consecutive valence states have to be involved. We have already seen that very strong colors appear if two valence states are mixed, for example, Fe^{2+} and Fe^{3+} in ink. There is a different reason for colorful substances if the disproportionation mechanism is realized. For that, we need a theory and such theory was provided by P. Day, K. Prassides, N. Hush, J. Reimers, and their collaborators.

As previously for MV-2 systems we introduce two diabatic, parabolic PES (H_{11} and H_{22}) in each of two equilibrium points, $Q = -Q_0$ and $Q = Q_0$, but this time the PES correspond to localization of an electron pair at two different sites, which we may write as M^+M^- and M^-M^+. In MV-3 systems we have to consider the MM configuration also, although it may not form a stable ground state. We notice that M^+M^- and M^-M^+ are different from each other by an electron pair.

The third diabatic PES (H_{33}) corresponds to MM. It can be assumed that its energy minimum occurs when the geometry is the same on both sites, for $Q = 0$. We assume that the force constants are the same for all three states. H_{33} is placed above or below H_{11} and H_{22} at $Q = 0$:

$$H_{11} = \frac{1}{2} k \left(Q + Q_0\right)^2$$

$$H_{22} = \frac{1}{2} k \left(Q - Q_0\right)^2 \tag{10.91}$$

$$H_{33} = \frac{1}{2} k\, Q^2 + U.$$

The eigenvalues corresponding to the PES are determined from the following secular equation:

$$\begin{vmatrix} H_{11} - E & 0 & H_{13} \\ 0 & H_{22} - E & H_{13} \\ H_{13} & H_{13} & H_{33} - E \end{vmatrix} = 0. \tag{10.92}$$

The interaction matrix element, H_{13}, is the coupling between the M^+M^- and MM states (or the M^-M^+ and MM states). The direct coupling between H_{11} and H_{22} vanishes in the MV-3 case, as will be shown later. The solution of Equation 10.92 is shown in Figure 10.18.

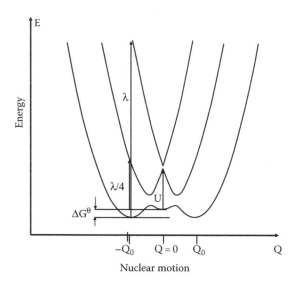

FIGURE 10.18 PES for three valence states (M^-M^+, M^+M^-, and MM). In this case, $\Delta G^\ominus < 0$, meaning disproportionation. M^-M^+ have a minimum close to $Q = -Q_0$ and M^+M^- close to $Q = Q_0$.

We may choose H_{13} and Q_0 to fit previously calculated PES along a breathing or half-breathing mode. The vertical energy U at $Q = 0$, which appears in Equation 10.90 also appears in the theory of condensed matter physics; therefore, to avoid unnecessary confusion, we use the notation employed there. U is the so-called *Hubbard U* parameter (see Chapter 17).

To establish the exact minima of the parabolas in Figure 10.18, it is possible to use reduction potentials, from which $\Delta G^\ominus$ is obtained. Unfortunately, reduction potentials are often unknown for "missing" valences. If the lower and higher valence states are accessible, the valence in between disproportionates, in a solution as well as in the solid state.

Thallium is known in its oxidation states +1 and +3, corresponding to the configurations $[Xe]6s^2 5d^{10}$ and $[Xe]5d^{10}$, respectively. The following oxidation potentials have been obtained for Tl:

$$Tl^+ + e^- \rightarrow Tl \quad E^\theta = -0.34 \text{ V}$$

$$Tl^{2+} + e^- \rightarrow Tl^+ \quad E^\theta = +2.22 \text{ V}$$

$$Tl^{3+} + e^- \rightarrow Tl^{2+} \quad E^\theta = +0.30 \text{ V}$$

$$Tl^{3+} + 2e^- \rightarrow Tl^+ \quad E^\theta = +1.25 \text{ V}.$$

(10.93)

The Frost diagram is thus very concave for Tl^{2+}. Valence state +2 may, in fact, be regarded as a "missing" valence state. In a solution with the average valence +2, half of the ions are Tl^+ and half are Tl^{3+}. In Tl^{2+}, the 6s orbital would be unfilled and such a state is always unstable. It has not been easy to measure all reduction potentials by the same method, but the values given above are consistent and may be checked by obtaining the reduction potential for the last from the second and third ($\Delta G^{\ominus} = 0.30 + 2.22 = 2.52$ from the second and third and $\Delta G^{\ominus} = 2 \times 1.25 = 2.50$ from the third).

In solids such as $TlBr_2$, the valence state of thallium is +2, judging from the stoichiometry. Reduction potentials do not favor the conclusion that this valence state exists in the ground state, however, so disproportionation into Tl^+ and Tl^{3+} must take place. This is confirmed by alternating TlO distances. In fact, two different valence states appear in a number of cases with Tl^{2+}, for example, in TlS. A spectrum that can be explained by Figure 10.18 can be seen and corresponds to $Tl^+Tl^{3+} \rightarrow Tl^{2+}Tl^{2+}$.

Antimony and its cousins in group V of the periodic table (N, P, As, Sb, Bi) are known in the oxidation states +3 and +5, but not in the oxidation state +4. The valence state +4 is "missing" in group V. It is quite easy to make the salt K_2SbCl_6, a salt referred to as Setterberg's salt. The (average) oxidation state for Sb in the salt is +4. Crystallography reveals that antimony disproportionates into Sb^{3+} and Sb^{5+}, with typical SbCl bond lengths. This absorption is ascribed to $Sb^{5+}Sb^{3+} \rightarrow Sb^{4+}Sb^{4+}$ by Day et al. The energy of this absorption is $E = \lambda/4 - \Delta G^{\ominus}$, as follows from Figure 10.18.

10.8   QUANTIZED NUCLEAR MOTION

The theory advanced so far assumes that the nuclei move by classical mechanics, on Born–Oppenheimer (BO) energy surfaces. This is an approximation, since nuclear motion should be treated quantum mechanically like electronic motion. One may think that as long as the temperature is as high as room temperature, a classical treatment of the nuclei should not cause any problem, except possibly for the very lightest one, the proton. But at low temperatures, even heavy nuclei show quantum correction. This is particularly the case for ET problems, where we know that the coupling between electrons and nuclei is particularly important.

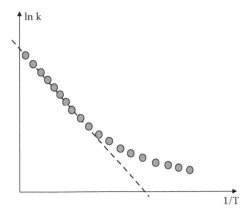

FIGURE 10.19   Typical $\ln k = E_a/kT$ plot for ET reactions. Circles are experimental numbers.

When the rate for low temperatures is explored, there is always one typical behavior at low temperature, that is, when 1/T tends to infinity. This is shown in Figure 10.19 in a plot of the logarithm of the ln k, where k is the rate, as a function of 1/T. This plot should be linear and the slope should be $-E_a/k$. In ET reactions (and many other reactions), the slope bends up for large values of 1/T, that is, as T → 0. This is the typical sign of tunneling. The activation energy we are talking about is the activation energy in the Marcus curves, due to nuclear motion.

Next, two models designed to deal with nuclear tunneling will be discussed: the PKS model and the Bixon–Jortner model. The Bixon–Jortner model (1968) provided the conceptual basis for the understanding of radiationless transitions in excited electronic and vibrational states.

10.8.1 PKS MODEL

This model was originally used along with the Day–Hush disproportionation model, mentioned in Section 10.7.2, and got its name from the pioneers S. Piepho, E. Krausz, and P. N. Schatz (PKS). The PKS model is a useful and consistent model for the treatment of spectra of mixed valence systems. The model includes the vibrational levels for MV-2 and MV-3 systems. It is assumed that the coupling is small so that the harmonic approximation can be used.

In the PKS model, four parameters are used to characterize the mixed valence system: (1) a vibronic coupling parameter, related to the reorganization energy λ; (2) the electronic coupling parameter Δ; (3) the frequency of the promoting vibrational mode ν; and (4) the Hubbard parameter U. The model describes well the transition from a localized to a delocalized system. Large reorganization energy λ compared with electronic coupling Δ means that the system is strongly localized. This is the case even if the Hubbard parameter U decreases to zero in the MV-3 system.

10.8.2 NUCLEAR TUNNELING

Various models that take this problem into account exist. Here, we will mention a model by Kestner et al. based on the *Fermi golden rule*. It is possible to show (see Section 7.3.2) that the probability per unit time that a system in an initial vibronic (electronic and vibrational) state, n_v, will undergo a transition into a set of vibronic levels {mw} is given by

$$W_{nv} = \frac{\pi}{2}\Delta^2 \sum_{m,w} \left|\left\langle \chi_{nv} \middle| \chi_{mw} \right\rangle\right|^2 \delta\left(\varepsilon_{nv} - \varepsilon_{mw}\right), \tag{10.94}$$

where Δ is the electronic coupling matrix element defined previously. Equation 10.94 represents a weighted density of final states, ε_{nv} and ε_{mw} are the unperturbed energies of the vibronic levels, and δ is the Dirac delta-function, which ensures energy conservation.

Assuming a Boltzmann distribution over the vibrational energy levels of the initial electronic state n, the thermally averaged probability per unit time of passing from the initial set of vibronic levels {nv} to a set of vibronic levels {mw} is given by

$$k = \frac{1}{Q} \sum_{n,v} \exp\left(-\frac{\varepsilon_{nv}}{k_B T}\right) W_{nv}$$

$$= \frac{\pi}{2} \frac{\Delta^2}{Q} \sum_{n,v} \sum_{m,w} \exp\left(-\frac{\varepsilon_{nv}}{k_B T}\right) \left|\langle\chi_{nv}|\chi_{mw}\rangle\right|^2 \delta(\varepsilon_{nv} - \varepsilon_{mw}),$$

(10.95)

where Q is the partition function:

$$Q = \sum_{n,v} \exp\left(-\frac{\varepsilon_{nv}}{k_B T}\right).$$

(10.96)

In Equation 10.95, there are contributions to the rate even if the thermal energy is not sufficient to ascend the barrier. This is seen if the logarithm of the rate is plotted as a function of 1/T and is referred to as *nuclear tunneling*. Without nuclear tunneling, a straight line is obtained, as is obvious from Equation 10.51 with κ equal to unity. Typical for nuclear tunneling is larger values of k for low temperature than those corresponding to the straight line. For very low temperatures, Equation 10.95 suggests a straight line parallel to the x-axis.

It may be shown that the equations for nuclear tunneling converge to the ordinary rate equation for a large activation barrier and high temperature. In the case of low temperature or a low activation barrier, the ordinary rate equation based on classical nuclear motion should not be used.

10.8.3 Vibrational Model for ET in the Limit of Low Barrier

ET may be treated as a vibrational quantum mechanical problem in a symmetric BO potential of Marcus type, where the distance between the energy minima is obtained from the reorganization energy λ and force constant k. The interaction is characterized by a gap Δ at the avoided crossing. An alternative way to account for the avoided crossing is to use the correction terms of the BO approximation. The energy splitting $\Delta E_{12} = E_2 - E_1$ between the two lowest energy eigenvalues is related to the rate of ET in a wave packet model. For large Δ and λ, ΔE_{12} becomes the frequency of the promoting vibrational mode, independent of Δ. The theory is illustrated by internal ET in symmetric positive molecular ions with two double bonds, separated by single bonds.

Completely delocalized ionization is obtained in the conjugated case when only one single bond separates the double bonds. More than one separating bond leads to mode softening and partial localization, whereas a completely localized, ionized double bond is obtained if many single bonds separate the double bonds.

In the Marcus model, it is assumed that the vibrational energy of the activating mode is small compared with the activation barrier. However, very often, the coupling is larger than the reorganization energy. Let us assume that we have a symmetric system. In Equation 10.29 we find that the barrier disappears if $|\Delta| > \lambda$. The eigenfunctions of the eigenvalues in Equation 10.20 may be written as

$$\phi_1 = \cos\theta \; \phi_a + \sin\theta \; \phi_b$$

$$\phi_2 = -\sin\theta \; \phi_a + \cos\theta \; \phi_b \tag{10.97}$$

θ is a function of Q and both eigenfunctions are normalized, assuming ϕ_a and ϕ_b are normalized.

We may assume that the diabatic wave functions are

$$E_{\pm} = \frac{k}{2}\left(Q^2 + Q_0^2\right) \pm \sqrt{k^2 Q_0^2 Q^2 + \left(\Delta/2\right)^2}. \tag{10.98}$$

These are the ordinary Marcus curves for a symmetric system. The ground state forms a symmetric double-well potential. The BO approximation does not work here because if we are at one minimum, the electron is at the donor D and if we are at the other minimum, the electron is at the acceptor A.

Previously, we treated the spectrum as being either localized or delocalized (single molecule). To solve the problem accurately, it is necessary to go beyond the BO approximation:

$$\Psi(q,Q) = \sum_i \Phi_i(q,Q) \sum_k \chi_{ik}(Q)C_{ik} \tag{10.99}$$

χ_k is a set of harmonic functions centered at $Q = -Q_0$, and $Q = Q_0$ (lower) and $Q = 0$ (upper). Equation 10.21 gives the BO energy surfaces. From the BO approximation, we also obtain equations for the vibrations. We insert Ψ from Equation 10.99 into these equations and obtain

$$H\Psi(q,Q) = \sum_{i,k} C_{ik}\left[-\frac{1}{2M}\nabla_Q^2 + E_i(Q)\right]\Phi_i(q,Q)\chi_{ik}(Q)$$

$$= \sum_{i,k} C_{ik}\left[-\frac{1}{2M}\nabla_Q^2\chi_{ik} + E_i(Q)\chi_{ik}(Q)\right]\Phi_i(q,Q)$$

$$- \sum_{i,k} C_{ik}\frac{1}{2M}\left[-2\nabla_Q\Phi_i(q,Q)\nabla_Q\chi_{ik}(Q) + \chi_{ik}(Q)\nabla_Q^2\Phi_i(q,Q)\right] \tag{10.100}$$

$$= \sum_{i,k} C_{ik}\Phi_i(q,Q)\chi_{ik}(Q).$$

A matrix eigenvalue problem is obtained by multiplying by $\Phi_j^* \chi_{jm}^*$ and integrating over electronic and vibrational coordinates. Note that the electronic functions form an orthonormal set for a given Q. For the evaluation of the integrals, see the original paper by Klimkāns and Larsson (2000). The vibrational functions are a set of harmonic functions centered at $Q = -Q_0$, $Q = 0$, and $Q = Q_0$. The eigenvalue problem is therefore of the form:

$$HC = SCE. \tag{10.101}$$

S is an overlap matrix. The eigenvalue problem is evaluated by *symmetric orthogonalization*.

11 Biological Electron Transfer

11.1 INTRODUCTION

The existence of electron transport in living systems has been known since approximately 1935. In 1961, Peter D. Mitchell proposed the *chemiosmotic hypothesis*. One important aspect of his theory is that protons are transported against the gradient to a more acid locality. The driving force provides coupling to an electron transfer (ET) chain. The free energy of ET originates from nutrients. The created proton gradient is used in the synthesis of adenosine triphosphate (ATP) in a process called *oxidative phosphorylation*. ATP is the energy "currency" of the living system.

In this chapter, we will discuss the fundamental aspects of electron transport of life processes. The chemiosmotic model is fully consistent with the Marcus model for ET and proton transfer (PT). ET takes place between π-systems as well as between metal ions, or between metal ions and π-systems. A biological chemical reaction that occurs perhaps a thousand times per second must work in a conserved structure and ET reactions are therefore well suited, since bond lengths, bond angles, and solvent structures are only slightly modified during the reaction. Most ET steps occur in a chain on the time scale of microseconds (μs), and no reaction step needs to be faster than a sufficiently fast rate-determining reaction step. Among biological ET reactions, those related to the charge separation in photosynthesis are a million times faster. But the latter reactions are photoreactions, where the competition with side reactions in the excited state demands high rates for the "useful" reactions. Chlorophyll molecules are used, probably because the change of structure when it is excited or ionized is very small. Photosynthetic ET will be discussed in Chapter 15.

11.2 THE LIVING SYSTEM

Living systems have existed for more than a billion (10^9) years. ET reactions may have been of importance from the early beginning, since they are possible in proteins at a quite large distance (~15 Å) and hence there are no serious restrictions due to diffusion limitations for reactants and products. Efficient ET and PT systems must have developed spontaneously at an early stage of a chemical evolution (the *prebiotic* evolution), which preceded the biological evolution.

The ability to reproduce is an important property of a living system. In modern life, this happens with the help of a genetic code stored in a DNA molecule. Sometimes, errors occur in the transcription of the code, called *mutations*. In rare cases, a more efficient system may result and survive in the natural selection. After many mutations, a new species has been formed that is more competitive than the original species in the *evolution*. This is mentioned here to remind the reader that natural biological ET and PT are optimized in one way or another.

11.2.1 Formation of Life

Several billions of years passed between the Big Bang and the appearance of life. During this time, while the earth cooled down, a "molecular evolution" took place. Living systems are immensely complicated and can hardly have started from a single cell just by chance. Molecules with very special functions have appeared and have been stable enough to avoid destruction. The original processes may have been simple chemical steady-state reactions. A steady-state reaction stops when it has run out of supplies. Only the reactions with a continuous supply "survive." We may speculate on the following steps in the molecular evolution toward life:

1. On our energy-rich but desolate planet, thermal or photochemical excitation leads to chemical reactions that form molecules when the temperature decreases. Products are formed that take part in more complicated reactions. The reaction products are plentiful and ubiquitous if the supply is sufficient. A most important point in the evolution of life is reached when the temperature on the planet goes below the melting point of silicates. Probably, a crust had already formed on the earth, 4 thousand million years ago.

   Another important step happens when the temperature goes below the boiling point of water, allowing water to condense. Ponds, lakes, and seas are now formed. Solid surfaces are available in contact with water. Aggregates of larger molecules are formed and precipitate on stones and rocks at the water front.

2. Water (H_2O), ammonia (NH_3), methane (CH_4), and nitrogen (N_2) are present in the early atmosphere. The supply of cyan gas ($(CN)_2$), hydrogen cyanide (HCN), carbon monoxide (CO), and hydrogen (H_2) is also adequate. Amino acids and lipids are formed in large amounts. Subsequently, proteins are formed from amino acids and colloidal systems from lipids. Vesicles bounded by phospholipid bilayers turn out to be strong and reproducible.

3. About 2.5 thousand million years ago, the isotopic ratio of sulfur changed, and this has been ascribed to the presence of oxygen on the earth. Oxygen may have originated from some primitive photosynthetic reaction. It is hard to imagine any source other than atmospheric carbon dioxide. Later, 1.7 thousand million years ago, it was certain that O_2 existed as one of the major components of the atmosphere.

   The appearance of oxygen in the atmosphere changed the conditions on the earth considerably. New structures were formed, such as carbohydrates and certain amino acids. Three-dimensional structure appeared. Nitrogen-containing molecules bind to sugar and to phosphates. ET processes are already ubiquitous and carry out chemical reactions at a distance.

4. Colloidal particles serve the purpose of increasing the concentration of reactants inside, and allow photoreactions to take place. Large aggregates

of molecules form by polymerization, and when a certain size has been achieved, the aggregate is ruptured and the pieces move away to new locations and continue to grow. Some primitive form of mitosis develops, making reproduction possible. The competition for sunlight starts a primitive evolution.

5. The spark of life might have come when layers are formed on top of others, efficiently excluding less "successful" processes from light. Stable colloids survive and eventually become the origin of cells. Energy transport systems evolve, organized in cell membranes. Primitive leaves develop with a favorable geometry for maximizing absorption of sunlight. Simple trunks and roots are formed to take care of the supply of water and other chemicals.

6. Eventually, systems that carry "instructions" for chemical reactions for the formation of cell-like aggregates are formed. Biological evolution is now possible. A few processes dominate, but are eventually replaced by more efficient ones in the struggle for efficient energy transport and a place in the sun. The "purpose" is just to be plentiful. What is plentiful "exists" as a primitive molecular species. Different processes may exist side by side and even cooperate.

7. Life as we know it starts. Nucleotides, including phosphates with one (AMP), two adenosine diphosphate (ADP), and three adenosine trisphosphate (ATP) phosphorous atoms are formed spontaneously. RNA and DNA and other molecules, which are common in the reproduction of living systems of today, are formed.

The first hypothesis of this type for explaining the appearance of life on our planet was due to the Russian biochemist Alexander I. Oparin and independently by the British evolutionary biologist John B. S. Haldane, and is more than 50 years old. The basic idea in the Oparin–Haldane hypothesis is that the atmosphere contained only reducing molecules mentioned previously under (2), during the first few thousand million years of its existence.

Facts supporting the Oparin–Haldane hypothesis have now accumulated. The ancient atmosphere has been reconstructed in closed volumes. After some weeks of discharge between electrodes, new molecules were discovered, some of which occur in living systems (Stanley Miller and Harold Urey). The conclusion is that a "molecular evolution" may have started in a "primordial soup" of this type, where some large and stable molecules are created in relatively high concentrations.

The probability for *drastic* changes is small. RNA or photosynthetic reaction centers cannot have been created in seconds, just because all constituent atoms and molecules happened to meet at a certain time. Life may have started from a single cell, when such a cell finally existed after millions of years of prebiotic evolution. Still, there have been many possibilities for variation during the biological evolution. On another planet, plants may be red or black and ducks may be able to talk and use boats and sails. There is no reason to believe that present life is perfect.

11.2.2 Cells, Mitochondria, and Cell Membranes

One-cell organisms such as bacteria have a diameter of 1–2 μm. Cells in multicellular organisms are usually ten times larger. The inside of the cell, the *protoplasm*, contains *enzymes*, RNA, molecules of the degrading process (*metabolism*), *mitochondria*, and cell nucleus. Enzymes are proteins in union with active molecules called *coenzymes*.

Most cells have a nucleus containing DNA with reproductive information (the *genome*). The cell is separated from the outer world by a *plasma membrane*. The latter consists of a double layer of *phospholipids*, arranged perpendicular to the membrane, with the hydrophobic end turned inward. The hydrophilic end is turned to the polar water solutions inside and outside the cell. The hydrophobic parts of each layer attract each other, probably due to great structural flexibility, increasing the entropy at room temperature.

Cells with a well-defined nucleus contain mitochondria, enclosed by the plasma membrane (Figure 11.1). Cells in multicellular organisms may have very different functions, but are similar in structure and content and operate on their own. Together with many other cells, they may accomplish a particular task, for example, muscle contraction or signal processing (in the nerve cells). All biochemical reactions, including all energy transfer systems, are reproduced in each cell.

Mitochondria have membranes but no cell nucleus. Their size is that of small bacteria. Mitochondria generate most of the supply of ATP (Figure 11.2). There may be hundreds of mitochondria in a cell. Each mitochondrion is a subcell within the cell and contains two membranes. The inner membrane is wrinkled to increase its surface area. The inner compartment of the mitochondria is called the *matrix* and contains a concentrated aqueous solution of enzymes and other molecules. The inner membrane contains trapped membrane proteins (Figure 11.1) where cell respiration and photosynthesis take place. Here, ET and PT reactions and photosynthetic reactions take place.

The free energy of easily mobile electrons is used to pump protons from the matrix across the inner mitochondrial membrane into the intermembrane space. There, an acidity gradient is built up, giving free energy for the protons to transfer back through the inner membrane and form ATP from ADP with the help of the protein enzyme *ATP synthase*. Apparently, a mechanical rotation first takes place in

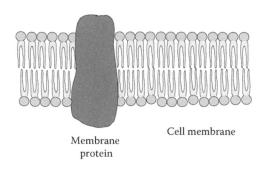

Membrane
protein

Cell membrane

FIGURE 11.1 Cell membrane with protein.

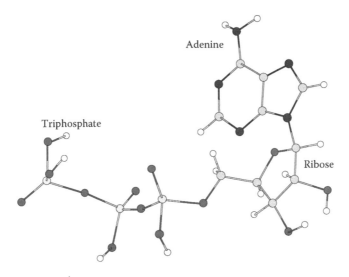

FIGURE 11.2 (See color insert) ATP. The adenine group has four nitrogens (dark). The sugar group is a pentose with three oxygens (dark). In ATP, there are three phosphorous atoms (grey) bonded to four oxygen atoms (dark); in ADP only two and in AMP only one phosphorous atom.

a protein called F_0F_1-ATPase, followed by conformational changes and trapping of ADP. There is sufficient free energy to synthesize ATP. The details of this synthesis are not yet known.

Since conserved structures with a very long lifetime have a great advantage in the evolution, a biological system is used over and over again. The result is cyclic reactions where reactants are fed in along a pathway and products are given away. There are several examples in Nature, for example, the citric acid cycle and the Calvin cycle.

11.2.3 MEMBRANE PROTEINS

Cell membranes are perforated by proteins extending through the membrane space (Figure 11.1). These proteins can be removed by detergents and studied in biochemical experiments. Even crystallization of these large molecules has been possible. Since 1985 and particularly in the new millennium, it has become possible to determine the structures of membrane proteins by x-ray crystallography methods. Every new structure needs the attention of large research groups during one or several years of its determination. Quantum chemical calculations of structural features have proven to be helpful in determining local bond distances and bond angles.

Given time, a cell membrane can be directly penetrated by quite large molecules. In a more efficient system, membrane proteins have evolved to simplify and control the transport through the protein itself. Membrane proteins form pores that act as a filter for transportation across the membrane. Several membrane proteins form channels for the penetration of potassium ions, K^+, but do not allow the passage of smaller ions, such as Na^+, Li^+, and Mg^{2+}. Since the latter ions are small, the water molecules form stronger ionic bonds with them than with the larger K^+ ions. In this case, the membrane protein has evolved to bind stronger than the water ions in the case of K^+. The smaller ions cannot rid themselves of the water solvent shell and cannot pass into the membrane channel.

ATP (Figure 11.2) belongs to a group of molecules called *nucleotides*, although nucleotides are normally thought of as molecules of the cell nucleus where they build DNA and RNA. The three different types of molecules forming a nucleotide have different tasks. Ribose and other saccharides are capable of forming polymers such as carbohydrates. The phosphate groups, finally, serve as the actual energy currency and at the same time have the property of easily binding to other molecules to make energy transfer possible.

Of particular interest to us are the membrane proteins, numbered *I–V*. Four of them are part of the *electron transport chain* in accordance with the chemiosmotic model. ET in the chains is driven by the energy of food or by photosynthesis. Protons are pumped across the membrane to a more acid location. This is done in *Complex I*, *Complex III*, and *Complex IV*. *Complex II* is used in reduction of ubiquinone to ubiquinol. Another molecule of this type is *Complex V*, an ATP synthase where ATP is synthesized from ADP and P_i, as just mentioned. This complex does not have any electron transport chain.

In the mitochondrial electron transport chain, electrons move from an electron donor (NADH, FADH, or QH_2) to a terminal electron acceptor (O_2) via a series of redox reactions. Complex III is called the *cytochrome bc$_1$* complex. Complex IV is the end station where oxygen molecules are accepting electrons and protons and are reduced to water. Complex IV is also called *cytochrome c oxidase* (CcO).

The mitochondrial membrane in photosynthetic organisms also contains proteins for harvesting energy from sunlight. The oxidation in the citric acid cycle provides energy for ATP synthase.

11.3 ELECTRON CARRIERS AND OTHER FUNCTIONAL GROUPS

An electron carrier is a subsystem that has accepted an electron in the lowest unoccupied molecular orbital (LUMO) or donated one from the highest occupied molecular orbital (HOMO). Usually, the carrier is a metal complex or a π-system with low reorganization energy. We will now discuss molecules that are involved in biological ET. PT reactions have been discussed already in Chapters 9 and 10, and not much needs to be added here. Proteins, also discussed in Chapter 9, can be electron carriers in the side groups of some of the peptides, for example, tyrosine. Next we will discuss some important molecules and molecular groups.

11.3.1 FUNCTIONAL GROUPS

Great variety in living systems is possible due to the diversity of proteins, folded in two-dimensional and three-dimensional structures. Proteins are one-dimensional chains of peptides, as we have seen in Chapter 9. These chains are created in RNA according to the DNA blueprint. The folding takes place in fractions of a second. Hydrogen bonds are formed to stabilize the final structure. In the folding pattern, the peptide backbone may be recognized as planar regions, called β-sheets, and helical regions, called α-helices. The end result is the geometry of the protein molecule in its electronic ground state. It is possible that other lower states exist with a different folding, and can be reached under different circumstances. Any sequence of the peptide chain can be artificially created in the test tube, folding into a unique protein.

After folding, the space is almost filled and only small molecules, primarily water, are able to penetrate into the protein. The structures occurring in nature are referred to as *wild type*. There are few cavities and these can be tested by *mutation* of the protein. A new protein is created by exchanging a peptide in the sequence and checking the folded structure afterward by crystallographic methods. Very often, the structure is the same, except slight changes around the exchanged peptide. Also, other properties can be tested by artificial mutation. In the case of ET proteins, the rate of ET has been compared to the wild type rate. In most cases, mutations lead to a protein that works less well, since the "best" structure has already been selected in the evolution.

The side groups of proteins may be important for the functioning of the protein. This is also the case for other molecules where the side group corresponds to a molecular group, covalently bonded to the macromolecule. We refer to these groups as *functional groups*. In Figure 11.3, some common ones have been shown. The alkyl group (methyl, CH_3-; ethyl, C_2H_5-; propyl, C_3H_7-; butyl, C_4H_9-) can be replaced by the covalent connection to the macromolecule.

The disulfide group (Figure 11.3) is spontaneously formed in a protein between two cystein peptides by two-electron reduction. Two fused cystein amino acids are shown, fused by removing two protons and two electrons as in the protein. The peptide dimer is called *cystin*.

Compounds or ions bound to a protein and important for the biological activity of that protein are called *cofactors*. These molecules are not proteins themselves, but function together with the proteins, and are termed as enzymes. Even metal ions such as Mg^{2+}, which is used in chlorophyll, are termed cofactors. Loosely bound cofactors are termed *coenzymes* and tightly bound ones, *prosthetic groups*.

The given functional groups may bind to metal ions and together form cages for important metal ions that are involved in ET. The copper ion, Cu^{2+}, for example, forms a metal center with a very intense blue color called a *blue protein*. The copper ions are bonded by the side groups of methionine, which is a thioether, the side group of cystein, which is a sulfhydryl group, and two side groups of histidine, which are imidazols. It might be added that the concentration of blue proteins in living matter is so low that its blue color cannot be seen.

Some functional groups or cofactors may be involved in ET by themselves, but this is not the general rule. Of the groups given in Figure 11.3, carboxyl and ester groups do not have favorable reduction potentials for biological ET. For formic acid, HCOOH, a reduction potential may be determined. Oxidation of HCOOH gives the same result as reduction of CO_2 to CO_2^- in the presence of two protons. CO_2^- is a radical and is written as $\cdot CO_2^-$.

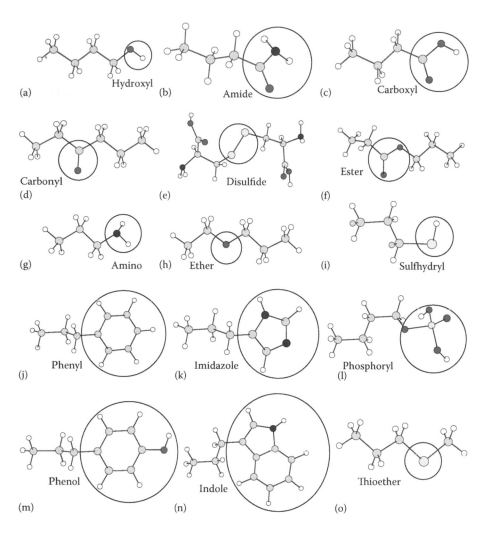

FIGURE 11.3 Functional groups: (a) hydroxyl $-OH$; (b) amide $-CONH_2$; (c) carboxyl $-COOH$; (d) carbonyl $=CO$; (e) disulfide $-S-S-$; (f) ester $-COO-$; (g) amino $-NH_2$; (h) ether $-O-$; (i) sulfhydryl $-SH$; (j) phenyl $-C_6H_5$; (k) imidazole $-C_3N_2H_3$; (l) phosphoryl $-OPO(OH)_2$; (m) phenol $-C_6H_4(OH)$; (n) indole $-C_8(NH)$; (o) thioether $-S-$.

11.3.2 CARBOHYDRATES AND LIGNIN

The general formula of a carbohydrate is $C_mH_{2n}O_n$ (usually m ≈ n). The net formula can also be written as $C_m(H_2O)_n$ and in this sense, carbohydrates are hydrates of carbon, as the name indicates. Structurally, however, it is more accurate to view them as polyhydroxy aldehydes and ketones. A synonym of carbohydrate is *saccharide*.

A monosaccharide is a ring-shaped molecule with five carbon atoms and one oxygen atom, or with four carbon atoms and one oxygen atom in the ring. A disaccharide is a condensation product of two monosaccharides. Structural formulas for two monosaccharides ($C_6H_{12}O_6$) and a disaccharide ($C_{12}H_{22}O_{11}$) are shown in Figure 11.4.

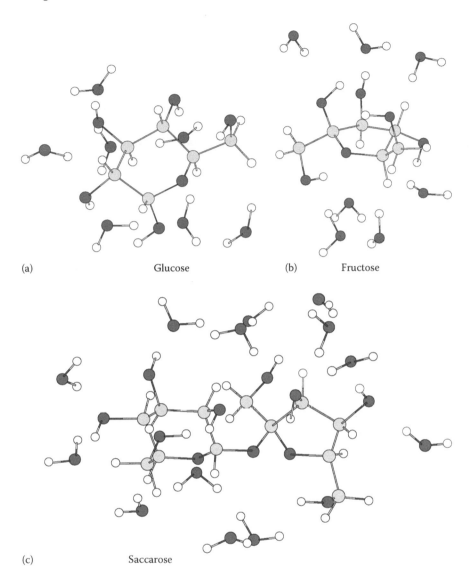

(a)   Glucose   (b)   Fructose

(c)   Saccarose

FIGURE 11.4 Structure of the monosaccharides: (a) glucose and (b) fructose. The condensed disaccharide sucrose (ordinary sugar) is seen on (c). Water molecules have been added to simulate the influence of the solvent.

The carbon atoms have hydroxyl and CH_2COH groups as covalently bonded substituents. There are many possibilities for hydrogen bonds, even between the monomers in a dimer. In a disaccharide, an oxygen atom connects two monosaccharide rings. Monosaccharides and disaccharides end with -ose. Blood sugar and fruit sugar are examples of monosaccharides, called glucose and fructose, respectively (Figure 11.4). Disaccharides are table sugar or cane sugar (sucrose) (Figure 11.4), milk sugar (lactose), and sugar in beer production (maltose). Maltose obtained from starch is a disaccharide formed from two glucose monosaccharides. Lactose is formed from glucose and another monosaccharide called galactose.

Disaccharides can be classified into two types: reducing and nonreducing disaccharides. Glucose can reduce $Cu(II)$ to $Cu(I)$ in copper hydroxide. This used to be a sensitive check for sugar in the urine of diabetes patients. If there is no sugar, which is normal, the color becomes black on heating, due to CuO. If sugar is present, red Cu_2O is formed instead.

Starch and cellulose are polysaccharides. Sugar is easily solvable in water, where it forms hydrogen bonds with the water molecules. In Figure 11.4, the structures have been optimized by quantum chemical methods (PM3). Water molecules are added to all possible "cavities" to simulate the solvated structure. Hydrogen bonds are formed between the saccharide and the water molecules. If optimization is done in open space, on the other hand, hydrogen bonds are formed between and in the monosaccharides. This structure appears in a water-free crystal. In the aqueous solution simulated in Figure 11.4, the final structure is relevant for aqueous solution.

Lignin consists of polymers of different lengths interconnected with cellulose in wood. Lignin may be isolated as a brown powder during the manufacturing of cellulose. In addition to sugar-like groups, lignin also contains phenol groups in the structure. The structure is very strong, making it possible for trees and other plants to build trunks and thereby compete for sunlight.

11.3.3 LIPIDS

Lipids are the molecules of fat. The lipids we find in higher animals are esters between glycerol and hydrocarbon chains. Most common is the chain of palmitine acid with 15 carbon atoms in the chain. Butter fat has a much shorter chain. Liver oil has a chain with unsaturated CC bonds. These molecules form an amorphous mixture where the attraction between the molecules is *hydrophobic* (meaning "water refusing") and is due to entropy. The opposite to hydrophobic is hydrophilic (water attracting). The attraction is particularly strong when hydrogen bonds are formed.

If pieces of fat are strongly stirred in water, a colloidal solution may be formed (for example, shaking a salad dressing). This colloid may be stabilized by surfactants, as occurs when pieces of fat are washed away in the laundry. The surfactant consists of salts of fatty acids.

Lipids build up the cell membrane. Chemically, lipids are esters between fatty acids and the threefold alcohol *glycerine*. The latter can form three ester bonds. The third is an ester with phosphoric acid (Figure 11.5). In the membrane, the hydrophobic parts are parallel. The polar head groups are hydrophilic and point outward.

Most of the fat that we eat does not contain any phosphate groups. Hydrocarbons are energy-rich, as are molecules with hydrocarbon tails. Biosystems use hydrocarbon tails in a controlled way. Oxidation takes place at the end of the ET chain, but the energy is extracted at the beginning of the chain. We cannot go into details here of

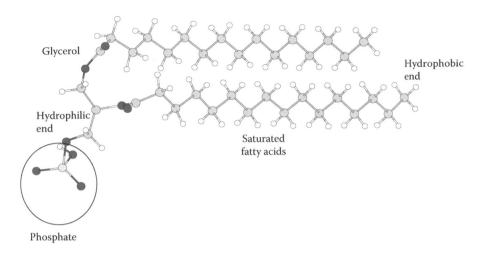

Glycerol

Hydrophobic
end

Hydrophilic
end

Saturated
fatty acids

Phosphate

FIGURE 11.5   Negative ion of a lipid with a short, saturated hydrocarbon chain.

how the energy is transferred with the help of a coenzyme A (CoA), which is a chain
molecule with many functional groups.

11.3.4   NICOTINAMIDE ADENINE DINUCLEOTIDE (NAD⁺)

Nucleotides are constructed in the same way as ATP/ADP/AMP (Figure 11.2) with
adenine at the end, followed by a sugar group (ribose), a diphosphate, and (in this
case) another ribose to which the nicotinamide group is attached in the same way as
adenine (Figure 11.6).

The precise function of adenine in the many reactions where this type of molecule
is involved is not yet known. The redox features belong to the nicotine amide group.
The connection point in nicotine amide and adenine is the nitrogen atom of the ring.
Nicotine amide is capable of accepting two electrons if, at the same time, one proton
is bonded. NADH is then formed. Despite the fact that NADH, contrary to NAD^+,
is not a conjugated ring system, its geometry is slightly bent but its structure is still
almost planar and very similar to that of NAD^+.

In the oxidation of lipids, NAD^+ is used and NADH is formed. Since two electrons are added in the LUMO, and since this partly breaks the π-bonding, much
energy is stored in NADH.

The NAD^+ and NADH molecules (including sugars, phosphate, and adenine) are
coenzymes in the enzyme *dehydrogenase*. Dehydrogenase oxidizes a substrate by removing two electrons and a proton to the nicotine amide acceptor or, alternatively, to a flavin.
The proton and two electrons originate from foodstuff, for example, lipids, as mentioned
previously. NADH dehydrogenase is the first enzyme in the electron transport chain.

The process indicated in Figure 11.6 is a coupled electron–proton transfer reaction. For every two electrons received by NAD^+, the complex transfers four protons to the inner mitochondrial membrane, thereby contributing to building the free
energy to produce ATP.

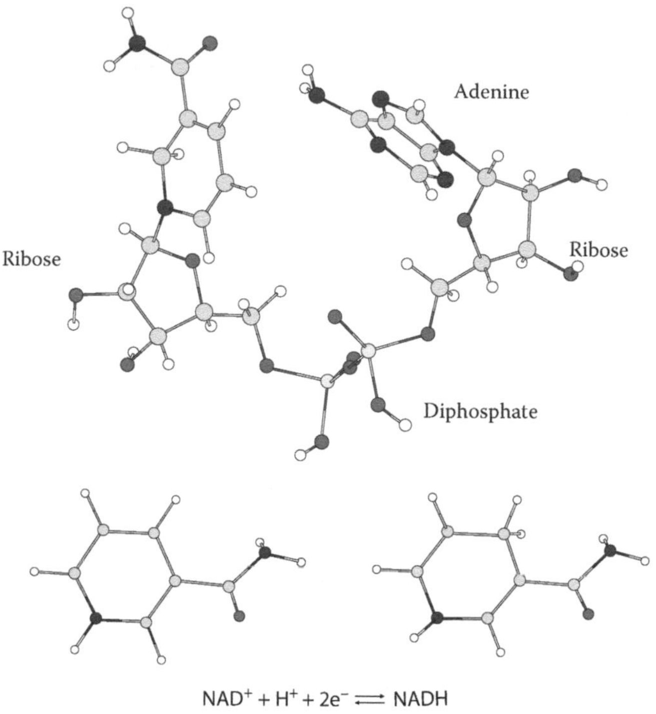

$$NAD^+ + H^+ + 2e^- \rightleftharpoons NADH$$

FIGURE 11.6 NADH (above) and (below) prosthetic groups of NAD$^+$ (left) and NADH (right).

11.3.5 FLAVINS

Other molecules that can accept two electrons are flavins: FAD (flavin adenine dinucleotide) and FMN (flavin mononucleotide). The former coenzyme is formed by adenine–ribose–diphosphate, followed by a linear sugar-type molecule and, finally, the aromatic ring (isoalloxazine) of Figure 11.7. FMN is missing adenine, ribose, and one of the phosphates at the beginning of the chain. Flavins are prosthetic groups in flavoproteins involved in the ET chain of Complex II.

As is seen in Figure 11.7, two protons are transferred at the same time as the electrons. The intermediate one-electron acceptor is called a semiquinone. Notice that the fully reduced form is not a conjugated π-system. Nevertheless, the structural changes between the two groups seen in Figure 11.7 are very small, and this makes ET between them possible at a quite high rate.

The primary task for FADH$_2$ is to carry high-energy electrons obtained in the oxidation of nutrients to a location where a proton gradient for oxidative phosphory-lation is built up. FAD is a prosthetic group in the protein succinate dehydrogenase (Complex II). The latter oxidizes succinate (HOOC−CH$_2$−CH$_2$−COOH) to fuma-rate (HOOC−CH=CH−COOH) in the citric acid cycle. FAD itself is reduced to FADH$_2$, which donates the electrons to the electron transport chain. The energy in FADH$_2$ is thus used to produce ATP by oxidative phosphorylation. Any oxidoreduc-tase protein that uses FAD as an electron carrier is called a flavoprotein.

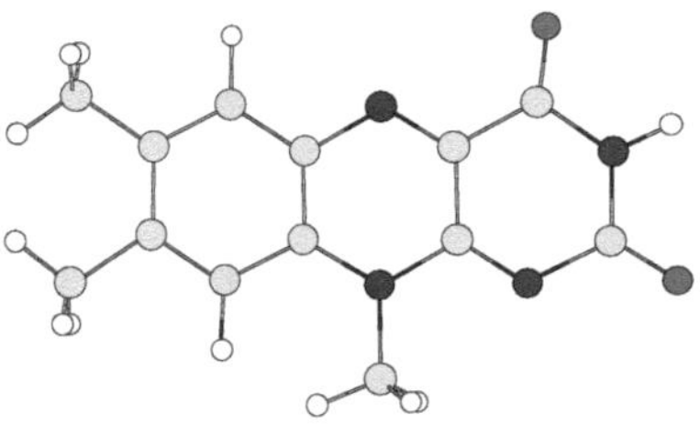

Flavin—oxidized form   Flavin—reduced form

FIGURE 11.7 Two-electron reduction of flavin. The attachment to the larger part of the FAD or FMN molecules is at the lower methyl group.

In Complex 1, NAD^+ receives two electrons and one proton and forms NADH. FMN receives two electrons and two protons and forms $FMNH_2$. NAD^+ and FMN cannot receive single electrons. Eventually, the electron pair and protons are given off to quinones, which are capable of single ET.

11.3.6 QUINONES

Quinones are mobile, lipid-soluble carriers. They are capable of accepting two electrons if two protons are also accepted at the O atoms. The reduced form of a quinone is seen in Figure 11.8. Quinones may take over two electrons from NADH or FADH, shuttle them in the membrane, and deliver them one after the other.

Other quinones are *phylloquinone* (vitamin K_1), *menaquinone* (vitamin K_2), *ubiquinone*, and *plastoquinone*. The latter two have partly unsaturated chains. Ubiquinone and phylloquinone contain a varying number of isoprene units of the type $-CH_2-CH=C(CH_3)-CH_2$ (Figure 11.8), but in phylloquinone only the first unit is unsaturated. The others are saturated $-CH_2-CH_2-CH(CH_3)-CH_2$ units, with a much higher flexibility than the partly unsaturated ones. The vitamin K quinones promote coagulation of the blood.

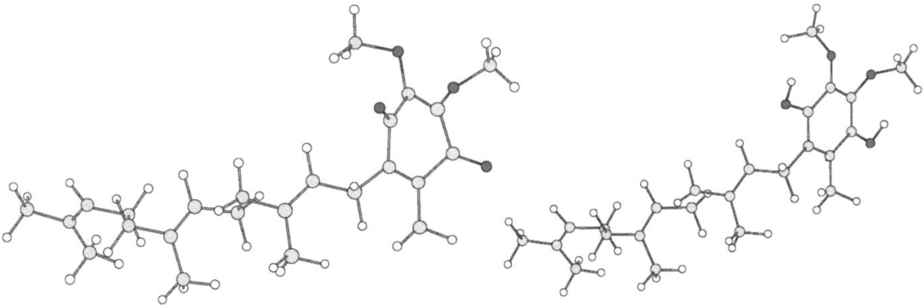

Übiquinon + $2e^-$ + $2H^+$ → Übiquinol

FIGURE 11.8 Ubiquinol, the doubly reduced form of ubiquinone. The chain on the left contains three double bonds and hence three $CH_2-CH = C(CH_3)-CH_2$ units. The molecule shown is thus Q3.

Ubiquinone occurs everywhere in living systems, thus giving rise to its name. Its most important role is as a transporter of electrons and energy in the membrane. Ubiquinone is also an antioxidant, as a part of several membrane proteins. One reason for aging of a living system may be sagging supplies of ubiquinone. Ubiquinone is produced in the body and can also be found in many foods. Q10 is available as a vitamin for elderly people. Ubiquinone is not fully planar. The structure is not much different from that of the π-system ubiquinol. Furthermore, protons are easily accepted at the oxygen lone pairs.

In the electron transport chain, ubiquinone receives two protons and two electrons from Complex I and Complex II and delivers them at Complex III. In between Complexes I and III and Complexes II and III, Q and QH_2 travel between the forest lipids. Q and QH_2 are highly hydrophobic and do not leave the mitochondrial membrane.

11.3.7 HEMES AND CYTOCHROMES

Heme is the prosthetic group of the protein hemoglobin. It binds sufficiently strongly to the oxygen molecule to transport this molecule in the body. Hemoglobin, its protein, makes up most of the red blood cells. Every cell in the body needs oxygen molecules to function, and every oxygen molecule has to be transported by a heme (B) group (Figure 11.9).

Hemoglobin also transports carbon dioxide and nitric oxide (NO), a molecule with important regulatory tasks in the body. The poisonous molecule carbon monoxide (CO) binds stronger to the heme group than oxygen, carbon dioxide, and NO, and this explains why CO is dangerous.

Hemes consist of a porphyrin-derived ring with an iron atom in the middle. The axial ligand is usually an imidazole group belonging to histidine.

Heme B is attached to the surrounding protein through a bond between the heme iron and an amino acid side chain. Other, fixed hemes are involved in ET chains in some photosynthetic bacteria.

Cytochromes are other proteins that contain a heme A group as a prosthetic group. They are found in different environments and have strong colors due to the heme group. Some cytochromes are water-soluble carriers. Cyt c (Figure 11.10) shuttles electrons between Complex III and Complex IV outside the mitochondrial membranes.

11.3.8 IRON–SULFUR PROTEINS

The peptides cystein and methionine contain sulfur in their side groups. Sulfides bind easily to metal ions, such as Fe^{2+}, Cu^{2+}, Zn^{2+}, and Pb^{2+}. In the case of Fe^{2+}, Fe–S complexes are formed, which are capable of easy reduction or oxidation. Pb^{2+} is poisonous and binds strongly at Zn^{2+} sites, but the structure is very different from the tetrahedral zinc complex.

The iron–sulfur complexes shown in Figure 11.11, bind to four cystein peptides. Iron–sulfur complexes are denoted according to the number of Fe and S atoms in the complex. In Figure 11.11, we see iron sulfide complexes with 1Fe (a), 2Fe–2S (b), and

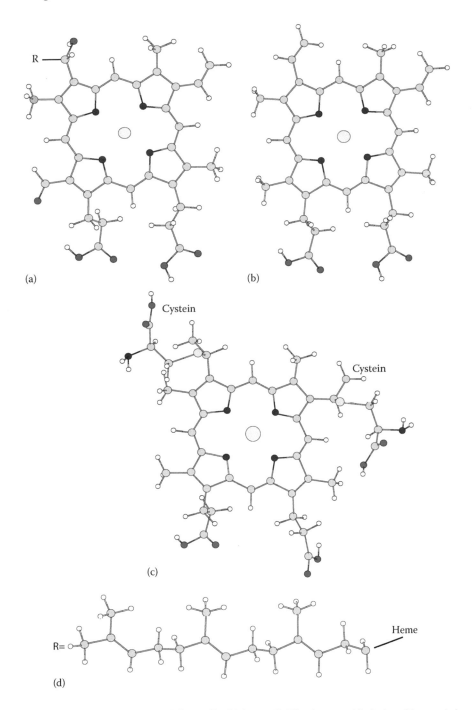

FIGURE 11.9 (a) Heme A; (b) heme B; (c) heme C. The isoprenoid chain of heme A is shown in (d).

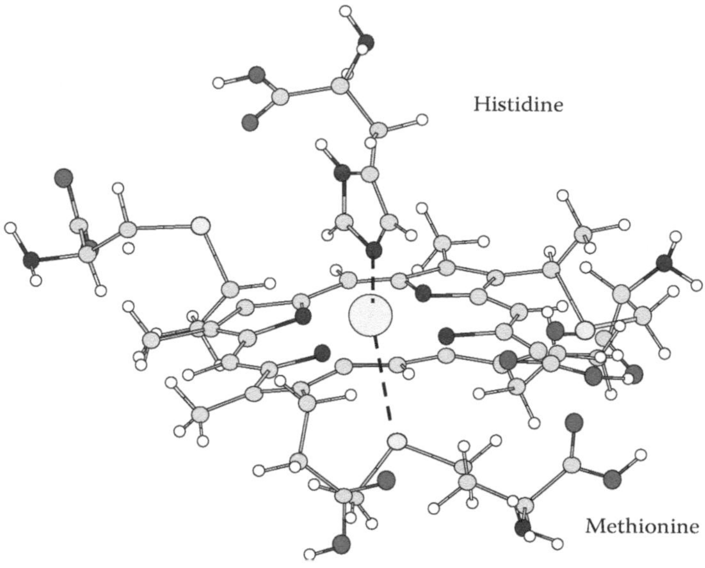

FIGURE 11.10   Cytochrome (cyt) c cofactor.

4Fe–4S (c) complex. The Fe atoms are connected to the sulfurs of the cys peptide. All iron atoms are surrounded by four sulfur ions in a tetrahedral geometry. Other iron-sulfur proteins have three iron ions. The oxidation state is varying. Iron and sulfur are strongly coupled and form a unit that can easily accept or donate electrons *without* great structural changes.

Complex I is a very large enzyme, also called NADH dehydrogenase or NADH quinone reductase. The electrons of NADH and two protons form $FMNH_2$ from FMN. The electrons are transferred through iron–sulfur proteins and then further on to ubiquinone. Many ET steps in Complex I have been studied down to atomic detail by Stuchebrukhov et al. Other detailed studies of iron–sulfur proteins have been carried out by L. Noodleman.

Iron–sulfur proteins are also found in Complex III and Complex IV and in ferrodoxins, cyt c reductase, nitrogenase, and a number of other systems. In most cases, they have a role in mitochondrial redox reactions. The iron–sulfur complex is easily oxidized or reduced. In fact, the HOMO–LUMO gap is small, so that Fe–S complexes work as small pieces of metal. Solid iron sulfide may become a metal if high pressure is applied.

11.4   BIOLOGICAL ELECTRON TRANSFER

Biological systems convert enormous amounts of energy-rich products (food) and oxygen to carbon dioxide. The chemical reactions have to take place at a low temperature between 10°C and 40°C without destroying the "reaction vessel". Biological processes are controlled in the sense that oxygen is not allowed to perform oxidation until the final step at a metal complex called Cu_B, which is part of a large membrane protein called

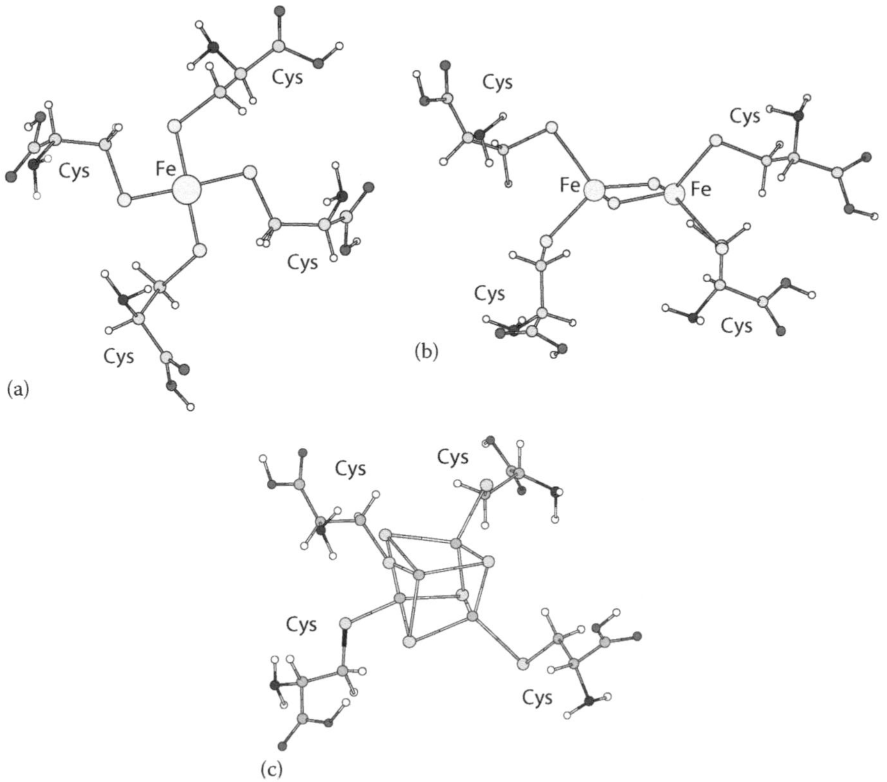

FIGURE 11.11   (See color insert) FeS centers and their connection to cystein (Cys) peptides in the protein, with one (a), two (b), and four (c) iron atoms.

cytochrome c oxidase (CcO) or Complex IV (Figure 11.12). This can only be achieved with the help of ET and PT processes. In CcO, the last proton is pulled over the edge to the proton pool of the intermembrane space, with the help of the remaining amount of free energy of the food molecules, and, of course, with the help of oxygen molecules. The oxygen molecules receive four electrons and four protons and become water.

11.4.1   ELECTRONS IN THE ELECTRON TRANSPORT CHAIN

Burning of organic compounds generates heat and CO_2, but hardly anything else. When foodstuff is oxidized in biological processes, the available free energy is used to build up a proton gradient and make ATP, which is later used as "money" to drive various life processes. In fire, oxygen is reduced to carbon dioxide in the flame. In life processes, on the other hand, oxygen is reduced at cytochrom c oxidase, separated in space from the oxidation of foodstuff. The degrading process of nutrients starts in various enzymes. The processes are very complicated and most details have to be left out here. One common end product that enters the electron transport chain on the left in Figure 11.13 is *succinate*.

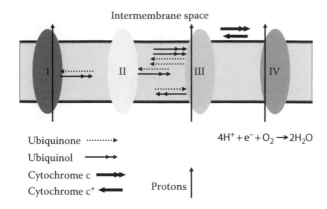

FIGURE 11.12 Electrons transported from Complexes I and II, through the mitochondrial membrane, to Complex III (bc$_1$). Cyt c transports electrons to Complex IV (cytochrome c oxidase) outside the membrane.

Succinate is a very simple molecule, a dicarboxylic acid where the two −COOH groups are held together by −CH$_2$−CH$_2$−. There is no place for an electron to localize, unless a π-system is formed. In Complex II (succinate dehydrogenase), succinate is oxidized to fumarate (HOOC−CH=CH−COOH) in a series of complicated, not fully known processes. In these processes, quinone (Q) is reduced to quinol (QH$_2$). The electrons are transported to Complex III through the membrane. The protons are transported perpendicular to the membrane to the intermembrane space, in order to build up a proton gradient for the production of ATP.

The distance between the central carbon atoms is 1.34 Å in succinate and 1.52 Å in fumarate. This leads to a large activation barrier, and this is why the reaction in Figure 11.11 cannot take place without enzymes—therefore, the complicated reactions in Complex II.

In Complex III, PT to the intermembrane space again takes place, driven by ET. Furthermore, the carriers are exchanged from quinones to cyt c. Cyt c moves outside the membrane and donates the electrons to Complex IV. Here the net reaction is

$$4 \text{ cyt } c^{2+} + 8H^+_{matrix} + O_2 \rightarrow 4\text{cyt } c^{3+} + 4H^+_{i.s.} + 2H_2O \tag{11.1}$$

The protons are again taken from the matrix and donated to the intermembrane space.

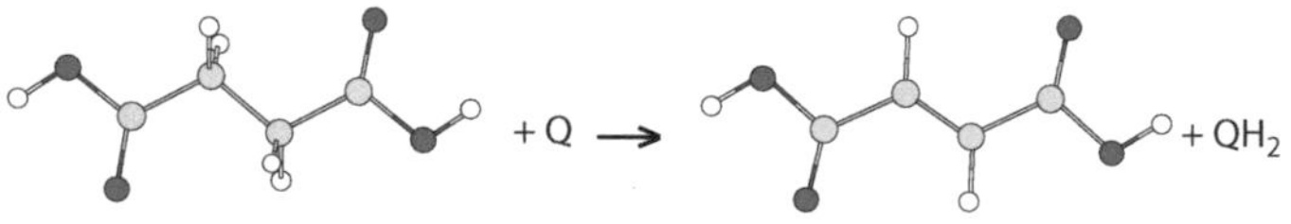

FIGURE 11.13 Net reaction of succinate and quinone.

11.4.2 ELECTRON TRANSFER STEPS

During the last century, an increasing number of structures of ET enzymes have become known by crystallographic means. Pathways for ET have been identified. The gross mechanisms follow the Marcus model.

The pathway concept generally refers to pathways for nuclei in the Born–Oppenheimer approximation. The name has been used alternatively to denote electron pathways between, say, two metal centers (Figure 10.16). It should be remembered, of course, that there we are not dealing with pathways in the classical sense, but with "the best orbital contact" between two metal centers. The electron cannot travel as a point-like classical particle, of course. The uncertainty relations spread out the electron to an orbital.

An immediate question in the studies of coupling is whether certain types of protein structure, for example π-systems, provide faster pathways than others? Has some structure been selected to promote efficient ET? Quite generally, this is not the case. The sequence and folding of a certain protein are of decisive importance for whether a particular protein has been selected for the ET chain. This is of importance for holding the coenzyme in the structure and to expose the prosthetic groups, possibly keeping the reorganization energy low, and adjust the reduction potential. In very few cases, there is evidence that pathways have also been selected. It is even doubtful that the selection of pathway is meaningful, unless the distance between the donor and the acceptor is also changed. The important thing is that the overlap between the atom orbitals is as large as possible along the pathway. A hydrogen bond, for example, may be important to pull the atoms closer together.

Electron carriers pass electrons to the electron transport chain of the membrane protein. Each step reaction within a membrane protein in the electron transport chain is a nonadiabatic ET reaction, meaning that the coupling between the metal ions in most cases is very small compared to the structural reorganization as electrons are donated or accepted. In a theoretical treatment, we always include two ET centers, one donor (D) and one acceptor (A). In the weak coupling limit, the equation for the rate, assuming that the nuclei move classically (the *semiclassical* ET model), may be written as in Equation 10.68:

$$k = \frac{\Delta^2}{4\pi\hbar} \sqrt{\frac{\pi^3}{\lambda k_B T}} \exp\left(-\frac{\Delta G^*}{k_B T}\right). \qquad (11.2)$$

In this equation, ΔG^* depends implicitly on the reorganization energy λ. The explicit dependence of activation energy on reorganization energy λ reads as in Equation 10.53:

$$\Delta G^* = \frac{\lambda}{4}\left(1 + \frac{\Delta G^\ominus}{\lambda}\right)^2. \qquad (11.3)$$

The explicit dependence on λ under the square root sign in Equation 11.2 is of minor importance.

From Equation 11.3, it follows that if λ is much larger than the driving force for the ET reaction, $-\Delta G^\ominus$, the ET reaction rate k is small due to a high activation barrier in the exponential of Equation 11.2. A high activation barrier is equivalent to trapping of the electronic charge, and this must not happen in biological ET. In biological ET reactions, it can be expected that $\Delta G^\ominus$ is small (to avoid energy loss) and λ is therefore also small, to avoid trapping. Since biological ET reactions take place at kT $\approx$ 0.03 eV, the activation energy ΔG^* should not exceed this value too much. If $\lambda = -\Delta G^\ominus$, ΔG^* is equal to zero, as follows from Equation 11.3.

By a "controlled" ET reaction, we mean one where the electron jumps with close to 100% probability between carriers, not losing too much free energy in the separate steps. This is not accomplished just by low ΔG^*. Coupling is also involved and may be more selective. However, coupling depends on distances rather than selected molecules in the ET overlap pathway between the carriers. As we will see next, $|\Delta|$ decreases exponentially in most cases, and hence distance is a likely part of the controlled ET. In other words, the intended carriers have to be well insulated. This may be achieved by large side chains not taking part in ET.

Vibrational energies smaller than kT = 0.03 Å (250 cm^{-1}) correspond to nearly classical motion. These motions may be counted to the *dynamics of the protein*. For example, motion of side groups, librational motion, and PT belong here. Some of the groups involved in the motion are charged dipoles. Therefore their motion causes oscillations of the local electric field. During these motions, $\Delta G^\ominus$ varies considerably. In fact, $\Delta G^\ominus$ may be negative only for a fraction of the time. In conclusion, slow vibrations determine the ET rate to a significant extent.

In a protein, the ET rate can hardly be faster than the characteristic time for protein dynamics. In reality, they are at least a thousand times slower in the electron transport chain. Transfer in the picosecond regime turns out to be sufficiently fast for the photosynthetic reactions in the excited state. By "sufficiently fast," we mean so fast that the many side reactions are avoided in the excited state. In the normal reactions of the ET chain of the membrane proteins, this rate is unnecessarily high. Instead it is possible to have sufficiently fast ET at a slightly larger distance. In other words, $|\Delta|$ in Equation 11.3 can be smaller than in the photosynthetic case, thereby permitting a longer distance for ET. This is important for the transfer of electrons in the membrane proteins. It is also important in the docking problem, when ubiquinone exchanges electrons with groups in the membrane proteins. We therefore find distances of about 15 Å between metal ions and aromatic systems exchanging electrons, distances that give very small coupling.

11.4.3 MORE ON ACTIVATION ENERGY

Equation 11.2 shows that if $\lambda = \Delta G^\ominus$, the activation energy ΔG^* is equal to zero. Since λ is often more than 1 eV for metal complexes, wasting of free energy to eliminate the barrier in this way would be costly for the organism. The most energy-saving mechanism for the system is $\Delta G^\ominus = 0$. The activation barrier is then equal to $\lambda/4$. Since kT at room temperature is about 0.03 eV, a reorganization energy of the magnitude $\lambda = 0.12$ eV is not seriously restricting the rate. At $\lambda = 1$ eV, the exponential

Arrhenius factor is still $\exp(-0.25/0.03) \approx 4 \times 10^{-4}$, which in most cases gives a sufficiently fast rate for ET.

Large reorganization energy, say 1 eV, can only be compensated by large electronic coupling. For metal ions, such as $Cu^{+/2+}$ or $Fe^{2+/3+}$, one often finds λ of this magnitude. The only way to bring down the activation energy is that coupling is large too. But then the metal ions have to be very close to each other. There must be substantial overlap between their orbitals.

In contradistinction to transition metal ions, π-systems have quite small reorganization energies, usually smaller than 0.2 eV. The coupling can be allowed to be smaller in this case.

Metal ions with bulky π-systems as ligands also have small reorganization energies. In heme, for example, the HOMO and LUMO have very much ligand character, thus "spreading out" the bond orders on many bonds. Warshel calculated the outer reorganization energy to about 0.1 eV, small because of the large size of the molecule and hence the solvent sphere.

In a pioneering work, De Vault and Chance studied the reaction center of *Chromatium vinosum* at very low temperatures. They looked for changes in the spectrum with photometric methods. The ln(rate) versus 1/T plot was far from linear. We remember from Chapter 10 that the logarithm of the rate at low temperature does not decrease exponentially as a function of 1/T, if nuclear tunneling is relatively large. In *C. vinosum*, the rate of the ET process studied by De Vault and Chance remains constant between 200 and 4 K. This means that ET takes place between the lowest vibrational levels of the precursor state and the successor state, and this is a signature of nuclear tunneling. This paper and others by De Vault and Chance had a large impact on the development of the theory for biological ET during the last decades of the twentieth century.

In summary, since energy is lost due to reorganization in each ET step within the membrane proteins, the number of steps is very likely minimized during the evolution. Efficient transport therefore means that the distance for the electron leap is maximized, under the condition that the rate should be faster than a rate-determining step ($\tau_0 \approx 1\mu s$). To minimize the reaction barriers and minimize the increase of the reduction potential, reorganization energy should be minimized and tuned with the free energy of reaction.

11.4.4 MORE ON COUPLING

The original Marcus model, although correct as an explanation of the activation energy for ET, did not contain anything about distance or the structural features of possible importance for ET. The same can be said about the Hush model, which is very similar to the Marcus model, and the vibrational model of Jortner and Bixon, originating from Levich and Dogonadze. The latter models account for nuclear tunneling, but treat the distance problem with the help of an arbitrary parameter called "coupling." Nowadays, this structural dependence can be calculated with reasonable accuracy and subsequently used within the Marcus–Hush and Jortner models.

The textbook tunneling model was originally invented by Gamow. Exponential decrease of rate with distance is obtained. This model cannot be used in proteins since it does not account for the thermal activation process and since the tunneling barrier cannot be calculated. The early paper by H. McConnell also does not include the thermal activation process. McConnell derived an equation for *symmetric* ET systems, which, with drastic approximations, also results in exponential decrease with distance. In the Gamow and McConnell models, only exponential decrease with distance is possible. Is nonexponential decrease possible? Yes, it is and the simple reason is that coupling is strongly related to the magnitude of distant chemical bonding. In chemical bonding, we are used to the fact that orbital overlap is the most important factor. Bond length is not of primary importance.

Are π-systems different from σ-systems in transferring electrons? Such questions could not be answered until the beginning of the 1980s. It turns out that the difference between σ- and π-systems is rather small. However, if a π-system is ordered in the optimum way, it has slower decrease with distance than exponential decrease.

Today, there is a well-defined set of accepted models for ET in proteins, all consistent with a generalized Marcus–Jortner model. Unfortunately, the name "superexchange model" is also used for exactly the same set of models. The word "superexchange" is highly confusing and even misleading. ET in proteins almost always involves one electron at a time, transferring to another site. Until 1983, the term "superexchange" was used exclusively for coupling of magnetic moments. This magnetic coupling problem, first discussed by Kramers, is very different since it involves one odd electron on each of two adjacent sites, in total an even number of electrons, coupled ferromagnetically or antiferromagnetically. In connection with ET, the term "superexchange" was first used by Jortner in 1983, as a new name for the model of McConnell, but McConnell himself did not use the term "superexchange" in connection with ET.

Actual measurements of rate between metal sites in proteins were performed by O. Farver and I. Pecht, and S. S. Isied et al. The former authors also showed that aromatic rings in the pathway between the donor and the acceptor do not seem to make any great difference to the rate and this was theoretically confirmed by A. Broo.

Harry B. Gray along with J. Winkler and others carried out a very enlightening experimental work on ET in proteins. Not only did they study natural ET, but also introduced artificial donors or acceptors on the surface of the protein. An important fact is that they most often obtained exponential decrease of the tunneling rate with distance, but a somewhat slower decrease factor for natural ET than in artificial ET steps. A theoretical calculation by A. Broo confirmed this. In the artificial case, the π-systems are not correctly located to promote ET.

C. C. Moser and P. L. Dutton carried out other studies. They concluded that donor–acceptor distance is of primary importance, but that protein structure between two electron exchanging aromatic groups plays an unimportant role.

Aside from the mentioned confusion, the ET field is now a quite mature field. The rate of ET reactions, contrary to the rates of most other chemical reactions, may be calculated using *ab initio* calculations in combination with molecular dynamics simulations. There should be no doubt that the physics of the ET process in proteins is correctly understood.

11.4.5 bc$_1$ COMPLEX

The structure of Complex III (bc$_1$ complex) has been crystallographically determined. It contains a 2Fe–2S complex and three cytochrome subunits (Figure 11.14).

Complex III (bc$_1$) is docked with ubiquinol (QH$_2$), which is oxidized to ubiquinone (Q) at the site Q$_0$, close to the intermembrane space. Q$_0$ delivers one electron to the Fe–S complex and another to cyt b$_L$. On the matrix side of the membrane, the opposite reaction takes place, but only half the amount can be produced. There is a net delivery of four protons to the intermembrane space, for each two electrons delivered to cyt c.

The ET steps involve a 2Fe–2S iron–sulfur complex as well as cyt c$_L$ and cyt c$_H$.

11.4.6 CYTOCHROME C OXIDASE

Cyt c delivers its electron to a Cu$_A$ center, which contains two coppers in a symmetric arrangement, and is located close to the intermembrane space. From the Cu$_A$ center, the electron moves on to low-spin heme A, and finally to a binuclear heme a$_3$–Cu$_B$ complex, where oxygen reduction takes place. The two metal ions in this binuclear center are only 4.5 Å apart. CcO is a large protein that contains 13 protein subunits with several metal sites: two heme irons and two copper centers, Cu$_A$ and Cu$_B$. Cu$_A$, located in subunit II, is a binuclear complex with two equivalent Cu^{2+} ions covalently bonded to two −CH$_2$S groups of cystein peptides and two nitrogen atoms of imidazole groups belonging to histidine peptides. Cu$_A$ absorbs at 830 nm and was first recognized by its spectral signal.

The electron leaps across the membrane from Cu$_A$ near the intermembrane space in three ET reactions to the binuclear site of heme a$_3$ and Cu$_B$, where dioxygen is reduced to water. Cu$_B$ (Cu$^+$/Cu^{2+}) has three histidine ligands and one OH$^-$.

In 1992, Babcock and Wikström clarified the ET in CcO (Complex IV). In the Wikström model, as updated by H. Michel, the reduction of dioxygen to water takes place in several steps. Four electrons are obtained from Cu$_A$ at arbitrary times. All intermediate steps have to be stable. Dioxygen binds to Fe at the binuclear complex. We have to remind ourselves of the orbital structure of the oxygen molecule. There is a strong σ-bond and two strong π-bonds. However, there are another two valence electrons, and these electrons enter the antibonding π* orbital. The spins are parallel

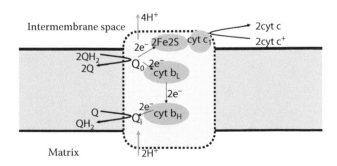

FIGURE 11.14 Complex III (bc$_1$ complex).

according to Hund's rule, making dioxygen paramagnetic. However, the two final valence electrons also cancel one of the two π-bonds. In dioxygen, the oxygen atoms are thus double-bonded.

Since the π^* orbital that can accommodate four electrons is only half occupied, two electrons can be added. Since π^* is antibonding, the bond length of O_2 is increased significantly, in this case by 0.3 Å. The first electron does not need any charge compensation by protons. O_2^- is formed. The dot stands for "radical"; in other words, an odd number of electrons. Because of the electroneutrality principle, very likely a proton is attracted to the outer oxygen atom, forming HO_2^-, before the next electron is accepted from the binuclear center. When protons are added without changing occupancy of orbitals, there is almost no change of interatomic distance.

The second electron enters the antibonding π^* orbital. There is now only the σ-bond between the two oxygen atoms left. The outer oxygen receives another proton whereby H_2O_2 is formed. Due to repulsion from the positive Fe ion of heme A, it is quite possible that the outer oxygen carries both protons.

In the next step, the σ-bond is broken by transfer of first one electron and then a second electron from the iron ion to the previously unoccupied, strongly antibonding σ^* orbital. The bond between the oxygen atoms is now fully canceled and one water molecule leaves the site. Another two protons are added and bind to the remaining oxygen atom.

Of the eight protons entering the binuclear center, four continue to the more acid intermembrane space, while four protons are used in the $4e^-$ reduction of O_2.

Due to the large size of CcO, it is hardly surprising that there are small differences between different species. There are three classes of CcO, denoted as A, B, and C. All three have the same heme–Cu_B binuclear center, but are quite different in other ways. One may speculate that the binuclear center, as a primitive CcO, developed already during the molecular evolution and was later mutated to improve the survival probability in different species.

In summary, CcO is an enzyme where diatomic oxygen (O_2) is reduced to two molecules of water (H_2O). For this, four electrons and four protons are needed and are provided by cyt c molecules moving on the outside of the membrane.

12 Photophysics and Photochemistry

12.1 INTRODUCTION

Photophysics and photochemistry deal with the interaction between light and matter. This is of immediate interest to us, since the nutrients that we need originate in natural photosynthesis. Furthermore, it has become very important to develop solar cells for electric energy production.

Small organic molecules usually absorb in the UV region or in the blue and violet of the visible region. The compound is white or transparent, or yellow if the absorption is in the blue. The emitted fluorescence light is usually blue. Organic molecules with large π-systems absorb in different parts over the whole visible region. A good example is provided by the chlorophyll molecule that absorbs light energy, drives photosynthesis on earth, and makes leaves green. The interaction between light and matter is studied in photophysics and photochemistry. Photosynthesis will be discussed in Chapter 15. Here, we will discuss more fundamental aspects of photochemistry and photophysics.

12.2 PHOTOPHYSICS

The impressive progress of photophysics in recent decades has been possible because of the *laser*, which provides short and intense light pulses. The basic laser mechanism is summarized in Section 12.2.4. Later, the basic rules for *absorption* and *emission* of light (or electromagnetic [EM] radiation in general) are derived from time-dependent quantum mechanics.

12.2.1 ABSORPTION AND REFLECTION OF LIGHT IN MATTER

EM radiation interacts with matter according to the Maxwell equations, published already in 1865. At this time, EM radiation had not yet been produced in the laboratory (not counting EM waves that drive alternating currents). Radio waves were first created by Hertz in 1888. Since light moves with the velocity that Maxwell derived for EM waves in general, it was assumed that light consists of EM radiation.

The molecular process of the absorption of light and excitation to a higher energy level in a single molecule is very fast—less than a femtosecond. Therefore, it may be assumed that the nuclei are not moving during the absorption process. The *Franck–Condon principle* expresses that electronic absorptions are *vertical* in a diagram where nuclear motion is plotted along the x-axis.

If light is absorbed at a certain frequency, it may later be emitted at the same or a lower frequency as *spontaneous emission*. It may also be emitted immediately as *stimulated emission*, while the light pulse that caused the absorption is still present.

The Bohr frequency relation determines the absorption of light by an atom or a molecule. To an observer, this looks like a certain amount of radiation is colliding with the molecule. The amount of energy and momentum absorbed or emitted is referred to as a *photon*. However, there is no obvious quantization of radiation that forms photons of limited extension. The photon should *not* be understood as an indivisible particle or a wave packet of atomic dimensions with given energy and frequency that enters the probe. Rather, the observable is an energy density as a function of frequency $\rho(\nu)$. Any field, no matter how weak, may produce a photon with some probability in space and time, proportional to its density. And any field may be divided up in a way that would contradict the particle nature of the photon. The photon concept is just a pedagogical construction to express that "here"— "now"— "with this frequency," a molecule has been excited with excitation energy $h\nu$. But the photon is absent as a particle before it is observed, just like a cat may be regarded as absent from a house before it has vomited on the floor or made itself observable in some other way.

When light hits the surface of a crystal, the electrons in the upper atomic layers try to follow the alternating radiation field. If the crystal is a metal, the oscillations are strong and become new radiators, creating EM radiation at all frequencies that compose the incoming EM radiation. If the surface is flat, coherent currents are induced at the first few atomic layers. The velocity of the light perpendicular to the plane is reversed, while the velocity component parallel to the plane remains the same. This is experienced as *reflection*. Practically all incoming light is reflected in a metal. A metallic, flat surface is a *mirror*.

If there is no absorption, as is the case for rays of visible light in window glass, water, benzene, and many other *transparent*, organic solvents and solids, the ray continues through the probe. The interaction with the solvent molecules slows down the ray in the medium. Because light has a smaller velocity in a medium, the light beam is bent or *refracted* according to *Snell's law* when it passes into the medium:

$$\frac{c}{v_{med}} = \frac{\sin \alpha}{\sin \beta} = n, \tag{12.1}$$

where α and β are defined in Figure 12.1. n is called the *refractive index*.

If there is absorption, the reflected and transmitted light will be missing the frequencies that are absorbed. If blue light is absorbed, the reflected light color will be the complimentary one, which is yellow. The leaves of plants are green because red light is absorbed in excess by chlorophyll molecules.

If the surface is irregular and rough and there is no absorption, incoherent and diffuse light is emitted in all directions as white light, if not absorbed. If all light is absorbed, the surface appears black.

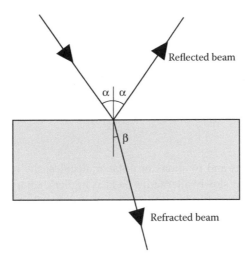

FIGURE 12.1 Refraction and reflection.

12.2.2 REFRACTION AND DIFFRACTION

Dipoles in the medium, consisting of water molecules, for example, are put in motion by radiation, and this motion will contribute to the EM field. Usually, EM waves of light have a much higher frequency than the eigenfrequencies of water molecules or other dipoles of that kind. Therefore, if light is propagated in a solid or in a solution, the molecules are not much affected, simply because they are too slow to follow the alternating field of the light wave. Similarly, the nuclear vibrations in the bonds are too slow to follow the field.

The speed of light in a medium (v_{med}) varies slightly with the wavelength of light. White light is *dispersed* into different wavelengths. This phenomenon is referred to as *refraction*. The bending angle depends on the refractive index n and hence on the wavelength of the light. The most common illustration of this is a glass prism, as seen in Figure 12.2. Another is refraction in spherical droplets of water producing a rainbow.

Two-slit diffraction is shown in Figure 1.4. A grating of many paralell slits acts as a dispersive element.

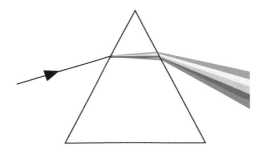

FIGURE 12.2 (**See color insert**) Refraction: Light ray passing a prism.

The dielectric properties of water and other polar substances are measured by the relative dielectric constant, ε_r, valid in a static electric field. For a propagating EM field, the *dielectric constant* depends on the frequency of the radiation, for the reasons just given. At light frequencies, the oscillation of dipoles or the vibrations of nuclei are unimportant. Then, as follows from Maxwell's equations, the dielectric constant for a substance is equal to the square of the refractive index (not proven):

$$\varepsilon_r = n^2. \tag{12.2}$$

Due to the vibrations and rotations of polar molecules, the relative dielectric constant, ε_r, in a static or slowly alternating field is higher than n^2. In water, depending on temperature, ε_r is roughly equal to 80, but n^2 is only ≈ 2.

When light propagates, every point on a wavefront may be considered as a source of a new wavefront. When light interferes after having passed a double slit (Figure 1.4) or a dispersion grating, the direction depends on the wavelength. A dispersion grating is therefore another possibility to disperse white light into its components.

12.2.3 LAMBERT–BEER'S LAW

Absorption is the difference between the incoming radiation intensity and the transmitted intensity. A light wave excites molecules in its way through a solution in a cuvette or through a slice of glass. We need to know the fundamental rules for how light is absorbed and how this *absorbance* can be measured. The concentration of the substance in the cuvette is [Y], and the thickness of a small slice is dx. The fraction dI/I absorbed by the solution should be proportional to the concentration [Y] and to the thickness dx:

$$\frac{dI}{I} = -\kappa[Y]dx, \tag{12.3}$$

where κ is a constant. Integration of Equation 12.3 gives

$$\ln I - \ln I_0 = -\kappa[Y](x - x_0), \tag{12.4}$$

where I is the intensity after the distance $(x - x_0)$, if I_0 is the original intensity. From Equation 12.4 follows

$$\frac{I}{I_0} = e^{-\kappa[Y](x-x_0)} = 10^{-\alpha[Y](x-x_0)}, \text{ where } \alpha = \kappa \log e. \tag{12.5}$$

α is called the molar extinction coefficient. Equation 12.5 shows how the intensity decreases along the pathway of the ray through the probe and is called the *Lambert–Beer's law*. The dimensionless quantity $\varepsilon[Y](x - x_0)$ is the *absorbance*. I/I_0 is the intensity that remains relative to the initial intensity after passing the distance $(x - x_0)$ and is therefore called the *transmittance*.

12.2.4 LASER RADIATION

In modern spectroscopy, it is necessary to have access to good sources of light, ultraviolet (UV), and infrared (IR) radiation. The laser is useful since it emits coherent EM radiation at a very high intensity and, if necessary, in a narrow frequency range.

The idea of amplification of EM radiation by stimulated emission was first realized for microwaves. The MASER (microwave amplification by stimulated emission of radiation) was developed in the Soviet Union in the early part of the 1950s, particularly by N. Basov and A. Prokhorov. At the same time, it was developed by C. H. Townes, A. L. Schawlow, and others in the United States. Originally, the laser was called "light-maser," but the name was changed to LASER (light amplification by stimulated emission of radiation).

The first laser, the ruby laser, was constructed by T. Maiman in the Hughes Laboratories in California. The same year, it was considerably improved in the Bell Laboratories. It consists of a corundum (Al_2O_3) crystal with quite sparse substitutions of Al^{3+} with Cr^{3+}. The corundum device is shaped in the form of a long cylinder.

The laser consists of a light source placed between two mirrors, where one of them can be penetrated by light. The first light source used for this purpose is a ruby crystal, which consists of a corundum, Al_2O_3, with Cr^{3+} impurities. This *gain medium* is excited from the sides. The Cr^{3+} ions are excited (*pumped*) to the lowest spin-allowed crystal field excited state. This level is called level 3 (Figure 12.3). Between the ground state (level 1) and level 3 there is another state called level 2, which has a different total spin from the ground state.

One wants to achieve population inversion, meaning that level 2 has a higher occupation than level 1. When the electrons start to deexcite and reoccupy the ground state, the intensity starts to increase in the frequency range of the emission. The deexcitation is now a stimulated emission and its intensity increases manifold. The laser is *fired* by itself and a light pulse is emitted. The emitted light penetrates the semitransparent mirror.

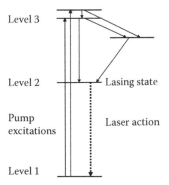

FIGURE 12.3 Three-level ruby laser. The levels shown are ligand field levels. Level 1 is the ^4A ground state; level 2 is the $^4T_1$ and $^4T_1$ excited states, and level 3 is the ^2E excited state.

All laser action in a ruby involves the ligand field transitions between the ground state $^4A_2$ ($t_{2g}\uparrow^3$), the excited $^4T_1$ and $^4T_2$ states ($t_{2g}\uparrow \rightarrow e_g\uparrow$) of the same spin multiplicity, and the excited ^2E ($t_{2g}\uparrow^2 t_{2g}\downarrow$) state of different spin multiplicity. The ruby is thus a three-level laser (Figure 12.3). Level 3 consists of the excited $^4T_1$ ($\lambda 407$ nm) and $^4T_2$ ($\lambda 551$ nm) states. Level 2 is the ^2E state, the lowest electronically excited state. The ^2E state has the same valence electrons occupied as the ground state. The transition $^2E \rightarrow {}^4A_2$ is spin forbidden. The population of the level 2 state is replenished from the level 3 states and becomes even larger than that of the ground state, just before the laser fires.

In the level 3 states, a strongly antibonding e_g orbital is occupied, which means that at least some of the bonds are considerably longer than in the ground state. The original excitation to the level 3 states (the *pump* transition) is therefore very wide due to vibrational broadening. Level 2, on the other hand, has almost the same equilibrium geometry as the ground state, since the occupied orbitals are the same as in the ground state. There is hardly any change of equilibrium geometry and the $^2E \rightarrow {}^4A_2$ transition is therefore very sharp.

An advantage with laser light is that the ray can be easily focused. The spread in space can be very small and the wavelength well-defined. Laser light may be used to measure distances with a high degree of accuracy.

When the laser is fired, the frequencies with the highest intensities are stimulated more than other frequencies. The laser light is therefore even sharper than the normal $^2E \rightarrow {}^4A$ spectral line. The narrowing is, in this case, a self-sharpening property. This gives a possibility to *tune* the laser using mirrors.

A great number of other sources of light exist. A modern laser is the YAG laser, where YAG stands for yttrium aluminum garnet ($Y_3Al_5O_{12}$). Nd^{3+} substitutes Y^{3+} at some sites. In contradistinction to the Y^{3+} ion, the Nd^{3+} ion has the 4f level occupied by three electrons, all with the same spin. In the case of lanthanides, the interaction of the 4f subshell with the ligands may be ignored and the states regarded as atomic states. In the Nd^{3+} ion, there are three 4f electrons with the same spin direction. The largest possible M_L state has $M_L = 3 + 2 + 1 = 6$, corresponding to an I state. Relativistic splitting makes the $^4I_{9/2}$ state the ground state (level 1), followed in energy by $^4I_{11/2}$, $^4I_{13/2}$, and $^4I_{15/2}$, which all belong to level 2 together with the higher vibrational levels of the ground state. Level 3 consists of $^4F_{5/2}$ and $^4F_{3/2}$ states. The transition is at $\lambda 808$ nm, thus in the near IR. Lasing is at 1064 nm, but also at other frequencies.

Other YAG lasers have been constructed where ytterbium, erbium, thulium, or holmium is used instead of neodymium.

A laser with wide absorption is the sapphire laser (ruby is red corundum and sapphire is blue corundum). In sapphire, some Al^{3+} sites in Al_2O_3 are replaced by Ti^{3+}. The latter ion has a single electron in the 3d shell of Ti^{3+} with a very broad spectrum in the visible region, depending on the large change in bond lengths when the 3d electron is excited (Figure 12.4). This laser also emits in a wider frequency region than the YAG laser. Level 2 is the same electronic state as the ground state, but the vibrational level is high.

By up-conversion techniques, low-energy radiation can be used to create higher-energy radiation, so that the pump energy can be larger than that of the pump source. For example, IR light may be used to emit visible pump light.

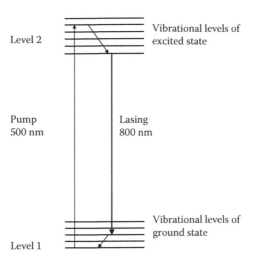

FIGURE 12.4   Energy levels for the sapphire laser.

12.2.5   Absorption of Radiation in Atoms and Molecules

The Hamiltonian for an electron in an EM field is derived from classical mechanics and electrodynamics:

$$H = \frac{1}{2\mu}\left(\vec{p}+e\vec{A}\right)^2 - e\Phi, \tag{12.6}$$

where μ is the rest mass of the electron, $\vec{A}$ is the vector potential, and Φ is the scalar potential field. In a pure field, we may assume that the vector and scalar potentials satisfy equations derived from the Maxwell equations:

$$\nabla^2\vec{A} - \frac{1}{c^2}\frac{\partial^2\vec{A}}{\partial t^2} = 0. \tag{12.7}$$

We assume the so-called Coulomb gauge to define the vector field $\vec{A}$, and the electron is in a zero scalar field:

$$\nabla\cdot\vec{A} = 0 \quad \text{and} \quad \Phi = 0. \tag{12.8}$$

Equation 12.6 may be written, using $\vec{p} = -i\hbar\nabla$, $\nabla\cdot\vec{A} + \vec{A}\cdot\nabla = 2\vec{A}\cdot\nabla$, and neglecting $\vec{A}^2$:

$$H = H_0 + V = -\frac{\hbar^2}{2\mu}\nabla^2 - i\hbar\frac{e}{\mu}\vec{A}\cdot\nabla, \tag{12.9}$$

where the first term is the ordinary kinetic energy and the second term may be regarded as a perturbation.

We assume that the wave propagates in the z-direction. The components of $\vec{A}$ are then in the xy-plane. We further assume that the radiation field may be written as a Fourier transform of a function $\vec{B}$ that depends on the frequency ω:

$$\vec{A}(z,t) = \int_{-\infty}^{\infty} \vec{B}(\omega) \exp\left[-i\omega\left(t - \frac{z}{c}\right)\right] d\omega, \tag{12.10}$$

where c is the velocity of light (= 137 a.u.). The perturbation operator V (Equation 12.9) may thus be written

$$V = \frac{e}{\mu} \int_{-\infty}^{\infty} \exp\left[-i\omega\left(t - \frac{z}{c}\right)\right] \vec{B}(\omega) \cdot \nabla(-i\hbar) d\omega. \tag{12.11}$$

In Chapter 7, we calculated the wave function under the assumption that the system is originally in the state n [$C_n(-\infty) = 1$; $C_m(-\infty) = 0$ for $m \neq n$]. Equation 12.11 is substituted into Equation 7.33 and we obtain, using $\omega_{mk} = (\varepsilon_m^0 - \varepsilon_k^0)/\hbar$,

$$C_m(t) = -\frac{i}{\hbar} \int_{-\infty}^{t} \exp(i\omega_{mn}\tau) \langle \psi_m^0 | V | \psi_n^0 \rangle d\tau$$

$$= -\frac{i}{\hbar} \frac{e}{\mu} \int_{-\infty}^{t} \exp(i\omega_{mn}\tau) \left\langle \psi_m^0 \left| \int_{-\infty}^{\infty} \exp\left[-i\omega\left(\tau - \frac{z}{c}\right)\right] \vec{B}(\omega) \cdot \nabla(-i\hbar) d\omega \right| \psi_n^0 \right\rangle d\tau. \tag{12.12}$$

If the perturbation dies out after some time, the upper integration limit may be replaced by $\tau = \infty$. The final result is thus

$$C_m(\infty) = -\frac{i}{\hbar} \frac{e}{\mu} \int_{-\infty}^{\infty} \int_{-\infty}^{\infty} \exp\left[i(\omega_{mn} - \omega)\tau\right] \left\langle \psi_m^0 \left| \exp\left(\frac{-i\omega z}{c}\right) \vec{B}(\omega) \cdot \vec{p} d\omega \right| \psi_n^0 \right\rangle d\tau. \tag{12.13}$$

We now make use of the Fourier integral theorem:

$$f(x) = \frac{1}{2\pi} \int_{-\infty}^{\infty} dt \int_{-\infty}^{\infty} f(\omega) \exp\left[it(\omega - x)\right] d\omega, \tag{12.14}$$

and obtain

$$C_m(\infty) = -\frac{2\pi i e}{\hbar\mu} \left\langle \psi_m^0 \left| \exp\left(\frac{i\omega_{mn} z}{c}\right) \vec{p} \right| \psi_n^0 \right\rangle \cdot \vec{B}(\omega_{mn}). \tag{12.15}$$

This equation shows that only the Fourier component $\vec{B}(\omega_{mn})$ is involved in the excitation to the state m, in agreement with the Bohr frequency condition. At absorption, the field loses the energy $\hbar\omega_{mn}$, corresponding to a photon.

The exponential function may be expanded:

$$\exp\left(\frac{i\omega_{mn}z}{c}\right) = 1 + i\omega_{mn}\frac{z}{c} + \cdots. \tag{12.16}$$

Keeping only the first term is justified as long as ω_{mn} is smaller than, say, 1 a.u., since $c = 137$ a.u. and z during the integration takes on values corresponding to one or a few atomic units. The first term gives

$$C_m(\infty) = -\frac{2\pi ie}{\hbar\mu}\left\langle\psi_m^0\left|\vec{p}\right|\psi_m^0\right\rangle\cdot\vec{B}(\omega_{mn}). \tag{12.17}$$

This first term in the expansion in Equation 12.17 is called the *electric dipole term*. There is a contribution only for the projection of $\vec{p}$ on $\vec{B}$.

In the N electron case, the momentum $\vec{p}$ is given by

$$\vec{p} = \vec{p}_1 + \vec{p}_2 + \cdots + \vec{p}_N. \tag{12.18}$$

Using

$$i\hbar\vec{p}/\mu = \vec{r}H_0 - H_0\vec{r} \quad (\mu = 1 \text{ a.u.}) \tag{12.19}$$

we obtain

$$\left\langle\psi_m^0\left|\vec{p}\right|\psi_n^0\right\rangle = \frac{\mu}{i\hbar}\left\langle\psi_m^0\left|\vec{r}H_0 - H_0\vec{r}\right|\psi_n^0\right\rangle$$
$$= \frac{\mu}{i\hbar}\left(\varepsilon_n^0 - \varepsilon_m^0\right)\left\langle\psi_m^0\left|\vec{r}\right|\psi_n^0\right\rangle = i\mu\omega_{mn}\left\langle\psi_m^0\left|\vec{r}\right|\psi_n^0\right\rangle. \tag{12.20}$$

The matrix element

$$\vec{d} = \left\langle\psi_m^0\left|e\vec{r}\right|\psi_n^0\right\rangle \quad (e = 1 \text{ a.u.}) \tag{12.21}$$

is called the *transition dipole moment* for the transition from state n to state m. If there are N electrons, we have instead

$$\vec{d} = \left\langle\psi_m^0\left|e\vec{r}_1 + e\vec{r}_2 + \cdots + e\vec{r}_N\right|\psi_n^0\right\rangle. \tag{12.22}$$

The dimensionless quantity

$$f_{mk} = \left(E_m - E_n\right)d^2 \frac{2\mu}{3\hbar^2} \tag{12.23}$$

is called the *oscillator strength* for the transition from n to m, and is used to express the strength of the transition.

We will now show the *Thomas–Reiche–Kuhn sum rule*:

$$\sum_i f_{ik} = N, \tag{12.24}$$

where N is the number of electrons.

We use the following quantum mechanical relations:

$$\left[x, H_0\right] = xH_0 - H_0x = \frac{i\hbar}{\mu} p_x$$
$$\left[p_x, x\right] = p_x x_0 - xp_x = -i\hbar, \tag{12.25}$$

which implies

$$\left[\left[x, H_0\right], x\right] = \frac{\hbar^2}{\mu}. \tag{12.26}$$

At the same time,

$$\left[\left[x, H_0\right], x\right] = \left[x, H_0\right]x - x\left[x, H_0\right] = 2xH_0x - H_0x^2 - x^2H_0. \tag{12.27}$$

We may now make use of the fact that the eigenfunctions of H_0 form a complete set:

$$f = \sum_i \langle f|i\rangle\langle i| \quad \text{for all f} \quad \text{since} \sum_i |i\rangle\langle i| = 1. \tag{12.28}$$

The latter relation is sometimes called the *resolution of the identity*. The expectation value of the operator in the left member of Equation 12.21 for the state n is then

$$\langle n|\left[x, H_0\right]x|n\rangle = \langle n|2xH_0x|n\rangle - 2E_n\langle n|x^2|n\rangle$$
$$= 2\sum_i \langle n|xH_0|i\rangle\langle i|x|n\rangle - E_n\langle n|x|i\rangle\langle i|x|n\rangle \tag{12.29}$$
$$= 2\sum_i \left(E_i - E_n\right)\langle n|x|i\rangle^2.$$

Equations 12.26 and 12.29 lead to the identity:

$$\frac{2\mu}{\hbar^2}\sum_i \left(E_i - E_n\right)\langle n|x|i\rangle^2 = 1. \tag{12.30}$$

For N particles, the operator x has to be replaced by $x_1 + x_2 + \cdots + x_N$ and Equation 12.30 by

$$\frac{2\mu}{\hbar^2}\sum_i \left(E_i - E_n\right)\langle n|x|i\rangle^2 = 1. \tag{12.31}$$

It is important that all excited states are included in the summation and that they form a complete set. This means that for continuous spectra, the summation has to be replaced by an integration.

12.2.6 RATE OF SPONTANEOUS EMISSION

The transition probability for emitting light is the same as for absorbing light. Continuous EM radiation of a frequency that matches a transition according to the Bohr relation would lead to the same population in the ground state and in the excited state. At the same time, the population in an excited state follows Boltzmann's statistics, according to which there is always a larger population in the ground state. Therefore, spontaneous emission of radiation must occur, independent of the EM field. Spontaneous emission is much slower ($\approx 10^{-9}$ sec) than stimulated emission. The latter emission occurs when the light pulse is still present and when the excited state has not yet started to deexcite in the vibrational manifold (the energy of transition to the ground state must be the same as in the light pulse). However, stimulated emission is a separate process from excitation. As a comparison, in the case of Raman scattering, excitation and deexcitation constitute the same process (=scattering of radiation).

The calculation of the rate of spontaneous emission from quantum electrodynamical theory is a rather complicated problem. We will use instead a treatment based on statistical equilibrium, originally derived by Einstein. We assume that the number of molecules in the ground state is N_0 and in the excited state N_1. For simplicity, we assume that both states are nondegenerate. The Boltzmann distribution ratio at equilibrium is equal to

$$\frac{N_1}{N_0} = \exp\left(\frac{-\Delta E}{k_B T}\right). \tag{12.32}$$

We assume that the radiation field is so weak that its presence does not require any modification of Equation 12.32. The rate of the induced absorption may be written as

$$\frac{dN_1}{dt} = N_0 B\rho(\omega_{01}) \quad (\hbar\omega_{01} = \Delta E = E_1 - E_0), \tag{12.33}$$

where ρ is the Planck radiation density:

$$\rho(\omega) = \frac{2\hbar\omega^3}{\pi c^3 \exp(\hbar\omega/k_B T) - 1}. \tag{12.34}$$

The rate of induced emission is derived in the same way as induced absorption and can be rewritten with the help of an expression equivalent to Equation 12.31:

$$\frac{dN_1}{dt} = -N_1 B\rho(\omega_{01}). \tag{12.35}$$

Obviously, if there were no spontaneous emission, we would have $N_1 = N_0$ at equilibrium, which would contradict Equation 12.32. The spontaneous emission rate must be proportional to N_1 and thus Equation 12.35 has to be replaced by

$$\frac{dN_1}{dt} = -N_1 \left[B\rho(\omega_{01}) + A \right]. \tag{12.36}$$

A and B are known as the *Einstein transition probability coefficients*.

At equilibrium, the sum of the absorption rate and the emission rate must be the same; therefore, using Equations 12.32 and 12.36, we obtain

$$N_0 B\rho(\omega_{01}) = N_1 \left[B\rho(\omega_{01}) + A \right]. \tag{12.37}$$

Using Equation 12.32, we obtain from the latter equation

$$\frac{N_1}{N_0} = \exp\left(-\frac{\Delta E}{k_B T}\right) = \frac{B\rho}{B\rho + A}, \tag{12.38}$$

hence

$$A = B\rho \left[\exp(\Delta E/k_B T) - 1 \right]. \tag{12.39}$$

Using Equation 12.34, we obtain

$$A = 2B\hbar\omega^3/\pi c^3. \tag{12.40}$$

The rate constant of induced absorption, $B\rho$, may be written in terms of the absorption rate, W_{01}, for the transition $0 \rightarrow 1$:

$$B\rho = W_{01}. \tag{12.41}$$

We assume that a monochromatic EM pulse starts at $t = 0$ and is directed along the z-axis. The vector field may then be written as

$$A_x = A_0 \cos\omega\left(t - \frac{z}{c}\right), \quad A_y = 0, \quad A_z = 0. \tag{12.42}$$

The electric and magnetic fields may be written in terms of $\vec{A}$ as

$$\vec{E} = -\frac{\partial\vec{A}}{\partial t} \quad \text{and} \quad \vec{H} = c\nabla x\vec{A}. \tag{12.43}$$

Thus

$$E_x = \omega A_0 \sin\omega\left(t - \frac{z}{c}\right), \quad E_y = 0, \quad E_z = 0,$$

$$H_x = 0, \quad H_y = \omega A_0 \sin\omega\left(t - \frac{z}{c}\right), \quad H_z = 0. \tag{12.44}$$

The electric and magnetic fields are at right angles to each other, and to the direction of propagation, and move with the velocity c. We are thus dealing with a plane-polarized wave with a well-defined frequency. Using the expression for V of Equation 12.9 in Equation 7.29, we obtain

$$\frac{d}{dt}C_1 = -\exp(i\omega_{10}t)\left\langle \psi_1^0 \left| \frac{\varepsilon}{\mu} \right| \psi_0^0 \right\rangle. \tag{12.45}$$

The wavelengths of interest to us exceed 100 nm, which is much more than the size of the molecule, and it is then a good approximation to consider A_x as constant over the molecule. We may then write A_x as

$$A_x = A_0 \cos\omega t = \frac{A_0}{2}\left[\exp(i\omega t) + \exp(-i\omega t)\right], \tag{12.46}$$

hence

$$\frac{d}{dt}C_1(t) = \frac{ie}{2\mu\hbar}A_0\left[e^{i(\omega_{10}+\omega)t} + e^{i(\omega_{10}-\omega)t}\right]\left\langle \psi_1^0 \left| p_x \right| \psi_0^0 \right\rangle. \tag{12.47}$$

The second term of Equation 12.47 contributes to C_1 (for $\omega \approx \omega_{10}$). Neglecting this term, we obtain C_1 by integration from 0 to t (*cf.* Equation 7.33):

$$\left|C_1(t)\right|^2 = \left|\frac{e}{2\mu\hbar} A_0 \left\langle \psi_1^0 \left| p_x \right| \psi_0^0 \right\rangle \int_0^t \exp i(\omega_{10} - \omega)d\tau\right.$$

$$= \frac{1}{2\hbar^2} \left|\frac{e}{\mu} A_0 \left\langle \psi_1^0 \left| p_x \right| \psi_0^0 \right\rangle\right|^2 \frac{\left[1 - \cos(\omega_{10} - \omega)t\right]}{(\omega_{10} - \omega)^2}. \tag{12.48}$$

We further obtain

$$\frac{d}{dt}\left|C_1\right|^2 = \frac{1}{2\hbar^2} \left|\frac{e}{\mu} A_0 \left\langle \psi_1^0 \left| p_x \right| \psi_0^0 \right\rangle\right|^2 \frac{\left[\sin(\omega_{10} - \omega)t\right]}{(\omega_{10} - \omega)}. \tag{12.49}$$

This expression cannot be used directly in Equation 12.41, since we need to integrate over ω from $-\infty$ to ∞. We obtain

$$W_1 = \frac{\pi}{2\hbar^2} \left|\frac{e}{\mu} A_0 \left\langle \psi_1^0 \left| p_x \right| \psi_0^0 \right\rangle\right|^2. \tag{12.50}$$

Using Equation 12.20, we may write this as

$$W_1 = \frac{\pi\omega^2}{2\hbar^2} A_0(\omega)^2 \left\langle \psi_1^0 \left| e_x \right| \psi_0^0 \right\rangle^2. \tag{12.51}$$

In EM theory, one can show that

$$A_0(\omega)^2 = \frac{8\pi}{\omega^2} \rho(\omega), \tag{12.52}$$

hence, from Equation 12.41,

$$B = W_1 / \rho(\omega) = \frac{4\pi^2}{\hbar^2} \left\langle \psi_1^0 \left| e_x \right| \psi_0^0 \right\rangle^2. \tag{12.53}$$

Using Equation 12.40,

$$A = \frac{8\pi\omega^3}{\hbar c^3} \left\langle \psi_1^0 \left| e_x \right| \psi_0^0 \right\rangle^2. \tag{12.54}$$

One may show that there is a contribution also from magnetic dipole and electric quadrupole radiation (and higher terms). The electric quadrupole term arises because the field is not uniform over the molecule. It is most important for wavelengths of the same magnitude as the width of the molecule. Such wavelengths are too short to be of interest here. Anyway, an improved Equation 12.54 may be written as

$$A = \frac{8\pi\omega^3}{\hbar c^3}\left\{\left\langle\psi_1^0\left|e_x\right|\psi_0^0\right\rangle^2 + \left\langle\psi_1^0\left|\frac{e}{2\mu c}(zp_y - yp_z)\right|\psi_0^0\right\rangle^2 + \left\langle\psi_1^0\left|\frac{3\pi e\Delta E^2}{40\hbar c^2}x^2\right|\psi_0^0\right\rangle^2\right\}.$$

(12.55)

The relative order of magnitude can be estimated. The expectation value of the dipole terms is of molecular dimensions, say 10 a.u. The magnetic dipole term contains $(2c)^2$ in the denominator and its magnitude is then $(2 \times 137)^2$, or about 10^5 times smaller than the electric quadrupole term. ΔE is smaller than 1 a.u. and because of the factor $1/c^2$, the electric quadrupole term should be even smaller than the magnetic dipole term. Thus, the contributions from the magnetic dipole and electric quadrupole terms are negligible compared with the electric dipole term, except when the matrix element of the latter is equal to zero for symmetry reasons.

Due to the low rate of spontaneous emission, a number of processes appear before emission. The processes depend strongly on the molecule; therefore, the content of the next section is referred to as molecular photophysics.

12.3   MOLECULAR PHOTOPHYSICS

Molecules or parts of molecules with transitions in the visible spectrum are referred to as *chromophores*. The first excited singlet state (S_1) is reached directly from the ground state or via one of the higher excited singlet states (S_2, S_3, ...). The Franck–Condon condition is satisfied, meaning that interatomic distances and bond angles immediately after excitation are the same as in the ground state. After the excitation pulse has passed, no stimulated emission can take place. The molecule can now be expected to remain in the electronic excited state during several picoseconds. During the first picosecond, relaxation of the molecular structure takes place. The vibrationally relaxed state is lower in energy than the state first accessed (=the Franck–Condon state). The decrease of energy at relaxation is one half of the *Stokes shift*. The Irish physicist and mathematician George G. Stokes showed already in 1852 that fluorescing substances become radiant when light is absorbed, but with a change in refractive properties such as wavelength.

12.3.1   FLUORESCENCE: STOKES SHIFT

Transition from an excited potential energy surface (PES) back to the ground state is referred to as *fluorescence*. The majority of the photoactive molecules have a spin-singlet ground state. The latter is referred to as the S_0 state in photophysics. There are several excited singlet states, of course (S_n, n > 0). The transition $S_n \rightarrow S_0$ is spin allowed, but it was early discovered (1950) that only the lowest excited state, S_1, makes any significant contribution to the fluorescence (Kasha's rule, after an American spectroscopist and theoretical chemist).

The name fluorescence is related to CaF_2 (fluorspar; latin: *fluor mineralis*), which emits light when it is irradiated by UV light. This emission appears to be caused by traces of europium, substituting at the Ca^{2+} sites as Eu^{2+}, and has nothing to do with what we call fluorescence today. The latin word *fluere* means to flow and refers to the use of a mineral to increase the fluidity of a mineral smelt. Fluor became the name for the element F and also for an emission that ceases as soon as the UV light source is removed (fluorescence). In the childhood of photochemistry (before 1945), fluorescence emission indeed appeared to be instantaneous. We know now that the timescale is nanoseconds; in other words, fluorescence is a million times slower than absorption.

The system relaxes to the lowest vibration state on the S_1 PES immediately after absorption. There may be an avoided crossing with the ground state PES, making *nonradiative* or *radiationless transitions* to the ground state possible (Figure 12.5). Another possibility is that the system emits radiation in a (vertical) fluorescence transition to the ground state. The system hits the PES away from the equilibrium position. The energy of the fluorescence radiation is smaller than the original excitation energy by $SS = SS1 + SS2$. SS is the Stokes shift. The emission frequency, v_1, is smaller than the absorption frequency, v_0. The change in energy $h(v_0 - v_1)$, corresponding to a red shift, is the Stokes shift.

Figure 12.5 shows the vibrational levels. Since the equilibrium geometry is different in S_0 and S_1, the transition moments are different from zero for a great number of vibrational levels (*vibrational broadening*). Since all absorption occurs from the lowest vibrational level of the S_0 state (at ordinary temperature), the *lowest* absorption energy corresponds to the lowest vibrational level of the S_1 state. Since all emission occurs from the lowest vibrational level of the S_1 state, the *highest* emission energy corresponds to the transition to the vibrational ground state of S_0. The vibronic transition between the lowest vibrational levels of the S_0 and S_1 states, and only this transition, is seen in both absorption and emission spectra at the same wavelength.

The next important step is to realize that the width of the absorption spectrum on *the red side* is equal to the free energy that drives structural reorganization in the S_1 state SS1 in Figure 12.5. In other words, it is the reorganization energy for the transition $S_0 \rightarrow S_1$. In the same way, the width of the emission spectrum on *the blue side* is

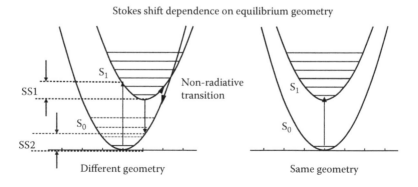

FIGURE 12.5 Potential energy surfaces for the ground state and the excited state.

equal to the reorganization energy of the $S_1 \rightarrow S_0$ transition SS2 in Figure 12.5. If we assume that the force constants for the S_0 state and the S_1 state are the same, the two reorganization energies are the same. The overlap between the common 0-0 absorption and emission lines is then in the middle between the maximum of the absorption and the maximum of the emission. The shift from the absorption maximum to the emission maximum, which is equal to the Stokes shift, is also equal to the sum of the reorganization energies for absorption and emission of a molecule. Under the given assumption, the Stokes shift is thus equal to the sum of the reorganization energies of absorption and fluorescence emission.

If there is no change of the equilibrium geometry (Figure 12.5 right part), there is overlap only with the lowest vibrational level, since the vibrational wave functions may be assumed to be the same in the two electronic states. There is a sharp absorption. In the case of very different geometry, the width of the absorption spectrum is large. Typical absorption and emission spectra are shown in Figure 12.6. Solution effects will lead to further broadening. The vibrational resolution often disappears in a solution spectrum.

The molecule of natural green photosynthesis, chlorophyll, has a small Stokes shift because the equilibrium geometry is hardly changed at excitation. This depends on the fact that the excitation scarcely changes the bond orders because of the distribution on many bonds. The corresponding reorganization energy is only about 0.1 eV. There is no nonradiative way back to the ground state because the S_0 and S_1 PES are little shifted and run almost parallel. The excitation energy is saved and becomes useful as chemical energy in natural photosynthesis (see further in Chapter 15).

In polyenes, single bonds go to double bonds and vice versa at excitation. This means that a *cis–trans*-isomerization takes place around double bonds (that is, double bonds in the ground state). A polyene called *retinal* is involved in the vision

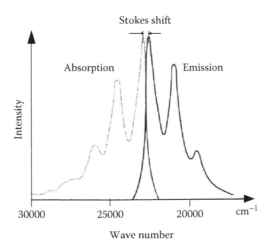

FIGURE 12.6 Absorption spectrum of an organic π-system and the corresponding fluorescence spectrum beside each other.

of most living organisms, since it *cis–trans*-isomerizes at the absorption of light. However, it must be said that the "double" bonds are weaker than in ethylene. The π double bonds contain less π bond order than the single bond in C_2H_4, while the "single" bonds also contain some π bond order.

12.3.2 INTERNAL CONVERSION

The excited state has the same geometry of the nuclei as the ground state a short time after excitation (Franck–Condon principle). The molecule immediately starts to relax toward the new equilibrium geometry by descending the vibrational levels without radiation and, at the same time, heating the probe. This process is referred to as *internal conversion*. Internal conversion also includes decay from higher excited electronic states down to the first excited singlet state (S_1).

Molecules such as melanin and DNA have fast rates of internal conversion, which means that they quickly arrive at the lowest vibration level of S_1, from where they deexcite. Formation of triplet states, deformed molecules, or free radicals is avoided. Molecules of this sort protect against radiation damage, since they deexcite quickly, before harmful chemical reactions have taken place.

In this context, it may be of interest to mention ZnO, which is a white crystal with absorption in the UV region with a band edge at 3.37 eV. This is likely due to 3d $\rightarrow$ 4s excitation. ZnO is used in radiation protection and in solar cells. It also appears to be a promising material for lasers. Remarkably enough, the emission spectrum is not yet well understood. A factor not taken into account so far in proposed mechanisms is the Jahn–Teller effect in the local 3d $\rightarrow$ 4s excited state, leading to a great distortion of the local geometry around the Zn^{2+} ion; in other words, a great Stokes shift. In pure ZnO, the emission has been reported as missing in the visible UV region, which is consistent with a large Stokes shift. The radiationless transitions are fast and triplet and singlet radical states are avoided.

12.3.3 SPIN–ORBIT COUPLING AND INTERSYSTEM CROSSING

If the Hamiltonian operator had no term containing spin, then spin would be conserved in the electronic transitions and the curve crossings. Transitions between singlet and triplet states would be forbidden. However, the Hamiltonian does contain spin terms, such as spin–orbit coupling. "Forbidden" becomes "slightly allowed" as long as there are no heavy atoms. The heavier the atoms, the more the spins mix. The electronic spin quantum number is no longer a good quantum number if very heavy atoms are involved (Figure 12.7).

In the Dirac relativistic equation, spin is naturally included. It is possible to identify the energy terms in the Dirac equation and include them in the ordinary Hamiltonian as "magnetic" terms. In particular, the spin–orbit coupling term is important. This term is physically due to interactions between the electron spin and its motional spin around the atomic nucleus. Spin–orbit coupling increases in importance for heavy atoms. Transitions or curve crossings are no longer spin forbidden.

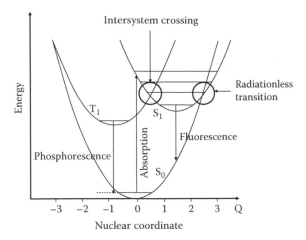

FIGURE 12.7 Photo-processes in the excited states. S_0 is the ground state PES; S_1 is the excited singlet state PES; T_1 is the excited triplet state PES.

In organic molecules without very heavy atoms, spin is almost conserved. The triplet state has a lower energy than the corresponding singlet state, according to the discussion in Section 3.5.6. Crossing of the singlet surface with the triplet surface is slightly avoided and makes it possible for the excited state to reach the lower triplet state. This process is referred to as *intersystem crossing* (ISC). In many cases, ISC can compete with fluorescence.

12.3.4 PHOSPHORESCENCE

Phosphorescence, together with fluorescence, is called *luminescence* is emission from long-lived triplet excited states. The intensity depends on whether faster processes occur in the excited state. The triplet states are, in most cases, the lowest energy excited states.

Phosphorescence has its origin in the Greek word for "carrier of light." At slow oxidation, white phosphorous emits light. This emission has nothing to do with what we today mean by phosphorescence. All phenomena where high enthalpy states, created in chemical reactions, decay with the emission of light, are nowadays referred to as *chemiluminescence*. Phosphorescence means the emission of light that continues after the radiation of the source has ceased. When measurements of fast processes became possible, it turned out that fluorescence also remains after some time. Phosphorescence now became the notation for a spin-forbidden transition, for example, one between a singlet state and a triplet state, while fluorescence became the notation for spin-allowed transitions.

The triplet state is considerably more long-lived than the singlet state; roughly speaking, the lifetime is on the microsecond rather than on the nanosecond scale. In any case, phosphorescence is an important part of the emission.

12.3.5 Types of Spectra

The wavelength (λ) range for absorption spectra take us from some hundred meters of radio waves and centimeters of microwaves over to x- and γ-rays with wavelengths smaller than atomic dimensions. Light absorption is measured as described in Section 12.2.3. A beam of radiation is directed at a sample and the intensity before and after the sample has been passed is measured. Dispersion may be achieved by a prism or a diffraction grating.

The absorption occurs for frequencies (ν) that fit the Bohr relation, $h\nu = E_m - E_n$. The energy levels E_m and E_n are characteristics of the probe molecule and are primarily due to electronic and vibrational quantum states. The vibrational energy component leads to a broadening of the electronic level as seen in the spectrum. In a solution, the discrete character is wiped out.

In the case of x-rays, the inner shell of the atoms is probed. The molecules of importance in photochemistry involve the valence electrons.

There are other broadening mechanisms. The shape may be Gaussian or Lorentzian or a mixture. Lorentzian broadening is the quantum mechanical broadening that was discussed in Chapter 7. Gaussian broadening occurs when the cause of the broadening has a statistical distribution, for example, due to solvent effects.

In an *emission spectrum*, the light intensity emitted from a probe after absorption is recorded (Figure 12.8). The emission is measured perpendicular to the incoming ray. Due to processes on the excited state energy surfaces that are faster than the emission, the intensity pattern is very different from the absorption spectrum. The *fluorescence spectrum* is recorded in the energy region where fluorescence can be expected due to the Stokes shift. Other emissions may be due to phosphorescence or charge transfer.

In an *excitation spectrum*, the amount of emitted light is measured at a certain low frequency. The incident light frequency is scanned from a maximum value over

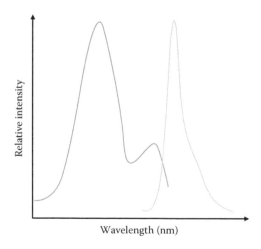

FIGURE 12.8   Excitation spectrum (dark) and emission spectrum (gray).

a certain absorption frequency region higher than the emission frequency. The more that is absorbed, the more is registered at the emission frequency. The excitation spectrum thus corresponds closely to what is measured in the absorption spectrum.

The x-axis in a spectrum is graded in wavelengths, frequencies, or energy units. It may be practical to remember that $\lambda = 1000$ nm (10^{-6} m) corresponds to $\tilde{\nu} = 10{,}000$ cm^{-1} (10^6 m^{-1}), 3×10^{14} s^{-1}, or 0.046 H (1.24 eV).

12.3.6 SPECTRAL NARROWING

Spectral narrowing simply means that an absorption in the spectrum is surprisingly narrow. In Chapter 6, we mentioned that some spin-forbidden transitions can be seen as very sharp absorptions in transition metal systems. The reason in that case is that the same spatial orbitals are occupied as in the ground state and therefore, the bond lengths are almost the same in the ground state and in the excited state. The Stokes shift is close to zero and therefore, there is no vibrational broadening.

It is important to remember that the Stokes shift is the same as the reorganization energy in the limit of zero coupling. If the coupling is large enough to offset the reorganization energy, the Stokes shift is equal to zero.

Narrowing of a more dynamic type occurs in laser emission where the emission with the highest intensity stimulates emissions with the same wavelength. These wavelengths are therefore intensified compared with emissions with less intensity. The line is sharpened.

Another type of narrowing occurs for long chains of molecules, called *J-aggregates* (a particular J-aggregate has been studied, for example, by Willig et al., see Chapter 14). In the ground state, one may think of small charges that keep the molecules lined up. The monomer (that is, a single molecule in the solvent, not interacting with the others) has the normal broadening of the absorption spectrum expected in a solvent for a charged species. If the molecules form J-aggregates, on the other hand, they are oriented in a way that is already "prepared" for the excited state. In the excited state, the molecules have almost the same geometry as in the ground state. The reorganization energy is small. If the coupling is large between the molecules, the lowest excited state is delocalized and therefore the transition energy is sharp. This will be further discussed in Chapter 14.

12.4 RATE MEASUREMENTS

Optical methods have been used to monitor chemical processes for a long time. In a photochemical reaction, a continuous flow of light is used to drive the reaction. In rare cases, light may be *produced* by a chemical reaction (chemiluminescence). A photochemical reaction may be initiated on the molecular level by a short pulse. By studying the absorption characteristics shortly after absorption, details on what happens on the molecular level can be obtained. We will first go through photokinetics, which is different from ordinary kinetics only in the appearance of radiation as a reactant.

Continuous improvements in time resolution finally led to "femtochemistry," the science of chemical processes that take place on the timescale of 10^{-15} sec. Among the many pioneers in this field, Ahmed Zewail should be mentioned in particular. Thanks to these efforts, we are now able to study natural and artificial photo-processes in detail.

12.4.1 PHOTOKINETICS

The spontaneous emission time for fluorescence is around 10^{-9} sec (=1 nsec). This is a million times faster than the time for excitation, which is less than 10^{-15} sec. Therefore, a number of fluorescence *quenching processes* can occur in the meantime. One of them is ISC. A most important quenching mechanism that will be discussed in the next chapter is charge transfer or *charge separation*. Other processes are chemical reactions.

The output of various processes relative to the original absorption rate is referred to as *quantum yield*. When the quantum yield of fluorescence is measured as less than 100%, some quenching takes place.

Just as in the case of reduction or oxidation reactions, bond lengths and bond angles change when a molecule is excited. In fact, in a highest occupied molecular orbital (HOMO) $\rightarrow$ lowest unoccupied molecular orbital (LUMO) transition, the total bond length change should be the sum of the bond length change for oxidation (removing an electron from HOMO) and the bond length change for reduction (adding an electron to LUMO).

One mole of excitations (photons) is called 1 Einstein. The reaction rate can be measured photophysically by time resolving the absorption spectrum. The actinometer is an older device for such measurements. It contains a substance that photochemically reacts with a predetermined quantum yield.

In the calculation of the kinetics, a steady-state assumption may be assumed for the short-lived intermediates on the excited PES, which we call A^S and A^T for singlet and triplet, respectively. The rate of absorption is set equal to unity. The rate constants are denoted as

Radiationless deactivation of singlet	k_s
Fluorescence	k_f
Intersystem crossing	k_{st}
Chemical reaction with molecule B	k_{rs}
Phosphorescence	k_p
Radiationless deactivation of a triplet	k_t
Chemical reaction between triplet and B	k_{rt}

The following steady-state equations hold:

$$\frac{d\left[A^S\right]}{dt} = I - \left[A^S\right]\left(k_s + k_f + k_{st} + k_{rs}\left[B\right]\right) = 0, \tag{12.56}$$

$$\frac{d\left[A^T\right]}{dt} = k_{st}\left[A^S\right] - \left[A^T\right]\left(k_t + k_p + k_{rt}\left[B\right]\right) = 0, \tag{12.57}$$

which alternatively may be written as

$$\left[A^S\right] = \frac{I}{k_s + k_f + k_{st} + k_{rs}\left[B\right]}, \tag{12.58}$$

$$\left[A^T\right] = \frac{k_{st}\left[A^S\right]}{k_t + k_p + k_{rt}\left[B\right]} = \frac{I}{k_s + k_f + k_{st} + k_{rs}\left[B\right]} \cdot \frac{k_{st}}{k_t + k_p + k_{rt}\left[B\right]}. \tag{12.59}$$

It is now possible to express the quantum yields for the different processes as functions of the rate constants and concentrations. The fluorescence quantum yield is

$$\Phi_S = \frac{k_f\left[A^S\right]}{I} = \frac{k_f}{k_s + k_f + k_{st} + k_{rs}\left[B\right]}. \tag{12.60}$$

The triplet quantum yield may be written as

$$\Phi_T = \frac{k_{st}\left[A^S\right]}{I} = \frac{k_{st}}{k_s + k_f + k_{st} + k_{rs}\left[B\right]}. \tag{12.61}$$

Both A^S and A^T may take part in chemical reactions and give rise to the same or different products. The greatest difference in behavior is related to the very different lifetimes. While A^S even in the absence of a chemical reaction has the lifetime

$$\tau_S = \left(k_s + k_f + k_{st}\right)^{-1}, \tag{12.62}$$

which is almost always less than 10^{-6} sec, and very often less than 10^{-10} sec, the lifetime for the triplet, A^T,

$$\tau_T = \left(k_t + k_p\right)^{-1}, \tag{12.63}$$

may be 1 msec or more. It is important that the solution is free of oxygen gas. An oxygen molecule easily reacts with the excited triplet state to form a singlet oxygen while the molecule decays to the singlet ground state.

Using Equation 12.58, we find the quantum yield Φ_{RS} for a chemical reaction with A^S:

$$\Phi_{RS} = \frac{k_{rs}\left[A^S\right]\left[B\right]}{I} = \frac{k_{rs}\left[B\right]}{k_s + k_f + k_{st} + k_{rs}\left[B\right]}. \tag{12.64}$$

Using Equation 12.59, we find the quantum yield for a reaction with A^T:

$$\Phi_{RT} = \frac{k_{rt}\left[A^T\right][B]}{I} = \frac{k_{st}}{k_s + k_f + k_{st} + k_{rs}[B]} \cdot \frac{k_{rt}[B]}{k_t + k_p + k_{rt}[B]}$$

$$= \Phi_T \frac{k_{rt}[B]}{k_t + k_p + k_{rt}[B]}.$$

(12.65)

From Equations 12.64 and 12.65, it follows that a large value of $k_{rs}[B]$ is required so that Φ_{RS} is not small. Since k_t and k_p are small, $k_{rt}[B]$ can be smaller and still have a large Φ_{RT}. It is difficult to drive chemical reactions on the singlet PES, since a number of deexcitation mechanisms occur.

12.4.2 FEMTOCHEMISTRY

A most important way to study kinetics is *spectrophotometry*. In the reaction $H_2 + Br_2 \rightarrow 2HBr$, only Br_2 has absorption in the visible range. The decrease in intensity is measured at different times and this is directly related to the concentration of Br_2. In *flash photolysis*, a *pump pulse* is used to start a unimolecular chemical reaction. A part of the same pulse is delayed and is used to measure the spectroscopic conditions at short time intervals after the probe pulse has started the reaction. Manfred Eigen in Germany and R. G. W. Norrish and George Porter in the United Kingdom started the development of fast measuring techniques. Today, extremely fast reactions may be studied using similar techniques. The time resolution is in the picosecond or femtosecond range, and there is motivation for using the terms *pico-chemistry* or *femtochemistry* for this type of extremely fast chemistry. Recently, even the *attosecond* range (10^{-18} to 10^{-15}s) has been opened for studies.

12.4.3 LASER LIGHT IN CHEMISTRY

Laser light is used in photochemistry to start a chemical reaction and to monitor what happens afterward. Time resolution is achieved by splitting the laser beam and letting one part of it "measure." The latter keeps track of time by passing a longer way between mirrors before it interferes with the other part of the same original pulse that passed through the probe. For example, if the way for the measuring beam is longer by 1 mm, the delay time is 3.3 psec. Whatever is registered at the reemergence depends on what happened 3.3 psec earlier to the probe part of the beam. Technically, it has been possible to create intense pulses of 10 fsec long or even shorter. This has given rise to a new science called *femtochemistry* (Figure 12.9). The technique is referred to as *pump-probe spectroscopy*.

Laser techniques have developed rapidly. Picosecond pulses became possible at the beginning of the 1990s. The time resolution of very fast photochemical reactions was improved considerably. Here, we are talking about time constants of the same order of magnitude as vibration cycles. Fast biochemical reactions, for example, the

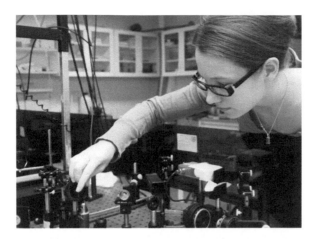

FIGURE 12.9 In the femtolaboratory. The laser ray is reflected on a certain wave path with the help of mirrors.

primary absorption process in vision, and photosynthesis including fast electron transfer and excitation energy transfer, can be studied in great detail.

12.4.4 TRANSIENT ABSORPTION SPECTROSCOPY

Short-lived excited states (transient states) of the relevant molecules may be studied by pump-probe spectroscopy. The probe is hit by a strong pump pulse that excites 0.1%–20% of the molecules. The probe pulse is weaker and delayed with respect to the pump pulse. Further excitation caused by the probe pulse can be ignored and the probe pulse thus acts as a measuring device for the changes in the absorption spectrum, $\Delta A(\lambda,t)$. The original laser pulse should have a rather wide wavelength range so that the absorption spectrum can be studied for several wavelengths (λ).

The difference in absorption $\Delta A(\lambda,t)$ compared with the ground state absorption is measured for various delay times at different λ. $\Delta A(\lambda,t)$ contains information on how the absorption spectrum is changed due to the initial excitation. The time of absorption corresponds to $t = 0$. These changes have different origins. The excited molecules may undergo internal conversion, ISC, or chemical reactions.

The experiment is repeated for different delay times. As the time proceeds, old absorptions at a certain wavelength may die out, while another will start at a different wavelength. From the absorption pattern, it is possible to deduce what dynamics takes place in the system.

12.4.5 TIME-RESOLVED RESONANCE RAMAN SPECTROSCOPY

Time-resolved Raman spectroscopy is a powerful tool for studying the structure and dynamics of transient species. Raman spectra have been briefly described in Chapter 4. Resonance Raman (RR) is different from the original Raman only by the fact that the incoming radiation is laser radiation tuned to fit ("be in resonance with") a single vibronic excitation in the probe. In most cases, the lowest vibrational state is involved.

An important example is resonance with a $\pi\pi^*$ transition. As we will see in Section 12.5, the bond distance changes much in $\pi\pi^*$ transitions. In particular, the stretching modes are now enhanced in the RR spectrum. The excited vibronic state is quantized together with vibrations that are typical for the excited state but slightly different from those of the ground state. The equation

$$\Theta = \sum_i C_i \Theta_i^0, \tag{12.66}$$

must be satisfied. Θ is the vibronic wave function originating in the S_1 state. It is "measured" in the ground state, but needs to include a number of vibronic levels in the excited state. The Stokes–Raman scattered lines have intensities proportional to C_i^2. For this reason, the vibrations associated with bond lengths or bond angles that change much have a high RR intensity.

In a biosystem, such as a photosynthetic reaction center, electron transfer leads to stretching of some bonds and shortening of others. The vibrations of these bonds are seen in the RR spectrum and are helpful in identifying the mechanism of photoinduced ET.

What is seen in the RR spectrum is automatically time-resolved in the same way as the excitation pulse. It is thus possible to study what happens to a vibration on a very short timescale and this bears witness to which bond is broken.

12.4.6 TIME-RESOLVED EMISSION SPECTROSCOPY

In many cases, it is possible to time-resolve emission spectra or excitation spectra. It is important to understand the properties of the excited states and the intermediate states during the geometric equilibration process on the excited state PES and afterward. There are different ways to do the time resolution. In some cases, it is possible to use a delay technique. If the lifetime of the fluorescing state is relatively long, one may decrease the radiation density so that "one-photon counting" becomes possible. It is then possible to obtain the time resolution with "ordinary" electronics, by letting the pump pulse start and letting the emission finish the time counting. Electronic devices are used to measure time.

12.5 PHOTOCHEMISTRY: MECHANISMS

Photoexcitation may be useful to drive chemical reactions, for example, photoinduced electron transfer and *cis–trans* isomerization. In the great majority of cases, aromatic π-systems are excited or deexcited, but the reaction may involve a rotation around a single bond. Not surprisingly, the products are often different from the products of a thermal reaction. These differences can be explained by the wave functions (orbitals) and states.

12.5.1 π-Systems as Absorbers of Light Energy

Planar π-systems are of great importance in electron transfer and excitation energy transfer systems. In π-systems containing more than 20 carbon atoms in the plane, the absorption is in the visible and lower UV regions.

In Chapter 3, we presented the character of the lowest few excited states of cyclic and linear π-systems. For both cyclic and linear π-systems, we found that the lowest excited state is formed either from the HOMO → LUMO substitution with some configuration interaction (CI) with the HOMO − 1 → LUMO + 1 substitution (linear systems), or from a CI between the HOMO − 1 → LUMO and the HOMO → LUMO + 1 states (cyclic systems). In cyclic systems the intensity is low in the lower states of each symmetry, but high for the second excited state.

The spectrum for naphthalene is shown in Figure 3.14. The intensity is along the y-axis. The first absorption for naphthalene is at about 30,000 cm^{-1}. This transition is from the ground state to a B$_{3u}$ state (or L$_b$ state) and has a very low intensity. The next transition is to a B$_{2u}$ state (or L$_a$ state) and starts to absorb at about 35,000 cm^{-1} with an intensity more than 10 times larger than the lowest energy transition. The upper states of B$_{2u}$ and B$_{3u}$ symmetry are referred to as B$_a$ and B$_b$, respectively. The B$_b$ state is lowest in energy and highest in intensity, but both transitions are more than 10 times as strong as the L$_a$ state.

The larger the π-system, the smaller the distance between the orbital energy levels and the smaller excitation energies. Benzene, naphthalene, anthracene, and tetracene, with one, two, three, and four 6-cycles merged along the bonds, respectively, absorb in the UV region, Tetracene though also absorbs in the visible (blue) and is pale yellow.

In the linear polyenes, there is an allowed ^{1}B$_u$ state and a forbidden ^{1}A$_g$ state at low energy. The former is a typical HOMO → LUMO state and the latter is a CI between the HOMO − 1 → LUMO and HOMO → LUMO + 1 states.

12.5.2 Photochemical Reactions

If organic molecules are ππ* excited, the π bond order (Equation 3.24) is much changed. This means that the excited molecule has very different properties compared with the ground state. An activation barrier of 1 eV in the ground state is almost insurmountable; however, in the excited state, the energy of the system is lifted far above the barrier and a number of reactions become possible. Visible light is in the energy region 1–3 eV and UV light 3–10 eV.

Very important laws for thermochemical and photochemical reactions were discovered by Roald. Hoffmann and are referred to as "Rules for conservation of orbital symmetry." Theoretically, one may predict by simple methods the energy order of the orbitals and the existence of an energy barrier for rotation around a bond. The same can be done for the excited molecule. It is then possible to predict whether *conrotatory* or *cis-rotatory* motion is favored.

A good example is given by polyenes that often have reaction insurmountable barriers for twisting around the double bonds equal to 2–4 eV (see below). When electrons are removed from the HOMO and added to the LUMO, the double bond character is transferred to the single bonds, and almost free rotation is possible in the previously strong bonds (see further on retinal below).

12.5.3 *CIS–TRANS* ISOMERIZATION

In Chapter 3, we have seen that aromatic π-systems are planar, stabilized by π-bonding. Out-of-plane motion of the carbon atoms is therefore an energy consuming process and the ground state is thus stable. In a $\pi \rightarrow \pi^*$ excited state, part of the π-bonding disappears, however, and a number of processes become possible. We will be interested particularly in the *cis–trans* isomerization process.

An ordinary CC single bond varies between 1.53 Å (ethane) and 1.54 Å (diamond) and has the bond energy 347 kJ/mol. The double bond in ethylene (ethene, C_2H_4) is considerably shorter (1.33 Å) and stronger (611 kJ/mol). Hydrocarbons with a single CC double bond are called *olefins*.

If the double bond is twisted so that BD are rotated compared with AC in Figure 12.10, the overlap between the two carbon π-orbitals is decreased. When the two planes are perpendicular, the π-bond disappears and the stabilization caused by π-bonding, equal to $2\beta \approx 4$ eV, also disappears. Turning the bond from the planar form to the perpendicular form thus causes a great increase in potential energy. Thermal breaking of the double bond is therefore out of question.

Excitation of an electron from the HOMO to the LUMO (both are π-orbitals) means that the π-bond is completely canceled. The Hückel π MOs in ethylene have the coefficients $(\sqrt{1/2},\sqrt{1/2})$ and $(\sqrt{1/2},-\sqrt{1/2})$ in the HOMO and LUMO, respectively. The bond order (Equation 3.24) in the ground state is $1/2 + 1/2 = 1$ and in the excited state $1/2 - 1/2 = 0$. The π-bond is thus canceled after the HOMO $\rightarrow$ LUMO excitation. The rotational barrier is close to zero. Therefore, the only way to carry out the rotation in Figure 12.10 is to excite the molecule in a $\pi\pi^*$ excitation. This transition is strongly allowed and takes up most of the visible absorption spectrum.

12.5.4 POLYENES

In cyclic or linear polyenes, every second CC bond is a double bond. Since the double bonds are conjugated, the single bonds are slightly shorter than typical CC single bonds and the double bonds are slightly longer than the ethylene double bond. It is still useful to refer to them as just single and double bonds, respectively. *Photoisomerization* is possible in the excited state. However, if the polyene

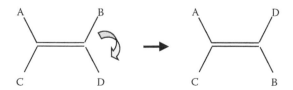

FIGURE 12.10 *Cis–trans* isomerization of ethylene.

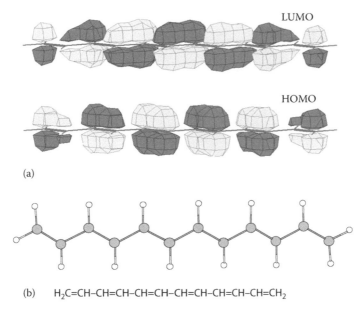

(a)

(b) $H_2C=CH-CH=CH-CH=CH-CH=CH-CH=CH-CH=CH_2$

FIGURE 12.11 The HOMO and the LUMO (a) for dodecapentane (b).

is substituted by methyl or larger groups near a double bond, photoisomerization is sterically hindered.

The spectrum of polyenes is dominated by the $\pi \to \pi^*$ excitation. The ground state is a ^1A_g state and the first excited state, with a high spectral intensity, is a ^1B_u state. However, there are also other low-lying excited states. The wave function and character of these states can be quite easily described. Transitions to several of the states are dipole forbidden and very hard to discover in the absorption spectrum. A transition to an excited state composed of the HOMO $\to$ LUMO + 1 and the HOMO $-1 \to$ LUMO configurations is forbidden. It has the same symmetry as the ground state (^1A_g) and is therefore denoted as 2^1A_g. For butadiene, the ^1B_u state is slightly lower in energy than the 2^1A_g states, but for the higher polyenes, the 2^1A_g state is the lowest excited state.

Isosurfaces of the HOMO and the LUMO are given in Figure 12.11. In the HOMO, the $C2p_z$ orbitals are overlapping wherever there is a π-bond. The LUMO, on the other hand, has a node in the "double" bonds. Excitation from the HOMO to the LUMO therefore weakens the double bonds and strengthens the single bonds considerably. The short bonds decrease in π character, and *cis–trans* isomerization becomes possible. This is the case for all molecules with a π-bonding orbital as the HOMO.

It must be remembered though that other occupied π-orbitals than the HOMO also contribute to the bonding. The short bonds for long polyenes are about 1.35 Å and the long bonds are about 1.45 Å. An isolated CC double bond is 1.32 Å and a typical CC single bond is about 1.53 Å. In the ionized state, the bond lengths become almost equal over 20 carbon atoms. In the excited state, there is even a swapping of single and double bonds.

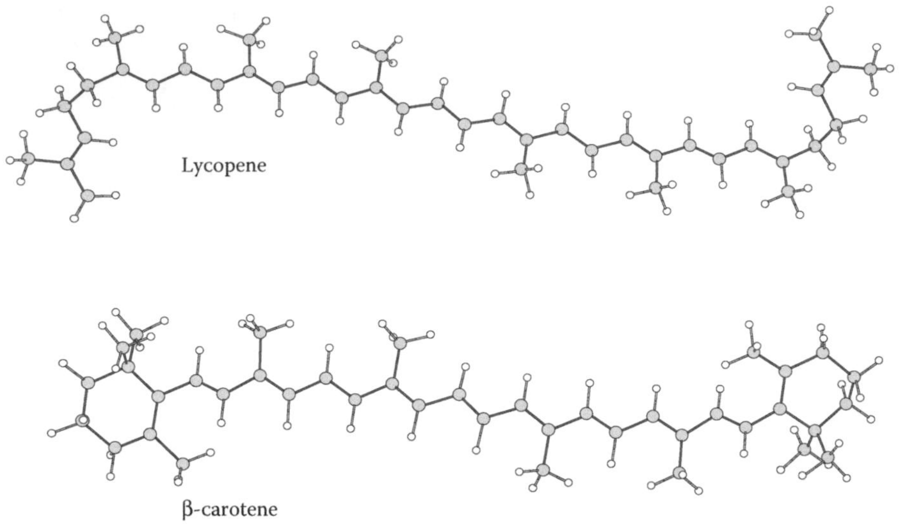

FIGURE 12.12   The carotenoids lycopene and β-carotene.

12.5.5   CAROTENOIDS

β-carotene is a polyene system (Figure 12.12), consisting of 40 carbon atoms of which 22 appear in a polyene chain. β-carotene exists in green leaves and plays a great role in the photosynthesis of green plants and in other types of photosynthesis. The color is not seen until the green chlorophyll disappears in wintertime. Carotene is the substance that makes carrots orange. Carrots are gene-modified wild carrots. The Stone Age people managed to develop carrots with a high concentration of carotenoids in the root (!), thereby improving the taste and increasing the content of the useful vitamin A.

β-carotene is an antioxidant that prevents singlet oxygen radicals from causing damage in living organisms, particularly in photosynthetic reaction centers and antenna systems.

$$^1\Delta_g(O_2) + {}^1A_g(\text{carotene}) \rightarrow {}^3\Sigma_g(O_2) + {}^3A_g(\text{carotene}) \qquad (12.67)$$

In this reaction, carotene is excited to the lowest triplet state at the same time as the excited singlet oxygen state decays to the ground triplet state. The excited carotene triplet state is not particularly reactive, but decays within nanoseconds to the ground state. The oxygen molecule is much less reactive in the ground state than in the singlet state.

β-carotene also takes part in the harvesting of sunlight. It absorbs in parts of the solar spectrum where chlorophylls absorb very little. The excitation energy is transferred into the chlorophyll and subsequently leads to electron transfer.

Lycopene is another carotenoid (Figure 12.12) that gives the red color to tomatoes. Contrary to β-carotene, it cannot be used by the body to produce vitamin A,

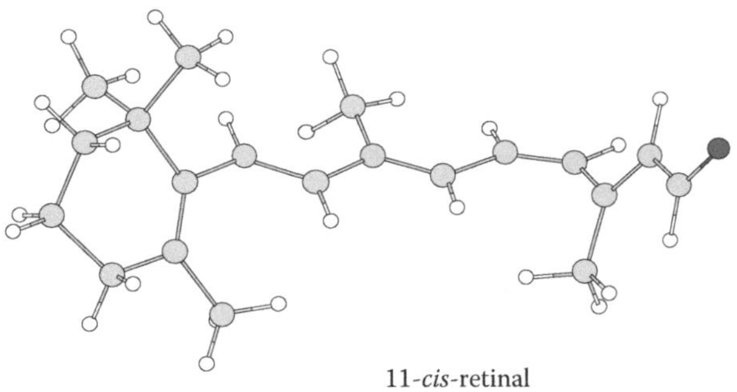

11-*cis*-retinal

FIGURE 12.13   11-*cis* isomer of retinal.

which is, in fact, easy to understand from its structural formula. On the other hand, it is useful to the body as an antioxidant.

12.5.6   RETINAL AND VISION

The polyene chromophore *retinal* or vitamin A aldehyde is the active molecule in the primary process of vision in almost all animals. Its 11-*cis* form (Figure 12.13) easily inserts itself into a well-defined cavity in between seven parallel α helices in the protein *opsin*. With retinal present in this cavity, the protein is called *rhodopsin*.

Retinal (Figures 12.13 and 12.14) is roughly half of the β-carotene molecule (Figure 12.12). Most animals get vitamin A from leaves, while carnivores (meat eaters) get vitamin A from other animals. All vertebrates (animals with a backbone) use retinal, while invertebrates use the hydroxylated forms of retinal. We are interested in the primary process of vision, which takes less than a millionth of the time for the whole process of seeing, that is, noticing and becoming aware of light.

The American George Wald knew that isomerization can only occur around double bonds, but only around CC bonds without bonds methyl groups because of steric hindrance. In the 1930s, Wald found that the *trans* form of retinal (Figure 12.14) does not bind with opsin but remains in the water solution. The *cis* form, on the other hand, enters the protein as soon as opsin and *cis* retinals are brought into contact.

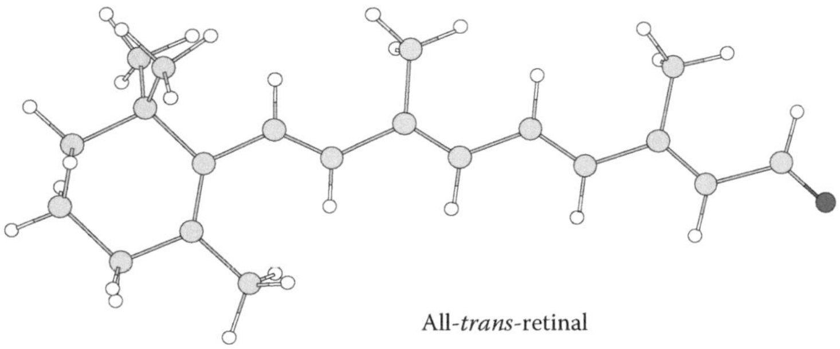

All-*trans*-retinal

FIGURE 12.14   All-*trans* isomer of retinal.

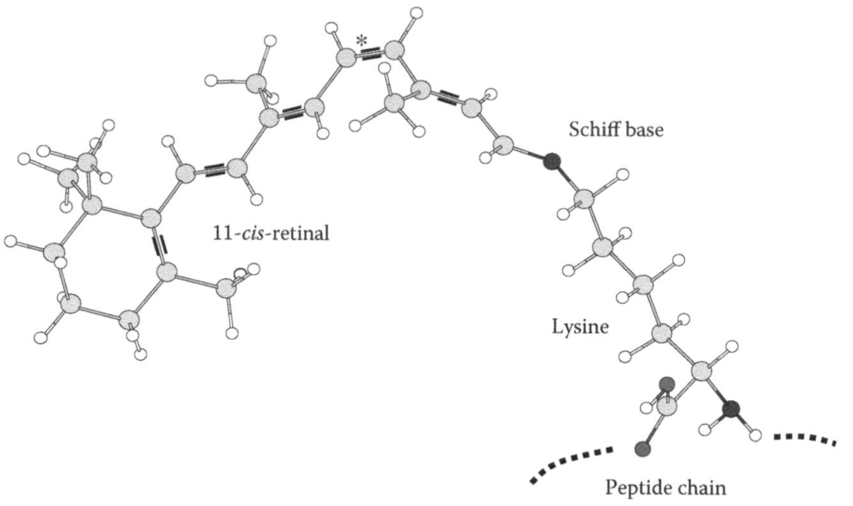

FIGURE 12.15 The attachment of 11-*cis*-retinal to the protein is via a nitrogen atom (dark, on the far right), called a Schiff base, belonging to the side group of the peptide lysine. Short bonds are dashed (=). The bond involved in the *cis–trans* isomerization is marked by a star.

Wald had good reasons to believe that a *cis–trans* isomerization triggers the signal that reports to the brain that light has been registered.

We now know, not least from crystallographic studies, that the *cis* form of the molecule fits well into its cavity and is attached to the protein via the nitrogen atom (Schiff base) of the side chain of the peptide lysine. The cavity does not fit into *trans*-retinal. But how is it possible to carry out a *cis–trans* isomerization inside the protein cavity? It is well known that protein matter does not allow void space, so it might be expected that the retinal fits exactly. One possibility is that the protein adjusts itself to the changed structure of the molecule, but in that case it must be a process that requires a time of the order of protein folding, about 1 μsec.

The time for the photoisomerization process is a very good object for time-resolved spectroscopy studies. As the time resolution was still moderate during the 1970s and 1980s, one was able to conclude only that the *cis–trans* isomerization process happened faster than the time resolution (<1 ps). It was not until the beginning of the 1990s that it could be established (by Callender et al. and Mathies et al.) that the *cis–trans* isomerization time constant is only 200 fsec! This is a surprisingly short time. Not even the vibrational cycle time of slow vibrations is faster. The retinal molecule is attached to the protein via the Schiff base (Figure 12.15) and the cyclohexene ring of the retinal molecule (to the left in Figures 12.13 through 12.15). Such a large group is too heavy to be set in motion during 200 fsec.

By geometry optimization using the CASSCF method, it was found that the geometry-optimized structure at a 90° rotation around the 11-bond gives a very different molecule from the *cis* form, which also does not fit into the cavity. As was customary at the time, the calculation was carried out in "open space." To include the hindrance from the protein would require too much computational resources.

Geometry optimization is no problem, but its purpose is just to find the structure corresponding to the energy minimum. It gives no information about the dynamics of the *cis–trans* isomerization in the protein.

It was important to find out whether a pathway exists without a large activation barrier, which leaves the retinal in a shape that still fits into the opsin cavity. Such a pathway was found by minimizing the motion of the fixed end points of the retinal molecule (F. Blomgren). The structure of the molecule 200 fsec after light absorption is not the *trans* form of the molecule. Instead, a different structure is adopted where the end points of the retinal are not changed, but where the two hydrogen atoms close to the 11-bond are moved corresponding to a 90° rotation. Only hydrogen atoms can move so far in such short a time.

After 200 fsec, the molecule remains in what looks like the 11-*cis* isomer, but it is, in fact, an intermediate called *bathorhodopsin*. The intermediate is still in a high-energy state. During the next few picoseconds, this intermediate starts to relax slowly (nanoseconds) to the *trans* form. As said before, the *trans* form does not fit into the protein. It is fair to assume then that it is the relaxation of bathorhodopsin that forces the retinal molecule to leave the protein. The successive "dark" processes are partly known but have nothing to do with photochemistry.

13 Photoinduced Electron Transfer

13.1 INTRODUCTION

In the future, we may be able to harvest solar energy in parts of the world where there is no natural photosynthesis, in a way that is even more efficient than natural photosynthesis. The primary reaction in photosynthesis is a *photoinduced electron transfer* (PIET) reaction.

Transfer of an excited electron to another molecule is an important mechanism for quenching of fluorescence. The high-energy electron may be stabilized on the other molecule, in a *charge-separated* state for a long time, or it may leap back to the original molecule in a *charge recombination* process, depending on energy conditions and coupling. The lifetime of the charge-separated state may be long enough for useful chemical reactions to take place. Such reactions will be discussed in Chapter 15. In this chapter, we will discuss PIET in general.

13.2 CHARGE TRANSFER TRANSITION IN SPECTRA

It is customary to call a molecule that is photo-oxidized the *donor* (D) and a molecule that receives the electron the *acceptor* (A). The orbital scheme is shown in Figure 13.1. Both molecules may be excited by radiation, but only the one with the highest excited state can be the donor. Except the ground state (DA), two other electronic states are involved, the locally excited state (D*A) and the charge-separated state (D$^+$A$^-$).

13.2.1 CHARGE TRANSFER STATES AS EXCITED STATES

Let us assume that the lowest unoccupied molecular orbital (LUMO) of D is higher in energy than the LUMO of A (Figure 13.1). More correct is to say that the lowest excited state on D has a higher free energy than the lowest excited state on A. Excitation directly from the highest occupied molecular orbital (HOMO) on D to the LUMO on A is very unlikely, since the orbitals belong to different molecules and have small overlap. Cases with orbital mixing will be discussed in Section 13.2.2. For the same reason, radiative charge recombination of the charge-separated state to the ground state is a slow process in most cases.

From Chapter 6, we recall that dd transitions in transition metal complexes with inversion center are forbidden. These transitions are in the visible region and may become slightly "allowed" by various mechanisms, giving some color to the complex. Since most oxides or halides are *not* black or strongly colored (as are sulfides

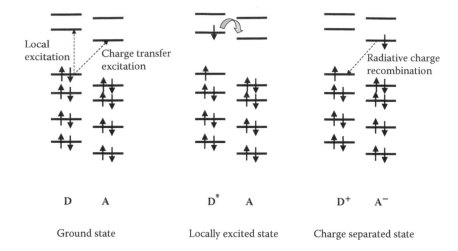

D   A   D*   A   D+   A−

Ground state   Locally excited state   Charge separated state

FIGURE 13.1 Orbital scenario of the PIET process. The charge-separated state may recombine radiatively or nonradiatively.

and selenides), one may conclude that the allowed spectrum starts well into the UV region for oxides. The allowed spectrum in a transition metal complex consists of charge transfer (CT) transitions L → M, and as we have seen in Chapter 6, some of them absorb strongly. Well-known examples are $KMnO_4$ (violet) and other permanganates, and $K_2Cr_2O_7$ (orange) and other chromates or bichromates. Here, we are talking about CT *within* a transition metal complex. Another type of CT is *intermetallic charge transfer*, that is, between two metal ions.

A CT transition in an organic system, with the possible exception of the colored Mulliken CT complexes (see below), is a transition from one molecule to the next, possibly separated from the first one by solvent molecules. The mixing of orbitals is very small, at least much smaller than between metal atom and ligand in a metal complex. CT transitions in the case of organic systems are weak and cannot be easily distinguished behind the local excitations in the spectrum. To photochemists active in the organic area, a CT transfer transition therefore means a very *weak*, almost forbidden transition. But CT also has another meaning. CT in the organic case usually refers to the subsequent PIET process, taking place after excitation.

To inorganic chemists, solid-state physicists, and chemists working in the field of metal complexes, a CT transitions is synonymous with an intense transition.

13.2.2 Mulliken Charge Transfer Complexes

If two aromatic compounds, A and D, are mixed, there is high probability that the different molecules are attracted to each other because of weak binding forces. The HOMO on A will mix with the LUMO on D, and the LUMO on A will mix with the HOMO on D. The system of molecules A and D is best treated as a single system. As long as the mixing is small, the intensity of the transition tends to be low. The width of a CT transition is large, and this depends partly on the weak bonding between A and D, permitting many different geometrical arrangements between the two molecules, and partly on the great width associated with all CT transitions in the Marcus model, with average energy equal to λ.

This type of organic complex was studied theoretically by R. S. Mulliken and W. B. Person. In this case, the ground and excited states have an even number of electrons and are spin-singlet states. On the other hand, the Creutz–Taube complex and its many cousins of almost symmetric, bimetallic complexes are odd electron systems where there are degenerate ground states rather than one ground state and one excited state. The situation with the Mulliken CT complexes is different and sometimes very complicated. We are talking about CT both as (1) wave function mixing in the ground state and (2) in excitations. The charge is differently localized in the ground state and in the excited state.

13.2.3 EMISSION FROM CHARGE-SEPARATED STATES

Spontaneous emission from higher excited singlet (S_2, S_3, etc.) states is rare, since the system quickly finds the *lowest* excited singlet state by nonradiative transitions (internal conversion, Kasha's rule). Usually, emission spectra are formed either from the lowest excited singlet state of a molecule (fluorescence) or from the lowest triplet state (phosphorescence). In some other cases, low-lying CT states, singlets or triplets, are deexcited radiatively.

The lowest CT state is reached in an electron transfer (ET) process on the lowest excited potential energy surface (PES). As in other cases of ET, the rate may be calculated if we know the free energy, reorganization energy, and electronic coupling. The final CT state is usually reached and stabilized within a nanosecond. The time for this process depends on the solvent. If the solvent is nonpolar, polarization is electronic and rather fast. Only structural reorganization remains to slow down the speed of the process. The calculation of the PIET rate can be done by ordinary methods (see below).

In the case of a polar environment, on the other hand, the Stoke shift is time-dependent due to solvent dynamics. The calculation of rate may be complicated. Usually, molecules have emission only from the locally excited state. Others have *dual fluorescence* from both the locally excited state and the CT state.

A rather small molecule deserves particular mention as an example of dual fluorescence and solvent interactions. It was first discovered by E. Lippert et al., and is referred to as DMABN (Figure 13.2). L. Serrano-Andrés et al. studied DMABN in detail using *ab initio* calculations. In the 1970s, Grabowski et al. coined the term "twisted intramolecular charge transfer" (TICT).

The lower energy fluorescence band is clearly a CT band, since its position depends on the dielectric constant of the solvent.

13.2.4 TRIPLET FORMATION BY CHARGE TRANSFER

As we have seen in Chapter 3, there is at least one triplet state below the first excited singlet state in the Franck–Condon region. The triplet is formed by reversing spin on the excited electron. It is lower than the corresponding singlet state by $2 \cdot K_{ij}$, where K_{ij} is defined as

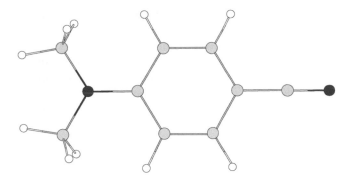

FIGURE 13.2   4-(N,N-dimethylamino) benzonitrile (DMABN).

$$K_{ij} = \left(ij|ij\right) > 0. \tag{13.1}$$

This integral is the Coulomb self-repulsion of the overlap charge $|\phi_i \cdot \phi_j|$. If ϕ_i and ϕ_j are on the same site, the exchange integral K_{ij} has a value in the range of 0.1–0.5 eV. The triplet state is considerably lower in energy than the singlet state (Figure 13.3). If the orbitals ϕ_i and ϕ_j are on different sites, on the other hand, the charge density $\phi_i \cdot \phi_j$ decreases exponentially with distance between the sites. At a large separation between ϕ_i and ϕ_j, the singlet and triplet states become almost degenerate ($K_{ij} \approx 0$).

We assume that the CT state is stable. Since K_{ij} is small in this state, the singlet and triplet have almost equal energy. The mixing of spin in the two states with M_S equal to zero [$S_1(CT)$ and $T_1(CT;M_S = 0)$] is therefore quite large.

$T_1(CT;M_S = 0)$ may still be above the locally excited triplet state. If the probability for CT is reasonably large, the scenario will be as follows: First, the Franck–Condon singlet excited state is formed. Subsequently, there is a CT to the charge-separated state. Here, the spin ($S = 0$) changes easily to triplet ($S = 1$). Charge recombination becomes possible at the original site for the triplet state (Figure 13.3). CT, therefore, speeds up transfer from the singlet state to the triplet state. The phenomenon was studied particularly by Kaptain and is referred to as *photo-CIDNP* (chemically induced dynamic nuclear polarization) or *RPM* (radical pair mechanism).

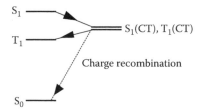

FIGURE 13.3   Intersystem crossing induced by charge transfer (photochemically induced dynamic nuclear polarization [photo-CIDNP]). The arrows indicate the CIDNP pathway. The full-drawn line from the charge-separated state ends on the local triplet and the dashed line on the ground state.

13.3 POLARIZATION ENERGY

If the charge separation takes place in a solvent, there is a polarization around the acceptor molecule during and after PIET. The relative dielectric constant, ε_r, appears in the equation for the stabilization energy. Immediately after CT, the relevant dielectric constant is n^2, which is the dielectric constant for electronic polarization. If the solvent molecules have a dipole moment, the polarization around the donated charge will further increase, given time.

13.3.1 REACTION FIELD

Whether CT is possible depends on the polarity of the solvent, as measured by the dielectric constant. There are essentially two important dielectric constants: one for slow processes or the static dielectric constant (ε_s) and the other for very fast processes (faster than any reorganization process), referred to as ε_∞ or the "dielectric constant at infinite frequency." ε_s of a compound can be obtained by measuring the capacitance of a condensator where the compound is used as a dielectricum. ε_∞ is obtained by measuring molecular polarizability. The higher the frequency of the applied electric field, the slower are the motions in the medium to follow the variations in the field. For example, a water molecule has a certain rotation time and when the field frequency is too fast, the water molecule no longer moves with the field. When the frequencies applied correspond to UV frequencies, ε_∞ (for all practical purposes) is measured. One may show that $\varepsilon_\infty = n^2$, where n is the refractive index.

13.3.2 REHM–WELLER EQUATION

The system consists of donor D and acceptor A. To obtain charge separation, the energy of the D*A state should be higher than that of the D$^+$A$^-$ state. The energy of the excited state D*A is the excitation energy between the two lowest vibration levels, E_{00}. In the gas phase, the energy of the charge-separated state is $I(D) - A(A)$, since the process means that one electron is transferred from D to A. In a solution phase, we have to use the oxidation potential for D and the reduction potential for A in the relevant solvent.

In the gas phase, the free energy of the charge separation process is

$$\Delta G(g) = I(D) - A(A) - E_{00}(g) - \frac{1}{R}. \tag{13.2}$$

We have assumed that the charge of one electron is moving and we use atomic units. The term $-1/R$ represents the lowering of the final energy due to the attraction between the separated charges at a distance R.

In the solvent with dielectric constant equal to ε, we have to use instead

$$\Delta G(s) = E_{ox}(D) - E_{red}(A) - E_{00}(s) - \frac{1}{\varepsilon R}. \tag{13.3}$$

(Notice that $4\pi\varepsilon_0 = 1$ in atomic units. To get ΔG in electron volts, we multiply by 27.21/1.89. To obtain ΔG in calories per mole, we multiply by Faraday's constant.) E_{ox} and E_{red} are absolute oxidation potentials and reduction potentials, respectively.

We see that the larger the distance for ET, R, the less negative is the free energy. The larger the distance R, the higher in energy is the CT state compared with the locally excited states.

$\Delta G(s)$ given in Equation 13.3 represents the situation after some time when the solvent has had time to relax around the new charges. It is of great interest to study also the situation immediately after excitation. The problem is that there are no reduction potentials for that situation. We first derive the relationship between $E_{ox}(D)$ and $I(D)$ with the help of the Born equation. The ionization energy I is the oxidation potential in a medium with dielectric constant $\varepsilon = 1$, while E_{ox} is the same oxidation potential in a medium with the dielectric constant ε after a sufficiently long time for the solvent to polarize (Figure 13.4). We have (a is a solvent radius around the charge)

$$E_{ox}\left(t = \infty\right) - I = \frac{q^2}{2a}\left(1 - \frac{1}{\varepsilon}\right),\tag{13.4}$$

where q is the charge in the sphere.

For the case with immediate polarization, we use instead $\varepsilon = n^2$:

$$E_{ox}\left(t = 0\right) - I = \frac{q^2}{2a}\left(1 - \frac{1}{n^2}\right).\tag{13.5}$$

The difference in Coulomb attraction energy between unit charges at a distance R in the gas phase and in a medium with dielectric constant ε is

$$U\left(g\right) - U\left(\varepsilon\right) = -\frac{q^2}{R}\left(1 - \frac{1}{\varepsilon}\right).\tag{13.6}$$

The Coulomb attraction in the medium, $U(\varepsilon)$, is thus less when R is larger.

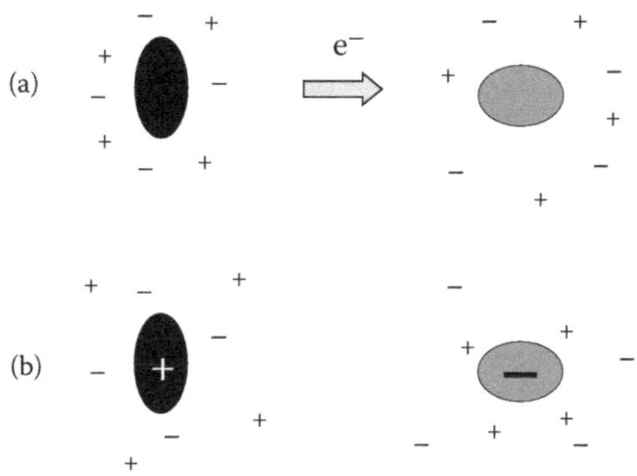

FIGURE 13.4   Charge transfer (a), followed by slower solvent reorganization (b).

If Equation 13.6 is written as

$$U(\varepsilon) = \frac{q^2}{R}\left(1 - \frac{1}{\varepsilon}\right) + U(g), \tag{13.7}$$

we see that as R is increased, the Coulomb attraction $U(\varepsilon)$ decreases, since $U(g)$ is constant. The charge-separated state thus increases its energy if R is large. The CT may be the lowest excited state for a short CT distance R; however, for longer distances, the locally excited states may be more stable.

We notice that in the Born polarization, if the charge is negative rather than positive, the sign before the right member is the same, since the Born equation has Z^2 for an arbitrary charge Z (we have $Z = 1$), while the ionization energy should be replaced by the electron affinity A.

The additional polarization energy due to the medium compared with the immediate polarization is obtained as the difference between Equations 13.4 and 13.5:

$$E_{ox}(t = \infty) - E_{ox}(t = 0) = \frac{q^2}{2a}\left(\frac{1}{n^2} - \frac{1}{\varepsilon}\right). \tag{13.8}$$

Suppose we do a quantum chemical calculation using some method that is at least able to treat the excited states in a configuration interaction between singly substituted Slater determinants. Among the ordinary locally excited states, we find the CT states. Usually, the energies of the CT states are calculated to be too high and part of the reason for this is that there is no polarization energy, since the calculation is done corresponding to empty space. Depending on the molecule, the CT state may become the lowest excited state at a certain distance R between D and A.

The reaction field is absent if the excited state is noncharged. The energy of locally excited states does not depend on polarization forces. The electronic polarization from the molecules themselves is accounted for in the quantum chemical calculation. It is important to include a few solvent molecules to fill the sphere with radii a_D and a_A, around D and A, respectively, with solvent molecules.

To account for the polarization outside the spheres, we have to add a *reaction field* that simulates the stabilizing polarization forces from the medium due to the charges created (the Coulomb attraction between the charges is already included in the calculation):

$$\Delta G^{\ominus}(t) = -\frac{q^2(t)}{\varepsilon(t)}\left(\frac{1}{2a_D} + \frac{1}{2a_A}\right). \tag{13.9}$$

Here, $q(t)$ is the charge transferred at time t. For example, $t = 0$ represents the situation when the dipoles have not yet started to adjust to the field created by charge separation, while $t = \infty$ represents the final state. $\varepsilon(0) = n^2$ and $\varepsilon(\infty) = \varepsilon_r$. The calculation has to be carried out iteratively, since the transferred charge has to be the final consistent charge.

In practice, a better result is obtained if experimental reduction potentials are used. Weller has derived the following equation, which follows from the previous equations:

$$\Delta G^{\ominus}(\varepsilon) = \left(E_D^{ox} - E_A^{red}\right)_\varepsilon - E_{00} - \frac{1}{\varepsilon R}\frac{27.2}{1.89}(eV). \qquad (13.10)$$

E_{00} is the excitation energy between the lowest vibrational states. The Coulomb term is added to account for the attraction of the charges due to PIET.

In practice, the reduction potentials are measured in a particular solvent, for example, acetonitrile ($CH_3C\equiv N$; $\varepsilon_r = 37$). In another solvent with relative dielectric constant equal to ε', the free energy for the CT reaction may be written

$$\Delta G^{\ominus}(\varepsilon') = \left(E_D^{ox} - E_A^{red}\right)_{\varepsilon'} - E_{00} + \left[\frac{1}{\varepsilon'}\left(\frac{1}{2a_D} + \frac{1}{2a_A} - \frac{1}{R}\right) - \frac{1}{37a}\right]\frac{27.2}{1.89}(eV), \quad (13.11)$$

where a_D and a_A are radii spheres around D and A, respectively.

13.4 INTERMOLECULAR AND INTRAMOLECULAR PIET

The rate of PIET may be obtained in the same way as the rate of thermal ET (TET) reactions, using the Marcus model. The locally excited PES replaces the ground state PES. The former has a high free energy, and the possible range of free energy values is therefore larger in PIET than in TET. It has been possible to test the functional dependence of rate as a function of free energy change, as discussed in Section 10.5.2 (Figure 10.11). In solution, the expected higher rate in the Marcus inverted region ($-\Delta G^{\ominus} = \lambda$) is not realized. For $-\Delta G^{\ominus} > \lambda$, the rate is about the same as the maximum rate. This is very common in intermolecular PIET in solution and we need to know why.

13.4.1 RATE OF PIET

We need the free energy of the reaction $-\Delta G^{\ominus}$, the total reorganization energy λ, the coupling $\Delta/2$, and the vibrational frequencies of the modes coupled to ET:

$$k = \nu_n \cdot \kappa \cdot exp\left(-\frac{\Delta G^*}{k_B T}\right). \qquad (13.12)$$

κ is the *electronic (transmission) factor* equal to the probability to pass the avoided crossing region on the lower PES.

We will first assume that κ is equal to unity. This is called *adiabatic ET* and physically means that the system moves exclusively on the lower PES (Figure 13.5). When it has passed the barrier, the character of the electronic state is one where the electron has moved to the other site.

The activation energy is calculated in the usual way as

$$\Delta G^* = \frac{\lambda}{4}\left(1 + \frac{\Delta G^{\ominus}}{\lambda}\right)^2, \qquad (13.13)$$

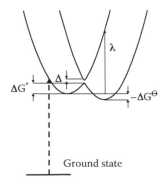

FIGURE 13.5 The Marcus model with the important parameters: reorganization energy λ, coupling $H_{12} = \Delta/2$, free energy of reaction $\Delta G^{\ominus}$, and activation energy ΔG^{*}.

where $\Delta G^{\ominus}$ is obtained from Equation 13.11.

The total free energy change is obtained as

$$\lambda = \lambda_{in}(0) + \lambda_{out}(0). \tag{13.14}$$

The inner reorganization energy, λ_{in}, is obtained by geometry optimization of the structure of the neutral and oxidized D and the bond length change of the neutral and reduced A. The outer reorganization energy is obtained from Equation 10.31. The spheres used are the same as was used for calculating the free energy of the reaction.

In Equation 13.12, v_n is the vibrational frequency that corresponds to the nuclear promotional motion (abscissa in Figure 10.10). In a metal complex, we are dealing with the so-called "breathing modes" for the ligands around the nucleus, usually at $\approx 500 \ cm^{-1}$.

It is thus possible to use Equation 13.11 and determine the free energy $\Delta G^{\ominus}$ of the reaction experimentally by measuring reduction potentials for the acceptor and donor in a solvent. If we assume $\Delta S^{\ominus} = 0$, we have ordinary PES based on $\Delta H^{\ominus}$. PES may be assumed to be parabolic. It is also possible to define an effective force constant.

It follows from Equation 13.13 that the activation barrier ΔG^{*} disappears if $\Delta G^{\ominus} = -\lambda$. If, on the other hand, $\lambda \gg \Delta G^{\ominus}$, ΔG^{*} depends linearly on λ. We may insert Equation 13.13 into Equation 13.12 and obtain Figure 10.11. The rate k of the ET reaction is increased with increased driving force until a maximum is reached for $\Delta G^{\ominus} = -\lambda$. If the driving force is further increased, the rate decreases with increasing driving force according to the Marcus model.

Equation 13.12 with $\kappa = 1$ is a typical expression for a rate constant if it is assumed that the nuclei are moving by classical mechanics with the ground state PES as potential. This is the simplest approximation for the rate. However, in ET reactions, we have to remember that the barrier is associated with an avoided crossing between two different PES, and that there is a probability that the system jumps to the upper surface. In that case, $\kappa < 1$ in Equation 13.12. Thus, the system is in a formally different electronic state, but the character of the wave function does not change, meaning that no ET takes place. The electronic factor κ was calculated in Section 10.5.4.

In conclusion, PIET is favored if the medium is polarizable so that $\Delta G^\ominus$ is negative. Equation 13.11 shows that PIET is also favored by a small distance between the charges. The distance for ET cannot be too large. This is for two reasons: coupling and free energy.

13.4.2 INTERMOLECULAR PIET IN SOLUTION AND IN GLASS

Rehm, Weller, and many others obtained the rates for intermolecular ET over large reaction free-energy regions. By using polar solvents, such as acetonitrile, Rehm and Weller easily found cases when the final polarization was so large that the system was far into the Marcus inverted region. The plot of logarithm of rate versus free energy measured from the Franck–Condon state, excited state ($\Delta G^\ominus$), increases with free energy until the inverted region is reached, where $\Delta G^\ominus = \lambda$.

However, the measured rates did not signal that the CT system was in the inverted region. The rate increased as the free energy decreased, as predicted by the Marcus model, but only until a maximum rate was obtained. A further increase of the driving force did not bring down the rate.

The apparent failure of the Marcus model can be explained if we look at the dynamics. During the first few femtoseconds after excitation, the system moves from the Franck–Condon region down on the energy surface determined by structural reorganization, toward the new equilibrium point. The kinetic energy increases at the same time and hence the system climbs up on the other side of the parabolic, vibrational energy surface. The parabolic PES of the CT state intersects the parabola of the local excitation in such a way that the avoided crossing is passed twice in each cycle. The PES of the CT state is now being populated. The solvent dipolar molecules start to rearrange around the transferred electron. The CT PES goes down in energy and the activation barrier decreases. Vibrational quanta are lost in the internal conversion. Finally, the CT PES has reached the point where it intersects the locally excited PES at the minimum. We are now at the Marcus maximum in Figure 10.11. The CT PES is fully populated before the final equilibrium, corresponding to the inverted region, is reached.

Since ET takes place before the Marcus inverted region is reached, it is unlikely that the latter will be observed. The reaction rate remains at its maximum. The greatest chance to find the reduction of rate typical for the inverted region is to use *intramolecular* ET reactions. Alternatively, it is also possible to use solids or frozen solvents (J. R. Miller experiment). When such experiments were carried out, the inverted region appears.

The dependence on R in Equation 13.10 is interesting. As R decreases, the free energy of the reaction becomes more negative. The explanation above may be used to explain the absence of any influence of the Marcus inverted region in the rate data. This also means that the Marcus inverted region is entered as the two molecules approach each other. The electron jumps in through the solvent when the Marcus maximum region is passed ($\lambda = -\Delta G^0$), and has already transferred when the inverted region is reached.

13.4.3 CHARGE RECOMBINATION

Charge recombination to the T_1 state has been discussed above. Charge recombination to the ground state may also occur if the PES are favorable (Figure 13.6). Depending on Stokes shift and geometry change, the charge-separated PES may intersect the ground state with its minimum well separated from or very close to the ground state equilibrium. In the first case, the activation barrier will be small and in the second case, large. The effect on the charge recombination rate is great.

13.4.4 INTRAMOLECULAR PIET

Another way to prove that the Marcus equation is correct is to carry out PIET at a fixed distance in the same molecule. There are a number of ways to do this. C. Creutz and N. Sutin (1977) carried out early intramolecular ET experiments at Brookhaven which tend to confirm Marcus inverted region. Later, two series of experiments were carried out. G. L. Closs and J. R. Miller at Argonne Labs near Chicago achieved ET through a saturated spacer by using electrons from radiolysis as already discussed in Section 10.5. J. Verhoeven and collaborators in the Netherlands obtained similar results using PIET. In both cases, *D-bridge-A* molecules were used.

13.4.5 INTRAMOLECULAR CHARGE TRANSFER

PIET from one π-system to another π-system separated by a covalently linked spacer (Figure 13.7) was carried out by Verhoeven et al. in the early 1970s. The lowest excited state is localized to one of the π-systems, metoxynaphthalene, that subsequently acts as a donor of the electron to the other π-system, the acceptor = $C(CN)_2$.

The rate of ET was measured. The main factor that influences the rate is the distance through the spacer and the chemical structure of the bridge. The notion of through space interaction could be ruled out. In this case, we are dealing with ET through σ orbitals.

Verhoeven was one of the first to show that electron interactions and ET can occur through several σ bonded spacers. He showed that electron coupling is an experimental reality. Earlier, it was believed that ET is only possible if there is a contact between two metallic electrodes, or possibly between two π-systems.

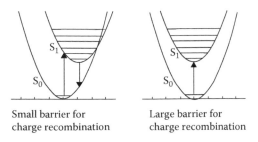

Small barrier for
charge recombination

Large barrier for
charge recombination

FIGURE 13.6 Possible PES for charge recombination to the ground state.

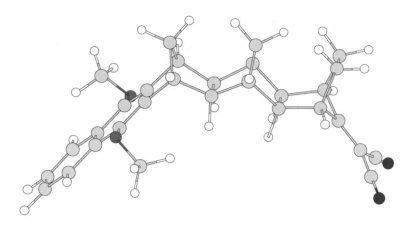

FIGURE 13.7 One of the molecules studied for PIET by Verhoeven et al.

Verhoeven et al. constructed a number of molecules of varying length of a similar type as the one shown in Figure 13.7. The distance dependence of the coupling is approximately exponential.

ET through aliphatic bridges with weak exponential decrease of the rate with distance, is possible provided the chain is *all-trans*. The dihedral of four carbon atoms is then 180°. If there are "*cis*-turns," the coupling decreases fast with distance, roughly proportional to the number of *cis*-turns. The Verhoeven molecules are built in a way that closely resembles the *all-trans* structure.

13.4.6 FULLERENE SYSTEMS

Fullerenes are excellent electron acceptors, since the electron affinity of C_{60} is exceptionally large (2.8 eV) and the LUMO is sixfold degenerate. Suggestions have been made to use fullerenes as "storage" of electrons, and indeed there is some sense in it. It has been possible to fill all six orbitals subsequently at an electrode, using cyclic voltammetry. The surface of the molecule is large, which permits solvent molecules to polarize and "neutralize" six electrons.

In Figure 13.8, one of the first fullerene systems that was studied in detail regarding PIET is shown (Devens Gust et al.). A substituted porphyrin is connected via phenyl groups with C_{60} on the right and a β-carotene on the left.

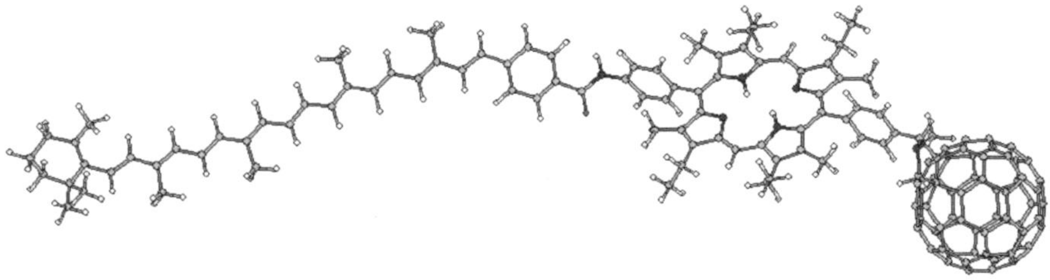

FIGURE 13.8 PIET fullerene system studied by Devens Gust et al.

The carotene or the porphyrin is excited to S_1. There is ET to the C_{60} ligand from the porphyrin. Later, the hole moves to the carotene ligand. There is now a rather stable charge-separated state. Finally, this state decays by ET from C_{60} to carotene. The triplet excited state of carotene is formed with long lifetime, before the whole complex returns to the ground state. This system holds the promise as an artificial photosynthetic system.

The molecular orbitals in the vicinity of the HOMO–LUMO gap are localized on either of the three π-systems. Two HOMOs are found on porphyrine and the next two on carotene. The seven LUMOs are found on C_{60}. However, it should be mentioned that information on the orbitals is *not* sufficient to explain the mechanism. Unfortunately, the system is too large to permit meaningful calculations of the excited states.

H. Lemetyinen, J. Andréasson, and many others have studied other large π-systems including the C_{60} molecule. After excitation, electrons or excitations can be "seen" in time-resolved transient spectroscopy jumping between the connected molecules. This type of work is useful in understanding natural photosynthetic processes. It deserves to be mentioned that ET or conductivity cannot occur without excitation.

13.4.7 OTHER INTRAMOLECULAR PIET EXPERIMENTS

Porphyrine systems of the type seen in Figure 13.9 are of interest since different parts of the π-bridges are perpendicular. The functioning of the bridge can be studied in great detail. The phenyl bridges are connected by an acetylenic group. The porphyrine π-electrons transfer in the σ-system of the phenyl groups across the π-bridge. The π-orbitals of phenyl are not used. B. Albinsson et al. have studied ET and excitation energy transfer using time-resolved spectroscopy.

Other work of this type, for example, by M. R. Wasielewsky et al., aim at finding π-bridges that permit ET over large distances. Even photoinduced *conductivity* over a large distance appears to be possible.

For some scientists, the aim is the oxidation of the water molecule to evolve oxygen. Nature has already succeeded in doing this in photosynthesis, by trial and error, over thousands of millions of years. Unfortunately there has been no convincing success so far. Perhaps it cannot be expected that modern research can improve on Nature in just a few decades.

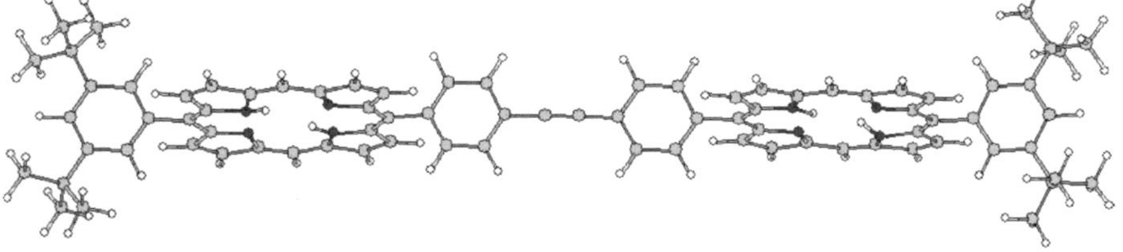

FIGURE 13.9 Diporphyrin complex studied by B. Albinsson et al.

13.5   MOLECULAR PHOTOVOLTAICS

Due to all the "progress" made by humankind, we now need not only natural photo-synthesis to survive, but also artificial photosynthesis, consisting of devices of various types to collect solar energy. Solar cells of the "silicon type" will be discussed in Chapter 16. Here, we will focus on *molecular photovoltaic systems*. A few attempts to create free energy from solar radiation for use in chemical reactions have already been mentioned. However, the systems mentioned thus far have not possessed any way to use the free energy produced as an electric voltage. In the development of photovoltaic devices, the aim has been to provide a useful source of energy in the form of a voltage.

A primitive photovoltaic cell may be obtained by shining light on the surface of one of the electrodes in an electrochemical cell. Electrons are lifted to the excited state, thereby creating a voltage between the two electrodes.

A useful electrode material is titanium dioxide (TiO_2). TiO_2 is an insulator at the temperature $T = 0$, and for that matter also at room temperature. No low-lying orbitals are involved in ET or conductivity. Ti^{4+} does not contain any 3d electrons. However, $O2p \rightarrow Ti3d$ excitation is possible. The excited electron has a nonzero probability to leap into the next Ti ion. The electron then continues to move by Ti^{3+}/Ti^{4+} electron exchange. In the $O2p \rightarrow Ti3d$ excitation, the electron leaves a nonbonding O2p orbital and ends up in a slightly bonding $Ti3d(t_{2g})$ orbital. The excitation energy is 3.4 eV. The reorganization energy is small. It is also small for Ti^{3+}/Ti^{4+} electron exchange. The coupling between the $Ti3d(t_{2g})$ orbitals is via an O2p orbital and, therefore, is quite large. Thus, $\Delta > \lambda$ and there is no activation barrier. TiO_2 is a good conductor of electrons in the lowest excited state (also called the conduction band). Because excitation in TiO_2 does not lead to any great structural changes (low reorganization energy), this oxide is able to endure repeated photoxidation.

Nevertheless, there is a problem with the mobilization of the excited electron. It is reasonable to assume that the O2p hole left behind is attractive, making it hard for

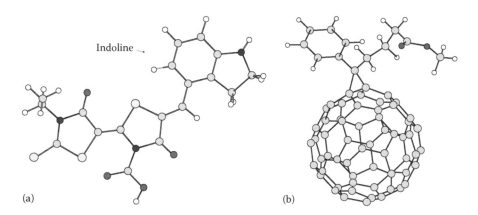

Indoline →

(a)   (b)

FIGURE 13.10   (See color insert) Examples of photosensitizers in Grätzel cells: (a) indoline-like molecule and (b) PCBM.

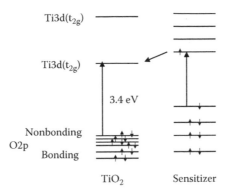

FIGURE 13.11   Orbital energies for the system TiO_2 and sensitizer.

the electron to proceed into the conduction band. Furthermore, there must be electron carriers in the electrolyte, for example, Fe^{3+}, which would be reduced to Fe^{2+} by the other electrode. The carrier could be any metal ion or molecule that is solvable in two adjacent valence states.

The greatest problem though is that the lowest excitations start in the UV region (at 3.4 eV). Therefore, the solar visible spectrum is left out from the production of voltage. The latter problem is solved by the so-called *sensitizer* (Figure 13.10). The sensitizer molecule can be excited in the visible region with a high quantum yield. The HOMO of the sensitizer is at a higher energy than the O2p level in TiO_2. The excitation energy is now in the visible region (Figure 13.11), while at the same time the LUMO is in level with the conduction band of TiO_2.

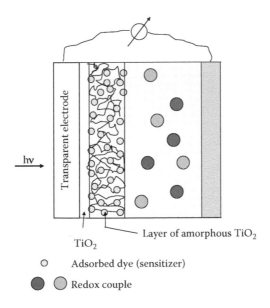

FIGURE 13.12   (See color insert) Grätzel cell.

In a dye-sensitized solar cell, organic molecules are adsorbed on the surface of TiO_2. The system is photoelectrochemical. In 1991, this cell was invented by Michael Grätzel and his collaborators in Lausanne, Switzerland. The TiO_2 layer consists of solid TiO_2 with a porous TiO_2 layer at the top to increase the surface area. This layer can be penetrated by the electrolyte. The sensitizer molecule is attached to the surface of TiO_2. For this reason, it needs groups (for example, $-COOH$) that bind to the TiO_2 surface.

The rate of ET from the sensitizer to TiO_2 has been measured in several cases using transient spectroscopy. Very high rates on the femtosecond scale have been found. The donation of electrons to TiO_2 is one of the fastest ET reactions known.

The modern Grätzel cell is sketched in Figure 13.12. Its efficiency is around 25%. It is important to bring down the price by using cheap sensitizers, electrode materials, and redox systems. One of the electrodes has to be transparent for light. The often-used indium-tin-oxide (ITO) is expensive. It is possible to use instead fluorine-doped SnO. The price of Grätzel cells is considerable lower than the price of ordinary silicon-based solar cells.

14 Excitation Energy Transfer

14.1 INTRODUCTION

In a crystal of identical chromophores, for example, a naphthalene crystal, local excitations may be transferred to other sites, just like electrons or holes. *Excitation energy transfer* (EET) is also called *electron-hole pair transfer* because the excited electron and remaining hole are transferred simultaneously to the other atom or molecule. Degeneracy appears in a finite system with two identical chromophores. In a repetitive system, the excitations may be delocalized, but this is not always the case. Devices for solar light harvesting, and natural antenna systems are examples of such repetitive systems. It is important to understand their properties.

14.2 EXCITED STATES OF BICHROMOPHORES

We will be interested in oligomers (dimers, trimers, etc.) and polymers with a large and undefined number of monomers. We first assume that there are two identical chromophores, related by translation, inversion center, or mirror plane. The excitation $\phi_i \rightarrow \phi_a$ on one corresponds to $\phi_j \rightarrow \phi_b$ on the other. The two excitations are assumed to be degenerate. If the chromophores are different, there is no degeneracy. The excitation with the highest energy tends to transfer to the site with the lower excitation energy. A comparison between electron transfer (ET) and EET is given in Figure 14.1.

14.2.1 CHROMOPHORES

A chromophore is the part of a protein or other molecule that is primarily responsible for its color. In the present chapter, we are particularly interested in chromophores that have so-called exciton spectra or are involved in EET. The spin-singlet states play the most important role in chromophores. The single Slater determinant approximation may be less accurate in some cases. In polyenes, for example, doubly substituted determinants give rise to low-lying excited states, and in porphyrins there is also a near degeneracy among the lowest excited states. After configuration interaction (CI), multiconfigurational electronic states are created, which are split in energy and may have very different transition moments in the excitation from the ground state.

Indole (Figure 14.2) is a side group in the protein tryptophan. Its spectrum has been studied in detail. In a protein that contains more than one tryptophan residue, the excited state can be transferred between them.

The most important chromophores in biological systems are chlorophylls and bacteriochlorophylls. They derive from chlorine and bacteriochlorine. Porphyrins derive from porphine (Figure 14.3). The two protons on the nitrogen

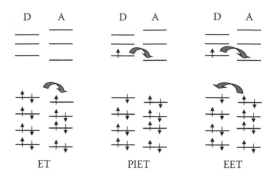

FIGURE 14.1 Orbital energy schemes for donor (D) and acceptor (A) molecules in the cases of ET (electron transfer), PIET (photoinduced electron transfer), and EET (excitation energy transfer).

atoms may be replaced by metal ions. The doubly negative ion missing the two protons is called a *free-base porphine*. These molecules are planar π-systems with conjugated double bonds. In chlorine, one and in bacteriochlorine, two pyrrole rings are reduced.

The electronic structure is rather complicated, but conforms to the rules mentioned in Chapter 3 for large cyclic π-systems. The important orbital substitutions are from the highest occupied molecular orbital (HOMO) and HOMO − 1 to the lowest unoccupied molecular orbital (LUMO) and LUMO − 1. CI leads to four excited states, the Q_x and Q_y states at lower energy and the Soret states at about 3 eV, on the border of the ultraviolet (UV) region.

If molecular groups are substituting the carbon hydrogens in the molecules in Figure 14.3, as in chlorophyll and bacteriochlorophyll, the mentioned laws for cyclic systems are only approximately valid. The Q band transitions tend to be more intense than in porphine, chlorine, and bacteriochlorine. For example, chlorophyll has quite a high absorption intensity in the Q region, for the transitions, which are mainly responsible for the harvesting of light energy.

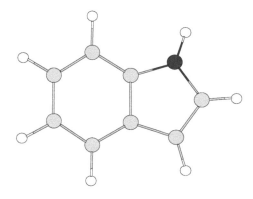

FIGURE 14.2 Indole is a typical molecule or side group (of the peptide tryptophan) for excitation energy transfer in proteins.

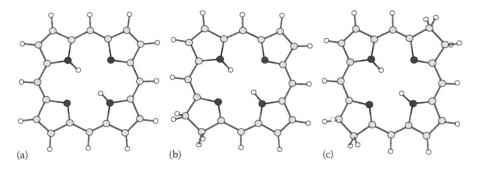

FIGURE 14.3   (a) Porphine, (b) chlorine, and (c) bacteriochlorine. Notice that the 5-ring (left-down) is reduced in chlorine and two 5-rings are reduced in bacteriochlorine.

As was shown in Section 3.5.6, a spatial orbital substitution, $\phi_i \rightarrow \phi_a$, results in three spin-triplet wave functions and one spin-singlet wave function. The triplet states are

$$^3\Phi_i^a = N!^{-1/2} \left| \prod_{\neq i} (occ)\phi_i \alpha \phi_i \alpha \right|, \tag{14.1}$$

$$^3\Phi_i^a = N!^{-1/2} \left| \prod_{\neq i} (occ)\phi_i \beta \phi_i \beta \right|, \tag{14.2}$$

and

$$^3\Phi_i^a = (2N!)^{-1/2} \left\{ \left| \prod_{\neq i} (occ)\phi_i \phi_a (\alpha\beta + \beta\alpha) \right| \right\} \tag{14.3}$$

and the spin-singlet state is

$$^1\Phi_i^a = (2N!)^{-1/2} \left\{ \left| \prod_{\neq i} (occ)\phi_i \alpha \phi_a (\alpha\beta - \beta\alpha) \right| \right\}. \tag{14.4}$$

The ground state energy is $\left\langle {}^1\Phi_0 \middle| H \middle| {}^1\Phi_0 \right\rangle$. The symbol Φ indicates that the theoretical treatment particularly concerns the approximation to a single Slater determinant. The expressions for the difference to the singly substituted ($\phi_i \rightarrow \phi_a$) expectation value are given by

$$\left\langle {}^1\Phi_i^a \middle| H \middle| {}^1\Phi_i^a \right\rangle - \left\langle {}^1\Phi_0 \middle| H \middle| {}^1\Phi_0 \right\rangle = \varepsilon_a - \varepsilon_i - J_{ia} + 2K_{ia} \quad \text{(for spin singlets)} \tag{14.5}$$

and

$$\left\langle {}^{3}\Phi_i^a \left| H \right| {}^{3}\Phi_i^a \right\rangle - \left\langle {}^{1}\Phi_0 \left| H \right| {}^{1}\Phi_0 \right\rangle = \varepsilon_a - \varepsilon_i - J_{ia} \quad \text{(for spin triplets)}. \quad (14.6)$$

The derivation of Equations 14.5 and 14.6 is simple, but is left out. The orbital energy is ε_i for the orbital occupied in the ground state and ε_a for the orbital occupied in the first excited state. The difference in the expectation values for singlet and triplet states is the exchange integral K_{ia}. We will use the following notation (the Mulliken notation):

$$\int \phi_i^*(1)\phi_j^*(1)\frac{1}{r_{12}}\phi_k(2)\phi_l(2)d\tau_1 d\tau_2 = \left(ij|kl\right) \quad (14.7)$$

and find that the Coulomb and exchange integrals may be expressed as

$$J_{ia} = \left(ii|aa\right) = \int \phi_i^*(1)\phi_i^*(1)\frac{1}{r_{12}}\phi_a(2)\phi_a(2)d\tau_1 d\tau_2 \quad (14.8)$$

and

$$K_{ia} = \left(ia|ia\right) = \int \phi_i^*(1)\phi_a^*(1)\frac{1}{r_{12}}\phi_i(2)\phi_a(2)d\tau_1 d\tau_2. \quad (14.9)$$

In a CI calculation, there is no interaction between the ground state Hartree–Fock Slater determinants and the singly substituted states, as we have seen in Chapter 2. A reasonable approximation for the excited states is obtained by including only singly substituted Slater determinants in the CI calculation. The matrix element between two substitutions, $\phi_i \rightarrow \phi_a$ and $\phi_j \rightarrow \phi_b$ on the other molecule, may be written (without proof)

$$\left\langle {}^{1}\Phi_i^a \left| H \right| {}^{1}\Phi_j^b \right\rangle = 2\left(ai|jb\right) - \left(ab|ji\right), \quad (14.10)$$

$$\left\langle {}^{3}\Phi_i^a \left| H \right| {}^{3}\Phi_i^a \right\rangle = -\left(ab|ji\right). \quad (14.11)$$

From Equations 14.7 and 14.8, it follows that the integral $(ai|jb)$ is an electrostatic repulsion between two charge distributions, $\phi_i\phi_a$ and $\phi_j\phi_b$, called transition charges, and from Equations 14.7 and 14.9 that $(ab|ji)$ is the electrostatic repulsion between $\phi_a\phi_b$ and $\phi_j\phi_i$. The latter are products between exponential or Gaussian functions that are located on different molecules and therefore tend to zero exponentially as the distance between the chromophores is increased. We conclude that $(ab|ji)$ can be neglected in comparison with $(ai|jb)$ in Equation 14.10. The absolute value of the matrix element in Equation 14.11 is very small in comparison with the absolute value of Equation 14.10. In other words, triplets do not interact except at a very short distance between the molecules.

14.2.2 WAVE FUNCTIONS AND MATRIX ELEMENTS OF BICHROMOPHORES

As above, the active orbitals are called ϕ_i and ϕ_a on one chromophore (A) and ϕ_j and ϕ_b on the other chromophore (B) (Figure 14.4). We assume that the two chromophores are identical, and related by symmetry operations. The ground state wave function contains symmetric and antisymmetric molecular orbitals (MO) of the type $\phi_i \pm \phi_j$ and $\phi_a \pm \phi_b$. Index i and a belong to center A, and j and b to center B. We include only the relevant orbitals and write the ground state wave function as

$$\Phi_0 = \frac{1}{4\sqrt{N!}}\left|...\left(\phi_i+\phi_j\right)\alpha\left(\phi_i+\phi_j\right)\beta\left(\phi_i-\phi_j\right)\alpha\left(\phi_i-\phi_j\right)\beta...\right|. \quad (14.12)$$

The excited state wave functions are obtained by substituting by the unoccupied MOs $\phi_a + \phi_b$ and $\phi_a - \phi_b$ with the same spin. There are eight such substitutions, but it is sufficient to do the substitution for one of the spins, α or β, and project out the correct spin function afterward. We, therefore, substitute only β spin orbitals and obtain four new functions, two (Φ_1 and Φ_2) with transitions polarized along the intermolecular axis (x), and two (Φ_3 and Φ_4) polarized perpendicular to it (y)

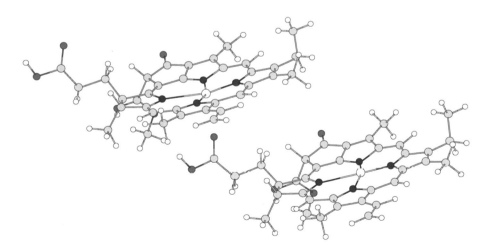

FIGURE 14.4 Dimer of chlorophyll (phytyl chains not shown). The shape of the dimer is due to the charges of opposite signs coming closer by skewing the two planes. The most important attraction is between the carboxyl group at the upper left of the lower monomer and the Mg^{2+} ion at the center of the upper monomer.

$$\Phi_1 = \frac{1}{4\sqrt{N!}} \left| ...(\phi_i + \phi_j)\alpha(\phi_i + \phi_j)\beta(\phi_i - \phi_j)\alpha(\phi_a + \phi_b)\beta... \right|, \qquad (14.13)$$

$$\Phi_2 = \frac{1}{4\sqrt{N!}} \left| ...(\phi_i + \phi_j)\alpha(\phi_a - \phi_b)\beta(\phi_i - \phi_j)\alpha(\phi_i - \phi_j)\beta... \right|, \qquad (14.14)$$

$$\Phi_3 = \frac{1}{4\sqrt{N!}} \left| ...(\phi_i + \phi_j)\alpha(\phi_i + \phi_j)\beta(\phi_i - \phi_j)\alpha(\phi_a - \phi_b)\beta... \right|, \qquad (14.15)$$

$$\Phi_4 = \frac{1}{4\sqrt{N!}} \left| ...(\phi_i + \phi_j)\alpha(\phi_a + \phi_b)\beta(\phi_i - \phi_j)\alpha(\phi_i - \phi_j)\beta... \right|. \qquad (14.16)$$

The Slater determinants of Equations 14.13 through 14.16 are added and subtracted to the wave functions of Equations 14.17 through 14.20:

$$\Phi_x = \frac{1}{\sqrt{2}}(\Phi_1 + \Phi_2) = \frac{1}{2\sqrt{2N!}} \left| ...\phi_i\alpha\phi_j\,\alpha\left(\phi_i\phi_b + \phi_j\phi_a\right)\beta\beta... \right| = \frac{1}{\sqrt{2}}\left(^1\Phi_i^a - {}^1\Phi_j^b\right),$$

$$(14.17)$$

$$\Phi_y = \frac{1}{\sqrt{2}}(\Phi_3 + \Phi_4) = \frac{1}{2\sqrt{2N!}} \left| ...\phi_i\alpha\phi_j\,\alpha\left(\phi_i\phi_b - \phi_j\phi_a\right)\beta\beta... \right| = -\frac{1}{\sqrt{2}}\left(^1\Phi_i^a + {}^1\Phi_j^b\right),$$

$$(14.18)$$

$$\Phi_{ctx} = \frac{1}{\sqrt{2}}(\Phi_1 - \Phi_2) = \frac{1}{2\sqrt{2N!}} \left| ...\phi_i\alpha\phi_j\,\alpha\left(\phi_i\phi_a + \phi_j\phi_b\right)\beta\beta... \right| = \frac{1}{\sqrt{2}}\left(^1\Phi_i^b - {}^1\Phi_j^a\right),$$

$$(14.19)$$

$$\Phi_{cty} = \frac{1}{\sqrt{2}}(\Phi_3 - \Phi_4) = \frac{1}{2\sqrt{2N!}} \left| ...\phi_i\alpha\phi_j\,\alpha\left(\phi_i\phi_a - \phi_j\phi_b\right)\beta\beta... \right| = -\frac{1}{\sqrt{2}}\left(^1\Phi_i^b + {}^1\Phi_j^a\right).$$

$$(14.20)$$

The wave functions Φ_x and Φ_y do not interact with the functions Φ_{ctx} and Φ_{cty} because of the vanishing overlap between the functions centered on A and B.

The wave functions of Equations 14.17 and 14.18 are sums of two Slater determinants in which the substitutions [(i → a) or (j → b)] are within the same molecule. The wave functions Φ_x and Φ_y thus correspond to two locally excited states. The wave functions of Equations 14.19 and 14.20, on the other hand, are sums or differences of Slater determinants where the excitations are charge transfer excitations, since the number of electrons on A and B is 3 + 1 and 1 + 3, respectively.

In an asymmetric system, the ground state will be mainly localized on one of the centers. If the sites are identical, asymmetry may be created if Φ_x and Φ_y are added or subtracted. The sum $\Phi_x + \Phi_y$ gives $^1\Phi_i^a$, which corresponds to an excitation on site A. Polarization will lower the energy of this state, and cause permanent localization of the excitation on A. The EET between A and B is described by the Marcus model. There may be an activation barrier. In that case, we are back to the case with ordinary local excitation.

However, if the coupling is large, the excited states are delocalized over two centers. The excited states without charge transfer, Φ_x and Φ_y, are referred to as *exciton states.*

Whether the locally excited states have a lower energy than the charge transfer states, $\Phi_{ctx} + \Phi_{cty}$, depends on the polarization effects. The latter lower the energy of the charge transfer states, even in nonpolar solvents. The smaller the distance between A and B, the larger is the lowering, as we have seen in Chapter 13.

According to Equation 14.10, the difference in energy expectation value is

$$\left\langle \Phi_x \middle| H \middle| \Phi_x \right\rangle - \left\langle \Phi_y \middle| H \middle| \Phi_y \right\rangle = 2\Big[\, 2\big(ia \big| jb\big) - \big(ij \big| ab\big) \Big] = 2W_S, \qquad (14.21)$$

for the wave functions of Equations 14.17 and 14.18. Later, this energy difference will be written as an "interaction" between transition moments.

14.2.3 Covalent Bonding in Ground and Excited States

There may be a coupling between the locally excited states, which is the main reason for EET. In some cases, this leads to an energy lowering of several hundreds of cm^{-1} in the excited state, even if the monomers are not covalently bonded. If the monomers are covalently linked, we are, in principle, talking about single molecules.

In Figure 14.5, we see a planar "dimer" of zinc porphyrine where the two monomers are linked by a butadiyne molecule (HC≡C–C≡CH). The ground state conformation is the perpendicular form on the right, but there appears to be almost free rotation around the central CC bond. The perpendicular form is probably preferred because of weak steric hindrance in the planar form between the adjacent hydrogen atoms on different monomers.

In the excited state, the LUMO is singly occupied and since the LUMO is bonding between the central carbon atoms, there is now some conjugation between the two

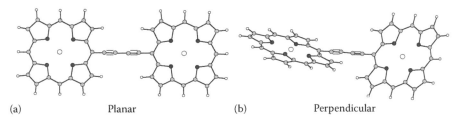

(a)   Planar   (b)   Perpendicular

FIGURE 14.5 Butadiyne-linked zinc porphyrins in planar (a) and perpendicular (b). The triple bonds are marked out geometry.

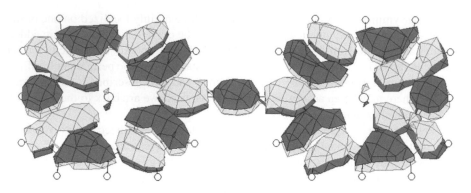

FIGURE 14.6 LUMO of the planar complex, showing the π-bonding contribution between the central carbon atoms of the butadiyne group.

π-systems. Figure 14.6 shows a symmetric LUMO. The first excited state is stable in the planar state and behaves as a state where symmetric monomers are interacting. The HOMO (not shown), on the other hand, has a node through the central CC bond. Taking one electron from the HOMO and putting it in the LUMO means that the bonding between the monomers is increased in the planar state.

Delocalization over the whole molecule is important. The reorganization energy is spread out over a large complex. Hence, the geometry change between the ground and excited states is small. The coupling is similar to the exciton coupling and it remains large. Therefore, $|\Delta| > \lambda$ and the system is delocalized.

14.3 TRANSITION MOMENTS

Van der Waals interactions are due to forces between molecules that are not bonded together by chemical bonds. Most important are the interactions between charges, charges and dipole moments, and quadrupole moments, and interaction between dipole moments and quadrupole moments. A particular kind of van der Waals force is London interactions, which decrease with distance as $1/R^6$, where R is the distance between the molecules or atoms (for example, inert gas atoms). In a way, this interaction is a correlation effect between electrons on different atoms. The physical picture is that a momentary charge deviation of the electrons in one atom causes a reaction of the same kind in the other atom. The interaction between these two deviations leads to a small energy lowering, which is responsible for condensation of the inert gas at low temperature. One may also talk about attractions due to mutual polarizability.

In the excited states, there is a similar interaction between molecules. This interaction is strongly spin dependent and is of particular importance for EET. Briefly, EET is possible by spin-singlet excited state, but almost impossible between triplet states.

14.3.1 TRANSITION DENSITIES

Electronic transition moments have been treated above in Section 3.5.3. The electronic transition moment is

$$\vec{\mu} = \left(\mu_x, \mu_y, \mu_z\right). \tag{14.22}$$

From this, the intensity of the absorption line (oscillator strength) is obtained, proportional to the square $|\vec{\mu}|^2$. A component of the transition moment is obtained as in Equation 3.50:

$$\mu_x = \int \phi_i^* \cdot x \cdot \phi_a \, dxdydz. \tag{14.23}$$

The transition density is the product:

$$\rho_t = \phi_i^* \cdot \phi_a. \tag{14.24}$$

ρ_t is obviously a mathematical function that is positive in some regions of space and negative in others. If ρ_t is integrated over all space, the integral is equal to zero, since the two wave functions, ϕ_i and ϕ_a, are orthogonal (nonoverlapping).

In the following, we will show that the transition density determines the energy splitting between the exciton states as well as the transition moments by Equation 14.23.

14.3.2 ENERGY ORDER OF DIMER EXCITON STATES

Flat dimes are oriented differently to each other, flat sides toward each other, or head to tail (Figure 14.7). Most often, dimers are oriented in the skewed way of Figure 14.4. To get an idea about the usefulness of the transition moment and its similarity with the dipole moment, we will derive the energy order of the excited states of the dimer from the transition moments.

Different orderings between the molecules and transition moments of a dimer are shown in Figure 14.7. The arrow is the dipole moment directed from a positive

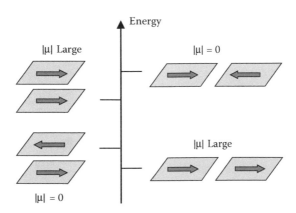

FIGURE 14.7 Orientation of dimers in two cases: (left) stacked and (right) head to tail.

transition charge to a negative transition charge. In the stacked case (left), the direction is the same or different. Because the energy can be obtained from the transition moments, we calculate the energy as a Coulomb energy. If the transition moments have the same directions, the charges of the same sign are close to each other. There is more Coulomb repulsion than attraction, so the energy is higher than if the directions are opposite to each other. The sum of the transition moments is higher in the upper case on the left. In the lower case, the total transition moment is close to zero.

On the right-hand side of Figure 14.7, we have the head to tail case. Obviously, the same direction case is lower in Coulomb repulsion than the different direction case. The total transition moment is larger in the same direction case.

Below, we will give the quantum mechanical proof. In classical mechanics, oscillating dipoles behave in the same way. In the Bohr–Sommerfeld picture ("old quantum theory"), the transition moments are also oscillating dipoles, but contrary to the classical case, only two quantized states are possible at the end.

14.3.3 DISTANT CHROMOPHORES INTERACT VIA TRANSITION CHARGES

Assume now that ϕ_i and ϕ_a are orbitals on one atom or molecule and ϕ_j and ϕ_b are orbitals on another atom or molecule, at some distance from the first one. We are interested in the matrix element (ai|jb) (Mulliken notation), which is the main term in the right member of Equation 14.10. We will show that this interaction matrix element can be described as an interaction between the two distant transition charge distributions, $\phi_a\phi_i$ and $\phi_j\phi_b$.

The charge distributions $\phi_a\phi_i$ and $\phi_j\phi_b$ may be replaced by a number of point charges $q_1, q_2, q_3, \ldots$, on each molecule, sufficiently densely disposed to represent the total, continuous charge distribution of the electrons. We first calculate the potential in the point $\vec{r}$ at some distance from the molecule.

The origin is located somewhere inside the molecule (Figure 14.8). The vectors $\vec{a}_1, \vec{a}_2, \vec{a}_3, \ldots$, are directed from the origin to the point charges. The potential φ in the point $\vec{r} = (x,y,z)$ from one of the point charges, for example, $\vec{a} = \vec{a}_1$ with the charge q, may be written as (in atomic units):

$$\varphi(\vec{r}) = \frac{q}{|\vec{r}'|} = \frac{q}{|\vec{r} - \vec{a}_i|}. \tag{14.25}$$

FIGURE 14.8   Calculation of the potential in the point $\vec{r}$ from a number of point charges in $\vec{a}_1, \vec{a}_2,$ and $\vec{a}_3$.

The potential φ is Taylor expanded at a long distance from the molecule ($|\vec{r}| \gg |\vec{a}|$). In three dimensions, δx of the ordinary Taylor expansion is replaced by $(-a_x, -a_y, -a_z)$, the derivative for $x = 0$ by the gradient, and the product $x \cdot (d/dx)_{x=0}$ by the scalar product grad $\varphi \cdot \vec{a}$, with the gradient calculated at the origin:

$$\varphi(\vec{r}') = \frac{q}{r} + q(xa_x + ya_y + za_z)\frac{1}{r^3} + \text{terms in } a^2 \text{ (quadrupole terms)}. \quad (14.26)$$

The second term of the right member is the derivative of the function $1/|\vec{r} - \vec{a}|$ with respect to a_x, a_y, and a_z:

$$\frac{\partial}{\partial a_x}\left[(x - a_x)^2 + (y - a_y)^2 + (z - a_z)^2\right]^{-1/2}$$
$$= \left(-\frac{1}{2}\right)\left[(x - a_x)^2 + (y - a_y)^2 + (z - a_z)^2\right]^{-3/2} \cdot 2(x - a_x)(-1). \quad (14.27)$$

For $(a_x, a_y, a_z) = (0,0,0)$, this is equal to

$$\left[x^2 + y^2 + z^2\right]^{-3/2} x = \frac{x}{r^3}. \quad (14.28)$$

In SI units, $4\pi\varepsilon_0$ has to be included in the denominator of Equations 14.26 and 14.28.

The first term in the right member of Equation 14.26 is a monopole term. It is simply the potential from the charge placed at the origin. The next term may be written

$$q(xa_x + ya_y + za_z)\frac{1}{r^3} = \frac{(\vec{r} \cdot \vec{\mu})}{r^3}. \quad (14.29)$$

The vector

$$\vec{\mu} = q(a_x, a_y, a_z) \quad (14.30)$$

is the contribution from the charge in point $\vec{a}$ to the dipole moment of the molecule. Summing over all discrete charges, we obtain

$$\sum_i q_i(xa_{ix} + y_{ix} + za_{iz})\frac{1}{r^3} = \sum_i \frac{(\vec{r} \cdot q_i\vec{a}_i)}{r^3} = \frac{(\vec{r} \cdot \vec{\mu})}{r^3}, \quad (14.31)$$

where $\vec{\mu}$ is now the total dipole moment from all discrete charges:

$$\vec{\mu} = \sum_i q_i\vec{a}_i. \quad (14.32)$$

For a continuous charge distribution, we have instead

$$\sum_i q_i \rightarrow \int \rho(\vec{a}) da_x da_y da_z, \tag{14.33}$$

$$\vec{\mu} = \sum_i q_i \vec{a}_i \rightarrow \int \vec{a} \cdot \rho(\vec{a}) da_x da_y da_z. \tag{14.34}$$

Neglecting the quadrupole terms, we obtain

$$\varphi(\vec{r}) = -\frac{1}{r} \int \rho(\vec{r}') dx' dy' dz' + \frac{1}{r} \sum_i Z_i$$

$$-\frac{\vec{r} \cdot \int \vec{a} \rho(\vec{a}) da_x da_y da_z}{r^3} + \frac{\vec{r} \cdot \sum_i Z_i \vec{a}_i}{r^3} = \frac{Q}{r} + \frac{\vec{r} \cdot \vec{\mu}}{r^3} + \cdots, \tag{14.35}$$

where Q is the charge of the molecule and $\vec{\mu}$ is its dipole moment.

If the molecule is neutral, the monopole term disappears and the Taylor expansion starts with $(\vec{r} \cdot \vec{\mu})/r^3$. This term gives rise to a dipole field. The simplest charge distribution that leads to such a field consists of two charges q with different signs, at a distance R from each other (this is the meaning of a "dipole"). The dipole is then a vector of length $\vec{\mu} = q\vec{R}$ directed from the negative to the positive charge. An "ideal dipole moment" is obtained when $q \rightarrow \infty$ and $R \rightarrow 0$ in a way that $\vec{\mu}$ remains constant. Far from the molecule, one may assume that all dipole moments are ideal dipole moments.

The electric field $\vec{F}$ outside an ideal dipole is the derivative of the potential φ with respect to x, y, or z: $(F_x, F_y, F_z) = (\partial\varphi/\partial x, \partial\varphi/\partial y, \partial\varphi/\partial z)$:

$$\frac{\partial \frac{\vec{r} \cdot \vec{\mu}}{r^3}}{\partial x} = \frac{\frac{\partial(\vec{r} \cdot \vec{\mu})}{\partial x} \cdot r^3 - (\vec{r} \cdot \vec{\mu}) \cdot \frac{\partial r^3}{\partial x}}{r^6} = \frac{\mu_x \cdot r^3 - (\vec{r} \cdot \vec{\mu}) \cdot 3rx}{r^6}. \tag{14.36}$$

Since $\vec{r} = (x, y, z)$, we may write

$$\vec{F} = \frac{\vec{\mu}}{r^3} - \frac{3(\vec{r} \cdot \vec{\mu})\vec{r}}{r^5}. \tag{14.37}$$

The potential energy for a dipole $\vec{\mu}_2$ in the electric field $\vec{F}$ is then ($\vec{\mu}$ in Equation 14.24 is called $\vec{\mu}_1$)

$$\vec{F} \cdot \vec{\mu}_2 = \frac{\vec{\mu}_1 \cdot \vec{\mu}_2}{r^3} - \frac{3(\vec{r} \cdot \vec{\mu}_1)(\vec{r} \cdot \vec{\mu}_2)}{r^5}. \tag{14.38}$$

Thus, if we let $\bar{F}$ be the field from the other molecule, we find that the matrix element (ai|jb) of Equation 14.10 can be represented as in Equation 14.38. This is the well-known expression for the interaction between two dipole moments. A quite remarkable thing is that this is also the Hamiltonian matrix element in the Schrödinger equation for the excited singlet states, if the dipole moment is to be replaced by the transition dipole moment. Next, we will find that the rate of transfer of the S_1 state to the site is proportional to the square of this matrix element.

14.4 FLUORESCENCE RESONANCE ENERGY TRANSFER

In Section 14.2, we found in Equation 14.11 that the interaction matrix element between two chromophores in an excited triplet state decreases exponentially with distance, while the corresponding singlet matrix element (Equation 14.10) decreases more slowly as $1/R^6$ with distance. These interactions were derived a long time ago from the old quantum theory by D. L. Dexter in the triplet case and Theodor Förster in the singlet case. The conclusions are valid also for the case of unequal chromophores.

Förster's theory is behind a very useful method for determining the absorption spectrum of a collection of identical chromophores within a separation distance of less than 100 Å, placed in a protein, DNA, or some other "scaffold." One example is the indole groups in a protein, where the interaction causes a broadening of the absorption spectrum.

Förster's theory is also behind the FRET method (Förster resonance energy transfer or fluorescence resonance energy transfer). In this case, excitation takes place with a narrow laser pulse, corresponding to the lowest excited state of a chromophore. The rate of EET is determined by time-resolved fluorescence.

14.4.1 INTERACTION BETWEEN SPIN-SINGLET EXCITATIONS (FÖRSTER)

The integral in Equation 14.10 is equal to the sum of twice the Coulomb repulsion (ai|jb) between the transition charge distributions $\phi_a\phi_i$ and $\phi_j\phi_b$, minus the Coulomb repulsion (ab|ji) between the transition charges $\phi_a\phi_b$ and $\phi_j\phi_i$. This is referred to as the Förster contribution. The second term is exponentially decreasing with distance and is referred to as the Dexter term. The original derivations were done by Th. Förster and D. L. Dexter, respectively.

Since the respective integrated charges are zero, the first non-zero term in (ai|jb) is a dipole–dipole term. The interesting thing is that the two dipole moments are identical to the dipole moments for the transition charges $\phi_i \rightarrow \phi_a$ and $\phi_j \rightarrow \phi_b$, respectively. It is easy to derive that an interaction between two dipole moments decreases as R^{-6}, where R is the distance between the atoms. The transition dipole–transition dipole interaction integral was first derived by Förster and is named after him.

The Coulomb integral (ab|ji) in Equations 14.10 and 14.11 are between the charge densities of orbital products where the two orbitals are on different centers. Hence, the two density distributions decrease as e^{-R}. This much faster decrease was originally derived by Dexter and the exponentially decreasing integral is named after him.

We may now guess what results are obtained if the excited states of several molecules are included in a CI treatment. If the diagonal elements are slightly different, the triplet states at a lower energy are localized on single chromophores. The interaction matrix elements between the chromophores are too small to permit extensive delocalization. In the case of singlet states at a higher energy, there may be a substantial interaction between excitations on different molecules. This directly applies to antenna systems in photosynthesis.

We write Equation 14.10 as

$$\left\langle {}^1\Phi_i^a \left| H \right| {}^1\Phi_j^b \right\rangle = 2\left(ai\left|jb\right.\right) - \left(ab\left|ji\right.\right) = V_{\text{Förster}} + V_{\text{Dexter}}. \tag{14.39}$$

The derivation of the Förster matrix element thus leads to the dipolar term

$$\left(ai\left|jb\right.\right) = \frac{\vec{\mu}_1 \cdot \vec{\mu}_2}{R^3} - \frac{3\left(\vec{R}\cdot\vec{\mu}_1\right)\left(\vec{R}\cdot\vec{\mu}_2\right)}{R^5}. \tag{14.40}$$

We use the Fermi golden rule (Equation 7.39) to derive the decay rate due to Förster interaction and obtain

$$W_{\text{Förster}} = 2\frac{2\pi}{\hbar}\left\langle m\left|V_{\text{dip}}\right|k\right\rangle^2 \rho\left(E_k^0\right). \tag{14.41}$$

In the first derivation, Förster applied the old quantum theory to derive Equation 14.40. In a later derivation, he used the methods described above where we are considering the two chromophores as a single quantum system. The Förster interaction is an electronic coupling between excited states in different parts of the system. There is a coupling matrix element H_{12} that determines the rate of transfer together with reorganization energy λ as always in the Marcus model. It is usually not realized that the nuclear coupling due to structural reorganization may be important, leading to temperature dependence.

Somewhat confusing is that we calculate the electronic coupling element as a Coulomb interaction between the transition densities on two different sites, as if two different excitations are involved. This is not the case, of course. There is only one excitation, exchanged between the sites.

The trivial picture of an excitation transfer as the emission of one quantum of light by the first molecule or molecules, followed by absorption by the second molecule is unsuitable. The probability of the latter process (emission and obsorption of light) would be extremely low. Furthermore, it would occur with equal probability over the whole crystal, independent of the distance between the chromophores. In reality, due to Equations 14.40 and 14.41, the probability for transfer is proportional to $1/R^6$.

An early experiment was carried out by Cario and Franck in 1922, where they used a mixture of mercury and thallium vapor. When irradiated with the light of a special frequency belonging to mercury, the emission spectrum shows the spectra of both atoms, although only mercury absorbs at the probing light frequency. Hence, the thallium atoms have been excited by an excitation transfer from the mercury atoms. In this experiment, the distance for excitation transfer cannot be decided. Similar observations of fluorescence were made with molecular vapors and in solution.

Eventually, it became clear that excitation transfer may occur between statistically distributed molecules over as much as 40 Å. To determine the distance dependence, it is important to use an array where the location of the molecules is known, as in proteins. The distance between the indole groups (Figure 14.2) can actually be determined by observing the transfer probability for EET.

14.4.2 The Mysterious Factor of Two

As is clear from the derivation above, that the coupling can be obtained as a dipole–dipole interaction between the transition moments for the two molecules. Förster, in his updated derivation, derived his matrix element from the Slater determinant wave functions. The factor of two in the right member of Equations 14.39 and 14.41 is therefore missing. The reason is that a Slater determinant is not an eigenfunction of spin, but rather an equal mixture of singlet and triplet states. For triplet states, the factor is equal to zero, since (ai|bj) is missing from Equation 14.10. In Equation 14.10 there is clearly a factor of two in front of (ai|bj). The average between the singlet and triplet is a factor equal to unity, as in Förster's incorrect derivation.

On the other hand, Förster showed a lot of insight by including the factor n^2 in the denominator. This factor is not included in our derivation. It depends on the polarizability of the medium, since n is the refractive index. In our derivation, we have not accounted for the fact that there are a number of other electrons present that are not excited. In electrodynamics theory, these electrons are approximately taken into account by the factor of $n^2 \approx 2$ in the denominator.

In summary, Förster included the factor of two in the denominator, but left it out in the numerator. The present author included the factor of two in the numerator, but left it out in the denominator. Our final result is thus a factor of four larger than Förster's result. The correct result is the geometrical average. This value is used by the majority of scientists, since they tend to forget about both of the important factors of two.

14.4.3 Dexter Coupling

In the triplet case, the large term (ai|bj) is missing in Equation 14.11. In the term (ab|ji), the function ab decreases as $\exp(-4\varepsilon R)$, where ε is the orbital energy measured in H. Assuming that ε is 0.25 H, the decrease factor of ab is $\exp(-R)$. The same is the case with the function ji. The integral (ai|bj) thus decreases as $\exp(-2R)$. The transition probability for a Dexter transition is proportional to the square of (ai|bj) and therefore decreases as $\exp(-4R)$. This means that triplet–triplet excitation transfer only takes place at a small distance between the chromophores. In practice, one talks about *sensitation*, meaning a collision between the molecules.

14.4.4 RATE EQUATIONS FOR EET

The rate expressions derived from the Marcus model in the previous chapters for ET may also be used for EET. We know already that the geometry of a molecule is slightly changed when it is excited. In the case of polyenes, strong bonds become weak bonds and vice versa.

The coupling matrix elements are given in Equation 14.10 for singlets and Equation 14.11 for triplets. The reorganization energy may be estimated or determined by quantum chemical calculations.

We have found that an excitation may be exchanged with another, identical chromophore. If the bond distance change is large at excitation, the thermal activation energy for such an exchange is large. For example, polyenes are known to have a large reorganization energy λ. It is not likely that polyenes exchange excitation with each other. In other π-systems, this may be possible. Chlorophylls exchange electrons with almost zero activation energy and this is used in nature in the antenna systems of photosynthetic organisms.

The $1/R^6$ dependence was experimentally proven by Stryer and Haugland in an experiment on an α helix in 1967.

15 Photosynthesis

15.1 INTRODUCTION

Green plants use sunlight energy in *natural photosynthesis*. The sun emits radiation particularly in the wavelength range of 400–700 Å and energy range $h\nu = 2 - 3$ eV or $h\nu = 200 - 300$ kJ/mol. In this range, molecules can be electronically excited and covalent bonds broken or formed. In a biological organism, the energy take-up has to be done in a controlled way, without breaking the chemical bonds. The *reaction center* (RC) of photosynthetic organisms is part of a membrane protein, localized in the *chloroplasts*. Light harvesting takes place in the membrane proteins of the thylakoid membrane. Photosynthetic electron transport chains (ETC) have many similarities to the oxidative chains discussed in Chapter 11.

In the competition for sunlight, plants have developed the root–trunk–leaf construction that we see in most green plants. The total surface area perpendicular to the sun's rays is important for the species to be competitive. The total power of the green plant photosynthesis on Earth is 100 TW (1 terawatt = 10^{12} W). This may be compared to the worldwide consumption of energy per second of 15 TW and to the total solar radiation power falling on the earth equal to 174,000 TW.

15.2 MOLECULES OF PHOTOSYNTHESIS

The most important molecule is *chlorophyll* (Chl), which structurally derives from chlorine (Figure 14.3). *Bacteriochlorophyll* (BChl) derives structurally from bacteriochlorin (Figure 14.3). Chl and BChl serve as acceptors of radiation, transporters of excitation energy, in charge separation, and for the transport of electrons. They are suitable for this purpose because of low reorganization energy.

Chl and BChl cover only a small region of the solar spectrum, however. Accessory pigments are used by nature to absorb in the spectral region between the Soret band at the border of the ultraviolet (UV) region and the Q bands of Chl and BChl.

15.2.1 Chl AND BChl

The pigment proteins of photosynthesis are located in the *chloroplasts*. The latter are "cells within cells," which, like mitochondria, reside in the cytoplasm. Mitochondria use chemical energy to produce adenosine triphosphate (ATP). Chloroplasts harvest solar energy. The chloroplasts contain *thylakoids*, which may be described as flat sacks, whose membranes contain the pigment proteins (Figure 15.1). Thylakoids have an inside (lumen) and an outside (stroma) and form stacks referred to as *grana*. A great number of grana and stroma are surrounded by the inner and outer membrane of the chloroplast. The four protein complexes in the thylakoid membrane of green plants are photosystem I (PSI), photosystem II (PSII), a cytochrome b6f complex, and ATP synthase.

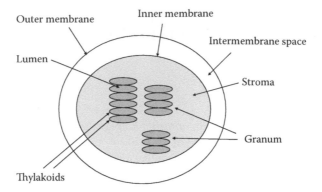

FIGURE 15.1   The chloroplast.

The most important pigments are Chl and BChl. Compared to the green Chl molecule, the purple BChl molecule is reduced in ring II (Figure 15.2). The outer double bond is replaced by a single bond. The vinyl group of ring I has been exchanged with a carbonyl group in BChl. The Mg atom is bonded to an imidazole side group of histidine on one side of the plane in the axial direction.

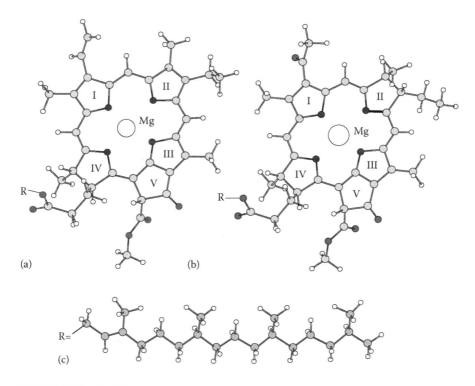

FIGURE 15.2   (a) Chl *a* (upper left) and (b) BChl *a* (upper right). The substituent R is the phytyl chain. The corresponding pheophytin molecules are identical except that they are missing the magnesium ion.

Pheophytins (Ph) and bacteriopheophytins (BPh) are different from the BChls only in that they are missing the Mg^{2+} ion at the center of the molecule. Ph and BPh are more easily reduced than Chl and BChl, respectively. Ph and BPh are therefore used as the end stations in the RC part of the electron transfer (ET) chain, immediately after the charge separation reaction.

Chl and BChl molecules have small differences in their side groups. Chl a and BChl a molecules are shown in Figure 15.2. For example, in BChl b, the $>CH–CH_2–CH_3$ group on ring II is replaced by $>=CH–CH_3$. In Chl b, the $–CH_3$ group is replaced by a formyl group, $–CH=O$.

The reorganization energies for electron transfer (ET) or excitation energy transfer (EET) are small for the Chls and BChls. Trapping of both electrons and excitations is shallow.

15.2.2 CAROTENOIDS

Other molecules of great importance in biological photosynthesis are carotenoids, discussed in earlier chapters (Figure 15.3). Carotenoids are active as light-harvesting pigments at considerably shorter wavelengths than the Chls (480–500 nm). Another very important role for carotenoids is to act as acceptors for triplet excitations occasionally formed in the photosystem. A third role in the oxygen-rich parts of the system is to react with excited singlet oxygen to form ground state triplet oxygen and triplet excited carotene.

Among the carotenoids, we find *xanthophyll*, an oxidized carotenoid with 11 conjugated double bonds. A hydroxyl or other oxygen-containing substituents have replaced one of the hydrogen atoms on a saturated carbon of the end groups (Figure 15.3). Xanthophylls are yellow and appear in egg yolk. When the green Chl color disappear from the leaves in winter, the yellow xanthophyll can be seen.

Ordinary xanthophyll absorbs in a region quite close to the Soret band (400 nm). Stacking moves the absorption closer to 500 nm. Other xanthophylls and carotenoids also absorb in the upper visible region. Thus, the accessory carotenoid pigments considerably improve the total absorption power of green plants.

15.2.3 PHYCOCYANOBILINS

Cyanobacteria, and green, red, and brown algae are responsible for about half of the photosynthesis on Earth. The different colors indicate that there are several different

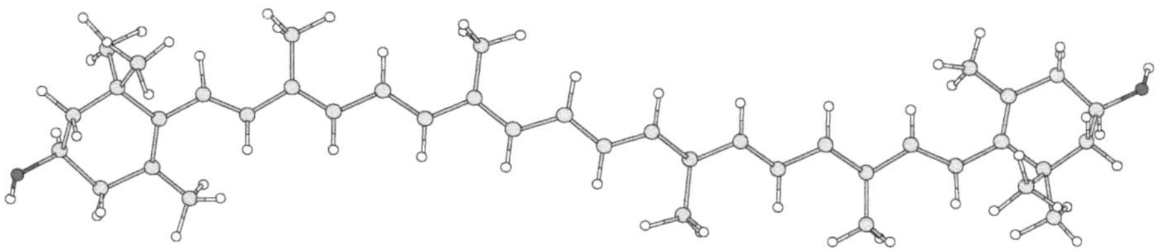

FIGURE 15.3 Xanthophyll, the yellow pigment of autumn leaves. The molecule is a planar π-system except at the ends.

active molecules involved as acceptors of light. Contrary to purple bacteria, cyanobacteria and algae produce oxygen. The pigment proteins phycocyanin, phycoerythrin, and allophycocyanin belong to the phycobiliprotein family of photosynthetic
proteins.

Early experiments on algae provided the basis for our present knowledge of photosynthesis. By illumination with very short pulses and measuring the production
of oxygen, Emerson and Arnold were able to show, in 1932, that antenna systems
exist as a large number of molecules, which together produce only a single oxygen
molecule in each flash.

Emerson and Arnold also showed that eight excitations (photons) are necessary
for each oxygen molecule produced. This is consistent with the fact that two electrons are necessary to raise the oxidation state of oxygen from -2 (as in H_2O) to
zero (as in O_2). There are two oxygen atoms per oxygen molecule and there are two
photosystems to be passed by the electron, thus $2 \times 2 \times 2$ excitations are necessary
to produce one O_2 molecule.

The name phycocyanin (phyco is Greek for algae) derives from blue-green algae
(cyanobacteria). Cyanobacteria and green plants derive their photosynthetic system
from primitive forms of a photosynthetic system that might be responsible for the
appearance of oxygen on Earth, 2.5 billion years ago.

The chromophores of the above-mentioned proteins are called phycocyanobilin,
phycoerythrobilin, and allophycobilin, respectively. Phycoerythrobilin is an accessory pigment chromophore in the green plant photosynthesis. The light energy har-
vested in the red region between 500 and 700 nm by phycoerythrin is passed on to
the PSI and PSII RCs. The structure of phycocyanobilin is shown in Figure 15.4.
This structure closely resembles structures determined together with proteins. It is
not excluded that there are a number of other stable structures for the molecule. The
one shown here is stabilized by the protein.

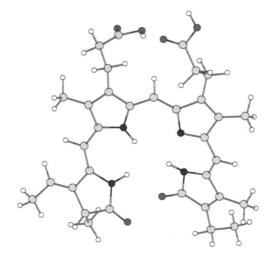

FIGURE 15.4 Phycocyanobilin, another pigment of photosynthesis. Notice the similarity
to the porphyrins.

15.3 ANTENNA SYSTEMS

The first antenna system (or *light-harvesting complex*) to be structurally determined belongs to purple bacteria (Cogdell et al., 1995). The remarkable fact with this antenna system is that it has evolved into circular shapes containing a certain number of identical chromophores.

15.3.1 PURPLE BACTERIA ANTENNA SYSTEMS

Biological antenna systems consist of a number of identical chromophores disposed in the membranes of chloroplasts. The amount of sunlight that can be absorbed is proportional to the number of molecules in the antenna system. In the antenna molecules, charge separation is energetically prevented due to the rather long distance between the chromophores (Figure 15.5).

The antenna system is thus a way for nature to increase the cross section for light harvesting of sunlight for the same RC. One single RC can take care of excitations from a large number of antenna chromophores, and carry out charge separation.

Charge separation (photoinduced electron transfer, PIET) is thus a competing mechanism to exciton formation by photoinduced excitation energy transfer (PIEET) only at a short distance between the chromophores. The longer distance for the electron to be transferred the larger the increase of free energy (the less the driving force). Therefore, charge separation is usually only possible within a sphere of a small radius around donor D.

Excitation energies of the locally excited states are the same for all chromophores in the antenna system (Figure 15.6). However, the excited states interact to form excitons, if the coupling is larger than the reorganization energy, as we have seen in Chapter 14. In a delocalized exciton, the reorganization (Stokes shift) tends to zero with the size of the delocalization region.

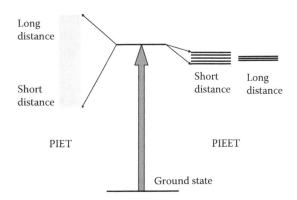

FIGURE 15.5 The significance of distance in the transfer of electrons (PIET) and excitations (PIEET) between identical chromophores. Charge separation (PIET) has a lower energy, the shorter the ET distance. PIEET has the same average energy for the final state, independent of distance.

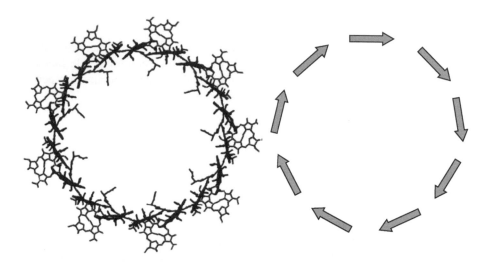

FIGURE 15.6 Part of the antenna system of purple bacteria. The figure on the right shows how the transition moments are coupled in the ground state.

In PIEET, the *average* final state energy is independent of the distance. The solvent or protein contribution to the Stokes shift is quite negligible. A high reaction rate may be obtained if $\lambda = -\Delta G^{\ominus}$. If we have PIEET between equal molecules ($\Delta G^{\ominus} = 0$), a small Stokes shift is beneficial for a high reaction rate. PIET, on the other hand, has a large solvent shift and is possible only at short transfer distance.

The lowest electronic state has the transition moments organized as in the right part of Figure 15.6. This state has the same symmetry as the ground state and seems to be delocalized in antenna systems. The transition to the ground state is forbidden. Possibly, the circular orientation serves as a "storage ring" for the excitation.

Another task of the storage ring may be to prevent the transfer of triplet states to the RC. Triplets cannot be transferred over long distances. If decay to spin triplet states takes place, the excitation may be considered as lost.

Transition to the ground state is symmetry forbidden, but the excitation is free to move to another storage ring. The inner ring seen in Figure 15.7 acts as a way into the RC situated in the middle of the outer ring.

The RC is the end station of the excitation. The very last site for excitation is the *special pair* (SP), where two chromophores are oriented so close to each other that the excitation is trapped. Charge separation becomes possible and occurs primarily within the SP. The excited electron subsequently continues to the accessory BChl. The time constant for this step is 1 ps. This time constant is thus a million times shorter than the typical time constants for ET in biological systems. However, it is considerably longer than in efficient photovoltaic cells.

The coupling, of course, is different in PIET and PIEET. In the latter case, as we have seen in Chapter 14, the coupling follows the Förster theory for singlet states, and may be written as an interaction between transition dipole moments. The probability for excitation transfer depends on the intensity of the absorption. The dependence

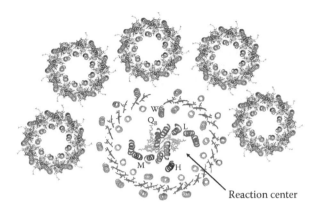

Reaction center

FIGURE 15.7   Antenna system of purple bacteria, determined by x-ray crystallography by R. J. Cogdell et al. The light-harvesting smaller circles (LH1) donate the excitation to the inner ring (LH2). LH2 donates the excitation to the special pair of the reaction center.

on distance decreases as $1/R^6$, thus much slower with distance than the exponential dependence in the case of PIET.

If the distance is about 10–15 Å, the excitations are usually mobile, even if not delocalized, with a rather large rate ($>10^6$ s^{-1}). At this large distance, charge separation is impossible. From the point of view of biological efficiency, it is appropriate if the antenna is made up of the same molecules as those that perform charge separation. The type of process that takes place thus depends only on the distance between the chromophores.

15.3.2   Green Plant Antenna Systems

The antenna pigments of PSII and PSI consist of Chl a, Chl b, and carotenoids. The antennas of green plants contain about 200 Chl molecules. The structure is known but complex. The *core antenna* contains Chl chromophores, some of which are also used in the charge separation and subsequent ET.

It has been possible to split off the core antenna system from the outer antenna and study it separately, both spectroscopically and by quantum chemical calculations. Eight chromophores are included in this split-off, six from the subunits containing the RC and two from neighboring subunits.

The spectrum of each chromophore may be calculated by using the so-called "time-dependent density functional theory (DFT) method," a method that has proven reliable and simple in previous calculations. The exciton coupling, the Förster coupling between the chromophores, may also be calculated. A calculation of the spectrum of the eight core chromophores is done subsequently. This calculation includes the eight chromophores and couplings in an 8×8 configuration interaction (CI) problem for the excited states. The previously calculated monomer energies are used as diagonal matrix elements in the Hamiltonian matrix. The calculated coupling matrix elements are used as nondiagonal matrix elements.

The diagonal matrix elements are different because of the different bonding angles of the side groups. The Ph excitation energies are different from the Chl excitation energies. One problem in such a calculation is that the coordinates are not well determined, since the crystallographic resolution is not yet sufficiently good. Another problem is that the excited states of the protein side groups in the vicinity of the chromophores also have to be included.

15.4   BACTERIAL REACTION CENTERS

Photoreactions that produce chemical energy by excitation of BChl or Chl molecules take place in RCs. The process is referred to as the *primary charge separation*. Purple bacteria use a type of photosynthesis that, to some extent, resembles green plant photosynthesis in PSII. In the 1980s, two purple bacteria, *Rhodopseudomonas viridis* and *Rhodobacter sphæroides*, reached a prominence that few had expected from species living at the bottom of ponds and similar places. Two German chemists, Johann Deisenhofer and Hartmut Michel, managed to dissolve the protein from the membrane, crystallize it, and determine its structure.

Earlier time-resolved spectroscopic work had allowed time constants to be determined. In 1978, the Russian photochemist V. A. Shuvalow determined a time constant for charge separation. At this time, no details were known about the geometry of the system where this photoreaction takes place.

The first determined structure of an RC was thus the bacterial one. In fact, this was the first membrane protein whose structure could be determined. The electron is transported in the RC with the help of chromophores of the same type used in the antenna system. The free energy is used to build up a proton gradient across the membrane.

15.4.1   STRUCTURE

The RCs of the purple bacteria, *R. viridis* and *Rb. sphaeroides*, turn out to be similar. In the former case, a cytochrome subunit is accompanying the subunits L, M, and H, which contain the chromophores. The cytochrome subunit contains four heme groups organized in such a way that the electrons can transfer through these heme groups and reduce the photoxidized chromophores. In other words: after excitation, there is a hole in the highest occupied molecular orbital (HOMO) and this hole is filled by an electron coming in from the cytochrome system.

The prosthetic chromophore groups show a high degree of local two-fold symmetry with the symmetry axis perpendicular to the membrane plane. It has been possible to determine that ET takes place in only one of the two almost equivalent sides, from the SP, via an accessory $BChl_A$ to BPh_A to a quinone on the active side. Index A is used for the active side and B for the inactive side. The A chromophores belong to the L subunit and the B chromophores to the M subunit.

The bacterial photosynthetic RC is called *RC type II*. To this group belongs the green plant RC of PSII. RC type II contains four BChl (Chl) and two BPh (Ph) chromophores, three protein subunits, and a number of molecules involved in ET. The latter cofactors are not covalently bonded to the protein. BChl (Chl) chromophores bind axially to the imidazol side group of histidine. BChl (Chl) and BPh (Ph) chromophores are anchored to the protein with the help of the phytyl chains.

Two of the BChl chromophores squeeze into a protein cavity and form a skewed SP dimer. The distance between the Mg ions of the monomers is 8 Å. The geometry is favorable for the formation of a low-energy excited state that contains both excitonic and charge transfer characters.

In *R. viridis*, the wavelength of the transition to the excited state is only 900 nm, corresponding to 11,000 cm^{-1} or about 1.4 eV. The absorption is quite wide, hinting at a large change in equilibrium geometry after absorption. The purple bacteria all absorb in the infrared and near visible region. Evidently, this decrease of the excitation energy is connected to the dimer formation. It is reasonable to conclude that bacterial RCs have evolved from the same ancient RC type II as PSII, and that the specialization of the dimer in bacterial RCs has improved the absorption of the stray sunlight deep in the water where the species is living.

In Figure 15.8, the protein structure is left out and only the six mentioned chromophores are shown, together with one ubiquinone molecule. *R. viridis* has menaquinone, a naphthoquinone, as Q_A. Phs have lower excitation energy and lower reduction potential than the other chromophores.

15.4.2 Charge Separation and ET

The SP donates the excited electron to the accessory BChl on the A side. The ways of avoiding the B side are very subtle. It seems to be related to a water molecule connecting the SP and the accessory BChl$_A$. In terms of evolution and reproduction, the system is created by dimerization of two monomers. However, the precise structure of the protein and the way the subunits are attached to each other is enough to cause great differences in ET properties.

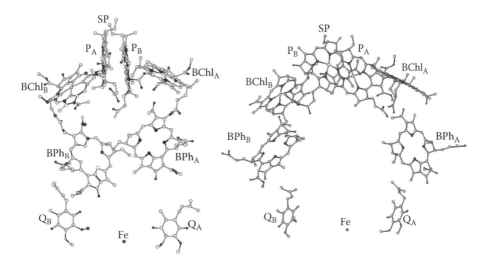

FIGURE 15.8 Bacterial RC (*Rb. sphaeroides*) from different angles. Only the cofactors are seen here. The protein structure was determined by Deisenhofer and Michel.

The BPh$_A$ chromophore is a safe haven for the electron. It cannot return to the accessory Chl because of a low Boltzmann factor. The electron could return to the SP, whose triplet state is at a lower energy. However, in this case, the electronic factor is too low because of a large distance. The average lifetime on BPh$_A$ is 200 ps. In the next step, it transfers to Q$_A$, a quinone molecule.

Between Q$_A$ and another quinone, Q$_B$, there is an Fe^{2+} ion. This ion is not involved in ET, but remains as Fe^{2+}. Instead, it appears to be active in connecting the imidazol side chains of the peptides located between the two quinones, Q$_A$ and Q$_B$.

The natural way of operation of a bacterial RC, permits only SP to be excited from the antenna system. It is possible to separate the RC from the antenna, however, and excite directly with higher energies. This way, the accessory BChl may become excited. Such an excitation also leads to charge transfer of type BChl$_A$$^+$–BPh$^-$ or SP$^+$ BChl$^-$ → SP$^+$ BChl BPh$^-$.

Normally, the electron excited in the SP moves quickly to BChl$_A$ (3 ps) and then to BPh$_A$ (1 ps). The short time compared to the time for the majority of biological ET steps (10^6 ps = 1 µs) is necessary to prevent side reactions such as the formation of triplets. The triplet state is, in fact, lower than the charge separated state when the electron is at BPh$_A$. Thus the triplet state of the SP can only decay to the ground state, meaning that the excitation energy would be lost.

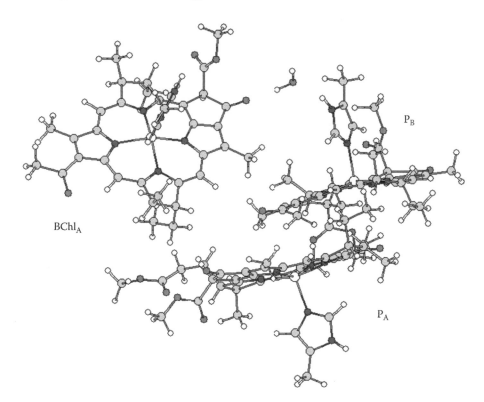

FIGURE 15.9 (See color insert) Detail of the bacterial RC. Notice the water molecule between BChl$_A$ and P$_B$.

The residence time for the electron on $BChl_A$ is very short (1 ps), very likely to avoid back transfer to the SP triplet state. The problem is the possible CDNAP process, mixing singlets and triplets, and making charge recombination to the triplet state possible.

There is orbital contact and therefore coupling directly between the BChls, between the SP and $BChl_A$, and between $BChl_A$ and BPh_A. The calculated couplings are consistent with the short residence times. The water molecule between the SP and $BChl_A$ (Figure 15.9) has nothing to do with electronic coupling, but may be helpful in the trapping on $BChl_A$. The SP and BPh_A can only couple via side groups and small molecules and this together with the large distance, lowers the magnitude of the coupling considerably.

There is thus very little coupling between the SP and BPh_A. The residence time on BPh_A is quite long (200 ps) and hence, if the coupling between the SP and BPh_A were large, the electron could back-transfer to the triplet of the SP and the excitation energy would be lost. Back-transfer via $BChl_A$ is impossible due to a large Boltzmann factor.

Both quinones are provided with an *isoprenoid chain*, which very likely functions biologically as an anchor of the molecule to the protein. In the same way, BChl and Chl molecules appear to be anchored by the *phytyl chain*. All four BChl molecules are attached to the molecule by an axial ligand, binding to the Mg ion.

15.5   GREEN PLANT PHOTOSYNTHESIS

In green plants and cyanobacteria, there are two different PSs, PSI and PSII, coupled in series. PSI and PSII are together capable of oxidizing water to O_2. This oxidation takes place in PSII. Typical for PSII is a manganese complex, called the oxygen-evolving complex (OEC), where water molecules enter and oxygen molecules exit.

15.5.1   PHOTOSYSTEM I

PSI contains six Chl a molecules and two phylloquinones in the RC. There is a C_2 symmetry axis. The symmetry is generally higher than in the bacterial RC and PSII. The two branches for ET, called the A and B branches, are both active as ET chains, but the rate of ET is a factor of ten faster in the B branch. The two quinones are symmetrically placed and function approximately as Q_A in the bacterial RC.

PSI and PSII both absorb at about 700 nm, corresponding to about 1.7 eV.

The light-dependent reactions begin in PSII. A Chl a molecule within the RC of PSII is excited and attains a higher energy level. In this state, an electron is transferred from one molecule to another. The electron flow goes from PSII to cytochrome b6f and further on to PSI. PSI fills the hole after a previous excitation and is excited to a state with a high free energy. The final electron acceptor is NADP. $NADP^+$, NADP, and NADPH are different from NAD^+, NAD, and NADH (Figure 11.6) in a third phosphate group bonded to one of the ribose groups, but not to the other phosphates. Two protons from OEC also take part in the formation of NADPH.

15.5.2 PHOTOSYSTEM II

In PSII, water is oxidized by O_2 in OEC. This type of photosynthesis is called *oxygenic photosynthesis*. OEC contains four manganese ions, in oxidation states varying between +3 and +4, and one calcium ion (Ca^{2+}). Oxygen may be regarded as waste product. In fact, if it happens to become oxidized to singlet oxygen, it threatens to destroy the system. The structure contains carotenes for protection. Nevertheless, the lifetime of PSII is fairly short (≈ 1 day).

The RC of PSII has great similarities with the bacterial RC (Figure 15.10). The monomers of the SP in PSII are almost as widely separated as in any pair of Chl molecules of the RC. Charge separation appears to take place in both the SP and in the Chl_A – Ph_A pair. In the latter case, the hole in the HOMO of Chl_A is refilled from the SP.

A tyrosine side group (Tyr_Z) is located only 5 Å from the OEC. When the SP is excited, the hole is refilled from Tyr_Z. The distance between Tyr_Z and the SP is almost 14 Å, which means that the ET is rather slow. It may be surprising that such a small molecular group as phenol is an electron carrier. The oxygen atom on phenol, however, is hydrogen bonded to one water molecule and two histidines (see Figure 9.6). There is coupling between ET and proton transfer in a way that is not yet fully understood.

The excitation moves by EET in the antenna system until it reaches the SP, which acts as a very shallow trap. Charge separation takes place within the RC, probably due to the slightly smaller distance between the chromophores. Subsequently, the electron moves to Phe_{D1} via the accessory Chl_{D1}. It seems that the ETs are only on one side (D1) of the dimeric RC, just as in the bacterial RC.

A carotene molecule is shown in Figure 15.10. Together with Cyt b559, this molecule is involved in the continued transport of the electrons.

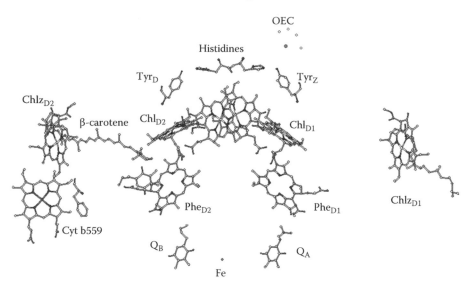

FIGURE 15.10 PSII reaction center. All protein structures and all hydrogen atoms are left out. One of the β-carotenes are shown.

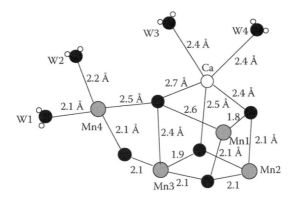

FIGURE 15.11 Oxygen-evolving complex (OEC). Wx are water molecules. Large white is Ca^{2+}. Dark is oxygen and grey manganese (Mnx). The picture is copied from the work of Umena et al.

In 2001, Zouni et al. obtained the first crystallographic structure. Later, Ferreira et al. (2004) improved the resolution. It should be added that the RC shown in Figure 15.10 is only a small part of the total structure determined. The structure of the OEC was not determined until 2011, by Umena et al. (Figure 15.11).

When the electron has left Tyr_Z, a neutral radical remains. It is believed that this radical picks up an electron and a proton from a water molecule and thus takes a direct part in the oxidation of water.

15.5.3 BINDING OF CARBON DIOXIDE: RuBisCo

The reduction of carbon dioxide takes place at a location different from the RC. Energy-rich ATP is used in the "dark reactions" of photosynthesis, which produce carbohydrates from CO_2. The photosynthetic net reaction may be written as

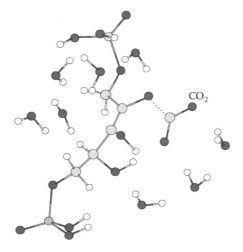

FIGURE 15.12 Ribulose-1-5-biphosphate forming a bond with carbon dioxide (dashed line).

$$CO_2 + H_2O \rightarrow H_2CO + O_2. \qquad (15.1)$$

H_2CO is an abbreviation for a carbohydrate. The reaction is a part of the Calvin cycle. The enzyme necessary to carry out the binding of CO_2 is called ribulose-1,5-biphosphate carboxylase oxygenase (RuBisCO). The mechanisms are not fully known. The cofactor ribulose-1,5-biphosphate, a rather small sugar phosphate molecule, binds to CO_2 after abstraction of a proton. A covalent bond is formed between RuBis and CO_2 (Figure 15.12).

The details of the full reaction are not fully known. ATP, an Mg^{2+} ion, and an arginine residue play active roles. The molecule formed decays into two phosphoglycerate molecules, which later form sugars and carbohydrates.

16 Metals and Semiconductors

16.1 INTRODUCTION

Metals are distinguished by their opacity, reflectivity (metallic shine, metallic mirror), and conductivity. All of these properties are related. Reflectivity depends on high conductivity. Conductivity is high since there is no energy gap at the Fermi level (HOMO–LUMO gap), between the occupied and unoccupied levels.

In amorphous form, most metals are black, and absorb light in the whole visible region. However, since most of the important excitations for conductivity have a much lower energy, it is possible to have a "white" metal, not absorbing at all in the visible region. Such metals exist and are important as an electrode material in photovoltaics. The reverse is also possible, of course, that a black body is lacking conductivity at a low temperature.

A piece of metal may be considered a very large molecule. The molecular orbital (MO) energies form a continuum in some energy regions and gaps in other regions. Normally, the energies of the valence orbitals form an energy band called the *valence band*, followed by an *energy gap* at a higher energy. Above the gap, there are unoccupied orbitals, forming the *conduction band*. In metals, the valence band is unfilled, so most of the conductivity takes place here.

In semiconductors, the valence band is filled. Conductivity appears only because of thermal activation across the gap. The conductivity is proportional to the number of carriers in the conduction band, according to the Boltzmann distribution law. Semiconductors are insulators at $T = 0$.

Theoretically, metals are best described in a model where the independent variable of the orbitals is the momentum of the electron rather than the position of the electron. Unfortunately, this difference often becomes a "language" difference between physicists and chemists. An attempt to bridge this difference will be made below.

Originally, "metal" was a term used to characterize certain elements in the periodic table. Here, we will adopt the physicist's point of view and call everything that is conducting, a metal or a semiconductor. A number of electron transfer (ET) systems are able to conduct electricity over a distance. Our primary concern is this interesting border region between ET and conductivity.

16.2 FREE ELECTRON MODELS AND CONDUCTIVITY

Wave packets are functions of either position or momentum. In the case of molecules and ordinary chemical phenomena, the momentum of a particle is of little interest.

"Local" descriptions are useful and fully sufficient. In the case of metals, on the other hand, a momentum space description is more useful and is the basis of concepts such as mobility and conductivity.

16.2.1 RESISTIVITY AND CONDUCTIVITY

According to Ohm's law from 1827, the current (I) is proportional to the voltage over the end points of the conductor:

$$V = IR. \tag{16.1}$$

In a wire with the cross-section A and length l, the resistance R is

$$R = \frac{\rho \cdot \ell}{A}, \tag{16.2}$$

where ρ is the resistivity.

Resistivity (ρ) or "specific electrical resistance" is a measure of how strongly an electric current is hindered in a material. *Conductivity* (σ) or "specific conductance" is the reciprocal quantity to resistivity and is a measure of the ability of a material to *conduct* electricity. Resistivity is measured in ohm meter (Ω m). Conductivity is therefore measured in ohm^{-1} m^{-1} (Ω^{-1} m^{-1}).

A typical feature of metals is that the lower the temperature the lower is the resistivity (Figure 16.1). For gold, silver, and copper metal, resistivity ranges between 10^{-6}–3×10^{-6} Ω cm (10^{-8}–3×10^{-8} Ω m) at room temperature. For graphite the resistivity is 0.3 Ω cm. The electron–nuclear (electron–phonon) interactions are responsible for the resistivity.

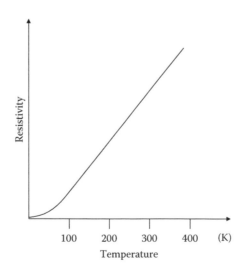

FIGURE 16.1 Conductivity of a metal as a function of temperature.

The resistivity of a metal tends to a value, known as the residual resistivity, as $T \to 0$. This value depends on the type of metal, the degree of purity, and the crystallographic structure. It has always been difficult to measure resistivity at very low temperatures, but as far as one can see, an ordinary metal does not completely lose its resistivity at very low temperatures.

Only *superconductors* lose all resistivity below a critical temperature (T_C). Superconductors are different in other properties too, such as perfect diamagnetism, and will therefore be treated in a separate chapter.

16.2.2 DRUDE MODEL

Soon after the electron was discovered in 1897, there appeared a model for metallic conductivity, developed by the German physicist Paul Drude. In this model, the electrons are moving as free particles. The positive ions left behind are treated as a homogeneous attractive field.

The most important property of a metal, that an electric field, no matter how weak, always causes conductivity, follows in the Drude model. However, there are other features of metals, particularly visible at a low temperature, where the Drude model fails. For example, resistivity as a function of temperature is incorrectly described.

In an electric field (E), the electrons move with an average velocity called the *drift velocity* (v_d). The electron *mobility* μ is defined as

$$v_d = \mu E, \tag{16.3}$$

where E is the electric field given in volt per meter, Vm^{-1}. The electron mobility is given in centimeter squared per volt-second, $cm^2(Vs)^{-1}$ (although the SI unit is meter squared per volt-second, $m^2(Vs)^{-1}$).

The conductivity is proportional to the product of mobility and carrier concentration. If we have n charge carriers per unit volume, the net flux of charge carriers per unit time is given by $n \cdot v_d$ and the current density $J = I/A$ by

$$J = nev_d, \tag{16.4}$$

where e is the electronic charge. The conductivity

$$\sigma = 1/\rho = I/A \cdot \ell/RI = J/E \tag{16.5}$$

is directly related to the mobility (μ), due to Equations 16.3 and 16.4.

$$\sigma = nev_d/E = ne\mu, \tag{16.6}$$

where e is the charge of the electron. Since $\sigma = J/E = I/EA$,

$$IR = JAR = Ae\sigma R = E\ell = V. \tag{16.7}$$

$I = J\,A$ is the current measured in amperes and V is the voltage over the conductor of length 1, measured in volts.

Even if we associate free electrons with wave packets, there are still problems with the Drude model. One is that the electrons in reality are subject to the Fermi–Dirac statistics. As $T \to 0$, the Drude behavior of conductivity is not the correct one. The Drude model, invented long before quantum mechanics, does not even distinguish between insulators and conductors. Inert gases would be conductors, since in classical models any small increase in energy should be possible. This is not the case in quantum mechanics, of course, and this is consistent with the fact that diamond and silicates are insulators at $T = 0$.

In 1933, Arnold Sommerfeld and Hans Bethe revised the Drude model. Their more complete, quantum mechanical theory goes under the name of the *free electron model* (FEM) for metals.

Before we leave the Drude model, we may derive the conductivity in terms of the collision theory. Let n electrons be moving in a certain direction at time $t = 0$; dn of these make a collision at time t, with constant probability P, within the time window dt. We obtain

$$dn = -P \cdot n \cdot dt. \tag{16.8}$$

Integrating this equation leads to

$$n = n_0 e^{-tP} = n_0 e^{-t/\tau}, \tag{16.9}$$

where $\tau = 1/P$. The average time between the collisions is given by

$$\frac{1}{n_0} \int_0^{\infty} n t\, dt = \tau. \tag{16.10}$$

The drift velocity of the scattered electron in the x-direction, v_{dx}, follows by integration of the equation of motion for the freely moving electron (m_e is the mass of the electron).

$$m_e \frac{dv_d}{dx} = -eE \Rightarrow v_{dx} = v_d - \frac{Eet}{m_e}. \tag{16.11}$$

The average velocity of all electrons in the x-direction is

$$v_{dx} = v_d - \frac{Eet}{\tau m_e} \int_0^{\infty} t e^{t/\tau} dt = -eE\tau/m_e. \tag{16.12}$$

The average velocity of v_d just after the collision is zero if we assume that the electron is moving in all possible directions after the collision.

To obtain the current density, we sum over all free electrons:

$$J_x = -\sum ev_{dx} = \frac{ne^2 E\tau}{m_e}.$$ (16.13)

The conductivity is thus given by

$$\sigma = \frac{ne^2\tau}{m_e}.$$ (16.14)

Not surprisingly, the conductivity is proportional to the average time between the collisions.

The remaining problem is to calculate τ. In an ET problem, τ is the average time for ET between the sites. The Drude model may thus be used to calculate the conductivity in an infinite ET system. This assumes that the transferring electrons are not strongly interacting with each other as in a metal, but this is usually not the case.

16.2.3 ATOMIC ORBITAL OVERLAP

To get a background for FEM, we may remind ourselves about how MOs are formed. Two atomic orbitals interact and form two new orbitals, one bonding at a lower energy and one antibonding at a higher energy. If another atomic orbital is added, there will be three molecular levels. If a great number (N) of atoms are lined up beside each other, there will be N MOs, with energies spread over a band (Figure 16.2).

The range over which the orbital energies spread is large for valence orbitals, but very small for core orbitals, since the overlap between the core orbitals on different sites is very small. The 3s and 3p orbitals overlap at a shorter distance (Figure 16.2).

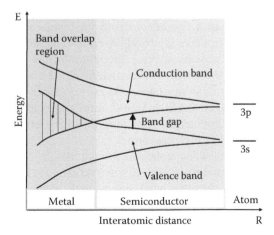

FIGURE 16.2 Interaction between the discrete energy levels on all atoms in a metal evolves into energy bands as the distance R between the atoms is decreased. (Model of J. C. Slater.)

This is the case, for example, for Mg. For Ca, the 4s and 4p orbitals overlap. There is a continuous energy band, but a band that is filled by only two electrons per atom. Mg and Ca are, therefore, metals. This case corresponds closely to the classical Drude model when electrons may take on any energy value. To make the simplest possible quantum mechanical model, we only need to treat the electrons as in the box model. We have to observe the Pauli principle and occupy the orbitals with two electrons until the actual total number has been reached. This is essentially the FEM for metals.

There is thus a wide, continuous, and unfilled band of valence orbital energies. If we have a crystal of noble gas atoms (He, Ne, Ar, Kr, Xe, and Rn) on the other hand, every orbital taking part in the interaction arises from an orbital that was filled in the atom. Since the number of MOs created by the interaction must be the same as the original number of orbitals, all orbitals will be filled. The orbitals above the gap receive no electrons at T = 0. For example, in the case of neon, the valence band created from the 2s and 2p orbitals will be filled with 12 electrons.

In the transition metals, for example, iron, the core orbitals (1s, 2s, 2p, 3s, and 3p) form narrow bands, which are all filled. The 3d, 4s, and 4p bands form wide bands. At the appropriate interatomic distance for iron metal, there is an overlap between the 3d and the 4s–4p orbitals on different sites. There is thus a set of orbitals of mixed 3d, 4s, and 4p character that forms a continuous band of orbital energies. This band is occupied only by the eight valence electrons per Fe site. The most antibonding orbitals of this valence band are empty, and this explains the cohesive forces of metals. The bonds, referred to as *metal bonds*, are type of covalent bond.

Within this band, there will be a possibility for excitation by any small amount of energy. Therefore, iron is conducting. The calculation of excitation energies is a very hard problem. The highest occupied molecular orbital (HOMO) – lowest unoccupied molecular orbital (LUMO) energy difference in Hartree–Fock (HF) does *not* correspond to excitation energy and the density function theory (DFT) gap has no meaning at all, in principle. The DFT gap tends to be smaller than the HF gap, and possibly even agrees with experiments if the expression for the exchange term is chosen "judiciously."

The only way to make a conductor from an insulator would be to apply extremely high pressure to make the bands overlap. The atoms would then come closer to one another and possibly overlap enough to form metals. Molecules with a sizeable energy gap to the lowest unoccupied level are also insulators at T = 0. For example, water (H_2O) is an insulator. The energy gap to the electrons that are antibonding between oxygen and hydrogen is simply too large and crystal formation causes only a small broadening. Condensed oxygen and sulfur, on the other hand, become conducting at high pressure.

Atoms that tend to form stable molecules are not metals, for example N, P, O, S, F, Cl, and Br. The condensed phase is insulating. Alkalis are metals under ordinary conditions. Hydrogen, on the other hand, has a single electron in its valence shell (1s), just like the alkali atoms Li, Na, K, Rb, and Cs. Two hydrogen atoms, form a very strongly bound hydrogen molecule and these molecules together have a much larger bonding energy than the same number of atoms would have had as a metal. If the pressure is increased, however, the hydrogen atoms are forced to be close to each other on all sides and start to resemble a metal.

Further down the periodic table, it is energetically favorable to form crystals even in the main groups V and VI. For example, bismuth (Bi) is a metal. In the large Bi atom, the enthalpy gained by forming a diatomic molecule is smaller than that gained by forming a crystal.

16.2.4 FREE ELECTRON MODEL IN ONE DIMENSION

FEM is a quantum mechanical treatment of the "electron in a box." The repulsive Coulomb interaction between electrons is assumed to be cancelled by a uniform positive background rather than by positive atomic nuclear point charges. This approximation is possible because we are only interested in the loosely bound valence electrons, among which we find the conduction electrons.

The Schrödinger equation (SE) is given by

$$-\frac{\hbar^2}{2m}\frac{d^2\psi}{dx^2} = E\psi. \tag{16.15}$$

The unnormalized eigenfunctions are determined by the boundary condition $\psi(L) = 0$:

$$\psi(x) = \sin kx; \quad k = \frac{n\pi}{L}, \quad n = 1,2,3,\dots; \quad E = \frac{\hbar^2}{2m}k^2 \tag{16.16}$$

We now instead let the electron move on a circular track, like a skier in a training tunnel. We have $L = 2\pi R$, where R is the radius. The relevant boundary condition is that the skier continues with the same speed when she arrives at the end of the tunnel, from where she started. The unnormalized eigenfunctions are

$$\psi = e^{ikR\varphi}; \quad k = \pm\frac{n}{R} = \pm\frac{n\cdot 2\pi}{L}, \quad n = 0,1,2,\dots \tag{16.17}$$

The eigenvalues are

$$E = \frac{\hbar^2}{2m}k^2; \quad k = \frac{m\pi}{L}, \quad m = 0,\pm 2,\pm 4,\dots\left(m = 2n\right) \tag{16.18}$$

thus the same expressions as in Equation 16.16 in terms of k. Since the allowed values of k are not the same in the two cases, we obtain slightly different eigenvalues and eigenfunctions. In a very large system, this difference is of no importance.

Since the energy is kinetic, we also have $E = p^2/2m$, which implies $p = \hbar k$. The momentum operator, Equation 1.13, is

$$\hat{p} = -i\hbar\frac{d}{dx}. \tag{16.19}$$

From this, it follows that only in the circular case are the eigenfunctions of energy also eigenfunctions of momentum. In the box case, we may define an expectation value over part of the phase space, but the expectation value over all space is equal to zero, since the particle moves back and forth. In the circular motion, the momentum is conserved.

The circular motion and the linear motion for $L \to \infty$ are analoguous. It is of great value to be able to describe metals in momentum space. Electrons are described as infinitely extended waves with a fixed momentum. Conductivity at $T > 0$ is hindered by scattering between electron waves and localized vibrational states.

Where is the HOMO–LUMO gap? The length of the box is L. Let us assume that we have N atoms and N electrons. The distance between the atoms is a. If we assume one electron per site, the equation

$$N = L/a \tag{16.20}$$

holds. The N electrons fill the N/2 lowest levels. According to Equation 16.18. the HOMO at $n = N/2$ has the energy:

$$E = \frac{\hbar^2}{2m}\left(\frac{N\pi}{2L}\right)^2 = \frac{\hbar^2}{2m}k^2 \quad \text{where } k = \frac{N\pi}{2L}. \tag{16.21}$$

The distance between the energy levels at $(N + 1)/2$ and $N/2$ is

$$\Delta E = \frac{(2N+1)h^2}{32mL^2}. \tag{16.22}$$

Since N is proportional to L, the difference in energy between the HOMO and the LUMO tends to zero as $1/L$.

FEM is a model, not an exact theory. One advantage with the simplification is that the Fermi–Dirac statistics may be applied to obtain the occupancy of the levels at any temperature. This is also true for the three-dimensional version of FEM, the Bethe–Sommerfeld model.

16.2.5 BETHE–SOMMERFELD MODEL

The potential for the electrons of the three-dimensional box is defined as

$$V(x,y,z) = 0 \quad \text{for } 0 \le x, y, \text{ and } z \le L \tag{16.23}$$

and $= \infty$ for x, y, or $z > L$. The SE in three dimensions is

$$-\frac{\hbar^2}{2m}\left(\frac{d^2\psi}{dx^2} + \frac{d^2\psi}{dy^2} + \frac{d^2\psi}{dz^2}\right) = E\psi. \tag{16.24}$$

The eigenfunctions are given by

$$\Psi(x,y,z) = \sin(k_x x)\sin(k_y y)\sin(k_z z). \tag{16.25}$$

The boundary conditions lead to the following energies:

$$E_{n_x n_y n_z} = \frac{\hbar^2 \pi^2}{2mL^2}\left(n_x^2 + n_y^2 + n_z^2\right) = \frac{\hbar^2}{2m}\left(k_x^2 + k_y^2 + k_z^2\right)m, \quad \text{where } k_i = \frac{n_i \pi}{L}. \tag{16.26}$$

The quantum numbers n_x, n_y, and n_z are natural numbers. We assume that there is one valence electron per metal atom and l atoms along the edge of the cube. The energy levels with

$$\left(n_x^2 + n_y^2 + n_z^2\right) \le R^2 \tag{16.27}$$

are occupied by two electrons each. Thus, all points with positive integer number coordinates inside a 1/8 sphere with the radius R ($n_x > 0$, $n_y > 0$, $n_z > 0$) are occupied. The number of points inside the 1/8 sphere is thus equal to the volume, divided by 8. Each point corresponds to an eigenvalue and two electrons. Since there are $N = l^3$ free electrons in total, the following equations must hold:

$$N = 2\frac{4\pi R^3}{3 \cdot 8} = l^3 \Rightarrow N = \frac{\pi R^3}{3} = l^3. \tag{16.28}$$

The energy difference between the HOMO and the LUMO may be calculated as the derivative dE/dN, where N is the number of electrons. For the HOMO, we have from Equation 16.26

$$E = \frac{\hbar^2 \pi^2}{2m(\ell a)^2}R^2, \tag{16.29}$$

where a is the bond distance (L = la) and R is the radius of the occupation sphere according to Equation 16.27. If we write dE/dN = (dE/dR)/(dN/dR) and use Equations 16.28 and 16.29, we obtain

$$\frac{dE}{dN} = \frac{\hbar^2 \pi}{Rm\ell^2 a^2} = \frac{\hbar^2 \pi^{2/3}}{3^{1/3}ma^2} \cdot \frac{1}{N}. \tag{16.30}$$

The HOMO–LUMO gap decreases inversely proportional to the number of electrons.

The *Fermi level* is located between the HOMO and the LUMO. In the present case, the Fermi level is thus localized in a continuous band of energy levels. The HOMO–LUMO gap is referred to as the gap at the Fermi level. In the case of a metal, the Fermi gap is equal to zero for all practical purposes.

The part of the valence band that is occupied at T = 0 is equal to E at the Fermi level, thus, according to Equation 16.29,

$$E = \frac{\hbar^2 \pi^2}{2ma^2} \left(\frac{3}{\pi} \right)^{2/3} . \tag{16.31}$$

The occupied band is extended between 0 and E.

16.2.6 CONDUCTIVITY IN A PERIODIC POTENTIAL AT T = 0

It is important to remember that metallic conductivity does not depend primarily on the delocalization of the orbitals, but on the fact that the Fermi level is in a continuous energy band. If orbitals are available at an infinitesimally small energy, a small electrical field causes an imbalance in the number of electrons moving with the field and against the field (we assume that the orbitals are eigenfunctions of momentum, as discussed above). More MOs are occupied if the electrons move with the field, and this leads to conductivity (Figure 16.3).

Crystalline insulators and semiconductors may also be said to have delocalized orbitals, but since there is a gap at the Fermi level, there are equal numbers of electrons with momentum in two opposite directions, even in the presence of a small electric field. The net conductivity is equal to zero. The only possibility to obtain conductivity in such systems is to lift the electrons across the Fermi gap to the conduction band. This costs thermal activation energy and hence the semiconductors are insulating at T = 0.

In molecular crystals, each molecule has doubly occupied orbitals (closed shells). The broadening of the energy bands is, in most cases, much smaller than the HOMO–LUMO gap. Electrons cannot move from one molecule to the next, so there is no conductivity.

The absence of a gap in the energy band at the Fermi level is a necessary but not sufficient condition for conductivity. The electrons or excitations may localize themselves if the reorganization energy is larger than the coupling between two neighboring sites. The orbitals are no longer eigenfunctions of the momentum and the band model cannot be used. In the following, we first assume that there is no localization

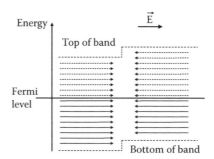

FIGURE 16.3 Energy levels in an electric field $\vec{E}$ for a metal. Dashed orbitals are unoccupied; full-drawn orbitals are occupied.

and that a description in terms of the delocalized orbitals is justified. Later, we will discuss the localization problem.

In a metal at $T = 0$, the electrons are thus moving as waves through the crystal, unhampered by nuclear motion. The delocalized electrons cannot cause any reorganization of the structure. We may assume that the energy is given by Equation 16.18 in one dimension and by Equation 16.21 in three dimensions and write

$$E = \frac{\hbar^2}{2m} \cdot k^2. \tag{16.32}$$

Equation 16.32 is the energy as a function of k for a free wave. The function $E(k)$ is a continuous band of energy levels. Models of this type are referred to as *band models*. From the above, it follows that the Fermi level is located at

$$k = \frac{\pi}{a}. \tag{16.33}$$

We have so far assumed that the electrons are completely free in the sense that the potential is flat (between the infinite walls). In a real case, there are atomic cores that may change the energy dependence on the wave vector. To account for the possibility that E is a function of k different from that given in Equation 16.32, we improve our FEM to account for the periodic atomic structure. Since this type of problem is studied in most textbooks on solid state physics, we will only sketch the results in Section 16.3.

In the Hückel model with constant β, which can be regarded as an improved FEM model, the Fermi level corresponds to the energy α. The top of the band for an infinite circular molecule is at $\alpha - 2\beta$ and the bottom of the band is at $\alpha + 2\beta$ ($\beta < 0$). K can be defined also for the Hückel model (see below). The orbital energy as a function of k is shown in Figure 16.4. As a comparison, the FEM parabolic eigenvalues are given in the same diagram.

We see that the modification is very large. However, since only the levels up to $k = \alpha$ are occupied, the main difference is an unimportant energy shift.

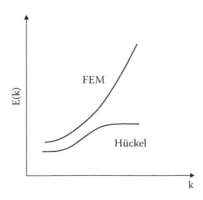

FIGURE 16.4   Comparison of the energy levels of FEM and Hückel.

16.2.7 CONDUCTIVITY AT ELEVATED TEMPERATURE

Resistivity is caused by electron waves scattering against the nuclei. The modes of the nuclei extend over the whole crystal and the energies form continuous energy bands. Whether they are delocalized or not makes no difference, however. In the ground state, all vibrations are in their lowest state. The electrons cannot be scattered unless vibrational modes are excited or deexcited. Both processes require some thermal agitation. The vibrational energy, $h\nu$, may assume any small value, however.

When an electron is scattered against a vibration it loses energy or gains energy. Just as in the case of the IR spectra, the selection rule is $\nu \to \nu - 1$ or $\nu \to \nu + 1$. The electron may also be assumed to have a continuous set of energy levels, as we see when we apply simple models, such as the FEM or the Hückel model, or more advanced ones for that matter.

It may be interesting to add here that in superconductors the majority of the conducting electrons have a gap to the first excited state. At low temperature, the electrons cannot be scattered against vibrations since the gap energy is too large. This is the important reason why superconductivity is possible.

Unfortunately, the FEM is not sufficiently accurate to derive an expression for the resistivity as a function of temperature. In an extended model, developed independently by Bloch and Grüneisen, the resistivity at temperature T, $\rho(T)$, is obtained as

$$\rho(T) = \rho(0) + C \frac{T^5}{\Theta^6} \int_0^{\Theta/T} \frac{x^5}{(e^x - 1)(1 - e^{-x})} \, dx, \qquad (16.34)$$

where Θ is the mean Debye temperature and C is a constant. This equation is referred to as the *Bloch–Grüneisen equation*.

At a high value of the temperature T, x is so small that the exponential of the integrand of Equation 16.34 can be Taylor expanded. The integrand is equal to x^3. Integration leads to T^4. Hence, $\rho(T) - \rho(0)$ is proportional to the temperature T.

At a low value of T, the upper integration limit tends to ∞, and the integrand tends to the function $x^5 e^{-x}$. The integral is finite (=5!) for $T \to 0$ and hence $\rho(T) - \rho(0)$ is proportional to the temperature to the power of five.

At high temperatures, the resistance of a metal thus increases linearly with temperature. As the temperature of a metal is reduced, the temperature dependence of resistivity follows a power law function of temperature. It may be shown that if defect scattering is included, the residual resistivity $\rho(0)$ tends to zero as T^n, where $n < 5$.

16.3 TIGHT-BINDING MODEL

The general description of a metal is in terms of orbitals, which are eigenfunctions of energy and momentum. The total wave function is, in principle, a Slater determinant. The Swiss physicist Felix Bloch developed the general band model.

16.3.1 ONE-DIMENSIONAL MODEL

Typical for atoms in metals is that they bond in many directions, even if only a few orbitals are available per atom. For example, in sodium metal there is only a single valence electron, which binds to the other atoms, primarily with the help of the 3s and 3p orbitals. We first limit ourselves to the one-dimensional case.

The one-dimensional system has a lot of similarities with the planar π-systems treated using the Hückel model in Chapter 3. In the case of solids, the Hückel model becomes the *tight-binding model*. Let us first assume that we have a one-dimensional metal containing a great number of atoms. Each atom provides one loosely bound electron, as in a linear π-system. Our experience tells us that instead of a linear system, we could use a cyclic system. The wave functions are different just at the end points, and our system is assumed to be so large that the end points do not matter.

The orbital energies are given by

$$\varepsilon_\ell = \alpha + 2\beta \cos \frac{2\pi\ell}{N}; \quad \ell = 0,\pm 1,\ldots,N/2, \tag{16.35}$$

for N even. The eigenfunctions corresponding to Equation 16.15 are

$$\phi_\ell = \sum_{\mu=0}^{N-1} N^{-1/2} \chi_\mu \cdot \exp\left(\mu i \frac{2\pi\ell}{N}\right); \quad \ell = 0,\pm 1,\ldots,N/2. \tag{16.36}$$

For benzene (N = 6) there is a single orbital for $1 = 0$. The next two, for $1 = \pm 1$, are degenerate. In Chapter 3, we added and subtracted them to get two real orbitals. In the present case, it is more convenient to keep the exponential form. The result agrees precisely with Equation 16.7 above and we may define momentum. Using

$$k = \frac{\ell\pi}{2L}; \quad \varepsilon_\ell = \alpha + 2\beta \cos \frac{2\pi\ell}{N}; \quad \ell = 0,\pm 1,\ldots,N/2. \tag{16.37}$$

The energy difference between the HOMO and the LUMO (after some calculations) is

$$\Delta E = 2\beta \sin \frac{\pi}{N} \approx 2\beta \frac{\pi}{N}, \tag{16.38}$$

which tends to zero as 1/N.

Since only the lower orbitals of the band are occupied, there is a gain of enthalpy equivalent to a *metal bonding*, creating the cohesive forces of the crystal.

16.3.2 Peierl's Distortion

One-dimensional crystals are subject to a distortion that leads to alternating short and long bonds. The half-filled valence band thereby develops a gap at the Fermi level. We have met this situation already for polyenes, The result is easily calculated using the Hückel model with bond-length-dependent coupling β.

The simplest polyene after ethylene is butadiene, with four carbon atoms. The simple Hückel model predicts two short bonds and a central long bond. The former have a slightly longer CC bond length than the double bond of ethylene. The central bond has almost the bond length of a single bond. The ordinary version of the Hückel model predicts less and less difference in bond lengths between short bonds and long bonds as the number of carbon atoms in the oligomer is increased. It we apply the bond-order-dependent coupling Hückel model the alternating bond-length pattern remains for any large polyene.

In Figure 16.5, a semi-empirical method has been used to optimize the structure. The bond-order-dependent β is automatically taken into account. With this method, there is alternation for the neutral molecule (1.449–1.344 Å in Figure 16.5). At the ends of the molecule, where the difference between long and short bonds increases. If even more accurate models are used, the bond length alternation is slightly less (0.08 Å).

Bond alternation in polyenes is an example of a more general theorem for one-dimensional crystals called Peierls' theorem. This theorem applies to systems like polyenes, where there is one orbital and one electron per atom, that is, a half-filled band. The theorem was first stated in 1955 by Rudolf Peierls. In the case of one-dimensional systems and half-filled bands, we use a proof by Lionel Salem. The Hamiltonian is expanded in a Taylor series for a geometry with equal bond lengths:

$$H = H_0 + \left(\frac{\partial U}{\partial Q}\right)Q + \frac{1}{2}\left(\frac{\partial^2 U}{\partial Q^2}\right)Q^2 + \ldots\ldots, \tag{16.39}$$

where Q is a small displacement of the nuclei (compare with the derivation of the Hellman-Feynman theorem in Section 4.2.5). First-order perturbation theory is used to obtain the total energy, using the truncated Hamiltonian of Equation 16.39:

$$E = E_0 + \left\langle \Psi_0 \left| \frac{\partial U}{\partial Q} \right| \Psi_0 \right\rangle + \frac{1}{2}\left\langle \Psi_0 \left| \frac{\partial^2 U}{\partial Q^2} \right| \Psi_0 \right\rangle Q^2 + \sum_{i \neq 0} \frac{\left[\left\langle \Psi_0 \left| \frac{\partial U}{\partial Q} \right| \Psi_i \right\rangle\right]^2}{E_0 - E_i} Q^2 + \cdots \tag{16.40}$$

The second-order Jahn–Teller term is the last of the Q^2 terms of the right member. This term may be calculated using the expressions given in Equation 16.36 for the wave functions. Since $E_i > E_0$, the denominator is negative, but has to compete

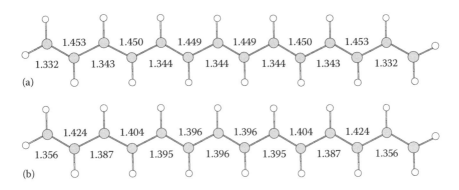

FIGURE 16.5 Oligomer of polyene. Geometry-optimized CC bond distances are given. (a) The neutral molecule and (b) The positive ion.

with the first Q^2 term, which is the ordinary parabolic term. We remember from Chapter 12 that the first excited state (i = 1) interacts strongly with the ground state and provides a large term, which is the essential energy gain due to the Peierls' theorem. Modification toward bond alternation thus gives a lowering of the total energy, as is seen in Figure 16.6. The period is doubled from a to 2a.

If there is one electron and one orbital per site, the band will be half-filled. We allow the period to be doubled (from a to 2a). We are free to move every second nucleus to decrease its distance to one neighbor and to increase the distance to the other neighbor. This leads to energy savings due to the second Q^2 term in Equation 16.40, based on the distortion of the bands in the vicinity of the new gaps.

Below k = π/2a, this distortion will cause a lower energy and above k = π/2a, a higher energy. This lattice distortion becomes energetically favorable when the energy gain by the second Q^2 term of Equation 16.40 outweighs the energy cost due to moving away from the previous equilibrium points. Energy savings due to the new band gap outweigh the elastic energy cost of rearranging the ions. The bond lengths thus become alternant and a gap is created in the band between the occupied and unoccupied energy bands.

Physicists call this appearance of short bonds and long bonds "dimerization," a notation that has very little to do with "dimerization" in chemistry. However, Peierls' theorem can also be applied to a stack of organic π-systems, for example, TCNQ. In this case, it is useful to talk about a multimer that forms dimers.

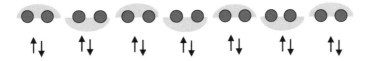

FIGURE 16.6 Peierl's distortion at half occupation of the band.

16.3.3   BLOCH BAND MODEL

From what we have learned above, particularly in Equation 16.36, a one-electron wave function in a crystal should be written as

$$\psi_k(\vec{r}) = u_k(\vec{r})\exp\left(i\vec{k}\cdot\vec{r}\right), \tag{16.41}$$

where u_k has the periodicity of the lattice:

$$u_k\left(\vec{r}+\vec{R}\right) = u_k(\vec{r}), \tag{16.42}$$

and $\vec{R}$ is a translation that leaves the system invariant. This form of Equation 16.41 was proven by Felix Bloch to be the correct form of the wave function and is referred to as Bloch's theorem.

In calculations, the solution ψ of a one-electron SE is approximated by a linear combination of atomic orbitals, $\psi_n(\vec{r})$:

$$\Psi(\vec{r}) = \sum_{n,\vec{R}} b_{n,\vec{R}}\psi_n\left(\vec{r}-\vec{R}\right), \tag{16.43}$$

where $b_{n\vec{R}}$ are variational coefficients and n is the index for the atomic orbital.

The eigenvalues appear as energy bands. The energy levels are functions of the wave vector $\vec{k}$:

$$\varepsilon_n = \varepsilon_n\left(\vec{k}\right) \tag{16.44}$$

The *density of states* is the number of states in a given energy interval.

Agreement with experiments, regarding the energies and shapes of the bands, is usually high. On the other hand, accurate band gaps cannot be expected to be well reproduced by a one-electron model such as the tight-binding model, which is, in principle, a model for the ground state.

Improved calculational schemes were originally developed by the American physicists John C. Slater and Conyers Herring, using plane waves as basis sets. Slater used a "muffin-tin potential," where the atomic wave functions inside the atomic spheres were fit to the plane waves. He was also the first person to use a viable local exchange approximation, which he called the Xα model.

J. C. Slater together with G. F. Koster developed a method called the Slater–Koster tight-binding method, which builds directly on Figure 16.1. This tight-binding model is a simplified MO model where the crystal orbitals are expressed in terms of atomic functions, as in the Hückel model. Later, basis sets of the same type as in quantum chemical calculations on finite systems have also come to be used in infinite systems.

16.3.4 EFFECTIVE MASS

In Chapter 1, we derived that the group velocity of a wave packet may be written as

$$v_g = \frac{1}{\hbar} \frac{d\varepsilon}{dk}, \tag{16.45}$$

where the band energy ε is given in Equation 16.44. In three dimensions, we have instead

$$\vec{v}_g = \frac{1}{\hbar} \nabla_{\vec{k}} \varepsilon(\vec{k}). \tag{16.46}$$

We want to write down Newton's equation for the electron in a metal and obtain

$$\frac{dv_g}{dt} = \hbar^{-1} \frac{d^2\varepsilon}{dkdt} = \hbar^{-1} \left(\frac{d^2\varepsilon}{dk^2} \right) \frac{dk}{dt}. \tag{16.47}$$

We now assume that a force F is acting on the wave packet. Newton's equation may be written as

$$F = \hbar \frac{dk}{dt} \tag{16.48}$$

and we obtain

$$\frac{dv_g}{dt} = \frac{1}{\hbar^2} \frac{d^2\varepsilon}{dk^2} F. \tag{16.49}$$

Thus

$$F = \hbar^2 \left(\frac{d^2\varepsilon}{dk^2} \right)^{-1} \frac{dv_g}{dt}. \tag{16.50}$$

We recognize this equation as Newton's law: force = mass × acceleration, provided that we define an effective mass m^* as

$$m^* = \hbar^2 \left(\frac{d^2\varepsilon}{dk^2} \right)^{-1}. \tag{16.51}$$

In the FEM model, the effective mass m^* is equal to the real mass of the electron. Narrow bands have a high density of states, a large absolute value of the second derivative, and a small value of the effective mass. It should be remembered, however, that in the narrow *filled* bands there is no electron motion at all.

Furthermore, if electron holes exist in narrow bands, there is often polarization around the hole. In that case, the band model cannot be used at all, in principle. An effective mass can still be defined, as in the Holstein model (see below). In the trapped case, the effective mass is very large.

16.3.5 CONDUCTIVITY IN ALLOTROPIC FORMS OF CARBON

Carbon is the atom of the periodic table that forms the greatest number of compounds. Pure carbon may also form large molecules and crystallize into a number of *allotropic* forms, such as C_{36}, C_{60}, and C_{70} from the large fullerene family, nano tubes (a tube-like form of a fullerene), graphite, and diamond. These substances have widely different conductivity properties.

A crystal of C_{60} molecules is an insulator. This is no surprise since the interaction between the atoms belonging to different C_{60} molecules is weak, while at the same time the HOMO–LUMO gap is large. The broadening of the bands due to intermolecular interactions is fairly small and hence the valence band is formed just by the HOMO, and thus completely filled. C_{60} belongs to the organic semiconductors that will be discussed in Chapters 17 and 18.

The next question is what happens if we oxidize a fullerene crystal. Will the hole (or electron if we reduce) be localized to a single C_{60} ion or will it be delocalized? This depends on the ratio between the coupling and the reorganization energy, as we have seen in Chapter 10. The reorganization energies (calculated by Duščesas) are of the order of 0.05 eV. The couplings are probably of the same magnitude or even larger (judging from quantum mechanical calculations). Fullerides of the type AC_{60} (A = alkli atom), therefore, be conducting. More electrons may be donated by making compounds of the type A_xC_{60} with $1 < x < 6$. Some of these compounds have proven to be metals at T = 0, while others are even superconductors.

Graphite occurs in pure form in nature. The name originates from the Greek to "draw" or to "write." Graphite is the "lead" in pencils, but, of course, it has nothing to do with lead. A layer of graphite may be regarded as "extended benzene," extended to a plane surface in all directions. The σ bonds of the plane are of the sp^2 type. The π bonds are perpendicular to the plane. The bonds are all of equal length. Graphite is therefore different from the one-dimensional polyenes. If the Hückel calculations with bond-order-dependent β are performed on finite systems, where the contour carbons are saturated by hydrogen atoms, the bond lengths converge to a single value, when the planar, hexagonal system is increased in size.

There is no energy gap at the Fermi level and this is consistent with the fact that graphite is an electric conductor. The $2p_z$ orbitals perpendicular to the molecular plane bind to form π bonds, as was described in Chapter 3. The bonds to other planes in a piece of graphite are noncovalent and weak. These bonds are of the London type, decreasing as $1/R^6$, where R is the distance between the planes. The London forces may physically be described as due to instantaneous electronic displacements that are responded to by the polarizability of the electron cloud from the next sheet. The London forces are of correlation type and depend on the polarizability of the systems that attract each other.

A single layer of graphite is referred to as a *graphene* sheet. Not surprisingly, graphene is conducting. Still, it is important to point out that closing the gap at the Fermi level does not *guarantee* conductivity. In principle, the removal or dropping of an electron could be localized. This is easy to check by calculations of a large graphite-like system. One way to do this is to replace one carbon atom with a nitrogen atom in the central part of the system. The HOMO represents the wave function of the additional electron. It is, in fact, delocalized over many atoms. The larger positive charge of the nitrogen as compared with the carbon nucleus prevents full delocalization. The LUMO is also delocalized. The delocalization remains if the system is structurally reoptimized. All these results are consistent with the fact that graphite is a metal.

Diamond is sp^3 hybridized and the valence band is filled. The band gap is large (5.4 ev) and there is no conductivity at T = 0. Diamond will be treated in Section 16.5.

16.4   LOCALIZATION–DELOCALIZATION

In the band model, the electron wave functions are *mathematically* written as delocalized wave functions. Most metals conform to this picture as far as metallic conductivity, optical properties, and magnetic properties are concerned.

The question of delocalization has nothing to do with the mathematical model used in the calculations of metals, but it is a *physical* problem. The test, as usual, is to remove or drop an electron and check whether it localizes or delocalizes over the whole system. If there is localization, the site geometry will deform. On the other hand, if it delocalizes, the geometry is exactly the same for all the sites.

An increase of the temperature may cause geometric changes and this may be enough to cause localization, as proposed in P. W. Anderson's model. However, metal–insulator (MI) transitions are not always caused by thermal disorder.

16.4.1   Metal–Insulator Transition

In the band overlap region (Figure 16.2), conductivity does not have to be activationless, since electron or hole trapping is possible and accompanied by a thermal barrier for ET. In a metal–insulator transition, the system goes from metal to insulator if certain parameters are changed, such as temperature, pressure, or doping level. The conductivity may change over several orders of magnitude. This means, of course, that the ground state wave function is not the same before and after the transition. If we look closely, site geometries also change significantly. The problem as far as the understanding of the electronic structue is concerned is that theoretical models used in the calculation of electronic structure may not tell whether there is metal–insulator transition or not. For example, if the band model is used the wave functions will always be delocalized, independent of possible metal–insulator transitions.

The lack of agreement with experiments is often blamed on the electronic correlation effects. However, this is not always justified. The wave functions for localized and delocalized systems are very different, but the difference is often in the mathematical form of the orbitals rather than in the number of configurations in the wave function. It is necessary to specify which type of correlation effects are present and this will be attempted next.

Perhaps the first-mentioned unexpected insulator was in a report by Boer and Verwey in 1937. Simple nickel oxide (NiO) was found to be a poor conductor of electricity. In fact, without impurities, NiO is an insulator at T = 0. This is no surprise to a chemist, who is used to local systems. NiO is surrounded by six O^{2-} ions. $Ni(H_2O)_6$ is also surrounded by six oxygen atoms, and even if the latter belong to water molecules, they may be regarded as O^{2-} ions for all practical purposes. NiO should thus have almost the same color as dilute solutions of Ni^{2+} salts and this is also the case (Figure 16.7). The color is the same: green-blue, certainly not black as would be expected for metals. The chemists make the right prediction, although this may be due to their disregard for the band model.

Boer and Verwey were committed to applying the band model. Obviously, the 3d of nickel is unfilled, and thus they expected a metal. After decades of futile attempts to improve the band model, John Hubbard took the pragmatic approach to simply add a positive number (U) to the unoccupied electronic levels, where the band gap appears experimentally (the Fermi level). Later, Neville Mott came up with an explanation. According to him, the system was indeed localized. The meaning of U is that energy is needed to transfer one electron from one site to the next. In NiO, we have to transfer one electron between two Ni^{2+} sites, to form Ni^{+} and Ni^{3+}.

It now appeared as justified to add U at the Fermi level. However, since the model was still the band model, not much was gained because the wave function is still incorrect.

Transition metal crystals with oxygen ligands (for example, MnO, FeO, CoO, NiO, CuO, nickelates, and cuprates with a single valence state +2 on nickel and

(a)   (b)

FIGURE 16.7   **(See color insert)** Nickel oxide (a) and aqueous solution of $NiSO_4$ (b). The green color is due to ligand field transitions in the NiO_6 complex.

copper) are referred to as "Mott insulators." Mott insulators do not disproportionate, but have a spin-coupled ground state.

16.4.2 POLARONS

Before we continue with the Mott insulators, we will introduce the concept of *polarons*. It is clear from earlier chapters that a localized electron deforms the structure around it. This deformation is a change of solvent structure and also bond length. If the electron jumps to another site, the same polarization occurs in the new site of course, while the old site goes back to the structure of all the unoccupied sites.

Physicists prefer to talk in other terms. A polaron is a *quasiparticle*, indeed it is an electron with an accompanying *polarization field*. One also speaks of *phonons* interacting with the electron. A quasiparticle, as the name indicates, is *almost* a particle. The phonon is the quasiparticle of the nuclear vibrations. The mass of the electron is replaced by an effective mass. The calculation of the effective mass is given in the Holstein theory, which will be discussed in Section 16.4.5.

It has been argued that polarization acts as a potential well. Unfortunately, such a treatment tends to obscure the difference between a potential well for the electron and the total energy potential that is part of the Marcus model. The nuclear reorganization of the nuclei appears to be omitted or built into the quasiparticle concept. Nuclear tunneling, leading to a larger rate at low temperatures, is easily confused with electron tunneling.

If the reorganization energy is large compared with the coupling, the mass of the quasiparticle is large. In an ordinary ET problem, this treatment becomes a bit clumsy. Chemists measure reaction rates or the mean time for ET. They are less interested in an equation of motion for a particle that stays at the same site most of the time.

Physicists also use the concept of a bipolaron, which is simply a pair of electrons with different spins, "dressed" by the accompanying polarization field. In chemistry, the bipolaron corresponds to a site where the metal ion has an oxidation state lower by two units. In a lattice of thallium sulfide (Tl_2S_3), an occasional Tl^+ ion with its different bond lengths, would be a bipolaron.

If carried through, the physicist's and chemist's theories should lead to the same results. Some concepts are definitely irreconcilable, however. Typical for ET reactions is the exponential dependence on the rate. At a decreasing temperature, nuclear tunneling leads to a weaker decrease of the rate and finally to a constant rate as a function of T^{-1}. This is interpreted as a tunneling through the Marcus barrier. In solid-state physics, this is referred to as "variable range hopping," which leads to a log(rate) dependence proportional to $T^{-1/4}$.

Nevertheless, the first theory of ET, which also recognizes the importance of structural reorganization and mark the beginning of the polaron theory, was due to two Soviet physicists, L. D. Landau and S. I. Pekar. In 1948, they calculated the effective mass of the polaron.

16.4.3 MOTT INSULATORS

In crystalline systems, the electrons localize if this is energetically favorable. In this case, the Bloch model should not be used. A well-known example is the transition metal oxides, for example, CoO or NiO. Since the 3d levels are incompletely filled in these oxides, the Fermi level is in the 3d band. Still, these substances are insulating at $T = 0$.

The wave function is very different between a localized case and a delocalized case. In the delocalized case, the eigenfunctions correspond to continuous band eigenvalues; however, in the localized ligand field case, there is an optical gap between the ground state and the excited state. There is thus no point in trying to calculate the wave function for a localized system using the band model, since the wrong wave function will be obtained as the end result.

Transition metal salts dissolved in water, for example, $NiSO_4$ or $CoSO_4$, have typical, quite weak ligand field absorption. We remember from Chapter 6 that ligand field transitions are dipole forbidden to first order. This is also the case for NiO. On the other hand, CuO and cuprates are black or brown, as are many other systems where sulfur or selenium are ligands. Traditionally, this is blamed on allowed metal–ligand transitions. However, at least in cuprates the absorption that covers the ligand field transitions has the properties of the Hubbard transitions and may, in fact, be ascribed to the Hubbard transitions.

Mott insulators are semiconductors since there are no electronic states at a low energy. The Mott insulators are antiferromagnetically coupled in the ground state. This means that the spins of the sites are alternating below a critical temperature, called the *Neél point*. In NiO, the Ni^{2+} ion is in a triplet state, corresponding to the high spin $3d^8$ configuration. The two triplet spins are correlated in the same way as the electrons in the case of H_2 (in the valence bond description). A description in terms of the Slater determinant of the band model is unsatisfactory.

A Mott insulator may become conducting after doping. If green NiO is doped with colorless Li_2O, a black substance appears with the formula $Li_xNi_{1-x}O$. In other words, Li^+ substitutes Ni^{2+} in some sites. To maintain electroneutrality, some Ni^{2+} sites have to be oxidized to Ni^{3+} (in air). Li^+ and O^{2-} cannot conduct electricity, so this has to be done by the nickel ions. In addition to the Mott reaction, where Ni^+ and Ni^{3+} are created, conductivity may now be obtained by electron exchange between Ni^{2+} and Ni^{3+} in an e_g orbital with the same spin. At room temperature, the conductivity increases proportional to the number of Ni^{3+} sites when the latter number is small. The blackness is obviously due to intervalence transitions. The latter transitions tend to be broad and apparently cover most of the visible spectrum.

If the system had become metallic, the hole at the Ni^{3+} site should be filled from a Ni^{2+} site. The new Ni^{3+} hole should be filled from other Ni^{2+} sites, etc., without any thermal barrier. Holes would have been created in the valence band, causing conductivity.

The doped NiO system remains a semiconductor, however, with a thermal activation barrier. This thermal barrier is the barrier for electron exchange of the Marcus model. The structural reorganization that takes place when an electron leaves the Ni^{2+} site and when an electron arrives at the Ni^{3+} site can only be accomplished by thermal agitation.

The same result is obtained for other doped Mott insulators. The conductivity is low if the rate of ET is low. In CuO and NiO, the number of antibonding e_g electrons is changed; hence, there is a large difference in some bond lengths between different oxidation states and, consequently, a large reorganization energy.

The simplest explanation for the optical properties of Mott insulators is that the excitation is localized. One Ni^{2+} ion with a surrounding O^{2-} is excited, while the others are left in their ground states. In the ligand field excitation of Ni^{2+}, one electron is lifted from a t_{2g} MO to an e_g orbital. The former is weakly antibonding and the latter is strongly antibonding between the metal ion and the ligands. Consequently, there is a substantial increase in some metal–ligand bond lengths. Large reorganization relative to the coupling leads to localization in the Marcus model.

The way to calculate a system of this type in principle is to vary the distances at a single site and obtain a local wave function. If localization remains after configuration interaction with the excited states in the ground state geometry, the system is localized. If, on the other hand, the difference in bond lengths disappears after configuration interaction with geometry optimization, the system is delocalized. This would be the case if we applied the same procedure to a nickel metal, for example. The lowest variational energy for Ni metal is for the delocalized wave function.

In cuprates, for example, La_2CuO_4, the copper atoms appear in a square grid called the CuO_2 plane. Doping may be performed by replacing some La atoms by Sr atoms, whereby some copper atoms in the CuO_2 plane are oxidized to Cu^{3+}. A metal–insulator transition appears at about 5% doping, but at low temperatures the system transfers to a superconductor rather than an ordinary metal. Cuprates will be further treated in Chapter 17.

16.4.4 SIMPLE MODEL FOR METAL–INSULATOR TRANSITION

In crystals, where some sites have unpaired electrons, these electrons may be localized or delocalized, depending on the influence of some external variable. If the system is localized, the electron has a certain residence time. The structure of a site with an electron is slightly different from a site without an electron. In Chapter 10, we learned that in this situation the reorganization energy λ is larger than twice the coupling, equal to $|\Delta|$. Suppose now that λ is a parameter that may be varied, while $|\Delta|$ remains constant. (Alternatively we may keep λ constant and vary $|\Delta|$.)

Let us assume a finite system with N identical sites and a single electron that enters one of the empty LUMOs on site i. Let us call this MO ψ_i. The eigenvalue problem looks like the eigenvalue problem of the Hückel model in Chapter 3, except that at site i where the electron is located, the diagonal element on site I is lowered by the reorganization energy λ, which represents the gain in energy by relaxation of the structure. The non-diagonal matrix elements are the couplings H_{12}. The next-neighbor coupling is the most important one.

The Hamiltonian matrix is diagonalized. The probability P_{1i} for the electron being on site i has now decreased, since the ground state wave function coefficients, C_v, on the other sites are different from zero. Thus, the remaining charge on site i after the first iteration is

$$P_{1i} = 1 - \sum_{v \neq i}^{N} C_v^2. \tag{16.52}$$

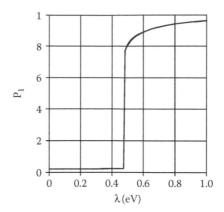

FIGURE 16.8   Metal–insulator transition, induced by varying λ.

In the next iteration, the diagonal elements have to be updated. The bond-length changes are proportional to the total occupation of the electron. The reorganization energy i is proportional to the bond-length change squared and hence the diagonal element corresponding to site i is scaled down by multiplication with P_{1i}^2. The vth matrix elements will be $\lambda \cdot C_v^4$. The non-diagonal matrix elements are the same as before. The eigenvalue problem is solved in the second iteration. The probability to be on site i is now further decreased. P_{2i} is obtained in an equation similar to Equation 16.52. A new Hamiltonian matrix is formed. The procedure is repeated until convergence is obtained. The result is shown in Figure 16.8 for different values of λ.

Figure 16.8 corresponds to an actual case. If the reorganization energy is decreased, the electron delocalizes over the sites for a certain λ. It is not possible to delocalize over only a few sites due to the C_v^4 power dependence in the matrix elements. The wave function is drastically changed at the phase transition (λ = 0.48 in Figure 16.8) and the equilibrium geometry is also changed.

If we go from the delocalized case and increase λ, nothing happens for small λ. After a further increase, *self-trapping* occurs on a certain site.

"Electron" may be replaced by "hole" or "excitation." A calculation of this type has to be carried out for the NiO system before it can be concluded if a hole is localized or an excitation is localized.

16.4.5   HOLSTEIN MODEL

The Holstein model is a model for structural reorganization just as the Marcus model. The Holstein model applies to a polaron localized in a lattice. The polaron is an electron moving with the lattice deformation caused by the electron. The polaron moves through the lattice with a certain effective mass. There are some similarities to the Marcus model, but also some differences. The Marcus model is, in principle, a many-electron model, since it assumes the existence of parabolic total energy surfaces. The Holstein model is a one-electron model that includes

FIGURE 16.9 Holstein model with a linear chain of diatomic molecules (N_2) with one added electron.

the reorganization energy in the effective mass of the electrons. The so-called molecular crystal model may be considered as a number of diatomic molecules lined up as in Figure 16.9 in the one-dimensional case. The molecule with the longer NN distance corresponds to N_2^-.

A certain deformation potential is assumed, but it is not clear how this potential should be parametrized to account for the actual reorganization energies when the bond lengths of the diatomic molecule are changed due to the presence or absence of an electron.

Contrary to the Marcus model, the Holstein model is expressed in k space. In principle, the correction terms in the Born–Oppenheimer approximation are calculated. The effective mass is proportional to J^{-1} where J is the electronic coupling between the sites.

16.5 SEMICONDUCTORS

Semiconductors may be said to be a similar type of system as metals in the sense that site localization does not occur. The band overlap region of Figure 16.2 has not been reached, however, and therefore the conductivity is activated. The inorganic semiconductors we describe here are different from the organic ones that will be mentioned in Chapter 18. In those discussed in this chapter, holes and electrons are delocalized and there is no or very small structural reorganization.

16.5.1 BONDING CONDITIONS IN DIAMOND

Diamond is an allotrope of carbon. The bonds are of sp^3 type, that is, each carbon is surrounded by four neighboring carbon atoms at equal bond length. The end atoms may be saturated by hydrogen atoms and the system geometry optimized.

The chemical bonding in methane was described in Chapter 3. The atomic C2s and C2p orbitals are sp^3 hybridized. The bonds are directed into four of the eight corners of a cube, with the carbon atom in the center. Bonds are formed with the H1s orbital. The bonding orbitals are a symmetric "a" orbital and three equivalent orbitals of t_2 symmetry at a higher energy. The corresponding antibonding orbitals have a node through the bond region and are found at a higher energy.

When the hydrogen atoms are replaced by four methyl groups, there are 32 valence MOs, of which 16 have a bonding character. The 32 valence electrons of $C(CH_3)_4$ occupy these orbitals and form a structure with a low enthalpy. There is still a large gap to the first orbitals, which are all partly or fully antibonding. As we continue to replace hydrogen atoms with methyl groups, the lower bonding orbitals and the upper antibonding orbitals form rather wide bands, but a gap remains between the lower and upper MO bands. Only the lower orbitals (the valence band) need to be occupied, as in all saturated organic molecules. The band gap is as large as 5.4 eV in diamond.

In Figure 16.10, one carbon atom has been replaced by a nitrogen atom. Compared to the case with just carbon atoms, there is now one additional electron. This electron cannot fit into the valence band, but is located in the Fermi gap. The corresponding MO is shown in Figure 16.10. It is important to notice that the HOMO and the LUMO are both delocalized. The HOMO is attracted by the positive core of the nitrogen atom. The HOMO electron is screening the N core to some extent and therefore the LUMO is considerably more delocalized than the HOMO. In any case, there is great delocalization in both the valence band and the conduction band.

Diamond has very high carrier mobility in the conduction band. This is consistent with the fact that, due to delocalization, there is very little reorganization energy.

The same electronic structure as for diamond is obtained for Si and Ge crystals, but the band gap is considerably smaller: 1.17 eV for Si and 0.74 eV for Ge. We replace Ge by its neighbors in the periodic table, to obtain gallium phosphide (GaP),

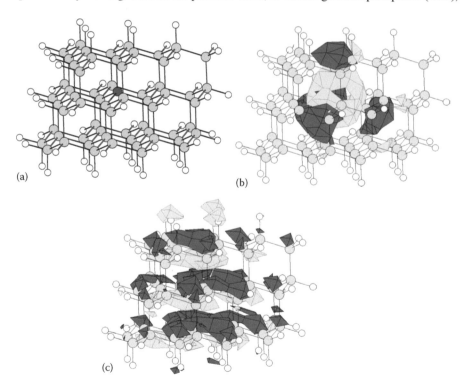

(a)

(b)

(c)

FIGURE 16.10 **(See color insert)** Piece of diamond saturated by hydrogen atoms and with a nitrogen atom (dark grey) replacing a carbon atom (a, left). In (b) we see the HOMO of a spin polarized calculation (center); the LUMO is seen in the right picture (c). Both the HOMO and the LUMO are considerably delocalized.

with a number of electrons per site. Some distortion is found among the MOs, but the band gap remains and is even larger than for Ge, or 2.32 eV. Other inorganic semiconductors are (band gap in electron volts in parenthesis) InSb (0.23), InAs (0.43), and AlSb (1.65).

16.5.2 CONDUCTIVITY AND DOPING IN SEMICONDUCTORS

Due to the band gap between the valence band and the conduction band (Figure 16.11), an applied electric field cannot make the electrons move, since all levels remain occupied in the valence band and all levels remain unoccupied in the conduction band. The only way is to raise the temperature to obtain a population of electrons in the conduction band. The lower electron levels, corresponding to electrons moving with the field, get the largest Boltzmann population. Since there are many empty orbitals in the conduction band (and some in the valence band), it is possible to form a wave packet that moves with the field.

The conductivity is proportional to the number of electrons in the conduction band, thus given by a Boltzmann factor. The conductivity (in a weak field) is thus proportional to

$$\sigma = C\exp\left(-\frac{\Delta E}{kT}\right), \tag{16.53}$$

where C is a constant.

Whether activation energy is necessary for the motion of the electrons in the conduction band depends on the electronic structure. In the inorganic semiconductors previously mentioned (silicon, GaAs, etc.), the electron is delocalized and no further thermal activation is necessary.

Semiconductors are used in the electronics industry. One important property is that the conductivity increases with the temperature (due to activation to the conduction band) while it decreases for metals. As can be understood from the band gaps, the conductivity at room temperature is very small for listed semiconductors. This

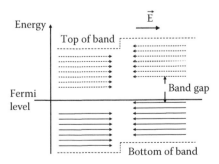

FIGURE 16.11 Energy levels in an electric field $\vec{E}$ of a semiconductor. Dashed orbitals are unoccupied; full-drawn are occupied.

can be changed by replacing the atoms in some sites with an adjacent atom in the periodic table. This is called *doping*. In Figure 16.10, doping has been performed by replacing one carbon atom with a nitrogen atom. The sp^3 valence band is filled and hence the additional electron of the N atom will appear just below the conduction band (since this orbital is singly occupied). This type of doping is called *n-doping* because there is a free particle (the hole) with a n(egative) charge.

Alternatively, diamond may be doped by boron. There is now one missing electron in the sp^3 valence band. The hole is quite delocalized, as the electron at n-doping. The positive hole is centered around the B atom, since the boron core is more negative than the carbon core, while the energy will be at the top of the valence band since the orbital is half filled. This type of doping is called *p-doping*, since the free particle (the hole) is p(ositively) charged.

The most useful semiconductor of this type (sp^3 hybridized diamond structure) is Si. Suitable doping atoms are phophorous, arsenic, and antimony in the case of n-doping, and boron, aluminum, or gallium in the case of p-doping. Silicium is considerably cheaper than diamond and has been used in transistors and other devices for many years. The most modern application is in solar cells. The absorption band is wide and in the visible region.

16.5.3   P–N JUNCTIONS

If p-type and n-type semiconductors are put into contact along a crystal plane, the n-type semiconductor acts like a donor and the p-type acts like an acceptor (Figure 16.12). This is called a p–n junction. The n-type electrons have a higher energy than the p-type holes and they fill the holes that are within reach. The p-doped semiconductor on the left is negatively charged and the n-doped semiconductor on the right is positively charged. There is a thin layer at the interface where the holes are filled and charges of different signs appear instead.

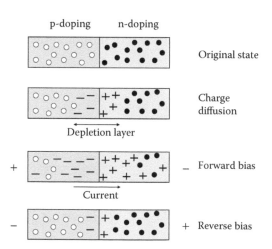

FIGURE 16.12   p–n junction with fields applied in different directions.

If an electric field is applied as in the third picture in Figure 16.12, the negative electrons are moved to the left. There is almost no trapping, so the charges move by fast electron exchange. In the n-doped part, the electrons fill the positive holes from the right, and when the depletion layer is reached, they fall into the hole on the p-side, which has become vacant after the electron moved away. Electrons thus move from right to left in the picture.

If the field is applied in the other direction, the electrons want to move the other way, but are stopped at the interface, due to the high energy that is necessary to reach the n-doped region. In the case of alternate currents, p–n junctions act as rectifiers (p–n junction diodes).

A light-emitting diode (LED) emits light when a voltage is placed over the electrodes, of which one is transparent. This happens in a p–n junction under forward bias, when the electrons are deexcited at the interface. The emission energy depends, of course, on the system. In the case of diamond, the emitted light is deep into the UV region at 235 nm.

In a boron p-doped semiconductor, there is a single electron in the HOMO, spread out around the boron impurity. This hole may become filled by an electron from another boron impurity. Thus, there are two possible states: the state where all boron centers are neutral and another state when the boron centers are alternatively positively and negatively charged. In Chapter 17, we will study the conditions for electron pair exchange between the B^- and B^+ centers. B^- is the carrier of an electron pair.

16.5.4   SOLAR CELLS

Semiconductors can be used as photovoltaic devices for harvesting of sunlight. The most common one is built with doped silicon (Figure 16.13). In the Si layer, electrons are photoexcited by sunlight into a broad conduction band. Electrons in a p-doped Si layer donate electrons to the holes created after the excitation. The electrode in

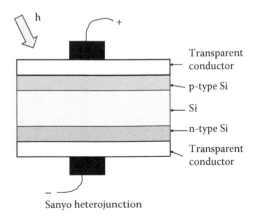

FIGURE 16.13   Simplified version of a so-called Sanyo heterojunction with 22% efficiency.

contact with this layer will therefore be positive. The excited electron is transferred to an n-doped Si layer. This electrode will be negatively charged.

Solar cells of this type can be used on the roof of a house. Coupled in a series, a high voltage over the end electrodes may be obtained. They are comparatively cheap to produce. In the future, it may become possible to harvest sunlight in deserts—where there is no photosynthetic activity anyway—on a very large scale.

16.6 PHONONS

The vibrations of the nuclei of a metal form delocalized waves in the same way as electrons or EM waves in open space. The word "phonon" corresponds directly to the word "photon" for EM radiation. "Phonon" means "sound" in Greek and was originally suggested by the Russian physicist Igor Tamm.

The eigenfunctions of the vibrational Hamiltonian are referred to as normal modes. The quantum mechanical normal modes are the same as the classical normal modes. Wave packets can be formed just as in the case of EM radiation. Heat is the source of both EM radiation and nuclear vibrations. The latter involve motion of atomic nuclei, heat motion, and also sound motion.

Sound has frequencies in the range of 10–10^4 s^{-1}, while vibrations in molecules are in the range of 10^{12}–10^{15} s^{-1}. In a crystal, the lower range of frequencies corresponds to *acoustic phonons* and the higher range to *optical phonons*. Optical phonons are similar to molecular vibrations. The smallest wavelengths have atomic or molecular dimensions. For both photons and phonons, the energy quantum is the Planck quantum, hv.

The size of the phonon is where the sound wave is, for example, a guitar string. The eigenfunctions are tones (or notes). A sound or a noise is a wave packet that can be written as a superposition of tones, called a Fourier series, involving different frequencies. The phonon is created as an excitation among the quantized energy levels and disappears in a deexcitation. Phonons play the major role in heat conduction, and the role of resistance in the case of electric conduction as we have previously seen.

17 Conductivity by Electron Pairs

17.1 INTRODUCTION

Electron pairs are as important in physics as in chemistry. Below a critical temperature, T_C, some metals become superconductors (SC) without resistance. Experiments show that the charge carriers are electron pairs. Calculated heat capacities using the Bose–Einstein statistics agree very well with the experimental results, and this is also consistent with pair formation among some of the electrons.

However, physicists and chemists tend to have different views on electron pairing. Physicists view electrons as free in metals, and feel obliged to find an attractive force that makes pairing between equally charged particles possible. To chemists, on the other hand, pairing is commonplace (bond pairs, lone pairs, disproportion). In most cases, electron pairs are strongly bound, and have large activation barriers for pair transfer, but there is no particular reason why this should be the case.

In this chapter, we will take the chemist's view, and use the same theory that we have used throughout this book. We will show that pair transfer is possible without activation energy under special circumstances, independent of the binding energy.

A conclusion will be that conductivity by electron pair transfer (superconductivity) is fundamentally different from ordinary metallic conductivity. The two lowest total electronic states are separated by an energy gap, contrary to the case in ordinary conductors.

17.2 SUPERCONDUCTIVITY

In 1911, the Dutch physicist Heike Kamerlingh Onnes discovered superconductivity. Onnes was also the first person to condense helium (1908). After he had made sufficient amounts of fluid helium, he used it to cool other substances to find out how they behave at very low temperatures. In particular, he was interested in the properties of metals at low temperatures. In 1911, it was known that the resistivity of metals decreases as the temperature is lowered. On the other hand, all motion should stop when the temperature reaches absolute zero, so the resistivity should be infinite. The latter view was held by one of Kamerlingh Onnes' friends, Lord Kelvin.

17.2.1 EXPERIMENT AND THEORY

The first experiment was carried out for mercury and the result surprised everyone (Figure 17.1). Not only did the resistivity decrease, but when the temperature went below a certain critical temperature, $T_C = 4.2$ K, all resistance disappeared

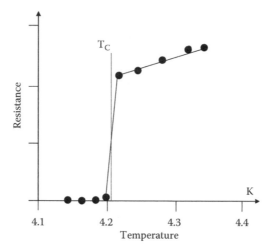

FIGURE 17.1   Resistivity as a function of temperature for mercury.

completely. This phenomenon became known as *superconductivity*. The critical temperature is the temperature at which superconductivity disappears on heating.

There appeared to be no explanation. To us, living 100 years later, the experimental results suggest some strange quantum effect. However, quantum mechanics had not yet been discovered in 1911, and even after 1926, the appearance of superconductivity remained mysterious for decades. Needless to say, superconductivity is an important test for theories, not least the theories presented in this book. In particular, it may be interesting to inquire about the importance of structural reorganization. The so-called *high T_C superconductors* have high chemical complexity and appear well chosen for the application of the Marcus model.

During the 1950s, it became clear in direct measurements that electrons conduct in pairs when the metal is an SC. In the case of fluid helium, Fritz London, whom we already know as the scientist who explained the chemical bond, pointed out (1938) that cooling involves the Bose–Einstein condensation in the case of superfluid helium, but he hesitated to extend this concept to superconductivity. In his impressive work on perfect diamagnetism, pair formation was not an issue. The first to point out an attractive interaction between the coupling of electrons in pairs and vibrations of the lattice was H. Fröhlich. We refer to books on superconductivity for details. The Bose–Einstein condensation was again mentioned by the chemist R. Ogg in connection with concentrated, highly conducting metal–ammonia solutions (1946), and by the physicists M. R. Schafroth and J. M. Blatt in the development of a molecular model of superconductivity (1954).

In 1957, Bardeen, Cooper, and Schrieffer (BCS) finally formulated a rough theoretical model for superconductivity. In this BCS model, the problem of electron pairing among free electrons was solved in a surprising way. The coupling between the electrons of the pair lies in the formation of the *Cooper pair*, which is held together by electron–phonon interactions. The BCS model explains the energy gap, since such a gap is a typical feature of the Cooper pair.

There were many experimental proofs of the BCS model. In one proof, N. E. Phillips (1959) compared the heat capacity of aluminum in the superconducting and nonsuperconducting phase at low temperature. In the latter case, superconductivity was destroyed by a strong magnetic field. Phillips found the expected heat capacity behavior as a function of temperature for the nonsuperconducting phase. In the superconducting phase, the heat capacity increased very rapidly from zero and reached a value much higher than normal, as the temperature approached the critical temperature from below. This behavior is typical for the Bose–Einstein condensation and depends on the rapid increase of the entropy as T approaches T_C from below.

The success of the BCS model continued. In connection with work on the so-called Josephson junction it was finally proven that the charge carriers consist of electron pairs.

17.2.2 MEISSNER EFFECT

Ordinary metals tend to be paramagnets. Each atom may be thought of as an isolated magnetic moment without electronic interaction with other magnetic moments. In 1933, the German physicists Walther Meissner and Robert Ochsenfeld carried out measurements showing that the magnetic field is equal to zero in an SC, already beyond a few atomic layers. This became known as the *Meissner effect*. Below T_C, an SC behaves as a *perfect diamagnet*. Fritz London showed that this is consistent with Maxwell's equations under a certain gauge.

In a diamagnetic material, there is magnetism only in an applied magnetic field. The direction of the induced field is opposite to the direction of the applied field, in contradistinction to the case for ferromagnets and paramagnets. In the applied magnetic field, an electric current is generated. Particularly in systems where such ring currents can be formed, for example, planar π-systems, the diamagnetic effect is large. This interpretation is consistent with shifts in the nuclear magnetic resonance (NMR) spectra. The currents are hard to measure in ordinary molecules, but can be calculated using accurate quantum chemical methods (D. Sundholm) using relativistic equations with the London gauge.

17.2.3 METAL–AMMONIA SOLUTION

At the beginning of the nineteenth century, the British chemist Davy first prepared alkali metal–ammonia solutions. Davy was also the first person to make alkali metals using electrochemistry. Ammonia condenses at −33.35°C and becomes solid at −77.7°C. Alkali and some other metals dissolve readily in anhydrous ammonia solutions. Already in 1908, on the basis of conductivity measurements, Kraus proposed that alkali ions A^+ exist in ammonia together with cavities containing a single electron (*solvated electrons*), in equilibrium with dissolved alkali metal atoms. At not too high concentration, alkali solutions are *all* deep blue, suggesting that the color arises from the electron cavities rather than directly from the metal ion.

At increased concentrations of alkali, the color changes to bronze-brown. In some temperature ranges, two phases exist. One is similar to the diluted phase and has the highest density of the two. The low-density phase has a higher concentration of metal. This is the bronze-colored phase.

The Australian chemist Richard Ogg cooled down the solution to liquid nitrogen temperatures and found indications that the solution was superconducting. At least, the conductivity was well *above* the conductivity of the metal ion.

At the same time, the magnetic susceptibility *decreases* as expected for electron pairs. This decrease was first believed to be due to pair formation in the cavities. Ogg suggested that some cavities contain two electrons. The German theoretical chemist Werner Bingel, on the other hand, suggested that species of type A^- are present (Figure 17.2). The difference to the Ogg model is that A^+ is present in the solution together with A^-.

Ogg went one step further. He reasoned that since the electrons are paired in the cavity, they follow the Bose–Einstein statistics at condensation, as earlier discussed by F. London for superfluidity. However, the idea of cavities with two electrons turned out to be unrealistic. Quantum chemical calculations, where the two electrons were assumed to move in a flat cavity potential with steep walls, show that the electron pair has a positive energy compared to the background and therefore is

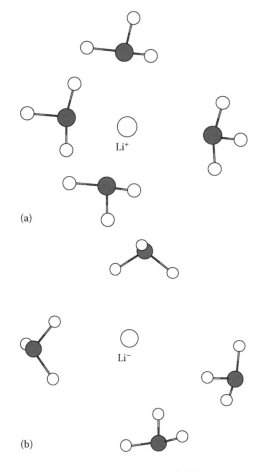

FIGURE 17.2 Bingel model. Geometry optimized $Li(NH_3)^+$ (a) and $Li(NH_3)^-$ (b).

unstable. The $2e^-$ cavity model is unrealistic from another point of view. The acceptance of a pair of electrons means that the cavity has to be built up from scratch every time an electron pair arrives and this means enormous reorganization of the energy for electron pair transfer. The Ogg model was deemed unrealistic.

The model of Bingel still appears as quite realistic (Figure 17.2). One could explain the high conductivity of concentrated alkali-ammonia solutions as due to the simultaneous presence of A^+ and A^-. The British chemist James Dye made similar conclusions in a comprehensive work on electrides and negatively charged metal ions.

In the Bingel–Dye model, the electron pair would form A^-. An important question is whether A^+ can receive an electron pair without great changes in the structure. If this is the case, the pair would move to A^+ and further on to new A^+ sites and the high conductivity could be explained. Semi-empirical calculations suggest that the size of A^+ and A^- is moderately different (Figure 17.2). However, this type of calculation is too primitive to make any firm conclusions. In any case, disproportionation in A^+ and A^- is realistic at a high concentration and this is in agreement with the susceptibility and conductivity measurements of Ogg.

17.2.4 COOPER PAIRS AND THE BCS MODEL

L. N. Cooper found on general grounds that there is a motional coupling between electrons and phonons in a metal. This coupling creates a gap between the two lowest energy levels. The ground state is one of concerted or coherent motion between electrons and nuclei. Cooper's paper is short and to the point and contains no details regarding the physical nature of the pairs. The lowest state above the gap is one where the electron pairs lose their coupling, and move independently with lost coherence.

Popular explanations of the Cooper pair formulate the coupling in the following way: "When the first electron passes, the nuclei are perturbed, and this simplifies for the second electron in its motion." However, many other possibilities exist for interpreting the Cooper pair, if such an interpretation is needed.

Why do electrons in some cases behave as single electrons and sometimes as a pair of electrons when the temperature is lowered? The mechanism for the Cooper pair explains some of the characteristics of superconductivity. If there is a gap, the pairing properties become visible because pairs of electrons obey the Bose–Einstein statistics and all condense into the lowest state. It costs energy to excite across the gap, therefore there is no resistance at low temperature.

The Cooper pair model is the central part of the BCS model. For a long time, the latter model was the only accepted microscopic model for superconductivity. The BCS model is a free-electron model that correlates the motion of electrons and nuclei. Electronic correlation is not in question.

The initial success of the BCS model has not continued in later years. Some elemental metals, for example, copper, silver, and gold, have so far refused to become supercomputing as $T \rightarrow 0$ K, and there is nothing in the BCS model to suggest why this is the case. But worse problems were to come.

New superconducting metals were discovered at a high rate, particularly by B. Matthias and his associates. In 1973, $T_C = 23.2$ K was reached for Nb_3Be and this appeared to be the record critical temperature for a long time. Theoreticians concluded that critical temperatures above 25 K can be regarded as impossible on the basis of the BCS model.

In 1986, Bednorz and Müller, and somewhat later Chu and Wu et al., discovered superconductivity at considerably higher temperatures than 25 K in cuprates. The latter reached 92 K in the cuprate $YBa_2Cu_3O_7$ (Figure 17.3), which physicists referred to as "123" or "YBCO." Twenty-five years later, there is still no accepted improvement of the BCS model that could explain high T_C superconductivity. Some scientists even believe that there is a new kind of superconductivity where pair formation is due to other interactions than phonon interactions.

A remarkable fact is that the nonactivated transfer of electron pairs occurs in compounds such as cuprates where structural reorganization is important in the description of electron transfer according to the Marcus model. This type of interaction is not considered at all in the BCS model. Further, there is no detailed and conclusive interpretation of the spectrum of cuprates. It therefore appears important to try to specify in more detail which type of electron–nuclear interactions might possibly be

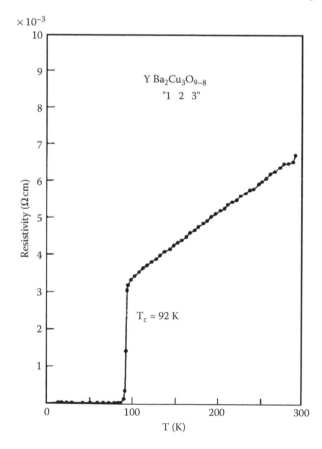

FIGURE 17.3 Resistivity of a single phase $YBa_2Cu_3O_7$ sample as a function of temperature. The picture is taken from the Nobel lecture of Bednorz and Müller in 1987.

of importance. We will start with a local description in terms of the Marcus model, and later study what is necessary for delocalization.

17.2.5 HIGH T$_C$ SUPERCONDUCTIVITY

Bednorz and Müller showed that a cuprate, La_2CuO_4, becomes superconducting at $T_C = 35$ K if some lanthanum atoms are replaced by barium atoms. This "doped" compound may be written as $La_{(2-x)}Ba_xCuO_4$. The oxidation state for La is +3, for Ba +2, and for oxygen −2. The oxidation state of copper, y, is thus determined from

$$(2-x)\cdot 3 + x\cdot 2 + y - 4\cdot 2 = 0 \Rightarrow y = 2 + x, \qquad (17.1)$$

where $y > 2$ corresponds to the presence of some Cu^{3+} ions.

The superconductivity appears in the so-called CuO_2 plane (Figure 17.4). In this plane, Cu^{2+} ions sit in the corners of squares, all with oxygen ions between them. Each copper ion belongs to four squares and every oxygen atom to two squares. Therefore, there are two oxygen atoms per copper atom. The CuO_2 plane has two negative charges per copper atom. In the undoped cuprates this is compensated by positively charged planes between the CuO_2 planes. The excess charge x, due to doping, appears as Cu^{3+} ions in the CuO_2 plane. The CuO_2 plane becomes superconducting at a critical molar concentration of Ba, $x_C \approx 0.05$. Ordinary copper oxide (CuO) remains a semiconductor at low temperatures, irrespective of doping.

Very soon after Bednorz and Müller's discovery, it was found that if Sr is used as a doping substance instead of Ba, the critical temperature is even higher. If the CuO_2 plane is absent, or if Cu^{2+} is replaced by Zn^{2+}, superconductivity disappears. Zn^{2+} has a full d shell, while Cu^{2+} is missing one d-electron for a full d shell. Superconductivity must be connected to the missing electron in the plane. Yet, the BCS model has nothing in it where "almost complete shells" or the "dopant concentration" are important. The BCS model can only be applied to systems where the

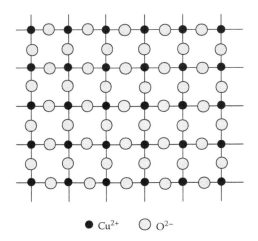

$\bullet$ Cu^{2+} $\bigcirc$ O^{2-}

FIGURE 17.4 CuO_2 plane.

electrons are free electrons with anonymous wave functions. A variation in chemical constituency is not an issue.

The term "high T_C superconductivity" is unfortunate since the decisive difference is not the critical temperature, but the type of system where superconductivity is found, for example, in oxides. In this sense, the cuprates were not the first ones. Reduced $SrTiO_3$ was known long before 1986, although with a very low critical temperature. Johnston et al. discovered superconductivity in ternary titanium oxides where the oxidation state of Ti is below +4. Sleight et al. found superconductivity in $BaBiO_3$, where some Pb^{4+} has replaced Bi ions. In both cases, the critical temperature was above 10 K. However, the highest measured critical temperatures are still in cuprates: 134 K under ambient pressure and 164 K under high-pressure conditions.

It is interesting that superconductivity appears for $T < T_C$ in doped cuprates, since ordinary cuprates are insulators, lacking conductivity at $T = 0$. Similar systems remain insulators or semiconductors. Doped CuO is semiconducting with a quite low activation barrier. NiO and nickelates with Ni in the +2 valence state are semiconductors, thus insulators at $T = 0$. In the Mott insulators, the Hubbard U parameter, already discussed in Chapters 10 and 16, is necessary to explain the spectra and the lack of conductivity. Is this phenomenological parameter hiding some important physics?

17.3 COUPLING AND CORRELATION IN ELECTRON PAIR TRANSFER

The theoretical problem is to understand electron pair transfer in the limit when Hubbard $U \to 0$ or even becomes negative. We will find that in this limit there is interaction between two many-electron states, the charged state, and the spin-coupled state. We will find that the motion of electron pairs is only possible in this "correlated" situation. We will also find that strong coupling to the vibrational states (phonons) is directly related to electron pair mobility.

17.3.1 MOTT'S JUSTIFICATION OF HUBBARD U

John Hubbard was the first person to note the inconsistency between theoretical band calculation and experimental conductivity data in the case of transition metal oxides. Without much ado, he simply added a band gap of the size U at the Fermi level. With $U = 2$ eV, the electronic transitions start at 2 eV instead of 0 eV, which, as it turns out, is a considerable improvement of the previous theoretical numbers.

N. Mott (later Sir Neville Mott) was rightly concerned and tried to justify Hubbard U. He came up with a model where the energy U is the energy needed to transfer one electron from one site to an equivalent site (thereby, in fact, assuming that the system is localized). It should cost in energy to move an electron to an equivalent site. In chemical terms, the "reaction" is as follows for NiO:

$$2Ni^{2+} \to Ni^+ + Ni^{3+} \quad \left(\Delta G^\ominus > 0\right). \tag{17.2}$$

The nearby Ni ions change their configuration from $3d^8$ to $3d^9$ and $3d^7$. Mott considered that U essentially consists of the energy $U = \langle 3d|1/r_{12}|3d \rangle$, the electronic repulsion between two electrons in a 3d orbital on an Ni atom, but this definition was later found to be less than precise.

If $|\Delta G^{\ominus}|$ is too large for thermal transition between the phases, optical transitions may take place between the spin-coupled states and the charged states. As Equation 17.2 shows, *Hubbard transitions* are, at the same time, *intervalence transitions*.

An important question is whether Hubbard U can be calculated or obtained experimentally, for example, from the spectrum. In almost all transition metal compounds, a "charge transfer edge" can be seen at 3–5 eV, but this refers to charge transfer transitions from the ligand valence subshell to the empty 3d type orbitals on the same site. The intermetal charge transfer transitions, however, start at an energy U, which is strongly dependent on the metal ion. It is interesting that only for cuprates, U appears to be lower than the energy of the site-local charge transfer transition.

Attempts to calculate U lead to diverging results. If U is calculated according to the original definition, $U = \langle 3d|1/r_{12}|3d \rangle$, very large numbers of the size U = 20 eV are obtained. In temperature, this corresponds to kT > 200,000 K, so U = 20 eV is obviously a meaningless number. Other methods have been used, but will not be commented on here.

17.3.2 APPLICATION OF THE MV-3 MODEL

Consider two different sites, M^+ and M^-, in a disproportionated system. We first assume symmetric geometry of the environment, so that M^+M^- has the same energy as M^-M^+. Since the spin-coupled state MM is unstable, two electrons may be transferred at the same time from M^- to M^+, forming M^+ and M^-. This is one of the rare situations when a transfer of electron pairs is possible (and, in fact, the only possibility) in a chemical system.

Disproportionated systems have been discussed in Section 10.7 above. Several elements have many accessible oxidation states, as well as "missing" oxidation states. Equation 17.2 simply expresses that a disproportionation reaction has the free energy $\Delta G^{\ominus}$. In the CuO case, however, there is no disproportionation ($\Delta G^{\ominus} > 0$). In Figure 17.5, we show the same Marcus parabolas as in Figure 10.18, but in this case $\Delta G^{\ominus} > 0$. The oxidation states Cu^+ and Cu^{2+} are well known. Cu^{3+} is known from $NaCuO_2$ and a few other similar compounds, and as impurities, as mentioned above. In most other cases, copper has the oxidation state +2 (or +1 or 0). Judging from spectral data, the Cu^{2+} disproportionation reaction may be written

$$2Cu^{2+} \rightarrow Cu^+ + Cu^{3+}; \quad \Delta G^{\ominus} \approx 2 \text{ eV}, \quad (17.3)$$

showing that Cu^{2+} is more stable than $Cu^+ + Cu^{3+}$ and does not disproportionate. In this case, there is a problem in measuring the reduction potential for this reaction, since Cu^{3+} does not appear together with the two other oxidation states in solution at an electrode.

Figure 17.5 may be constructed if force constants k, the reorganization energy λ, Hubbard U, and the coupling H_{13} are known. Figure 17.5 shows that U, in principle, can be defined alternatively as the free energy difference between the equilibria. The vertical energy is essentially the energy of a metal–metal transition. Since U is positive by several electron volts, this is consistent with the absence of conductivity.

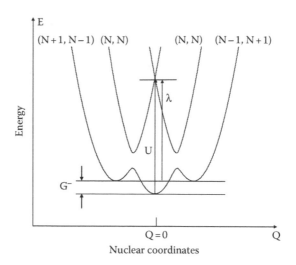

FIGURE 17.5   The MV-3 model for the case $\Delta G^{\ominus} > 0$ (no disproportionation).

Why is the ground state stable with equal valence? A very good reason is the antiferromagnetic spin coupling between the Cu^{2+} ions. In the same way, Ni^{2+} is a spin triplet with a large spin coupling between adjacent sites. Another reason is that it appears as energetically unfavorable to two electrons, repelling each other, on the same site. Still, this is not an unusual situation in reality.

Figures 10.16 and 17.5 are typical, fundamental Marcus models for electron pair transfer and the connected structural reorganization. The Holstein model can possibly be extended to include Hubbard U. This would not be an easy task, however. The two important obstacles are the inclusion of structural rearrangements more accurately than in the present Holstein models, and dealing with the correlation problem.

We may choose H_{13} and Q_0 to fit previously calculated PES. If the ground state is the MM state with $Q = 0$ as the equilibrium point, we are talking about a *spin-coupled* ground state, as for example in CuO, NiO, or in the cuprates. This is the anti-ferromagnetic state, sometimes referred to as the "spin density wave" (SDW) state.

In another chemical situation, the ground state may correspond to site occupation different in two electrons for example +3 and +1. In this case, the outer parabolas form the ground state. We call this state the *charged state* or charge density wave (CDW) state. In this case, U is small, but not necessarily negative.

If we start out from the situation shown in Figure 17.5 and decrease Hubbard U (for example, by special kinds of doping), the spin-coupled state MM is the ground state until the two parabolas M^-M^+ and M^+M^- are of the same level as MM, that is all three states have the same equilibrium energy. If this can be done, there is a strong interaction between the three degenerate states. This interaction continues to be important as long as U, measured from the MM parabola as in Figure 17.5, is positive. When $U < 0$, there is no longer strong interaction. The strongest interaction occurs for $\Delta G^{\ominus} = 0$. Below, we will show that the interaction between the M^-M^+, MM, and M^+M^- states is decisive for conductivity.

We call the region from $U = \lambda$ to $U = 0$ the *state overlap region*. Electron pair transfer is only possible under the additional condition that the charged state is the ground state.

17.3.3 INTERMETAL COUPLING

What is causing the coupling between the M^-M^+, MM, and M^+M states? There is also a coupling to the vibrations, but we suspect that this coupling may be derived from the correction terms to the Born–Oppenheimer approximation.

The direct coupling matrix element H_{12} between the two CDW states M^+M^- and M^-M^+ is normally equal to zero. This can be shown by just writing down the matrix element and calculating it. The book by Michl and Bonačić-Koutecky, listed in the references of Chapter 12, gives further details.

The interaction matrix element H_{13} between a charged state and a spin-coupled state is important. This is the matrix element for electron transfer between the M^+M^- state and the MM state, and between the M^-M^+ state and the MM state. Thus, H_{13} provides effective coupling between the two charged states, M^+M^- and M^-M^+, making it possible for electron pairs to be transferred between two adjacent sites. Outside the state overlap region, on the other hand, there is much less interaction and mixing due to nonmatching energies.

The wave functions of the charged state and the spin-coupled state are essentially the same as for the hydrogen molecule in Chapter 3. If the hydrogen 1s orbitals are called a and b, the wave function for the Heitler–London ground state is ab + ba. The wave function ab – ba is the triplet state and thus of no interest for electron pair transfer. The (unnormalized) wave function for the lowest charged state may be written as

$$\Psi_{C0} = \cos \varphi (aa + bb) + \sin \varphi (aa - bb). \tag{17.4}$$

In Figure 17.5, $\varphi = \pi/4$ ($\cos \varphi = \sin\varphi$) corresponds to $\Psi_{C0} = aa$ (left minimum) and $\varphi = -\pi/4$ ($\cos\varphi = -\sin\varphi$) to $\Psi_{C0} = bb$ (right minimum). For $Q = 0$, $\Psi_{C0} = (aa + bb)/\sqrt{2}$, while the excited state is $\Psi_{C1} = (aa - bb)/\sqrt{2}$.

The spin-coupled MM (unnormalized) wave function, corresponding to the ground state in cuprates, is

$$\Psi_S = (ab + ba). \tag{17.5}$$

The coupling between the spin-coupled state and the charged state (H_{13}) is hence calculated as

$$H_{13} = \langle (ab + ba)|H|aa \rangle \text{ or } H_{13} = \langle (ab + ba)|H|bb \rangle \tag{17.6}$$

(normalization factors are left out). The nature of H_{13} is a state interaction matrix element rather than an orbital interaction matrix element between overlapping orbitals from different centers, as in the case of electron transfer. This interaction may be extended between pairs over the whole lattice.

The interaction leads to the following orthogonal wave functions (C_1, C_2, D_1, and D_2 are coefficients):

$$\Psi_S = C_1 \left(aa + bb\right) + C_2 \left(ab + ba\right)$$

$$\Psi_A = aa - bb$$

$$\Psi' = D_1 \left(aa + bb\right) + D_2 \left(ab + ba\right). \tag{17.7}$$

The ground state is Ψ_S. Since the interaction between aa and bb is vanishing, the first excited state is Ψ_A. Using Ψ_S and Ψ_A, it is possible to form a time-dependent wave function that starts from aa at $t = 0$ and then oscillates between aa and bb, that is, between the sites. If $|C_2|$ is much larger than $|C_1|$, which is the case when the spin-coupled state is the ground state, this oscillation will be insignificant.

If the coupling matrix element H_{13} is large, the gap between Ψ_S and Ψ_A is large in magnitude (several millielectron volts or several tens of reciprocal centimeters). If there is a strong interaction between the charged state and the spin-coupled state, it is therefore possible that there is no longer any activation barrier between the minima in Figure 17.5. The above-mentioned electron pair moves as a free electron pair between the sites. The transition between Ψ_S and Ψ_A ought to be seen in the spectrum, but has relevance for the motion of the electron pair only if the ground state wave function is dominated by the aa + bb component. Furthermore, if the Ψ_A state is thermally populated, it should be possible to see the transition $\Psi_A \rightarrow \Psi'$.

Transitions between M^+M^- or M^-M^+ and the MM state are transitions in which the valence state is changed for adjacent metal ions and thus intermetal transitions. The energies depend strongly on the metal and may vary energy from typical small energy differences between the aa + bb and aa − bb states to Hubbard transitions of type ab + ba → aa − bb or aa + bb in the infrared, visible, or UV region.

The ground state is formed by the symmetric Ψ_S state. If $|H_{13}|$ is sufficiently large, this state delocalizes over the whole system as a coherent Ψ_S state. Ψ_S involves only a subset of the electrons of the system. Still, such a subset of freely mobile, bound electron pairs becomes visible in properties such as conductivity, diamagnetism, and heat capacity. The local Ψ_A state cannot be reached at a low temperature because of the gap. Local excitations of the type $\Psi_S \rightarrow \Psi_A$ can only take place as the temperature approaches the critical temperature. When this happens, the coherent Ψ_S state and the properties associated with it are ruined. The $\Psi_S \rightarrow \Psi_A$ excitations are two-electron excitations and create the Bose–Einstein scenario that we see. Single electrons follow the Fermi–Dirac statistics, of course, but since there are no single electron excitations available at a low energy, we do not see any trace of the Fermi–Dirac statistics.

17.3.4 STABLE CHARGED STATES

We may first study the limit case when the charged state is the ground state and no spin-coupled state exists below the crossing point of the M^+M^- state and the M^-M^+ state. One such case is the insulator $TlBr_2$. The difference in bond length clearly suggests that there are two different sites, corresponding to Tl^+ and Tl^{3+}. The Tl atom has the configuration $[Xe]5d^{10}6s^26p$, hence Tl^+ is $[Xe]5d^{10}6s^2$ and Tl^{3+} $[Xe]5d^{10}$. The difference between Tl^+ and Tl^{3+} is in the $6s^2$ electron pair. The $Tl^{2+}Tl^{2+}$ state is very high in energy and hence there is no coupling between the Tl^+/Tl^{3+} and Tl^{3+}/Tl^+ states. The $6s^2$ pair is immobile.

An often-heard argument is that no difference in density can be measured between Tl^+ and Tl^{3+} (as well as other similar systems, for example, Bi^{3+} and Bi^{5+}, Pb^{2+} and Pb^{4+}). Figure 17.6 shows the simplified picture in the case of Tl. The outer spherical Tl^+ shell contains roughly the $6s^2$ electrons. The inner shell contains the core electrons of the $[Xe]4f^{14}5d^{10}$ configuration. Tl^{3+} does not have any outer shell and its radius is therefore a bit smaller than the Tl^+ ion. The inner shells of the two ions are very similar. At the same time, it should be remembered that 6s electrons have a very small but nonzero electron density at the nucleus, which slightly increases the density at the Tl^+ nucleus. Which ion has the greatest density at the nucleus can only be determined by an accurate *ab initio* calculation. We conclude that the density in the core region is almost independent of the oxidation state.

Tl^{2+} would have a single electron in the 6s orbital. This electron would demand almost the same space as a doubly occupied 6s orbital. If the 6s orbital is empty, however, the ligands have a chance to come closer, decreasing the total energy. This is the main reason why Tl^{2+} spontaneously disproportionates into Tl^+ and Tl^{3+}. Tl^{2+} is a "missing" valence state. The stabilization of the Tl^+ and Tl^{3+} valence states has nothing to do with the Cooper pair formation (interaction with the phonons). Instead, it is related to the Born model, which prefers an unequal charge over the same charge (see below).

In the band model, the occupied 6s band lies at the top of the other bands. The lowest unoccupied band is also a 6s band, the one formed from the Tl^{3+} 6s orbitals. This means that the lowest electronic transitions go from $Tl^+(6s)$ to $Tl^{3+}(6s)$, in other words, the Hubbard transition, forming two Ti^{2+} ions. There is a question of how the bands are going to be calculated. Generally, the difference in energy between an occupied and an unoccupied orbital cannot be interpreted as excitation energy. It is highly questionable whether band models can reproduce the

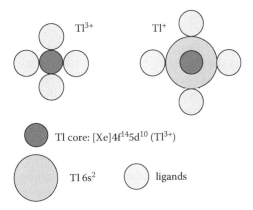

FIGURE 17.6 Schematic picture of the charge distribution in Tl^+ and Tl^{3+}.

spectrum with the help of the energy differences between the occupied and unoccupied bands.

The bismuth atom is another system with a missing valence state (+4). The ground state of $BaBiO_3$ is a charged state, since the average valence state is a missing valence state. The difference in BiO bond length is as large as 0.18 Å, signifying disproportionation, in Bi^{5+} sites and Bi^{3+} sites. Bi has the configuration $[Xe]4f^{14}5d^{10}6s^26p^3$. Consequently, Bi^{3+} has the configuration $[Xe]4f^{14}5d^{10}6s^2$ and Bi^{5+} has the configuration $[Xe]4f^{14}5d^{10}$. The case of $BaBiO_3$ is parallel to the $TlBr_2$ case just described.

The case isovalence-electronic with $BaBiO_3$ is K_2SbCl_6, a salt referred to as Setterberg's salt, which was discussed in Chapter 10. The average oxidation state for Sb in the salt is +4, but this is a "missing valence state." Absorption, giving the strong color, is ascribed to $Sb^{3+}Sb^{5+} \rightarrow Sb^{4+}Sb^{4+}$ by Day et al. The understanding of the intervalence spectrum is based on the local model.

17.3.5 SPIN-COUPLED STATES

In the spin-coupled state, the local spins are coupled antiferromagnetically in the same oxidation state. This is the ground state and the excitation energy to the charged state equal to U. However, the U transition is a broad transition and there must be some vibronic states at a much lower energy. There are no transitions other than those that have to do with the spin coupling and are mostly forbidden. Furthermore, the magnetic states cannot take part in the charge transfer.

The magnetic coupling discussed between sites has nothing to do with the coupling between magnetic moments, for example, between a compass and the magnetic field of the earth. The interaction between the magnets on different sites is a purely electronic interaction. P. W. Anderson termed this interaction *superexchange*. Typically, the electrons are localized and have a net magnetic moment on each site. For example, in MnO the five 3d electrons point in the same direction and form a spin sextet ($2S + 1 = 2 \cdot 5/2 + 1 = 6$). The spins on different sites are correlated in the sense that all spins on the neighboring sites point in the opposite direction. Anderson defined a coupling between the sites via the intervening, common ligand. This coupling has nothing to do with the electron transfer coupling.

Why doesn't Cu^{2+} disproportionate? One important reason is that the difference in occupation is not in the 4s orbital, but in the much smaller 3d orbital. In fact, the 4s and 5s orbitals do not get occupied as easily as the 6s orbital. This may have to do with the relativistic effects becoming important in the lower part of the periodic table. The valence states Zn^+ and Cd^+ are, in fact, "missing," while Hg^+ and Tl^+ exist. Ga^+ and In^+ are very "rare," meaning that it is hard to reduce Ga^{3+} or In^{3+} to Ga^+ and In^+, respectively. Ga^{2+}, In^{2+}, and Tl^{2+} are all missing, since a system with a single s orbital occupied is very unstable.

In the spin-coupled cases with equal valence, the coupling forces are of the same type as in ordinary chemical bonding, say between two nitrogen atoms. The only difference is that the bonding forces are smaller since there is usually an O^{2-} or S^{2-} in between the metal ions.

Cuprates are typical spin-coupled states. The ligand field transitions for Cu^{2+} are expected in the visible region of 1.5–2.5 eV. The ligand–metal (LM) charge transfer transitions are expected to appear at about 3–4 eV with a high intensity. In an undoped cuprate, there is additional intensity in the 1.8–2.5 eV region. This transition leads to photoinduced conductivity and is therefore very likely the U transition. Cu^{2+} behaves very differently from Ni^{2+} where the U transition is at a slightly higher energy, in a range where we also find the LM charge transfer transitions.

"Doping" of a cuprate effectively means that some Cu^{2+} ions are oxidized to Cu^{3+}. Also, a reduction to Cu^{+} leads to superconductivity. The Cu^{3+} sites are very likely localized. However, the Cu^{2+}/Cu^{3+} exchange may still have a quite low time constant (1 ps to 1 ns).

17.4 MV-3 SYSTEMS IN THE STATE OVERLAP REGION

The spin-coupled state and the charged state may become almost degenerate. In a series of compounds, such as Li_3C_{60}, Na_3C_{60}, K_3C_{60}, Rb_3C_{60}, and Cs_3C_{60}, the charged state (C_{60}^{2-}, C_{60}^{4-}) might be stable in one end of the series and the spin-coupled state (C_{60}^{3-}, C_{60}^{3-}) in the other end. Almost degeneracy somewhere in between leads to an interaction between the charged states and the spin-coupled states. This interaction will now be further investigated.

17.4.1 CALCULATION OF HUBBARD U: BORN EFFECT

As was mentioned above, Mott originally suggested

$$U = \langle 3d | 1/r_{12} | 3d \rangle \qquad (17.8)$$

as a definition of Hubbard U in NiO. It is assumed that the upper ligand field orbitals are pure Ni 3d orbitals. Since the ligand orbitals involved are fairly delocalized e_g^* molecular orbitals (MOs), the repulsion energy obtained should be considerably overestimated if atomic orbitals are used in Equation 17.8.

An improved interpretation is to write

$$U = I - A, \qquad (17.9)$$

that is, Hubbard U is the difference between the ionization energy (I) and the electron affinity A. This corresponds to the physics of the Mott process: Take one electron from one site, which obviously costs the ionization energy I, and add it to the neighboring site, whereby the electron affinity is regained. This process is acceptable in free atoms and molecules. In a liquid or solid, this is not possible and the definition has to involve polarization due to the surrounding molecules. This means that one should use reduction potentials, as has been done in Chapter 10.

In the case of C_{60}, I and A have been experimentally determined and we obtain $U = 7.6 - 2.7 = 4.9$ eV. Gunnarsson has subtracted the band gap $E = 1.6$ eV and obtained $U = 3.3$ eV. The magnitude of the polarization has been debated. We will use the Born equation, particularly since this equation is sensitive to the lattice constant, which is the distance between two C_{60} molecules in the lattice.

The site charges are originally the same, equal to Z. After disproportionation, the charges $(Z - 1)$ and $(Z + 1)$ alternate. The Born radius R can be taken as the lattice constant. In atomic units, the Born free energy (Section 6.2.2) difference between the spin-coupled state and the charged state is

$$\Delta G\left(Born\right) = -\frac{\left(Z-1\right)^{2}+\left(Z+1\right)^{2}-2Z^{2}}{2R} = -\frac{1}{R} \tag{17.10}$$

(neglecting the Born exponent). Thus, the Born contribution always favors disproportionation and more so the smaller the distance R.

R is about 20 Bohr in fullerides. ΔG(Born) is thus equal to $-1/20$ Hartree or -1.3 eV. This lowers the energy difference between the spin-coupled state and the charged state, thus Hubbard U decreases.

PbTe is a typical ionic crystal where Pb^{2+} alternates with Te^{2-}. What happens if some of the Pb^{2+} are replaced by thallium. The Pb^{2+} sites are "made" for a +2 ion, but Tl does not want to be Tl^{2+}. It is found that half of the sites are Tl^{+} and half of the sites are Tl^{3+}. The 6s electrons form a unit electron pair and populate alternant thallium sites. The $6s^{2}$ singlet may transfer between sites as an electron pair, provided that there is near degeneracy between the charged state and the spin-coupled case.

According to Equation 17.10 this comparison holds only for the low value of R that is relevant for TlTe. If the doping level is low, the distance R between the Tl sites is much larger. The Born contribution is now much smaller. The charged state is much less favored and near degeneracy is possible. Tl-doped PbTe is, in fact, superconducting below about 1 K in a certain.

17.4.2 FULLERENE SUPERCONDUCTIVITY

The discoveries of conductivity and superconductivity in $K_{3}C_{60}$ and $Rb_{3}C_{60}$ were great events in science during the 1990s. This happened within five years after the discovery of high T_{C} superconductivity and the discovery of fullerenes. Superconductivity was later discovered in a number of other $A_{3}C_{60}$ compounds, where A stand for any of the alkali atoms: Li, Na, K, Rb, or Cs, but never in any of the numerous alkali fullerenes with a different number of carbon atoms than 60.

$K_{3}C_{60}$ has $T_{C} = 28$ K. The highest critical temperature is obtained for $RbCs_{2}C_{60}$. In $Cs_{3}C_{60}$, superconductivity appears only under pressure.

The lattice constant depends on the size of the alkali ion. Generally, an increase of the lattice constant leads to a higher critical temperature. Some of the $A_{3}C_{60}$ compounds (A = Li, Na) with the smallest lattice constants are not superconducting. A larger lattice constant leads to a larger density of states at the Fermi level, which in the BCS model increases the critical temperature.

Looking into a greater detail the high pressure results are inconsistent with the BCS model, however. For $Na_{2}CsC_{60}$, the critical temperature increases from 10.5 to 29.6 K. In $K_{3}C_{60}$ ($T_{C} = 28$ K) and $Rb_{3}C_{60}$ ($T_{C} = 30$ K), on the other hand, superconductivity disappeares. $Cs_{3}C_{60}$, with the largest lattice constant, is an insulator at

T = 0 at ambient pressure, but already at the quite modest pressure of 3 kbar ($\approx$3000 atm) it turns into an SC with the highest critical temperature known for any fulleride.

It is more reasonable to use the Born effect. A larger lattice constant leads to a smaller Born effect according to Equation 17.10, whereby the charged state becomes more unstable compared to the spin-coupled state. In Rb_3C_{60}, the charged state is still slightly below the spin-coupled state. In Cs_3C_{60}, on the other hand, the spin-coupled state becomes the ground state and superconductivity disappears. High pressure decreases the lattice constant and lowers the charged state. When superconductivity appears, it is for the highest critical temperature of any alkali fulleride, or 38 K. In this case, the continuous tuning of pressure leads to a charged state and a spin-coupled state precisely in level.

The disproportionation model

$$2C_{60}^{3-} \rightarrow C_{60}^{2-} + C_{60}^{4-} \quad \left(\Delta G^0 \right) \tag{17.11}$$

leads to perfect agreement with the experiments. That Cs_3C_{60} is an antiferromagnet before a slight increase of pressure makes it superconducting, confirms the model.

17.4.3 Cuprate Superconductivity

As mentioned above, $La_{(2-x)}Ba_xCuO_4$ is superconducting below 35 K. The "mother compound" La_2CuO_4 is a cuprate and an insulator at T = 0 like CuO and all cuprates where copper has the well-defined oxidation state +2. CuO, the black, semiconducting oxide of copper, is formally not a cuprate (since a cuprate has two different metal ions). Cuprates, in contradistinction to CuO, can be made superconducting by oxidative or reductive doping. Cuprates are different from CuO by the presence of the CuO_2 plane in the structure.

There is no consensus on the mechanisms of high T_C superconductivity in doped cuprates. There is also no generally accepted assignment of the spectrum of pure as well as doped cuprates. Convincing experimental evidence shows, however, that the doping sites, consisting of Cu^{3+} ions, are local.

The Cu^{3+} sites with the neighboring oxygen ions are local and positive, since they miss one electron compared to the Cu^{2+} site. The energy needed for an intermetal transition where the electron ends on a site near a Cu^{3+} site must be much less than if it ends on a Cu^{2+} site for away from. We may conclude that the Hubbard gap is reduced in size locally in the CuO_2 plane close to the Cu(III).

The CuO_2 plane is the xy-plane. The metal–ligand covalent bonds are in the xy-plane because one electron is missing in the antibonding $O2p - 3d(x^2 - y^2)$ MO. The CuO distance is therefore smaller in the CuO_2 plane than in the perpendicular direction, where the $3dz^2$ orbital is fully occupied. This is the Jahn–Teller effect in action, ordering all the copper ions of the plane. The reorganization of the structure due to the number of electrons present at the site has little to do with the Jahn–Teller effect, however. When we apply the Jahn–Teller theorem, the number of electrons is constant.

The occurrence of superconductivity in the cuprates still had no accepted explanation when this book was written in 2011, 100 years after superconductivity was first discovered by Kamerlingh Onnes. One thing can be concluded, however. All attempts to squeeze any moving electron pair out of a spin-coupled CuO_2 plane, using fancy magnetic couplings of an unknown nature, must be regarded as futile. The Hubbard gap has to be eliminated in one way or another.

17.4.4 BISMUTHATES

If Bi ions are partly substituted with Pb ions, an SC with $T_C = 18$ K is obtained. $BaBiO_3$ is a typical CDW system, as is evidenced by alternant Bi sites (Bi^{3+} and Bi^{5+}). The BiO bond distances are different by as much as 0.18 Å. Doping with Pb or K, with compositions $BaPb_xBi_{1-x}O_3$ and $Ba_{1-x}K_xBiO_3$, respectively, leads to smaller distance differences, as determined by crystallographic means. Since the matrix element H_{12}, relevant for electron pair transfer directly from a Bi^{3+} site to a Bi^{5+} site, is vanishing, this can only be explained as being due to an *effective* H_{12} owing to the interaction with the $Bi^{4+}Bi^{4+}$ state. Doping lowers the energy of the SDW state, relative to the CDW states, and brings the states into interaction. This leads first of all to mobile electron pairs between the sites, and second to delocalization, with all the BiO bonds demonstrably equal in length.

17.5 PAIR CONDUCTIVITY IN THE GROUND STATE

An applied magnetic field induces electric currents in an atom or a molecule. In a molecular crystal or solid, these currents add up to a diamagnetic counter current, provided that the system in not paramagnetic, ferromagnetic, or antiferromagnetic. If the molecule is an aromatic molecule and the magnetic field is applied perpendicular to the aromatic ring, there are ring currents, which lead to a conspicuous NMR shift. The ring currents are moderately large even for normal aromatic systems, and can be calculated with the help of quantum chemical methods. The most up-to-date method is called GIAO (gauge including atomic orbitals) (Jusélius et al.).

There are situations where the ring currents are very large, for example, cyclobutadiene with equal bond lengths. cyclobutadiene with equal bond length is a hypothefical case, however, since cyclic C_4H_4 has two short and two long CC bonds. The situation of very large ring currents appears to be directly related to the superconducting currents. In this section, we will sketch how conductivity is possible in a nearly degenerate ground state of the kind that appears in MV-3 systems.

Our aim is to show that a superposition of the SDW and CDW wave functions leads to ground state currents. An effective intersite coupling $\beta < 0$ is assumed for neighboring sites. Otherwise, there is no coupling. We assume one "active" orbital (or hole) per site, as in an aromatic system. In our case, the active hole corresponds directly to the Cu^{2+} with one missing electron in a $b_{1g}*$ MO. The $b_{1g}*$ MO energies form a half-filled band.

17.5.1 CYCLOBUTADIENE WITH EQUAL BOND LENGTHS

The well-known Hückel eigenvalue problem for cyclobutadiene takes the form (see Chapter 3):

$$\begin{pmatrix} x & 1 & 0 & 1 \\ 1 & x & 1 & 0 \\ 0 & 1 & x & 1 \\ 1 & 0 & 1 & x \end{pmatrix} \begin{pmatrix} C_1 \\ C_2 \\ C_3 \\ C_4 \end{pmatrix} = \begin{pmatrix} 0 \\ 0 \\ 0 \\ 0 \end{pmatrix} \tag{17.12}$$

The eigenvalues and eigenfunctions may be written as

$$x_4 = -2\beta \quad \phi_4' = \frac{1}{2}(\chi_1 - \chi_2 + \chi_3 - \chi_4)$$

$$x_3 = 0 \quad \phi_3' = \frac{1}{2}(\chi_1 + \chi_2 - \chi_3 - \chi_4) \quad (Y)$$

$$x_2 = 0 \quad \phi_2' = \frac{1}{2}(\chi_1 - \chi_2 - \chi_3 + \chi_4) \quad (X)$$

$$x_1 = 2\beta \quad \phi_1' = \frac{1}{2}(\chi_1 + \chi_2 + \chi_3 + \chi_4) \tag{17.13}$$

where χ_i are atomic orbitals (Figure 17.7(a)).

In the ground state, ϕ_1' is always occupied. Since ϕ_2' and ϕ_3' are degenerate, there are a number of different possibilities for occupation and final many-electron states. In cyclobutadiene, we may occupy ϕ_2' by two electrons and obtain a short bond between atoms 2 and 3 and between atoms 1 and 4, since ϕ_2' is bonding between 2 and 3 and between 1 and 4. This is consistent with the experimental fact that two of the bonds are shorter in cyclobutadiene.

We are interested in the case with equal bond lengths and therefore may transform the eigenvectors as in Figure 17.7(b). The new eigenfunctions may be written as

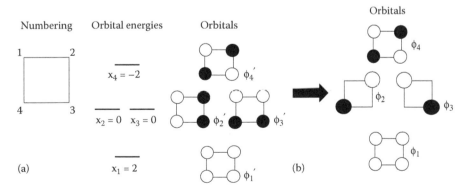

FIGURE 17.7   Four sites—four electrons (L = 1). (a) Ordinary Hückel orbitals and (b) transformed.

$$\phi_2 = \frac{\phi_3' + \phi_2'}{\sqrt{2}} = \frac{1}{\sqrt{2}}(\chi_1 - \chi_3) \quad (X + Y)$$

$$\phi_3 = \frac{\phi_3' - \phi_2'}{\sqrt{2}} = \frac{1}{\sqrt{2}}(\chi_2 - \chi_4) \quad (X - Y) \tag{17.14}$$

If ϕ_2 is occupied by a spin-up electron and ϕ_3 with a spin-down electron, we obtain (after symmetrization) the spin-coupled state. If ϕ_2 is occupied by two electrons or ϕ_3 is occupied by two electrons, we obtain the charged state. With these occupations the bond lengths remain the same, since in ϕ_2 and ϕ_3 two nonzero MOs are never on adjacent atoms.

17.5.2   WAVE FUNCTIONS AT THE VAN HOVE SINGULARITY (x = 0)

The Fermi level is in the middle of the valence band and in the middle of the eigenvectors with x = 0. The number of orbitals with x = 0 will increase linearly as the system size is increased. There is thus a singularity at the Fermi level that is called the *van Hove singularity*. We will now consider only the MOs with x = 0, beginning with the case with a single square (L = 1). The two degenerate MOs of Equation 17.14 are occupied by two electrons. We obtain the following possibilities for Slater determinants:

$$(X + Y)(X + Y) \leftrightarrow \left|...\phi_2^2 \alpha\beta\right|$$

$$(X - Y)(X - Y) \leftrightarrow \left|...\phi_3^2 \alpha\beta\right|$$

$$(X + Y)(X - Y) \leftrightarrow \left|...\phi_2 \alpha\phi_3\beta\right|$$

$$(X - Y)(X + Y) \leftrightarrow \left|...\phi_3 \alpha\phi_2\beta\right| \tag{17.15}$$

The short notation with X and Y is used for convenient X + Y corresponds to MO ϕ_2 and X − Y to MO ϕ_2. The following state wave function corresponds to the antiferromagnetic state:

$$(XX - YY) = \left[(X + Y)(X - Y) + (X - Y)(X + Y)\right]/2$$

$$\leftrightarrow \Psi\left({}^1B_{1g}\right) = \left|...\phi_2 \alpha\phi_3\beta\right| + \left|...\phi_3 \alpha\phi_2\beta\right| = \left|...\phi_2\phi_3\right|(\alpha\beta - \beta\alpha) \tag{17.16}$$

This wave function is correlated since it is easy to show that if one electron is at a given site with spin up, the other electron is at an adjacent site with spin down. If a minus sign is used, the spatial wave function is antisymmetric under an interchange of 1 and 2. The spin function has to be symmetric, corresponding to a triplet state.

$$\left(YX - XY\right) = \left[\left(X + Y\right)\left(X - Y\right) - \left(X - Y\right)\left(X + Y\right)\right]/2$$

$$\leftrightarrow \Psi\left(^3A_{2g}\right) = \left|...\phi_2\phi_3\right|\left(\alpha\beta + \beta\alpha\right) \qquad\qquad (17.17)$$

Finally, we may use a combination appropriate to the CDW wave functions:

$$\left(XY + YX\right) = \left[\left(X + Y\right)\left(X + Y\right) - \left(X - Y\right)\left(X - Y\right)\right]/2$$

$$\leftrightarrow \Psi\left(^1B_{2g}\right) = \left|...\phi_2^2\alpha\beta\right| - \left|...\phi_3^2\alpha\beta\right| \qquad\qquad (17.18)$$

$$\left(XX + YY\right) = \left[\left(X + Y\right)\left(X + Y\right) + \left(X - Y\right)\left(X - Y\right)\right]/2$$

$$\leftrightarrow \Psi\left(^1A_{1g}\right) = \left|...\phi_2^2\alpha\beta\right| + \left|...\phi_3^2\alpha\beta\right| \qquad\qquad (17.19)$$

Wave functions with angular momentum, corresponding to electron pair conductivity, may now be obtained:

$$\Psi_+ \leftrightarrow \frac{1}{2}\left[\left(XY + YX\right) + i\left(XX - YY\right)\right], \qquad\qquad (17.20)$$

$$\Psi_0 \leftrightarrow \frac{1}{2}\left[\left(X + Y\right)^2 + \left(X - Y\right)^2\right]/\sqrt{2}, \qquad\qquad (17.21)$$

$$\Psi_- \leftrightarrow \frac{1}{2}\left[\left(XY + YX\right) - i\left(XX - YY\right)\right] \qquad\qquad (17.22)$$

(i is the imaginary unit). It is possible to express Ψ_+ alternatively in terms of a Slater determinant of MOs:

$$\phi_+ = \frac{1}{2}\left(\chi_1 + i\chi_2 - \chi_3 - i\chi_4\right) = \left(\phi_2 + i\phi_3\right)/\sqrt{2} \qquad\qquad (17.23)$$

$$\phi_- = \frac{1}{2}\left(\chi_1 - i\chi_2 - \chi_3 + i\chi_4\right) = \left(\phi_2 + i\phi_3\right)/\sqrt{2} \qquad\qquad (17.24)$$

This is also the eigenvector form in the case of a perpendicular magnetic field.

A pair of electrons may be transported between the sites 1 and 3 to the sites 2 and 4, provided the activation energy disappears. This pair possesses an angular momentum and is held together as a state, in the sense that energy is required to promote it to an excited state where the pair is broken.

We now treat the case with two loops (L = 2). There are 2L = 4 copper atoms in each direction and 16 copper sites in all. A general row in the eigenvalue problem may be written as

$$xC_v + \sum_{\mu \neq v}{}' C_\mu = 0, \quad \text{for any } v. \tag{17.25}$$

The prime on the summation sign in Equation 17.25 indicates summation over the nearest neighbors only, since all other matrix elements between site v and other atoms is assumed to be zero. The matrix product of row v and the eigenvector column matrix has x from column v in the Hamiltonian matrix and multiplies the eigenvector coefficient C_v in row v. The remaining terms in the summation $\mathbf{HC}$ have 1 for the nearest neighbors of v in $\mathbf{H}$ and 0 otherwise. Thus, only coefficients for the nearest neighbors appear in Equation 17.25. The eigenvalue equation may be written as in Figure 17.8.

We are interested only in the eigenvectors for $x_i = 0$ (Figure 17.8). Equation 17.25 thus may be written for eigenvalues i:

$$\sum_{\mu \neq v}{}' C_{\mu i} = 0. \tag{17.26}$$

From Equation 17.25, it follows that the eigenvectors corresponding to the Fermi level with the eigenvalues equal to zero may be expressed in terms of a special class of orbitals with the property that the sum of the coefficients around an arbitrary point is always equal to zero. If the coefficient is different from zero in any point, all neighboring coefficients are equal to zero. Thus, the eigenvectors may be expressed in terms of a full set of orbitals with these properties. The latter set of orbitals is called "generating orbitals." In the case of $L = 1$, they are the two degenerate MOs of Figure 17.7. In the case of $L = 2$, the generating orbitals are given in Figure 17.9 and for $L = 3$ in Figure 17.10. Dark and grey indicate different signs in front of the site MO. The same procedure that was used to construct Figure 17.9 from Equation 17.26 may be used for any L, that is, for any number of squares inside each other. All

$$
\begin{pmatrix}
0100000000010000 \\
1010000000001000 \\
0101000000000100 \\
0010100000000000 \\
0001010000000100 \\
0000101000000010 \\
0000010100000000 \\
0000001010000010 \\
0000000101000001 \\
0000000010100000 \\
0000000001010001 \\
1000000000101000 \\
0100000000010101 \\
0010100000001010 \\
0000010100000101 \\
0000000010101010
\end{pmatrix}
\begin{pmatrix}
C_1 \\
C_2 \\
C_3 \\
C_4 \\
C_5 \\
C_6 \\
C_7 \\
C_8 \\
C_9 \\
C_{10} \\
C_{11} \\
C_{12} \\
C_{13} \\
C_{14} \\
C_{15} \\
C_{16}
\end{pmatrix}
=
\begin{pmatrix}
0 \\
0 \\
0 \\
0 \\
0 \\
0 \\
0 \\
0 \\
0 \\
0 \\
0 \\
0 \\
0 \\
0 \\
0 \\
0
\end{pmatrix}
$$

FIGURE 17.8   Eigenvalue problem $(H - \varepsilon)C = 0$ to obtain the eigenvectors for $x = 0$ in the $L = 2$ case.

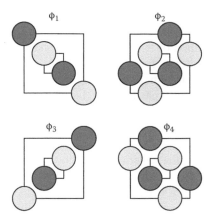

FIGURE 17.9 Orbital generators in the case with L = 2.

the orbitals shown have x = 0. Since the coefficient is equal to zero on every second atom, the bond lengths will remain unchanged in case the MOs are occupied.

Each new ring contributes another two eigenfunctions with x = 0. There are thus 2L such functions and this is also the number that can be constructed by observation of the rule that neighboring coefficients cannot both be nonzero.

The generating orbitals are nonorthogonal. Orthonormalization has to be applied among the orbitals of the upper row and separately among the orbitals of the lower row. The eigenfunctions φ_i are obtained. The latter MOs may be occupied with one or two electrons, but since electrons are available to occupy only half of the van Hove orbitals, the properties of the system will depend strongly on which MOs are finally occupied.

Summarizing, we have been able to calculate the generating orbitals from which the eigenfunctions corresponding to x = 0 can be obtained. The latter eigenfunctions

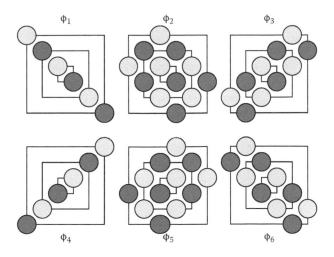

FIGURE 17.10 Orbital generators for L = 3.

may be occupied in different ways to give a charged state, a spin-coupled state, or the states that describe the motion of the electron pairs in the squares. The latter state is a linear combination of the two first mentioned. Depending on the system, the lowest state can be either of these states. What we have found is that one-electron per site systems are unstable if the site orbitals are not strongly interacting, as in the case of alkali metals or other typical metals. In the weak coupling situation, either a charged state, a spin-coupled state, or a mixture of these states develops into a state with electron pair currents.

A charged state is obtained if the upper row is occupied by two electrons each while the alternant sites in the lower row are left unoccupied. The total number of electrons with $x = 0$ is 2L, but there are $(2L)^2$ sites in the system. Apparently, the occupancy of the $x = 0$ states is just an indicator of a possible instability. We know from the BCS model that the phonon interactions contribute to pair formation. This may lead to an energetically favored charged state if the intermediate valence state is "missing."

Alternatively, we may occupy the upper orbitals in Figures 17.9 and 17.10 by one electron with spin up, while the lower orbitals are occupied by one electron with spin down. Adding the other possible wave function of this type with opposite spins leads to a spin-coupled state, since the spins are different on adjacent sites. Thus, if the intermediate valence state is present at a low energy, spin coupling occurs, as in the undoped cuprates. It should be noted that we are dealing with "chemical effects," which are not easily calculated and not easily understood. What is important is that in addition to these two possibilities, an electron pair current state is possible if the charged state and spin-coupled state are almost degenerate.

17.5.3 VIBRONIC WAVE FUNCTIONS

We now return to Figure 17.5 to determine the vibronic ground states in the MV-3 case. The wave function for the system is a function of electronic and nuclear coordinates. The electronic wave function is a product of an almost constant wave function for the $(N - 2)$ inactive electrons $\Theta(1,...,N - 2)$ and a two-electron function Φ describing the electron pair. We have seen above that if bonding and antibonding ligand field orbitals are fully occupied, the charge density does not change significantly. Only in Φ is there a great dependence on the metal–ligand distance, which leads to a charge transfer between the sites, which is the consequence of Figure 17.5. We indicate the implicit dependence on the nuclear position by Q and write $\Phi(1,2;Q)$.

The dependence on the positions of the other electrons is omitted. Φ is thus assumed to be a function of two electronic coordinates, 1 and 2:

$$\Psi(1,...,N;Q) = \Phi(1,2;Q)\chi(Q)\Theta(3,...,N), \qquad (17.27)$$

where Q represents the collective nuclear coordinate. From what we know so far, breathing and half-breathing modes are the most important nuclear coordinates for electron pair transfer. χ is the vibrational wave function that contains the important modes. As the nuclei on one site move to a position favorable for accepting a pair of electrons, the nuclei of the neighboring site move into a position for donating the same pair of electrons. There is a gain of energy involved in this type of motion that reminds us about the more generally defined Cooper pair. In both cases, there is an energy gap.

17.5.4 FINAL WAVE FUNCTION

We now have to write an expression for the total wave function that mathematically expresses both antisymmetry at the exchange of electrons and symmetry at the exchange of electron pairs. A general method exists that was mainly developed by the Canadian mathematician J. Coleman. A sum over all geminals (two-electron functions) is formed:

$$\Xi(1,2) = \sum_k \Xi_k(1,2), \tag{17.28}$$

which is a spin singlet wave function. The wave function is expanded as an antisymmetrized product of the singlet geminal (APSG) in the following way:

$$\Psi_{APSG} = A_N\left[\Xi(1,2)\Xi(3,4)...\Xi(N-1,N)\right]. \tag{17.29}$$

Because of the antisymmetrizer A_N, the wave function satisfies the Pauli principle. The product of geminal functions (note that there is only a single geminal function) ensures that we may exchange, say (1,2) with (3,4), and get the same wave function back. Whether the wave function appears to follow the Bose–Einstein statistics or the Fermi–Dirac statistics depends entirely on the experiment.

18 Conductivity in Organic Systems

18.1 INTRODUCTION

Organic conductivity is an extensive and rich field. Decades of scientific work has resulted in many interesting products, such as conducting organic polymers, organic light-emitting diodes (OLED), and organic superconductors. Yet, most organic crystals are semiconductors with a large band gap. Aromatic molecules with low excited states absorb in the visible or lower ultraviolet (UV) region at 2–4 eV. Crystals of anthracene and pentacene (Figure 18.1) have been used as model systems for studies of charge separation, electron mobility and exciton mobility.

18.2 ORGANIC SEMICONDUCTORS

Conductivity in organic systems is broadly divided into conductivity in crystals or films and conductivity in polymers. Single crystals without impurities are well suited for investigating the intrinsic charge transport properties of organic semiconductors.

18.2.1 ELECTRONS AND EXCITATIONS IN ORGANIC MOLECULAR CRYSTALS

Molecules with closed shells attract each other with London forces that are weak and proportional to $1/R^6$ as $R \to \infty$, where R is the distance between the molecules. The orbitals on different molecules overlap very little and therefore the energy bands are narrow. The band gap is almost the same as the lowest excitation energy for the single molecule.

Anthracene is useful as a model system in studies of solid-state properties. Single crystals may be grown from solutions of anthracene in organic solvents such as CS_2 and CCl_4. Recently, tetracene and pentacene have also received considerable attention owing to the high mobility of electrons. There appears to be many possibilities for their use in thin-film organic electronics. Electrons or excitations may be inserted into the conduction band by photoexcitation.

The low energy spectrum of anthracene (L_a) is shown in Figure 18.2. Benzene and naphthalene, with one and two aromatic rings, respectively, absorb at higher energy (shorter wavelength), and tetracene and pentacene absorb at lower energy. Judging from Figure 18.2, anthracene should be colorless since the absorption is in the UV region, but a pure anthracene crystal appears blue, probably due to fluorescence emission. The Stokes shift is quite small.

445

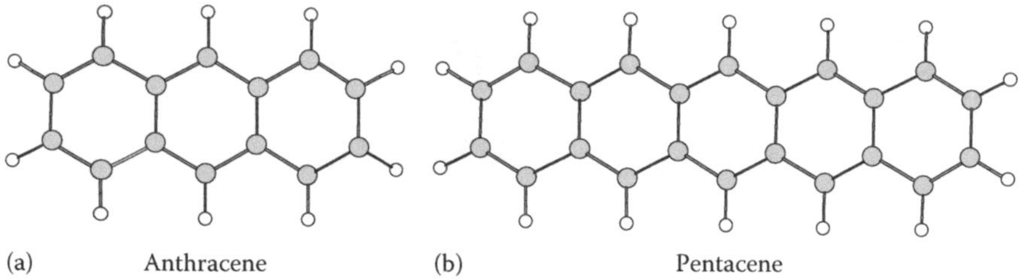

(a)   Anthracene   (b)   Pentacene

FIGURE 18.1   (a) Anthracene and (b) pentacene.

The spectrum depends on the solvent. The "solvent" may be a crystal, in which case there is a small red-shift (to longer wavelengths) of the absorption compared to the free molecule. Furthermore, there is a formation of excitons. Local excitons are referred to as "Frenkel excitons." Excitons extended over the whole crystal have been studied in great detail by Davydov and are called "Davydov excitons." The lowest exciton state is further red-shifted compared to the free molecule absorption. Local excitons polarize themselves and the medium (Figure 18.3). As usual, when the reorganization energy is less than the coupling, the excitons are delocalized.

Excitation of the crystal may alternatively result in charge separation. The negative charge has some probability to leave the positive hole, but remains localized (Figure 10.3) and travels around on its own, stepwise between the molecules, before it is captured by a positive hole. The residence time on a molecule of anthracene is in the region of 10^{-14}–10^{-13} sec. During this time, the electron causes a reorganization of the structure, since the vibrational cycle time is roughly of the same order of magnitude. An additional electron at a site may also cause some increase (or decrease) in the distance to the surrounding molecules. The time for displacement is more than a picosecond, however. Therefore, relaxation of the surrounding molecules should not be included in the reorganization energy.

Another type of induced polarization is polarization of the electronic charge cloud. However, this is an electronic effect and should not be counted in the reorganization energy.

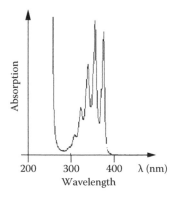

FIGURE 18.2   Absorption spectrum (mainly L_a) of anthracene.

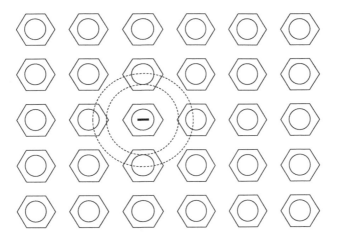

FIGURE 18.3   Polarization around a negatively charged molecule in the crystal.

18.2.2   Conductivity in Organic Systems

Before organic crystals began to be thoroughly investigated by, for example, the American physicists M. Pope, N. E. Geacintov, and C. E. Swenberg, and the Latvian physicist E. Silinsh, there was some confusion about the role of π-systems in conductivity. The π electrons have been regarded as free electrons. Why could they not just move around in the molecule and then jump to the next molecule? Others viewed organic crystals as systems with a large band gap, which would make conductivity virtually impossible. In any case, it was a great surprise when H. Akamatu, H. Inokuchi, and Y. Matsunaga in 1954 measured a rather high conductivity in perylene, containing some bromine. Perylene has a planar π-system (Figure 18.4), but how and why is the presence of a π-system important for conductivity?

Although the work of Akamata et al. did not involve photoexcitations, it may be of interest to lock at the spectrum. The energy of the ππ* transitions is generally

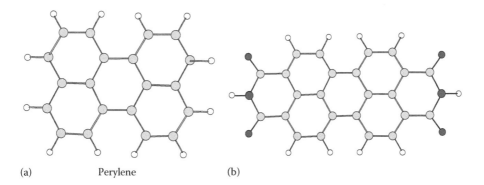

(a)   Perylene   (b)

FIGURE 18.4   A perylene molecule (a). The molecule on (b) is perylene diimide (PDI). The –NH group may be exchanged by other groups to obtain novel pigments with other hues and colors.

smaller than the gap of the $\sigma\sigma^*$ transitions. In small planar molecules, the lowest $\pi\pi^*$ transition is found high up in the UV region. Perylene (Figure 18.4) has a quite large π-system and the absorption is therefore at the border between the visible region and the UV region just below 3 eV (Figure 12.6). The highest intensity in the vibrationally split absorption is in the 0-0 state, indicating that the Stokes shift and thereby the reorganization energy for excitation is quite small. This can be blamed on the spreading of bonding characters over many bonds in a π-system, which leads to moderate bond order changes.

The large excitation energy for perylene is quite consistent with simple models such as the free electron model and the Hückel model. If we go to larger planar systems, toward graphite, the excitation energy finally tends to zero. In this case, we may rely on the simple Hückel model. The calculation of a spectrum by accurate methods is recommended, of course, but it may be hard to do the calculation for large molecules. In any case, the reason for the decrease of the excitation energy for larger molecules is easier to understand with the help of simpler models.

The absorption spectrum of perylene is roughly the same as for molecules where substitutions have been made in the perylene molecule. As usual in planar systems, the polarization (transition moment) is in the xy-plane. The substituents are not directly involved in the transition and only slightly modify the energy of the $\pi\pi^*$ transitions. This is enough to obtain pastel pigments with different hues.

In any case, the local band gap is of less interest for the conductivity in organic systems than in the inorganic ones (Si, GaAs, etc.). The large width of an organic band gap does not allow the conduction band to be thermally occupied. To have conductivity, we have to excite across a Hubbard gap in a charge transfer excitation. In an ordinary organic crystal, the molecules may be too far away to allow charge separation. The problem may be solved by having "special pairs" as in natural photosynthesis or simply by having different molecules as in the experiment of Akamatu et al., where the bromine molecule acts as the acceptor molecule.

18.2.3 CHARGE TRANSFER SPECTRA

Typical for perylene and many other π-systems is the relatively high energy of the highest occupied molecular orbital (HOMO). Typical for the bromine molecule, on the other hand, is the low energy of the lowest unoccupied molecular orbital (LUMO). Therefore, in the experiment of Akamatu et al., a Mulliken charge transfer complex is formed between perylene and bromine, where perylene is the electron donor (D) and bromine is the acceptor (A).

The theory for Mulliken charge transfer complexes is essentially the same as the Marcus theory applied to excited state electron transfer (ET). Typical for the absorption spectrum of a Mulliken charge transfer complex is additional absorption at quite low energy, absorption that does not exist in the pure substances (Figure 18.5).

In the band model, the semiconductivity property is interpreted as a gap at the Fermi level. The delocalized (one-electron) energy bands are used to obtain a theoretical gap. There are two problems with this interpretation. The first is that the energy splitting between the HOMO and the LUMO is not excitation energy in the Hartree–Fock method, but a difference between the ionization energy and electron

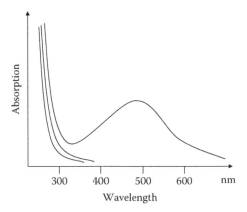

FIGURE 18.5   Charge transfer spectrum of a charge transfer complex between TCNE and TMB (Figure 18.6). The low energy absorption is the charge transfer transition defining the Hubbard gap.

affinity. In the density functional theory (DFT) method, there appears to be disagreement as to the meaning of the HOMO-LUMO gap.

Secondly, even if a band calculation of the gap leads to rough agreement with the excitation spectrum, the small value of $|\Delta G^{\ominus}|$ for ET to the bromine molecule cannot be explained within the same model. At wavelengths larger than 700 nm (~2 eV) in Figure 18.5, there is essentially no absorption to any electronic states of the TCNE–TMB charge transfer complex (Figure 18.6). On the other hand, there is actual charge transfer between D (TMB) and A (TCNE) at room temperature, as can be easily measured by conductivity.

The small value of $|\Delta G^{\ominus}|$ for ET and conductivity in a Mulliken charge transfer complex is because the complex gains energy after the electron has been transferred by simple Coulomb attraction. The donating "power" of D may be measured by its

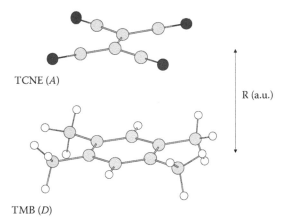

FIGURE 18.6   Optimized structure (PM3) for the charge transfer complex between tetramethylbenzene (TMB) and tetracyanoethylene (TCNE).

ionization energy E_I and the accepting "power" of A by the electron affinity E_A. Ignoring reorganization energies and other than ionic interactions between D and A, we may write approximately

$$E = E_I - E_A - \frac{1}{R_{DA}}(\text{a.u.}), \qquad (18.1)$$

where R_{DA} is the distance between D and A.

In a calculation, one may use the orbital energy of the HOMO as $- E_I$ and the orbital energy of the LUMO as $- E_A$, although the latter is usually connected with a rather large error (E_A is calculated too small). Equation 18.1 may be understood as a limit situation in open space with a large distance between A and D. Several terms are missing, particularly solvent interactions in the charge separated state. The energy of the charge transfer state (that is, after the electron has been transferred) is much lower than the vertical energy in the Marcus model. Furthermore, interactions between A and D, which are not included in the Coulomb term, are absent from Equation 18.1.

In Equation 18.1, it is important to include the reorganization energy λ, for the solvent and for the bonds. In the charge transfer complex between perylene and Br_2, there is large reorganization energy, particularly due to the extension of the BrBr bond when the electron is received in the strongly antibonding molecular orbital (MO). The final picture is seen in Figure 18.7. Therefore, Equation 18.1 must be updated to

$$\Delta G^{\ominus} = E_I - E_A - \lambda, \qquad (18.2)$$

where $-\lambda$ is a large negative term that includes the Coulomb attraction after charge transfer and polarization of the medium because of the charges created (mainly the Born term). Hence, $|\Delta G^{\ominus}|$ is much smaller than the excitation energy and this is the reason for conductivity.

If $|\Delta G^{\ominus}|$ is small in the Marcus model (Figure 18.7) there will be a Boltzmann distribution of ion pairs and stepwise ET may become possible. The important thing

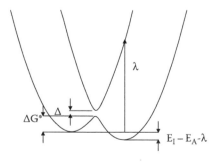

FIGURE 18.7 PES for a charge transfer. Reorganization energy λ, coupling $\Delta/2$, free energy of reaction $\Delta G^{\ominus} = E_I - E_A - \lambda$, and activation energy ΔG^*. E_I is the ionization energy of D and E_A is the electron affinity of A.

is that the ion pair state is reduced in energy by the Born contribution. Additional lowering arises because of a structural change when the electron jumps. ET and conductivity cannot be correctly described without taking structural changes into account.

Concerning solvent reorganization, it is important to remember that polarizability corrections, included via the refractive index, is a part of the electronic energy. The dependence on the refractive index can be expressed as in the Weller model or in other ways. Kuriyama, Ogilby, and Mikkelsen have shown that the dependence alternatively can be written in a dipolar form.

18.2.4 ORGANIC LIGHT-EMITTING DIODES

A display for a computer, a television set, or a handy walkie–talkie is filled with quantum devices. The mechanism of a light-emitting diode (LED) is, in a way, the opposite of the photoinduced ET mechanism. High potential electrons come in and radiation is emitted. A voltage is applied on the cathode (Figure 18.8). The transparent indium tin oxide (ITO) layer forms the anode. The voltage sucks out electrons from the HOMO (valence band) of the molecules of the organic LED (OLED). The electrons from the cathode enter the LUMO (conduction band). In the organic layer, light emission occurs spontaneously by deexcitation of the molecules, and light of a certain frequency is emitted through the transparent layers.

The transparent anode is made from In_2O_3, with some SnO_2 (ITO). For example, a crystal with two In_2O_3 molecules and three SnO_2 molecules forms a solid with the net formula $In_4Sn_3O_{12}$. O'Neil et al. have calculated the electron structure. There is nothing strange with a transparent metal or semiconductor. The only thing required is that there is no absorption in the visible spectrum, but an unoccupied band below the visible region, reachable at room temperature, where the electronic motion takes place. Unfortunately, ITOs of today are brittle and expensive. Much research is carried out to find better alternatives. In indium and tin oxide, the valence band is

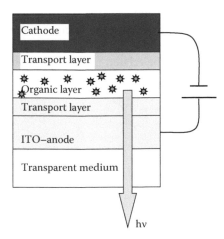

FIGURE 18.8 OLED device. The chemiluminescence light is emitted in the organic layer.

formed from the O2p orbitals with contributions from In4d. The band in the infrared consists mainly of O2p, In5s, and In5p. Above the visible region, there is another conduction band in the near UV region.

18.3   STACKED, CONDUCTING π-SYSTEMS

In the 1960s, a completely new form of organic conductors was found. Structurally, these new conductors turned out to be one-dimensional stacks of planar molecules. The conductivity is in the direction perpendicular to the π-system. In the 1980s, Klaus Bechgaard, and Denis Jérôme discovered organic, one-dimensional superconductors.

18.3.1   TTF–TCNQ

In 1964, the American chemist L. R. Melby and his collaborators synthesized a strong acceptor: tetracyanoquinodimethane (TCNQ) (Figure 18.9). In some compounds with the formula $M^{n+}(TCNQ^-)_n$, complete transfer of electrons from an organic molecule or metal atom to TCNQ takes place. The conductivity is quite high, possibly because the structural reorganization is moderate.

Since TCNQ forms one-dimensional stacks, it may be reasonable to suggest that Peierls distortion occurs along the stack and that this increases the activation barrier. In compounds with considerably less than complete ET from the other atoms or molecules, Peierls distortion should not occur. Here, the resistivity is, in fact, very small, suggesting that delocalization has occurred.

In 1972, F. Wudl et al. found a new donor, tetrathiafulvalene (TTF) (Figure 18.9). TTF is a semiconductor with a high conductivity at room temperature. TTF^+Cl^- is also a semiconductor with a band gap of 0.19 eV.

In 1973, it was discovered that solutions of TTF and TCNQ mix, resulting in a complex called TTF–TCNQ. Separate stacks of TTF and TCNQ are formed, adjacent to each other. A sharp conductivity peak appears at about 60 K. At this temperature, TTF–TCNQ conducts as a metal.

For elevated temperatures, the conductivity decreases slowly. The conductivity is thus activated with conductivity equal to zero at T = 0, but with a very small activation energy. The activation may be related to the formation of electrons and holes,

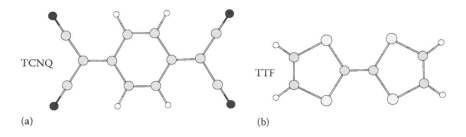

(a)   (b)

FIGURE 18.9   Acceptor tetracyanoquinodimethane, TCNQ (a), and donor tetrathiafulvalene, TTF (b).

due to ET between the stacks, or, alternatively, to the activation of motion along the stacks. These two activation mechanisms should be independent of each other.

TTF–TCNQ does not show any sign of Peierls "dimerization," which may depend on the fact that the band is much less than half filled, but with charge transfer occurring already in the ground state. The very small activation energy seems to be associated with ET within the stacks.

18.3.2 BECHGAARD SALTS: ORGANIC SUPERCONDUCTORS

SF_6 is a neutral, stable molecule. If the central atom is replaced by phosphorous, the atom preceding sulfur in the periodic table, the new molecule needs to have the same number of electrons and the same electronic structure; therefore, PF_6 has a large electron affinity. The Danish organic chemist Klaus Bechgaard made use of the strong electron accepting power of PF_6 to detach an electron from tetramethyl tetraselenafulvalene (TMTSF). The system obtained has a ground state in the form of a charge transfer salt (Figure 18.10). It should be observed that there are two TMTSF molecules per PF_6. At room temperature and ambient pressure, $(TMTSF)_2PF_6$ is a semiconductor with a small gap, 65 cm^{-1}. In the ground state, there are coupled spins between pairs of TMTSF molecules. Below the low critical temperature of 0.9 K and a pressure of 12 kbar, $(TMTSF)_2PF_6$ becomes a superconductor.

Bechgaard later used the even stronger electron acceptor molecule, perchlorate (ClO_4^-), instead of PF_6^-. He obtained a superconductor below $T_C = 1.4$ K at ambient pressure. All these organic conductors show magnetic ordering above the critical temperature up to several tens of kelvin, the transition temperature.

The molecule BEDT-TTF (bis-ethylenedithia tetrathiofulvalene; Figure 18.11) with acceptor molecule $Cu(NSC)_2$, was found to be conducting and even superconducting at a quite high critical temperature ($T_C = 10.4$ K) under ambient pressure. An insulator–metal transition temperature is observed at 135 K at ambient pressure.

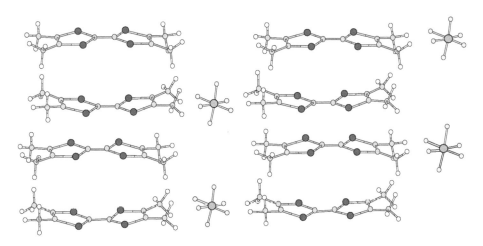

FIGURE 18.10 (See color insert) $8(TMTSF)^+$ and $4PF_6^-$ as a part of two stacks.

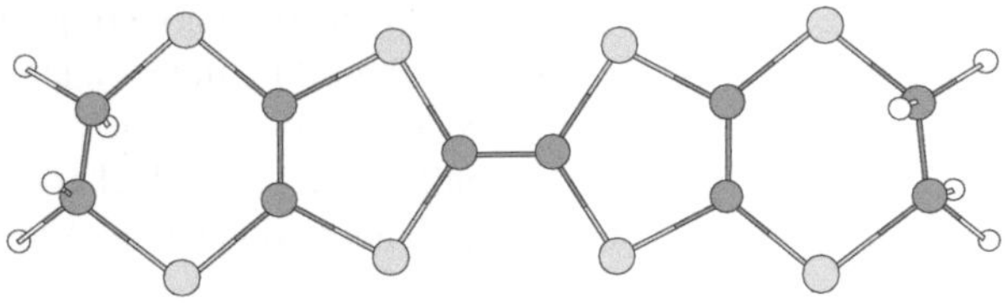

● Carbon   ○ Sulfur

FIGURE 18.11   BEDT-TTF.

As in the case with $(TMTSF)_2PF_6$, the oxidizing molecule receives one electron per two molecules.

In Bechgaard systems, the donor molecules of type TMTSF and BEDT-TTF are paired in the ground state. Each pair donates one electron to the oxidizing molecule. The HOMO is slightly split into weakly bonding and antibonding orbitals due to this weak pair formation. Removing one electron from the antibonding combination provides the small binding energy for the pair. In the whole crystal, the HOMOs form a band. This band is three-quarters filled, and the structure is not subject to any Peierls distortion.

Spin coupling occurs between two different pairs in some temperature regions. Apparently, this spin coupling provides the semiconductor gap in the non-conducting phases (for example, 65 cm^{-1} in the case of TMTSF).

18.4   CONDUCTING POLYMERS

The first known conducting polymer, *polyaniline*, was prepared from organic compounds during the 1800s. It took a long time, of course, before it was shown that it is a polymer and that it is conducting. In 1963, conductivity was documented in melanin and polypyrrole (PPY) by the Australian chemist D. E. Weiss and coworkers. The structure of melanin is complicated. It is created in our skin after UV radiation and appears to have something to do with the repair of damaged DNA. On the other hand, PPY is well defined and will be described in Section 18.4.4. Weiss found conductivity in PPY after electrochemical oxidation.

In 1975, a polymeric superconductor was found, $(SN)_x$ with $T_C = 0.26$ K. Apparently, this discovery had a greater impact on the physics community than the earlier efforts by the chemists in the field of conducting polymers.

18.4.1   (SN)x

In 1975, a remarkable inorganic polymer was found by R. L. Green and G. B. Street and their collaborators at the IBM Research Laboratory in San José, California. The crystalline polymer $(SN)_x$ (polysulfur nitride) (Figure 18.12) turned out to be a metal and superconductor below $T_C = 0.26$ K, the first polymeric material to exhibit superconductivity. This polymer is inorganic, since it does not contain carbon. However, there is no fundamental difference between organic and inorganic polymers of this type. We mention it in this section along with the organic, one-dimensional

(SN)$_x$

FIGURE 18.12 (SN)$_8$S according to the optimized structure.

conductors. In fact, the one dimensionality of (SN)$_x$ stimulated the development of conducting organic polymers and truly organic one-dimensional superconductors.

Geometry optimization by semi-empirical methods gives the bonding pattern seen in Figure 16.5. Each N atom corresponds to one CH unit in *polyacetylene* (PA). In PA, each CH unit contributes one π orbital and one electron. The valence band is half filled and the system PA is therefore subject to Peierls distortion. In (SN)$_x$, on the other hand, the sp^2 hybridized S atom contributes two electrons to the π-system. The π-system of (SN)$_x$ is therefore three-quarters filled and not subject to any Peierls distortion. However, three-quarter filling leads to other peculiarities, as we will see next.

18.4.2 POLYACETYLENE

Polyenes, $CH_2(=CH-CH)_n=CH_2$, have been discussed in previous chapters. The transitions to the first and second excited state are either strongly allowed ($^1A_g \rightarrow {}^1B_u$) or forbidden ($^1A_g \rightarrow 2^1A_g$) and go into the visible region for long chains (red, brown, and yellow autumn colors). If the number of carbon atoms tends to infinity, there appears to be a remaining gap of 1.5–2 eV, contrary to the case of graphene, where the gap has closed. This is directly connected to Peierls distortion to alternating bond lengths in the one-dimensional case. Peierls distortion cannot take place in the two-dimensional case.

Long chains without any specified number of carbon atoms are referred to as a polyacetylene polymer. Oxidative (I_2) or reductive (alkali) doping leads to conductivity. During the 1970s, Japanese chemists, including H. Shirakawa, were able to polymerize PA ($-CH=CH)_x-CH=CH_2$. In 1978, H. Shirakawa, A. G. MacDiarmid, and A. J. Heeger added I_2 as a "dopant" and found a high conductivity.

In calculations, it has been found that each alkali atom donates one electron to a PA chain. The electron delocalizes itself on about 20 carbon atoms. The alkali ion is attracted to the center of the distortion. There is no reason to believe that the electron has free mobility along the PA chain, at least not at low doping levels. The conductivity as a function of temperature shows a typical activated behavior for low doping levels. The conductivity probably arises when the electron jumps along a PA chain, or possibly between PA chains, from the neighborhood of an alkali atom to another without a near electron. The electron is not fully delocalized for low doping levels.

At higher temperatures, the mobility of the electron is stimulated by the nuclear vibrational motions. Already, the positive ion of an oligomer shows equal bond lengths and high intensity for vibrations that carry the electron forward along the chain. Very likely the high conductivity of PA at high doping levels is caused by bond length equalization and strong nuclear coupling.

H. Naarman and N. Theophilou improved PA in several ways. At high alkali doping levels, the conductivity was almost the same as that of copper ($10^5\Omega^{-1}$ cm^{-1}). As was found by M. Winokur et al., at high doping levels, three or four PA chains form a channel in which the positive alkali atoms are located. The bonding of one alkali atom to one PA chain is both ionic and covalent. A strong dipole is formed with a direction perpendicular to the PA chain. The dipole attracts another PA chain. Each doping alkali atom thus connects two PA chains. Next, an alkali atom may connect to another chain, and in this way the channels are formed.

In Figure 18.13, a geometry optimization has been carried out for three PA chains and six alkali atoms. The alkali atoms act as a zipper to hold the chains together. The channels consist of negatively charged walls of PA. The positive alkali ions (dark) are located in the channel. The alkali valence orbital appears to be partly involved in direct bonding to the closest π-system. A dipole is formed that attracts a second chain. Other alkali atoms bind to other chains in the same way.

By accepting electrons, the bond length alternation disappears. This does not mean metallic conductivity, however, since the charges are fixed to the alkali atom. Interaction with the zero-point motions of the alkali is apparently enough to make the electrons mobile. Heavily doped PA is conducting at T = 0. Contrary to the case for metals, the conductivity *increases* as the temperature is raised. This is very likely due to coupling between electron and alkali ion motion. For lower doping levels there is still some activation energy.

18.4.3 POLYANILINE

If one of the hydrogen atoms in benzene is replaced by an amino group, NH$_2$, we obtain aniline. Polyaniline (PANI) is easily polymerized from aniline. PANI is one of the oldest polymers known, first described in 1862 by Letheby. It gained a lot of new interest in the beginning of the 1980s because of its conducting properties.

The oxidation chemistry of PANI is complex. Three typical oxidation states are

1. Leucoemeraldine—colorless—fully reduced state
2. Emeraldine—green salt; blue base
3. Pernigraniline—blue, violet—fully oxidized state

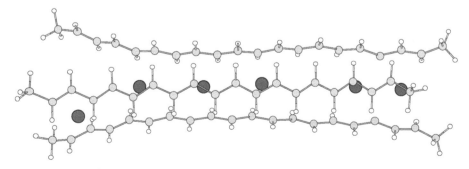

FIGURE 18.13   (**See color insert**) The result of geometry optimization of three PA chains and six alkali atoms (violet spheres).

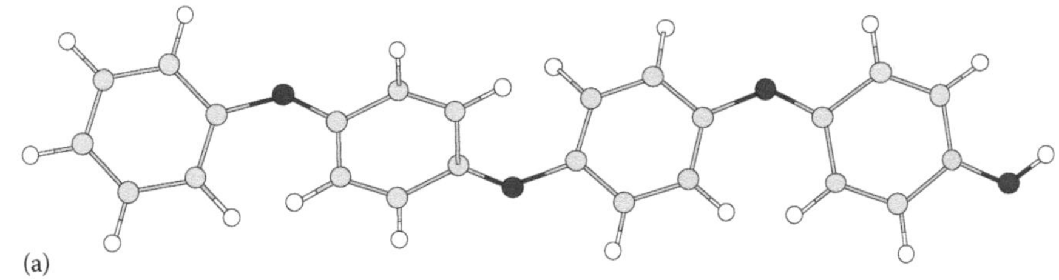

(a)

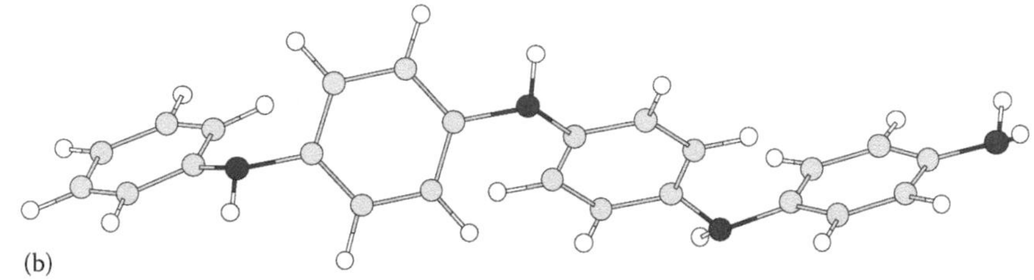

(b)

FIGURE 18.14 Oligomers of polyaniline (PANI): (a) fully oxidized state corresponding to pernigraniline and (b) fully reduced state corresponding to leucoemeraldine.

The emeraldine (Figure 18.14) base is the most stable form of PANI. If acid is added, emeraldine salt is obtained in a form that is electrically conducting. Leucoemeraldine and pernigraniline, on the other hand, are poor conductors. None of the systems are planar. This may be taken as proof that planarity, with "delocalized" π orbitals, is not necessary for high conductivity.

18.4.4 OTHER CONDUCTING POLYMERS

Polypyrrole (Figure 16.14) are chains of pyrrole connected at the carbon atoms that are neighbors of the nitrogen atoms. Two pyrrole molecules are joined together by removing two hydrogen atoms and joining the carbon atoms. In the chain containing nitrogen atoms, every fourth atom is a nitrogen atom.

As already mentioned, PPY (Figure 18.15) was synthesized and conductivity was measured already in 1963 by D. E. Weiss et al. PPY is electronically similar to PA and can be doped in the same way to become conducting.

In PPY, the conjugation is mainly in the chain of carbon atoms. This chain is wrinkled together by the nitrogen atoms (dark). If the chain is oxidized, the hole is extended over at least four pyrrole units (Figure 18.16). Figure 18.16 shows where

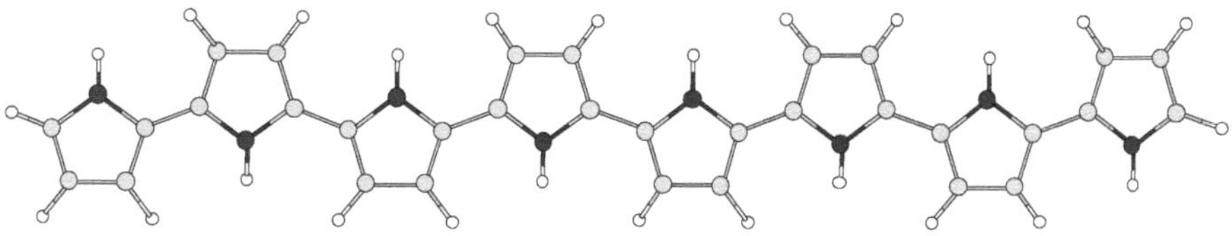

FIGURE 18.15 Octomer of polypyrrole (PPY).

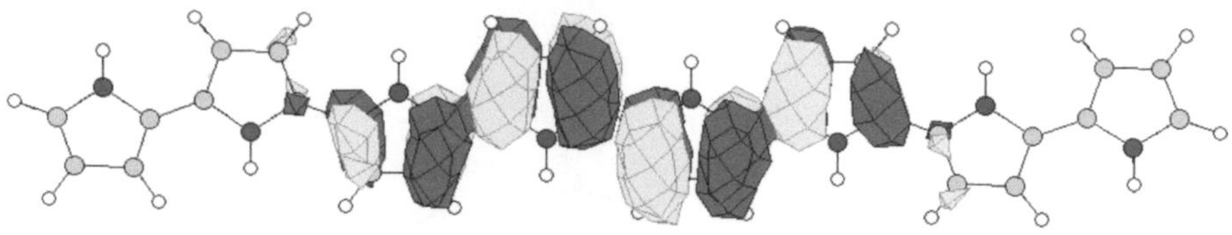

FIGURE 18.16   (See color insert) Hole state in oxidized PPY.

the bonding character is lost by oxidizing the molecule. The bond is increased from 1.400 to 1.443 Å at the center of the hole. The CC bond of the pyrrole group, where the hole orbital is antibonding, is decreased from 1.413 to 1.380 Å. The interpyrrole CC bond, where the hole orbital is also bonding, is decreased from 1.437 to 1.394 Å. These maximum changes are quite large and lead to a relatively large reorganization energy for the motion of the hole along the chain, or between the chains. Still, the hole is very much more delocalized than in the case of the saturated alkane chains, $CH_3(CH_2)_nCH_3$. The hole is moving with the help of stretching vibrations.

By replacing NH with isovalent-electronic sulfur, we obtain *polythiophene* (PT) (Figure 18.17) with similar properties as PPY. The PT chain is not fully planar, however. Apparently, this has little influence on the conductivity of the doped forms, since the bending angles are small.

Common to all organic polymers studied so far has been an unsaturated polyene chain where single and double bonds are formally alternating. In PT (as in PPY), the bond length alternation is in the CH chain. The bond opposite to the sulfur atom in each thiophene unit and the bond connecting thiophene units form the long bonds (1.43–1.44 Å), and the bonds in between the short bonds (1.375 Å). When an electron is removed in the case of an octomer the bond length alternation reverses. Adding or subtracting electrons from the chain leads to substantial structural reorganization.

Polyphenylene (PPE) and *polyphenylenevinylene* (PPEV) (Figure 18.18) also possess large bond length alternation in the ground state In PPE, the bond between the benzene rings is the longest. In PPEV, the benzene rings have bond lengths typical for benzene. In the connecting chains, the central bond is a short double bond, whereas the other two bonds are almost single bonds, thereby making the benzene rings "normal." As usual, the polymer chain has to be oxidized to create carriers. A conjugated π-system is not conducting in itself. The carriers cause changes in the structure that create quite large barriers for ET and conductivity.

PPE and other conducting polymers sometimes do not have the desired physical properties. One may then combine them with other nonconducting polymers with the desired properties. This is referred to as a *blend*, or a *polyblend*.

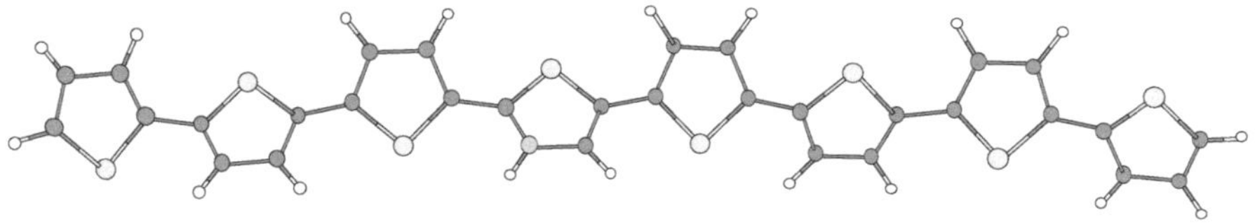

FIGURE 18.17   Octomer of polythiophene (PT).

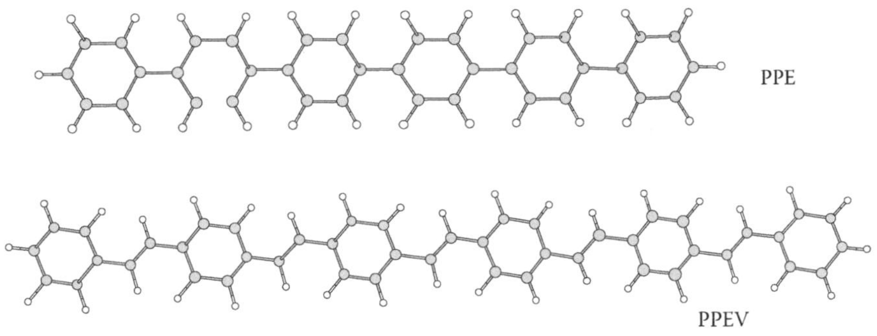

FIGURE 18.18 Hexamers of polyphenylene (PPE) and polyphenylenevinylene (PPEV).

The chemical name for styrene is vinyl-benzene (or phenyl-ethene) $C_6H_5C_2H_3$. *Polystyrene* (PS) (Figure 18.19) consists of an alkyl chain where on every second carbon atom, one hydrogen atom is substituted by a phenyl group (C_6H_5-). The phenyl groups repel each other and avoid stacking by turning away from each other. The structure is very flexible since the "skeleton" is an aliphatic chain.

PS is widely used as a hard plastic. In its pure form, it is colorless. PS is not conducting, contrary to all the other polymers we have studied so far, and one may wonder why. The simplest reason is that the polymer chain is saturated. Electrons or holes due to doping would appear in the π-system of the phenyl groups. The orbital overlap is too small to provide sufficient coupling between the nearest phenyl groups. $\lambda/\Delta > 1$, which means that the electron is almost immobile on its phenyl group.

Another question we may ask in connection with PS is why saturated chains are insulating (if holes can be created in a σ chain). Doping has to be carried out in any case, so what is the difference between carriers in a π-system and carriers in a saturated system? Very likely, the difference lies in the delocalization of the carriers that

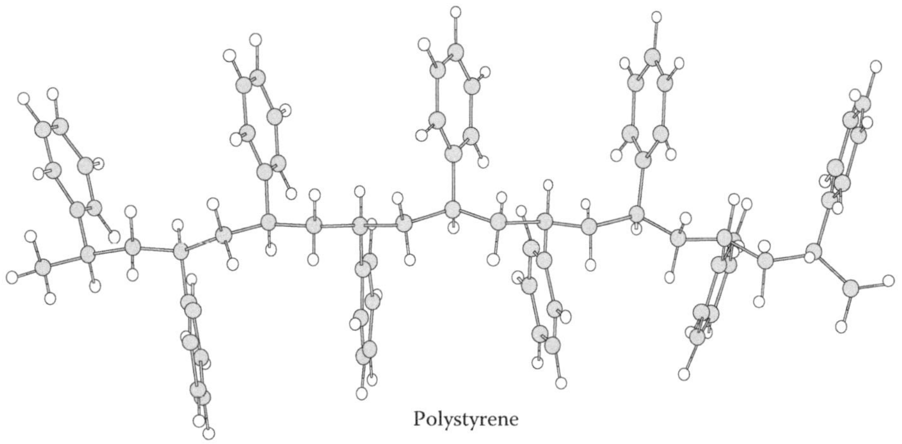

Polystyrene

FIGURE 18.19 Octomer of polystyrene (PS) containing a $-(CH_2)_n$ backbone with phenyl subunits.

takes place to a much larger extent in a π-system than in a σ-system. The π-system has shallow carriers that are easily mobile because of a small λ/Δ ratio.

18.5 ELECTRONIC STRUCTURE OF ONE-DIMENSIONAL CRYSTALS

What is the mechanism for conductivity in one-dimensional organic stacks or polymers? This depends on the electronic structure and how it is coupled to the nuclear vibrations. We want to treat these rather different systems within the same electronic model. In a stacked π-system, the principle is the same as for ET in general. There is always some vibration that cooperates with the electronic motion to carry electrons back and forth between equivalent locations. Complications arise in one-dimensional systems particularly due to the Peierls distortion.

18.5.1 SU–SCHRIEFFER–HEEGER MODEL

Immediately after the discovery of high conductivity in PA, a model was presented by Su, Schrieffer, and Heeger (SSH). This model goes back to earlier models on polyenes by Ooshika, Longuet-Higgins, Salem, Poplc, and Walmsley. The Peierls gap, related to alternating CC bond lengths, leads to insulation at T = 0, since the band gap in PA is a few electron volts. A new concept, called the *soliton*, was introduced in the SSH model. Originally, it was suggested that nonmatching alternancy could cause conductivity, and that single electrons or holes from the dopants should cause (one-electron) states in the gap. The model was extended in various ways. Direct calculations were carried out on doped systems, for example, by S. Stafström and J.-L. Brédas. It was beginning to be realized that structural deformations of the lattice are important for understanding the conductivity problem.

18.5.2 DELOCALIZATION MODEL FOR PA

Detailed calculations reveal the nature of the bonding between alkali dopants and the PA backbone. The bond-length-dependent coupling model, introduced in Chapter 3, appears to be the most logical extension of the SSH model. Bond length dependent coupling does not cause any significant additional complications in the simple Hückel model, so the model can be used on large chains in short calculation times. As far as possible, the results should be confirmed by using *ab initio* methods.

The Peierls distortion is confirmed in accurate models. If $|\beta_0|$ is given the value of 3 eV in a bond-length-dependent coupling model, the band gap has the value $|\beta| = 1.57$ eV.

As mentioned above, Winokur et al. have shown experimentally that threefold or fourfold channels are formed for high doping levels. The calculations behind Figure 18.13 support this interpretation. By examining CC bond lengths in the PA chains forming the channel, we find almost equal bond lengths over more than 20 carbon atoms. This means that the gap at the Fermi level disappears. This cannot be concluded from simple Hartree–Fock or DFT calculations. A method should be used (for example, CASSCF or TD-DFT) that is capable of giving correct excited states. In any case, the metallicity seems to be connected to almost delocalized electrons.

In the slightly alkali doped system, the electron is attracted to the positive alkali nucleus and forms a quite shallow carrier. Increased doping levels lead to a high density of such carriers. Vibrations along the chain promote mobility in the way that we are used to from the Marcus model. The localization of the carriers is directly coupled to the position of the nuclei as in all ET. This coupling already exists in the lowest vibrational state. Therefore, it is no longer surprising that there is conductivity at T = 0 K in the highly doped cases, and it is also not surprising that the conductivity increases for T > 0 K, as for a metal.

The activated conductivity for low doping levels is also easy to explain. In this case, there are no channels of the type seen in Figure 18.12. The electrons are trapped with greater binding energies than in the highly doped case. The reorganization energy is certainly also much larger than in the case of high doping levels. Thermal activation is therefore necessary to move the electrons around in the case of low doping levels. A quite large doping level ($\approx$5%) is necessary to create almost free electrons. At room temperature, the conductivity in PA is already quite high at low doping levels ($\approx$1%). The Pauli susceptibility remains small since the electrons are not free.

18.5.3 Behavior of Three-Quarter or One-Quarter Filled Bands

If the bond-length-dependent Hückel model is used for three-quarter or one-quarter filled bands, the result is surprising at first. A one-dimensional chain with 104 atoms, each bonded by a single bond is used. Three-quarter occupancy can be achieved by occupying the lowest 78 MOs by two electrons and leaving the rest empty. A band gap at the Fermi level opens up, but it is smaller than in the case of half-filled bands: $\beta = 0.12$ eV, if $|\beta_0| = 3$ eV.

If the occupancy is changed so that 77 orbitals are filled, while orbitals 78 and 79 have one electron each, we obtain a state that corresponds to an excited singlet or triplet state. After convergence, a completely different result is obtained compared to the previous occupation. All bonds are now of the same length (1.444 Å) and all sites have the same π charge (1.5|e|).

Similar wave functions are obtained when the semi-empirical PM3 method is run using RHF or UHF. It is possible to generate two different solutions. However, these both solutions are severe approximations of some fundamental many-electron state, since correlation is not included in the model. Nevertheless, the fact that there are two different wavefunctions already at the one-electron level, makes it possible to suggest Figure 18.20 for the conductivity.

Let us examine one-quarter occupancy of the valence band. The three-quarters case of (SN)$_x$ is similar, since the only difference is that electrons are replaced by holes. The solution with equal bond lengths is obviously the spin-coupled spin density wave (SDW) state seen at the top of Figure 18.20. The HOMO is shown in gray. A better analysis should include all the occupied orbitals. A complete analysis should be of an *ab initio* type and involve total energies. Here, the spin-coupled case was obtained by using the UHF method, which allows different spins on different atoms. In the solution, the bond order is equally distributed over the bond and therefore there are equal bond lengths.

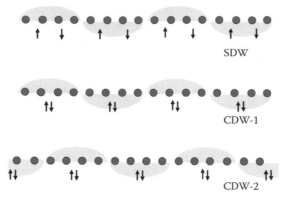

FIGURE 18.20   Degenerate wave functions for one-quarter filling.

The solutions, denoted CDW-1 and CDW-2, do not have a symmetric charge distribution (CDW = charge density wave). This result may be obtained using the RHF method, where the same spatial orbital is occupied both with spin-up and spin-down.

Three-quarters localization may be achieved by changing every second carbon atom in PA to a nitrogen atom. Optimization by the one-electron, unrestricted PM3 model leads either to an SDW or a CDW solution. The result is a deformed molecule with a repetition period of four atoms. It is clear that there are two possibilities for the ground state, and that correlation effects have to be included to obtain a correct result. The conclusion is that in the one-dimensional systems, one-half band occupation leads to a trivial Peierls dimerization, while the cases of one-quarter or three-quarters occupation may lead to interesting dynamics.

18.5.4   MOBILITY OF ELECTRONS

Weiss found that mobility is activated in oxidized or doped PPY (Figure 18.21). This is also the case in doped PA. The larger the activation energy, the smaller the doping level. This is readily explained by the fact that at high doping levels, the distance

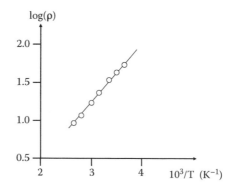

FIGURE 18.21   Activated behavior of resistivity in PPY as obtained by D. E. Weiss. T is absolute temperature and ρ is resistivity.

between the donor and the acceptor is so small that the delocalization regions overlap. The coupling between the donor and the acceptor is larger, and thus the activation energy is lower.

Conductivity is the carrier density multiplied by mobility. The Marcus model may be used to calculate the mobility residence time and the Drude model may be used to obtain conductivity. We recall that the two parabolas are total energy curves for nuclear motion along a reaction coordinate. There are two equal minima ($\Delta G^{\ominus} = 0$) since the electron can be localized at two equivalent sites, as long as the field is negligible. If a field is switched on, there is an imbalance in the Marcus curves that leads to less barriers to jump in the field direction. Metallic conductivity is obtained when the height of the barrier, ΔG^*, tends to zero. This happens when $\lambda < 2H_{12}$ (Chapter 10). $2H_{12}$ is directly related to the band width.

The connection between conductivity (σ) and mobility (μ) is as follows:

$$\sigma = n \cdot e \cdot \mu \tag{18.3}$$

where n is the number of electrons carriers and e is the electron charge. We may write the current as

$$J = n \cdot e \cdot \mu \cdot A; \quad n \cdot \mu = R \cdot k; \dagger \quad k = (k_+ - k_-), \tag{18.4}$$

where R is the distance between the metal ions in the direction of the electric field, and k_+ and k_- are the reaction rates for the forward and backward reactions, respectively. In the Marcus model, we may assume that the coupling is sufficiently small, so that the height of the barrier is not greatly affected by coupling. From Equation 10.29, we obtain the barrier height for forward (E_+) and backward transfer (E_-) as follows:

$$E_{\pm} = \frac{\lambda}{4}\left(1 \mp \frac{eER}{\lambda}\right)^2 = \frac{\lambda}{4} \mp \frac{eER}{2} + \frac{(eER)^2}{4\lambda}, \tag{18.5}$$

depending on whether ET is in the same or a different direction as the applied field. The ET rate between two molecules is

$$k_{\pm} = v_n \kappa \cdot \exp\left(-\frac{E_{\pm}}{k_B T}\right), \tag{18.6}$$

where v_n is the frequency for the vibrational mode that is coupled to ET, T is the temperature, and k_B is the Boltzmann constant. To know which vibrational mode is relevant, we have to find out which bonding distances are changed when an electron is removed or added. The connected modes induce ET. In the case of metal ions with ligands, the relevant mode is the breathing mode. If electrons are exchanged in between organic π-systems, the CC stretch frequency is the most important one. In the former case, the breathing mode is about 500 cm^{-1} and in the latter case it is 1600 cm^{-1}.

If the difference between the forward and the backward current is taken and the quadratic term is neglected in the exponent and the terms with E± are Taylor expanded, we obtain

$$k_+ - k_- = v_n\kappa\left[\exp\left(-\frac{E_+}{k_BT}\right) - \exp\left(-\frac{E_-}{k_BT}\right)\right] \approx v_n\kappa \cdot \frac{eER}{k_BT} \cdot \exp\left(-\frac{\lambda}{4k_BT}\right). \quad (18.7)$$

The total current may be expressed as

$$J = v_n\kappa \cdot \frac{ne^2ER^2}{k_BT} \cdot \exp\left(-\frac{\lambda}{4k_BT}\right). \quad (18.8)$$

In the case where $v_{el} \ll v_n$, the Taylor expansion, Equation 18.23, yields $\kappa = v_{el}/v_n$. If this is inserted into Equation 18.30, we obtain conductivity proportional to Δ^2 and independent of the nuclear frequency:

$$J = \frac{ne^2ER^2}{k_BT} \cdot \frac{\Delta^2}{4\pi\hbar}\left(\frac{\pi^3}{\lambda k_BT}\right)^{1/2} \cdot \exp\left(-\frac{\lambda}{4k_BT}\right). \quad (18.9)$$

The result of Equation 18.9 is shown in Figure 18.22. This curve is in good agreement with the experiment. It is worthwhile noticing that the same equation is used both in the activated region up to $T \approx 200$ K and in the higher temperature regions where the conductivity decreases again as the temperature increases.

18.5.5 CONDUCTIVITY IN DNA?

Ever since the discovery of the repetitive structure of DNA, it has been speculated that DNA is conducting. In fact, there have been a number of experiments that

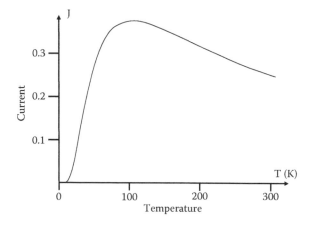

FIGURE 18.22 Calculated conductivity as a function of temperature in the case with localized charge carriers.

support that DNA is (1) conducting as a metal, (2) conducting as a semiconductor with activation energy, (3) superconducting below a certain critical temperature, and (4) conducting by hopping over a long distance. In addition to the fact that not all four conclusions can be valid at the same time, there seems to be no theoretical support for conductivity or fast ET. It is important to get to the bottom of this problem, for the simple reason that it is widely but incorrectly believed that everything repetitive and everything aromatic is conductive.

Figure 18.23 shows two bases along a DNA strand, both consisting of adenosine. The bases are cut from a pair and saturated to a molecular form that reminds us of FAD (Chapter 11), although FAD has two phosphates instead of one. The molecule shown in Figure 18.23 can be optimized with the simple HyperChem methods that have been used everywhere in this book. The problem is that in the fully optimized system, the adenosine rings tend to open up, in disagreement with the full crystallographic structure of a single strand. There may be many reasons for this. The calculational model may be inaccurate because it is too simple. It is more likely that the molecule that is cut from a long chain and saturated does not have the same equilibrium geometry as in the DNA strand. In any case, Figure 18.23, which is not fully optimized, quite closely represents the local structure between two bases in the chain.

Possible conductivity in DNA could occur vertically in Figure 18.23. A reason for confusing the results may be that immense experimental difficulties cannot be surmounted. A single DNA strand is picked out from its environment and inserted between two electrodes in a vacuum. The DNA chain may not "survive" that treatment.

Furthermore, theories for conductivity in fact require that the system is infinite for all practical purposes, and this may not be the case if conductivity is measured over a few layers of bases. In later and more relevant experiments, J. Barton et al. have attempted to use microscopic electrodes, built into both ends of a DNA string. This experiment is done either in the ground state as an ordinary ET experiment, or by photoinduced electron transfer. All ET is supposed to take place through the π-systems of the bases.

In the theoretical studies, the DNA strand is treated as an ET system. The reorganization energy (λ) has to be calculated for each of the bases. The coupling

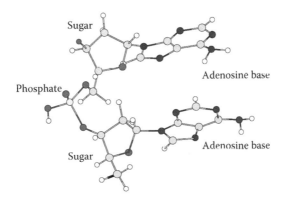

FIGURE 18.23 **(See color insert)** Part of a DNA strand with two adenosine bases.

($H_{12} = \Delta/2$) has to be calculated for each pair of base pairs. There appears to be no disagreement on this procedure. The time for one-step ET and multistep ET can be calculated, and the conductivity obtained using the Drude model.

The theoretical results are reasonably accurate and consistent between different calculations. ET takes place in single jumps, although a slight inaccuracy of the calculation is sufficient to allow multiple jumps in some cases. The most concise conclusion is that conductivity in DNA is caused by single leaps between the bases in the DNA strand, approximately, as in graphite, in the perpendicular direction.

The DNA bases are not particularly suitable for ET. In fact, they are avoided by the living systems in the transport of electrons, as we have seen in Chapter 11. In natural ET systems, other molecules, such as FAD or NADH, are used in the transport of electrons. The idea to use DNA as a basic structure in a molecular electronic device appears promising, however.

To be useful in ET, the DNA bases should be substituted with simpler organic moles of the type involved in the superconducting Bechgaard salts (Figure 18.24). PF_6 molecules, which are easily reduced, should be used to create positive carriers in the same way as in the case of organic superconductors.

The important thing is that the organic molecule is a π-system that is capable of giving away electrons without changing its structure. If we are looking for superconductivity, it is further necessary that the existence of one or two carriers on the same molecule is energetically possible.

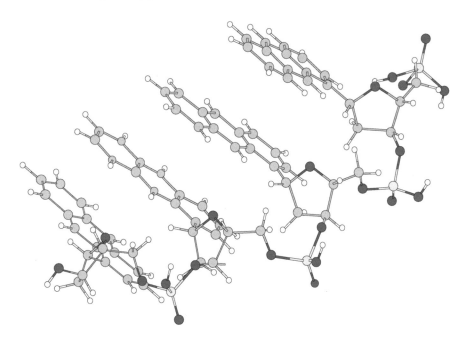

FIGURE 18.24 Part of a possible polymer system for organic conductivity and superconductivity. Organic molecules, staged by a DNA backbone, form a set of easily oxidizable molecules for the creation of carriers.

18.5.6 Conductivity at Low Temperatures

At normal temperature, there is an exponential decrease of conductivity versus temperature. In a diagram where conductivity is plotted as a function of $1/T$, where T is the absolute temperature, there is a linear dependence down to low temperatures. The power dependences of the temperature in the rate equation, as in Equation 18.8, or in the Landau–Zener term, are of very little importance in comparison with the exponential function.

At lower temperatures, the log k versus $1/T$ curve no longer decreases linearly but bends with a decreasing slope (Figure 10.19). This is due to nuclear tunneling, as described in the Bixon–Jortner model. What is stated here holds for all dimensions, of course, not just in the one-dimensional systems.

The inducing vibrational motion is mainly the breathing mode in metal complexes and the structural motions in organic π-systems of the type that change the lengths of the double bonds.

The tunneling behavior has mistakenly been ascribed to "variable range hopping" (VRH). The idea of the latter model is that when the temperature decreases and vibrational activation becomes more difficult, the electron decides to leap longer distances. However, most rate expressions are activated with an $\exp(-E_a/kT)$ factor and this dependence occurs independent of the leaping distance. A small fraction of the leaps may occur to the next near neighbor, but this fraction does not change very much with temperature. In fact, its dependence on temperature is the same as leaping to neighbors. The VRH model is thus of doubtful value, although it often leads to apparent agreement with experimental results.

In any case, nuclear tunneling cannot be avoided. The nuclear tunneling effect depends on the fact that at a low temperature the nuclei do not have enough energy to pass the activation barrier. Instead, tunneling through the barrier by the nuclei becomes relatively more important.

Bibliography

CHAPTER 1

PAPERS

M. Born, P. Jordan, W. Heisenberg, *Z. Phys. 34*, 858 (1925); *35*, 557 (1925).
L. Brillouin, *C. R. Acad. Sci. Paris 183*, 24 (1926).
L. de Broglie, *Compt. Rendus Acad. Sci. 177*, 507–510 (1923) (in French).
G. Gamow, *Z. Phys. 51*, 204–212 (1928).
W. Kutzelnigg, *J. Chem. Phys. 40*, 3640–3647 (1964).
P.O. Löwdin, *Phys. Rev. 97*, 1474–1489 (1955).
P.O. Löwdin, *Phys. Rev. 97*, 1490–1508 (1955).
P.O. Löwdin, *Phys. Rev. 97*, 1509–1520 (1955).
W.J. Meath, *Phys. Rev. A 41*, 1556 (1980).
M. Planck, *Ann. Phys. 309*, 553–563 (1901).

BOOKS

L.D. Landau, E.M. Lifshitz, *Quantum Mechanics* (Pergamon Press Ltd, Oxford, London, Paris, 1958, 1959, 1962). English translation by J.B. Sykes and J.S. Bell.
P.O. Löwdin, *Linear Algebra for Quantum Theory* (Wiley Interscience, 1998).
E. Merzbacher, *Quantum Mechanics*, 2nd edition (Wiley, New York, 1970).

CHAPTER 2

PAPERS

General

L. Brillouin, *Actualités Sci. Ind. 1933*, 71 (1933); *1934*, 159 (1934).
V.A. Fock, *Zeits. Physik 98*, 145–154 (1935).
E. Hylleraas, *Z. Phys. 48*, 469 (1928); *54*, 347 (1929); *65*, 209 (1930).
T. Koopmans, *Physica 1*, 104 (1934).
P.-O. Löwdin, *Revs. Mod. Phys. 36*, 966 (1964).
C. Møller, M.S. Plesset, *Phys. Rev. 46*, 618 (1934).
J.C. Slater, *Phys. Rev. 81*, 385 (1951).
J.C. Slater, *Adv. Quantum Chem. 6*, 1 (1972).
J.C. Slater, *Int. J. Quantum Chem. Symp. 9*, 7 (1975).

Local Exchange

A.D. Becke, *J. Chem. Phys. 98*, 5648 (1993).
A.D. Becke, *Can. J. Chem. 74*, 995 (1996).
A.D. Becke, *J. Chem. Phys. 117*, 6935 (2002).
R. Gáspár, *Acta Phys. Acad. Sci. Hung. 3*, 263 (1954).
P. Hohenberg, W. Kohn, *Phys. Rev. B 136*, 864 (1964).
W. Kohn, A.D. Becke, R.G. Parr, *J. Phys. Chem. 100*, 12974 (1996).
W. Kohn, L.J. Sham, *Phys. Rev. A 140*, 1133 (1965).
R.G. Parr, W. Yang, *Ann. Rev. Phys. Chem. 46*, 701 (1995).

470   Bibliography

J.P. Perdew, M. Levy, *Int. J. Quantum Chem. Symp. 49*, 539 (1994).
A. Savin, F. Colonna, R. Pollet, *Int. J. Quantum Chem. 93*, 166 (2003).
W. Yang, Q. Wu, *Phys. Rev. Lett. 89*, 143002 (2002).

Best Overlap Orbitals

S. Larsson, *J. Chem. Phys. 58*, 5049–5057 (1973).

BOOKS

P. Gombás, *Die Statistische Theorie des Atoms und Ihre Anwendungen* (Springer-Verlag, Wien, 1949).
R.G. Parr, W. Yang, *Density-Functional Theory of Atoms and Molecules* (Oxford University Press, New York, 1989).

CHAPTER 3

PAPERS

MOLCAS 7.4: F. Aquilante, L. De Vico, N. Ferré, G. Ghigo, P.-R. Malm-quist, P. Neogrády, T.B. Pedersen, M. Pitonak, M. Reiher, B.O. Roos, L. Serrano-Andrés, M. Urban, V. Veryazov, R. Lindh, *J. Comput. Chem. 31*, 224 (2010).
M.J.S. Dewar, E.G. Zoebisch, E.F. Healy, J.J.P. Stewart, *J. Am. Chem. Soc. 107*, 3902 (1985).
W. Heitler, F. London, *Z. Phys. 44*, 455 (1927).
E. Hückel, *Z. Phys. 60*, 423 (1930); *70*, 204 (1931); *72*, 310 (1931); *76*, 628 (1932).
H.C. Longuet-Higgins, L. Salem, *Proc. R. Soc. Lond. A 251*, 172 (1959).
P.-O. Löwdin, *J. Chem. Phys. 18*, 365 (1950).
R.S. Mulliken, *J. Am. Chem. Soc. 74*, 811–824 (1952).
B.O. Roos, in *Advances in Chemical Physics: Ab Initio Methods in Quantum Chemistry II*; Editor K.P. Lawley (Wiley, Chichester, England, 1987), Chap. 69, p. 399.
B.O. Roos, P.R. Taylor, P.E.M. Siegbahn, *Chem. Phys. 48*, 157 (1980).
J.J.P. Stewart, *J. Comput. Chem. 10*, 209–221 (1989).
J.D. Watts, R.J. Bartlett, *J. Chem. Phys. 96*, 6073–6084 (1992).

CHAPTER 4

PAPERS

R.J.M. Bennet, W.B. Sommerville, *Nature 223*, 489–490 (1969).
J.W. Cooley, *Math. Comput. 15*, 363 (1960).
K. Huang, A. Rhys, *Proc. R. Soc. Lond. A 204*, 406 (1950).
F. Hund, *Z. Phys. 46*, 805 (1927).
R.W. Nicholls, *Nature 219*, 151 (1968).
L. Nordheim, *Z. Phys. 46*, 833 (1927).
E.B. Wilson Jr., *J. Chem. Phys. 7*, 1047–1052 (1939).

BOOKS

R.G. Pearson, *Symmetry Rules for Chemical Reactions. Orbital Topology and Elementary Processes* (Wiley-Interscience, New York, London, Sydney, Toronto, 1976).
E.B. Wilson Jr., J.C. Decius, P.C. Cross, *Molecular Vibrations. The Theory of Infrared and Raman Vibrational Spectra* (Dover Publications, Inc., New York, NY, 1980). Originally published in 1955 by McGraw-Hill Book Co. Inc.

CHAPTER 5

BOOKS

D. Chandler, *Introduction to Modern Statistical Mechanics* (Oxford University Press, New York, 1987).

D.A. McQuarrie, J.D. Simon, *Molecular Thermodynamics* (University Science Books, Sausalito, CA, 1999).

CHAPTER 6

PAPERS

P.W. Anderson, *Phys. Rev. 79*, 350 (1950).

P.W. Anderson, *Phys. Rev. 115*, 2–13 (1959).

P.W. Anderson, H. Hasegawa, *Phys. Rev. 100*, 675 (1955).

I. Benjamin, *J. Phys. Chem. B 112*, 15801–15806 (2008).

M. Born, *Z. Phys. 1*, 45 (1920).

H.A. Jahn, E. Teller, *Proc. R. Soc. A 161*, 220 (1937).

H.A. Kramers, *Physica 1*, 182 (1934).

A.A. Rashin, B. Honig, *J. Phys. Chem. 89*, 5588–5593 (1985).

J.J. Van Vleck, *J. Chem. Phys. 3*, 803 (1935); *3*, 807 (1035).

BOOKS

C.J. Ballhausen, *Introduction to Ligand Field Theory* (McGraw-Hill Book Company, Inc., New York, San Fransisco, Toronto, London, 1962).

M. Born, K. Huang, *Dynamical Theories of Crystal Lattices* (Clarendon, Oxford, 1954).

S.A. Borshch, I.B. Bersuker, *The Jahn-Teller Effect and Vibronic Interactions in Modern Chemistry* (Plenum, New York, 1984).

C.K. Jørgensen, *Modern Aspects of Ligand Field Theory* (North Holland, Amsterdam, 1971).

J.H. Van Vleck, *Theory of Magnetic and Electric Susceptibilities* (Oxford University Press, London, 1932).

CHAPTER 7

PAPERS

L. Landau, *Phys. Z. Sow. 2*, 46 (1932).

M.S. Topaler, T.C. Allen, D.W. Schwenke, D.G. Truhlar, *J. Chem. Phys. 109*, 3321–3345 (1998).

J.C. Tully, *J. Chem. Phys. 93*, 1061 (1990).

C. Zener, *Proc. R. Soc. A 137*, 696 (1932).

BOOKS

E.E. Nikitin, *Theory of Elementary Atomic and Molecular Processes* (Oxford University Press, London, 1974).

CHAPTER 8

BOOKS

H. Eyring, J. Walter, G.E. Kimball, *Quantum Chemistry* (Wiley, New York, London, 1961).
A.A. Frost, R.G. Pearson, *Kinetics and Mechanism* (Wiley, New York, London, 1953, 1961).

CHAPTER 9

PAPERS

N. Agmon, *Chem. Phys. Lett. 244*, 456–462 (1995).
W.J. Albery, *Phys. Chem. 31*, 227–263 (1980).
W.J. Albery, *Faraday Discuss. Chem. Soc. 74*, 245–256 (1982).
G. Biczó, J. Ladik, J. Gergely, *Acta Phys. Acad. Sci. Hung. 20*, 11 (1966).
J.R. Bryant, T. Matsuo, J.M. Mayer, *Inorg. Chem. 43*, 1587–1592 (2004).
X. Duan, S. Scheiner, *J. Mol. Struct. 270*, 173–185 (1992).
R.M. Hochstrasser, H.P. Trommsdorff, *Chem. Phys. 115*, 1–6 (1987).
S. Iwata, C. Ostermeier, B. Ludwig, H. Michel, *Nature 376*, 660–669 (1995).
W.M. Latimer, W.H. Rodebush, *J. Am. Chem. Soc. 42*, 1419 (1920).
H.S. Lee, M.E. Tuckerman, *J. Phys. Chem. B 112*, 9917 (2008).
J. Lobaugh, G.A. Voth, *J. Chem. Phys. 100*, 3039 (1994); *104*, 2056 (1996).
P.-O. Löwdin, *Rev. Mod. Phys. 35*, 724–732 (1963).
R.A. Marcus, *J. Phys. Chem. 72*, 891–899 (1968).
M. Matsuo, *Phys. Rev. C 73*, 044309 (2006).
D.J. McLennan, *J. Chem. Educ. 53*, 348–351 (1976).
D.J. McLennan, *Aust. J. Chem. 32*, 1883 (1995).
K.R. Mitchell-Koch, W.H. Thompson, *J. Phys. Chem. B 112*, 7448–7459 (2008).
T.S. Moore, T.F. Winmill, *J. Chem. Soc. 101*, 1635 (1912).
D.E. Sagnella, K. Laasonen, M.L. Klein, *Biophys. J. 71*, 1172–1178 (1996).
M.E. Tuckerman, R. Laasonen, M. Sprik, M. Parrinello, *J. Phys. Chem. 99*, 5749 (1995).
R.J.P. Williams, *J. Solid State Chem. 145*, 488–495 (1999).
R.J.P. Williams, *Nature 376*, 643 (2002); *376*, 660–669 (1995); *378*, 235 (1995).
Y. Wu, B. Ilan, G.A. Voth, *Biophys. J. 92*, 61–69 (2006).

BOOKS

R.P. Bell, *The Proton in Chemistry* (Chapman and Hall, London, 1973).
R.P. Bell, *The Tunnel Effect in Chemistry* (Chapman and Hall, New York, 1980).
L. Melander, *Isotope Effects on Reaction Rates* (The Ronald Press, New York, 1969).
L. Melander, W.R. Saunders, *Reaction Rates of Isotopic Molecules* (Wiley, New York, 1980).

CHAPTER 10

PAPERS

General

D.N. Beratan, J.N. Onuchic, J.N. Betts, B.E. Bowler, H.B. Gray, *J. Am. Chem. Soc. 112*, 7915–7921 (1990).
B.S. Brunschwig, C. Creutz, N. Sutin, *Chem. Soc. Rev. 31*, 168–184 (2002).
B.S. Brunschwig, J. Logan, M.D. Newton, N. Sutin, *J. Am. Chem. Soc. 102*, 5798–5809 (1980).

A. Calhoun, M.T.M. Koper, G.A. Voth, *J. Phys. Chem. B 103*, 3442 (1999).
C. Creutz, H. Taube, *J. Am. Chem. Soc. 91*, 3988 (1969); *95*, 1086 (1973).
D.M. D'Alessandro, F.R. Keene, *Chem. Soc. Rev. 35*, 424–440 (2006).
S.K. Door, R.B. Dyer, P.O. Scoutland, W.H. Woodruff, *J. Am. Chem. Soc. 115*, 6398 (1993).
J. Halpern, L.E. Orgel, *Discuss. Faraday Soc. 29*, 32–41 (1960).
J.J. Hopfield, *Proc. Natl. Acad. Sci. USA 71*, 3640–3644 (1974).
S. Knapp, et al., *J. Am. Chem. Soc. 113*, 4010–4013 (1991).
S. Larsson, *J. Am. Chem. Soc. 103*, 4034–4040 (1981).
S. Larsson, *Chem. Phys. Lett. 90*, 136–139 (1982).
S. Larsson, *J. Phys. Chem. 88*, 1321–1323 (1984).
S. Larsson, *Chem. Scr. 28A*, 15–20 (1988).
S. Larsson, M. Braga, A. Broo, B. Källebring, *Int. J. Quantum Chem. QBS 18*, 99–118 (1991).
S. Larsson, K. Ståhl, M. Zerner, *Inorg. Chem. 25*, 3033–3037 (1986).
V.G. Levich, R.R. Dogonadze, *Dokl. Akad. Nauk SSSR 124*, 123–126 (1959).
V.G. Levich, R.R. Dogonadze, *Collect. Czech. Chem. Commun. 26*, 193 (1961).
C.H. Londergan, C.P. Kubiak, *J. Phys. Chem. A 107*, 9301–9311 (2003).
R.C. Long, D.N. Hendrickson, *J. Am. Chem. Soc. 105*, 1513–1521 (1983).
R.A. Marcus, *J. Chem. Phys. 24*, 966–978 (1956).
R.A. Marcus, *Annu. Rev. Phys. Chem. 15*, 155 (1964).
R.A. Marcus, *J. Chem. Phys. 98*, 7170 (1994).
G. McLendon, *Acc. Chem. Res. 21*, 160–167 (1988).
T.J. Meade, H.B. Gray, J.R. Winkler, *J. Am. chem. Soc. 111*, 4353–4356 (1989).
M.D. Newton, *Chem. Rev. 91*, 767–792 (1991).
M.D. Newton, N. Sutin, *Annu. Rev. Phys. Chem. 35*, 437 (1984).
D.C. Nocera, J.R. Winkler, K.M. Yokum, E. Bordignon, H.B. Gray, *J. Am. Chem. Soc. 106*, 5145–5150 (1984).
F.O. Raineri, H.L. Friedman, *Adv. Chem. Phys. 107*, 81 (1999).
M. Ratner, *Int. J. Quantum Chem. 14*, 675–694 (1978).
J.M. Saveant, *J. Am. Chem. Soc. 114*, 10595 (1992); *109*, 6788 (1987).
W. Schmickler, M.T.M. Koper, *Electrochem. Commun. 1*, 402 (1999).
J.D. Stong, J.M. Burk, P. Daly, P. Wright, T.G. Spiro, *J. Am. Chem. Soc. 102*, 5815 (1980).
H. Taube, H. Myers, R.L. Rich, *J. Am. Chem. Soc. 75*, 4118–4119 (1953).
M.J. Therien, et al., in *Electron Transfer in Inorganic, Organic and Biological Systems*; Editors J. Bolton, G.L. McLendon, N. Mataga (American Chemical Society, Washington, DC, 1991), pp. 191–199.
J. Ulstrup, J. Jortner, *J. Chem. Phys. 63*, 4358–4368 (1975).
G.C. Walker, P.F. Barbara, S.K. Boom, Y. Dong, J.T. Hupp, *J. Phys. Chem. 95*, 5712 (1991).
J.R. Wilier, J.V. Beitz, R.K. Huddleston, *J. Am. Chem. Soc. 106*, 5057–5068 (1984).
R.J.P. Williams, *J. Solid State Chem. 145*, 488–495 (1999).
X. Xuan, H. Zhang, J. Wang, H. Wang, *J. Phys. Chem. A 108*, 7513 (2004).

Reorganization Energy

P.G. George, J.S. Griffith, in *The Enzymes*; Editors P.D. Boyer, H. Lardy, K. Myrbäck, vol. 1 (Academic Press, New York, 1959), pp. 347–389.
N.S. Hush, *J. Chem. Phys. 28*, 962–972 (1958).
R.A. Marcus, *J. Chem. Phys. 24*, 966–978 (1956).
R.A. Marcus, *Faraday Discuss. Chem. Soc. 29*, 21 (1960).
R.A. Marcus, *J. Chem. Phys. 43*, 679 (1965).
R.A. Marcus, *J. Phys. Chem. 94*, 1050–1055 (1990) and further references therein.
R.A. Marcus, H. Sumi, *J. Chem. Phys. 84*, 4894 (1986).
J. Zhu, R. Ma, Y. Lu, G. Stell, *J. Chem. Phys. 123*, 224505 (2005).

J. Zhu, J.C. Rasaiah, *J. Chem. Phys. 101*, 9966 (1994).
J. Zhu, J. Wang, G. Stell, *J. Chem. Phys. 125*, 164511/1–7 (2006).

Coupling

P. Bertrand, *Chem. Phys. Lett. 113*, 104 (1985); *140*, 57–63 (1987).
M. Braga, A. Broo, S. Larsson, *Chem. Phys. 156*, 1–9 (1991).
M. Braga, S. Larsson, *Chem. Phys. 162*, 369–377 (1992).
M. Braga, S. Larsson, *Chem. Phys. Lett. 200*, 573–579 (1992).
M. Braga, S. Larsson, *Int. J. Quantum. Chem. 44*, 839–851 (1992).
M. Braga, S. Larsson, *J. Phys. Chem. 96*, 9218–9224 (1992).
A. Broo, S. Larsson, *Chem. Phys. 148*, 103–115 (1990).
A. Broo, S. Larsson, *J. Phys. Chem. 95*, 4925–4928 (1991).
A. Broo, S. Larsson, *Chem. Phys. 161*, 363–378 (1992).
S.S. Isied, *Progr. Inorg. Chem. 32*, 443–517 (1984).
A.D. Joran, et al., *Nature 327*, 508–511 (1987).
H. McConnell, *J. Chem. Phys. 33*, 115 (1961).
H. McConnell, *J. Chem. Phys. 35*, 508 (1961)
N. Sutin, *Acc. Chem. Res. 5*, 275 (1982).

Localized or Delocalized

G.C. Allen, N.S. Hush, *Progr. Inorg. Chem. 8*, 357–390 (1967).
C. Creutz, *Progr. Inorg. Chem. 30*, 1–73 (1983).
N.S. Hush, *Progr. Inorg. Chem. 8*, 391–444 (1967).
N.S. Hush, *Chem. Phys. 10*, 361 (1975).
N.S. Hush, in *Mixed-Valence Compounds in Chemistry, Physics and Biology*, NATO ASI Series; Editor D.B. Brown (D. Reidel Publishing Company, Dodrecht, The Netherlands, 1980), p. 151.
K. Prassides, P. Day, *Inorg. Chem. 24*, 1109–1110 (1985).
J.R. Reimers, N.S. Hush, *Chem. Phys. 299*, 79–82 (2004).

PKS Model

K. Boukheddaden, J. Linares, F. Varret, *Chem. Phys. 172*, 239–245 (1993).
S.B. Piepho, E.R. Krausz, P.N. Schatz, *J. Am. Chem. Soc. 100*, 2996–3005 (1978).
K. Prassides, P. Day, *J. Chem. Soc. Faraday Trans. II 80*, 85–95 (1984).
K. Prassides, P. Day, A.K. Cheetham, *J. Am. Chem. Soc. 105*, 3366–3368 (1983).
K. Prassides, P.N. Schatz, *J. Phys. Chem. 93*, 83–89 (1989).
K. Prassides, P.N. Schatz, K.Y. Wong, P. Day, *J. Phys. Chem. 90*, 5588–5597 (1986).
M.B. Robin, P. Day, *Adv. Inorg. Chem. Radiochem. 10*, 247–422 (1967).
P.N. Schatz, S.B. Piepho, E.R. Krausz, *Chem. Phys. Lett. 55*, 539–542 (1978).
H. So, *Bull. Kor. Chem. Soc. 13*, 385–388 (1992).
D.R. Talham, D.O. Cowan, *Organometallics 3*, 1712–1715 (1984).

Nuclear Motion

J. Jortner, *J. Chem. Phys. 64*, 4860–4867 (1976).
J. Jortner, M. Bixon, *Isr. J. Chem. 7*, 189–220 (1969).
N.R. Kestner, J. Logan, J. Jortner, *J. Phys. Chem. 78*, 2148–2166 (1974).
A. Klimkāns, S. Larsson, *Int. J. Quantum. Chem. 77*, 211–220 (2000).

Pathway Model

D.N. Beratan, J.N. Betts, J.N. Onuchic, *Science 252*, 1285–1288 (1991).
S. Larsson, M. Braga, *Int. J. Quantum. Chem. QBS 20*, 65–76 (1993).

Marcus Inverted Region

J.W. Beitz, J.R. Miller, *J.Chem. Phys. 71*, 4579–4595 (1979).

G.L. Closs, L.T. Calcaterra, N.J. Green, K.W. Penfield, J.R. Miller, *J. Phys. Chem. 90*, 3673–3683 (1986).

G.L. Closs, J.R. Miller, *Science 240*, 440–447 (1988).

R.K. Huddleston, J.R. Miller, *J. Phys. Chem. 85*, 2292–2298 (1981).

J.R. Miller, J.W. Beitz, *J.Chem. Phys. 74*, 6746–6756 (1981).

J.R. Miller, L.T. Calcaterra, G.L. Closs, *J. Am. Chem. Soc. 106*, 3047 (1984).

K.W. Penfield, J.R. Miller, M.N. Paddon-Row, E. Cotsaris, A.M. Oliver, N.S. Hush, *J. Am. Chem. Soc. 109*, 5061–5065 (1987).

Books

A.M. Kuznetsov, J. Ulstrup, *Electron Transfer in Chemistry and Biology: An Introduction to the Theory* (Wiley, Chichester, 1998).

W.M. Latimer, *Oxidation Potentials* (Prentice Hall, Inc., New York, NY, 1938).

S.I. Pekar, *Untersuchungen über die Electronentheorie der Kristalle* (Akademie Verlag, Berlin, 1954).

CHAPTER 11

Papers

G.T. Babcock, M. Wikström, *Nature 356*, 301–309 (1992).

A.M. Baptista, P.J. Martel, C.M. Soares, *Biophys. J. 76*, 2978–2998 (1999).

I. Belevich, M.I. Verkhovsky, M. Wikström, *Nat. Lett. 440*, 829–832 (2006).

M. Brändén, H. Sigurdson, A. Namslauer, R.E. Gennis, P. Ädelroth, P. Brzezinski, *Proc. Natl. Acad. Sci. USA 98*, 5013–5018 (2001).

M. Brändén, F. Tomson, R.E. Gennis, P. Brzezinski, *Biochemistry 41*, 10794–10798 (2002).

A. Broo, S. Larsson, *Int. J. Quantum Chem. QBS 16*, 185 (1989).

D.A. Cowan, F. Kaufman, *J. Am. Chem. Soc. 92*, 219–220 (1970).

D.A. Cowan, G. Pasternak, F. Kaufman, *Proc. Natl. Acad. Sci. USA 66*, 837–843 (1970).

V.L. Davidson, *Acc. Chem. Res. 41*, 730 (2008).

D. Devault, B. Chance, *Biophys. J. 6*, 825–847 (1966).

D. DeVault, J.H. Parkes, B. Chance, *Nature 215*, 642–644 (1967).

U. Ermler, G. Fritzsch, S.K. Buchanan, H. Michel, *Structure 2*, 925–936 (1994).

J.A. Garcia-Horsman, A. Puustinen, R.B. Gennis, M. Wikström, *Biochemistry 34*, 4428–4433 (1995).

M.R. Gunner, P.L. Dutton, *J. Am. Chem. Soc. 111*, 3400–3412 (1989).

A. Heller, *Acc. Chem. Res. 23*, 128–134 (1990).

J. Jortner, *Philos. Trans. R. Soc. B 361*, 1877–1891 (2006).

H.G. Khorana, *J. Biol. Chem. 263*, 7439–7446 (1988).

S. Larsson, *Int. J. Quantum Chem. QBS 9*, 385–397 (1982).

S. Larsson, *J. Chem. Soc. Faraday Trans. 2, 79*, 1375–1388 (1983).

R.A. Marcus, N. Sutin, *Biochim. Biophys. Acta 811*, 265–322 (1985).

H. Michel, *Biochemistry 38*, 15129–15140 (1999).

P. Mitchell, *Nature 191*, 144–148 (1961).

P. Mitchell, *Trends Biochem. Sci. 8*, 154–155 (1983).

C.C. Moser, J.M. Keske, K. Warncke, R.S. Farid, P.L. Dutton, *Nature 355*, 796–802 (1992).

C.C. Moser, C.C. Page, R. Farid, P.L. Dutton, *J. Bioenerg. Biomembr. 27*, 263–274 (1995).

W.W. Parson, Z.T. Chu, A. Warshel, *Biochim. Biophys. Acta 1017*, 251–272 (1990).

W.W. Parson, R.K. Clayton, R.J. Codgell, *Biochim. Biophys. Acta 387*, 265–277 (1975).

J. Quenneville, D.M. Popović, A.A. Stuchebrukhov, *Biochim. Biophys. Acta 1757*, 1035–1046 (2006).

R.A. Scott, A.G. Mauk, H.B. Gray, *J. Chem. Educ. 62*, 932 (1985).

S.S. Skourtis, D.H. Waldeck, D.N. Beratan, *Annu. Rev. Phys. Chem. 61*, 461–485 (2010).

A. Szent-Györgyi, *Nature 148*, 157–159 (1941).

M.I. Verkhovsky, J.E. Morgan, M. Wikström, *Biochemistry 34*, 7483–7491 (1995).

K.R. Vinothkumar, R. Henderson, *Q. Rev. Biophys. 43, 1*, 65–158 (2010).

H. Wang, S. Lin, J.P. Allen, J.C. Williams, S. Blankert, C. Laser, N.W. Woodbury, *Science 316*, 747–750 (2007).

M. Wikström, *Biochemistry 39*, 3515–3519 (2000).

P. Xiong, J.M. Nocek, J. Vura-Weis, J.-V. Lockard, M.R. Wasielewski, B.M. Hoffman, *Science 330*, 1075–1078 (2010).

Reorganization Energy

D.W. Brinkley, J.P. Roth, *J. Am. Chem. Soc. 127*, 15720–15721 (2005).

M.M. Crnogorac, N.M. Kostić, *Inorg. Chem. 39*, 5028–5035 (2000).

A. Warshel, A.K. Churg, R. Huber, *J. Mol. Biol. 168*, 693–697 (1983).

Cytochrome c Oxidase

M.R.A. Blomberg, P.E.M. Siegbahn, *Chem. Rev. 110*, 7040–7061 (2010).

M.R.A. Blomberg, P.E.M. Siegbahn, *Mol. Phys. 108*, 2733–2743 (2010).

O. Farver, S. Wherland, W.E. Antholine, G.J. Gemmen, Y. Chen, I. Pecht, J.A. Fee, *Biochemistry* (2011), DOI: 10.1021/bi100548n.

J.A. Fee, D.A. Case, L. Noodleman, *J. Am. Chem. Soc. 130*, 15002–15021 (2008).

M.-L. Tan, I. Balabin, J.N. Onuchic, *Biophys. J. 86*, 1813–1819 (2004).

CHAPTER 12

Papers

S.A. Antipin, T.B. Feldman, F.E. Gostev, O.A. Smitienko, O.M. Sarkisov, M.A. Ostrovsky, *Dokl. Biochem. Biophys. 396*, 127–127 (2004).

T. Baumert, V. Engel, C. Röttgerman, W.T. Strunz, G. Gerber, *Chem. Phys. Lett. 212*, 691 (1993).

H.A. Benesi, J.H. Hildebrand, *J. Am. Chem. Soc. 71*, 2703–2707 (1949).

F. Bernardi, M. Olivucci, I.N. Ragazos, M.A. Robb, *J. Am. Chem. Soc. 114*, 8211–8220 (1992).

S.A. Buzza, E.M. Snyder, A.W. Castleman Jr., *J. Chem. Phys. 104*, 5040–5047 (1996).

P.Y. Cheng, D. Zhong, A.H. Zewail, *J. Chem. Phys. 105*, 6216–6248 (1996).

A. Einstein, *Phys. Z. 18*, 121–128 (1917).

V. Engel, H. Metiu, R. Almeida, R.A. Marcus, A.H. Zewail, *Chem. Phys. Lett. 152*, 1 (1988).

T. Fiebig, C. Wan, S.O. Kelly, J.K. Barton, A.H. Zewail, *Proc. Natl. Acad. Sci. USA 96*, 1187–1192 (1999).

M.A. Fox, Editor, *Chem. Rev. 92*, 365–368 (1992).

J. Jortner, M. Bixon, *Collect. Czech. Chem. Commun. 63*, 1285–1294 (1998).

J. Jortner, M. Bixon, T. Langenbacher, M.E. Michel-Beyerle, *Proc. Natl. Acad. Sci. USA 95*, 12759–12765 (1998).

S.O. Kelly, J.K. Barton, *Science 283*, 375–381 (1999).

S.Y. Kim, S. Pederson, A.H. Zewail, *J. Chem. Phys. 103*, 477–480 (1995).

S. Larsson, B. Källebring, *Int. J. Quantum Chem. QBS 17*, 189–206 (1990).

F.D. Lewis, T. Wu, Y. Zhang, R.L. Letsinger, S.R. Greenfield, M.R. Wasielewski, *Science 277*, 673–676 (1997).

J.M. Nocek, J.S. Zhou, S.D. Forest, S. Priyadarshy, D.N. Beratan, J.N. Onuchic, B.M. Hoffman, *Chem. Rev. 96*, 2459–2489 (1996).

I. Schapiro, O. Weingart, V. Buss, *J. Am. Chem. Soc. 131*, 16–17 (2009).

G.G. Stokes, *Philos. Trans. R. Soc. Lond. 142*, 463–562 (1852).

M.H.B. Stowell, T.M. McPhillips, D.C. Rees, S.M. Soltis, E. Abresch, G. Feher, *Science 276*, 812–816 (1997).

M. Yoshizawa, Y. Hattori, T. Kobayashi, *Phys. Rev. B 49*, 13259–13262 (1994).

A.H. Zewail, *Science 242*, 1645–1653 (1988).

Q. Zhong, L. Poth, A.W. Castleman Jr., *J. Chem. Phys. 110*, 192–196 (1999).

O. Zhong, A.H. Zewail, *Proc. Natl. Acad. Sci. USA 96*, 2602–2607 (1999).

Rhodopsin

A. Altun, S. Yokoyama, K. Morokuma, *J. Phys. Chem. B 112*, 6814–6827 (2008).

A. Altun, S. Yokoyama, K. Morokuma, *J. Phys. Chem. B 112*, 16883–16890 (2008).

A. Altun, S. Yokoyama, K. Morokuma, *J. Phys. Chem. A 113*, 11685–11692 (2009).

P.O. Andersson, T. Gillbro, A.E. Asato, R.S.H. Liu, *J. Lumin. 51*, 11–20 (1992).

F. Blomgren, S. Larsson, *Chem. Phys. Lett. 376*, 704–709 (2003).

F. Blomgren, S. Larsson, *J. Phys. Chem. B 109*, 9104–9110 (2005).

L.M. Frutos, T. Andruniów, F. Santoro, N. Ferré, M. Olivucci, *Proc. Natl. Acad. Sci. USA 104*, 7764–7769 (2007).

T. Gillbro, P.O. Andersson, R.S.H. Liu, A.E. Asato, S. Takaishi, R.J. Cogdell, *Photochem. Photobiol. 57*, 44–48 (1993).

M.G. Khrenova, A.V. Bochenkova, A.V. Nemukhin, *Proteins 78*, 614–622 (2010).

J.E. Kim, D.W. McCamant, L. Zhu, R.A. Mathies, *J. Phys. Chem. B 105*, 1240–1249 (2001).

Y. Kurashige, H. Nakano, Y. Nakao, K. Hirao, *Chem. Phys. Lett. 400*, 425–429 (2004).

H. Nakamichi, T. Okada, *Angew. Chem. Eng. Ed. 45*, 4270–4273 (2006).

K. Nakayama, H. Nakano, K. Hirao, *Int. J. Quantum Chem. 66*, 157 (1998).

U.F. Röhrig, L. Guidoni, A. Laio, I. Frank, U. Rothlisberger, *J. Am. Chem. Soc. 126*, 15328–15329 (2004).

U.F. Röhrig, L. Guidoni, U. Rothlisberger, *Biochemistry 41*, 10799–10809 (2002).

R.W. Schoenlein, L.A. Peteanu, R.A. Mathies, C.V. Shank, *Science 254*, 412–415 (1991).

M. Schreiber, M. Sugihara, T. Okada, V. Buss, *Angew. Chem. Eng. Ed. 45*, 4274–4277 (2006).

R. Send, D. Sundholm, *Phys. Chem. Chem. Phys. 9*, 2862–2867 (2007).

L. de Vico, R. Lindh, M. Olivucci, *J. Chem. Theor. Comput. 1*, 1029–1037 (2005).

G. Wald, *Science 162*, 230–239 (1968).

Q. Wang, R.W. Schoenlein, L.A. Peteanu, R.A. Mathies, C.V. Shank, *Science 266*, 422 (1994).

M. Yan, D. Manor, G. Weng, H. Chao, L. Royhberg, T.M. Jedju, R.R. Alfano, R.H. Callender, *Proc. Natl. Acad. Sci. USA 88*, 9809–9812 (1991).

Books

P. Gaspardt, I. Burghardt, Editors, *Chemical Reactions and Their Control on the Femtosecond Time Scale* (Wiley, New York, 1997).

M. Klessinger, J. Michl, *Excited States of Organic Molecules* (VCH Publishers, Inc., New York, 1995).

J. Michl, V. Bonačić-Koutecký, *Electronic Aspects of Organic Photochemistry* (Wiley-Interscience, New York, 1990).

V. Sundström, Editor, *Femtochemistry and Femtobiology* (Imperial College Press, London, 1996).

CHAPTER 13

PAPERS

M. Bixon, J. Jortner, J.W. Verhoeven, *J. Am. Chem. Soc. 116*, 7349 (1994).

W. Brackman, *Recl. Trav. Chim. Pays-Bas 68*, 147 (1949).

A.I. Burstein, K.L. Ivanov, *Phys. Chem. Chem. Phys. 4*, 4115 (2002).

A.I. Burstein, A.A. Zharikov, *Chem. Phys. 152*, 23–30 (1991).

W.B. Davis, W.A. Svec, M.A. Ratner, M.R. Wasielewski, *Nature 396*, 60–63 (1998).

A. Ghosh, *J. Biol. Inorg. Chem. 16*, 819–820 (2011).

N.S. Hush, *J. Am. Chem. Soc. 109*, 3258 (1987).

B. Källebring, S. Larsson, *Chem. Phys. Lett. 138*, 76–82 (1987).

R. Kaptein, *Biol. Magn. Res. 4*, 145–191 (1982).

R. Kaptein, K. Dijkstra, K. Nicolay, *Nature 274*, 293–294 (1978).

J. Kroon, J.W. Verhoeven, M.N. Paddon-Row, A.M. Oliver, *Angew. Chem. Int. Ed. Engl. 30*, 1358 (1991).

S. Larsson, A. Volosov, *J. Chem. Phys. 85*, 2548–2554 (1986); Erratum: *J. Chem. Phys. 86*, 5223 (1987).

S. Larsson, A. Volosov, *J. Chem. Phys. 87*, 6623–6625 (1987).

R.A. Marcus, *Angew. Chem. Int. Ed. Engl. 32*, 1111 (1993).

R.S. Mulliken, *J. Phys. Chem. 56*, 801–822 (1952).

H. Oevering, M.N. Paddon-Row, M. Heppener, A.M. Oliver, E. Cotsaris, J.W. Verhoeven, P. Pasman, G.F. Mes, N.W. Koper, J.W. Verhoeven, *J. Am. Chem. Soc. 107*, 5839 (1985).

L. Onsager, *J. Am. Chem. Soc. 58*, 1486–1493 (1936).

J.D. Petke, G.M. Maggiora, *J. Chem. Phys. 84*, 1640–1652 (1986).

D.R. Prasad, G. Farraudi, *J. Phys. Chem. 86*, 4037–4040 (1982).

D. Rehm, A. Weller, *Ber. Bunsenges. 73*, 835 (1969).

D. Rehm, A. Weller, *Isr. J. Chem. 8*, 259 (1970).

A. Sautter, B.K. Kaletas, D.G. Schmid, R. Dobrawa, M. Zimine, G. Jung, I.H.M. van Stokkum, L. de Cola, R.M. Williams, F. Würthner, *J. Am. Chem. Soc. 127*, 6719–6729 (2005).

J.W. Verhoeven, I.P. Dirkx, T.J. de Boer, *Tetrahedron 25*, 4037 (1969).

M.R. Wasielewski, *Chem. Rev. 92*, 435 (1992).

A. Weller, *Z. Phys. Chem. Neue Folge 133*, 93 (1982).

R.M. Williams, M. Koeberg, J.M. Lawson, Y.Z. An, Y. Rubin, M.N. Paddon-Row, J.W. Verhoeven, *J. Org. Chem. 61*, 5055–5062 (1996).

R.M. Williams, J.M. Zwier, J.W. Verhoeven, *J. Am. Chem. Soc. 117*, 4093–4099 (1995).

Twisted Intramolecular Charge Transfer

Z.R. Grabowski, K. Rotkiewicz, W. Rettig, *Chem. Rev. 103*, 3899–4031 (2003).

N.S. Hush, M.N. Paddon-Row, E. Cotsaris, H. Hoevering, J.W. Verhoeven, M. Heppener, *Chem. Phys. Lett. 117*, 8 (1985).

E. Lippert, W. Lüder, F. Moll, W. Nägele, H. Boos, H. Prigge, I. Seibold-Blankenstein, *Angew. Chem. 73*, 695 (1961).

E. Lippert, W. Rettig, V. Bonačić Koutecký, F. Heisel, J.A. Miehé, *Adv. Chem. Phys. 68*, 1 (1987).

W. Rettig, *Angew. Chem. 98*, 969 (1986).

W. Rettig, V. Bonačić Koutecký, *Chem. Phys. Lett. 62*, 115 (1979).

K. Rotkiewicz, K.H. Grellmann, Z.R. Grabowski, *Chem. Phys. Lett. 19*, 315 (1973); *21*, 212 (1973).

L. Serrano-Andrés, M. Merchán, B.O. Roos, R. Lindh, *J. Am. Chem. Soc. 117*, 3189–3204 (1995).

A. Siemiarczuk, Z.R. Grabowski, A. Krowczyński, M. Asher, M. Ottolengui, *Chem. Phys. Lett. 51*, 315 (1977).

Intramolecular PIET

B. Albinsson, M.P. Eng, K. Pettersson, M.U. Winters, *Phys. Chem. Chem. Phys. 9*, 5847–5864 (2007).

C. Creutz, N. Sutin, *J. Am. Chem. Soc. 99*, 241 (1977).

W.B. Davis, M.A. Ratner, M.R. Wasielewski, *J. Am. Chem. Soc. 119*, 1405 (1997).

H. Imahori, N.V. Tkachenko, V. Vehmanen, K. Tamaki, H. Lemmetyinen, Y. Sakata, S. Fukuzumi, *J. Phys. Chem. A, 105*, 1750–1756 (2001).

B.K. Kaletas, R. Dobrawa, A. Sautter, F. Würthner, M. Zimine, L. De Cola, R.M. Williams, *J. Phys. Chem. A 108*, 1900–1909 (2004).

S. Larsson, J.M.O. Matos, *J. Mol. Struct. 120*, 35–40 (1985).

K.J. Smit, J.M. Warman, M.P. de Haas, M.N. Paddon-Row, A.M. Oliver, *Chem. Phys. Lett. 152*, 177 (1988).

Photoswitching

J. Andersson, S. Li, P. Lincoln, J. Andréasson, *J. Am. Chem. Soc. 130*, 11836–11837 (2008).

J. Andréasson, U. Pischel, *Chem. Soc. Rev. 39*, 174–188 (2010).

Fullerenes

G. Duščesas, S. Larsson, *Theor. Chim. Acta 97*, 110–118 (1997).

P.A. Liddell, D. Kuciauskas, J.P. Sumida, B. Nash, D. Nguyen, A.L. Moore, T.A. Moore, D. Gust, *J. Am. Chem. Soc. 119*, 1400–1405 (1997).

P.A. Liddell, J.P. Sumida, A.N. Macpherson, L. Noss, G.R. Seely, K.N. Clark, A.L. Moore, T.A. Moore, D. Gust, *Photochem. Photobiol. 60*, 537–541 (1994).

C.A. Reed, R.D. Bolskar, *Chem. Rev. 100*, 1075–1119 (2000).

BOOKS

V. Balzani, Editor, *Electron Transfer in Chemistry*, vols. 1–5 (Wiley-VCH, New York, 2001).

M.A. Fox, M. Chanon, *Photoinduced Electron Transfer* (Elsevier Science Publishers, Amsterdam, 1988).

G.J. Kavarnos, *Fundamentals of Photoinduced Electron Transfer* (VCH Publishers Inc., New York, 1993).

R.S. Mulliken, W.B. Person, *Molecular Complexes: A Lecture and Reprint Volume* (Wiley Interscience, New York, 1969).

CHAPTER 14

PAPERS

D.L. Dexter, *J. Chem. Phys. 21*, 836 (1953).

D.L. Dexter, *Phys. Rev. 108*, 707–712 (1957).

S. Eriksson, B. Källebring, S. Larsson, J. Mårtensson, O. Wennerström, *Chem. Phys. 146*, 165–177 (1990).

T. Förster, *Naturwissenschaften 33*, 166 (1946).

T. Förster, *Discuss. Faraday Soc. 27*, 7 (1959).

J. Franck, E. Teller, *J. Chem. Phys. 6*, 861–872 (1938).

J. Frenkel, *Phys. Rev. 37*, 17, 276 (1931).

J. Frenkel, *Phys. Z. Sow. 9*, 158 (1936).

E. Jelley, *Nature 138*, 1009 (1936).

M. Kasha, *Radiat. Res. 20*, 55 (1963).

G. McDermott, S.M. Prince, A.A. Freer, A.M. Hawthornthwaite, M.Z. Lawless, M.Z. Papiz, R.J. Cogdell, N.W. Isaacs, *Nature 374*, 517 (1995).
G. Scheibe, *Angew. Chem. 49*, 563 (1936); *50*, 51, 212 (1937).
L. Stryer, R.P. Haugland, *Proc. Natl. Acad. Sci. USA 58*, 719–726 (1967).
B. Trösken, F. Willig, K. Schwarzburg, A. Ebert, M. Spitler, *J. Phys. Chem. 99*, 5152–5160 (1995).
H. Wang, T.E. Kaiser, S. Uemura, F. Würthner, *Chem. Commun.* 1181–1183 (2008) (perylene bisimide: se J-aggregates Google).

BOOKS

A.S. Davydov, *Theory of Molecular Excitons* (McGraw-Hill, New York, 1962). Translation by M. Kasha and M. Oppenheimer, Jr.
T. Förster, *Molecular Exciton Theory*, Part III; Editor O. Sinanouğlu (Academic Press, New York, 1965).
T. Förster, *Organic Charge Transfer Complexes* (Academic Press Inc., New York, 1969).
T. Kobayashi, *J-Aggregates* (World Scientific, Singapore, 1996).

CHAPTER 15

PAPERS

J.P. Allen, G. Feher, T.O. Yeates, H. Komiya, D.C. Rees, *Proc. Natl. Acad. Sci. USA 84*, 5730–5734 (1987).
T. Arlt, S. Schmidt, W. Kaiser, C. Lauterwasser, M. Meyers, H. Scheer, W. Zinth, *Proc. Natl. Acad. Sci. USA 90*, 11757–11761 (1993).
R. Berera, R. van Grondelle, J.T.M. Kennis, *Photosynth. Res. 101*, 105–118 (2009).
M. Bixon, J. Jortner, *Chem. Phys. Lett. 159*, 17–20 (1989).
M. Bixon, J. Jortner, *J. Chem. Phys. 107*, 1470–1482 (1997).
M. Bixon, J. Jortner, M.-E. Michel-Beyerle, A. Ogrodnik, W. Lersch, *Chem. Phys. Lett. 140*, 626 (1987).
M. Chen, M. Schliep, R.D. Willows, Z.-L. Cai, B.A. Neilan, H. Scheer, *Science 329*, 1318–1319 (2010).
J. Deisenhofer, O. Epp, K. Miki, R. Huber, H. Michel, *J. Mol. Biol. 180*, 385–398 (1984).
J. Deisenhofer, O. Epp, K. Miki, R. Huber, H. Michel, *Nature 318*, 618–624 (1985).
J. Deisenhofer, H. Michel, *Science 245*, 1463–1473 (1989).
R. van Grondelle, et al., *Biochemistry 49*, 4300–4307 (2010).
N. Ivashin, B. Källebring, S. Larsson, Ö. Hansson, *J. Phys. Chem. 102*, 5017–5022 (1998).
N. Ivashin, S. Larsson, *J. Phys. Chem. B 106*, 3996–4009 (2002).
N.V. Ivashin, S. Larsson, *Chem. Phys. Lett. 375*, 383–387 (2003).
H. Michel, J. Deisenhofer, *Pure Appl. Chem. 60*, 953–958 (1988).
J.A. Potter, P.K. Fyfe, D. Frolov, M.C. Wakeham, R. van Grondelle, B. Robert, M.R. Jones, *J. Biol. Chem. 280*, 27155–27164 (2005).
J.P. Ridge, P.K. Fyfe, K.E. McAuley, M.E. van Brederode, B. Robert, R. van Grondelle, N.W. Isaacs, R.J. Cogdell, M.R. Jones, M.R. *Biochem. J. 351*, 567–578 (2000).
V.A. Shuvalov, A.V. Klevanik, A.V. Sharkov, Y.A. Matveetz, P.G. Krukov, *FEBS Lett. 91*, 135–139 (1978).
S. Spörlein, W. Zinth, J. Wachtveitl, *J. Phys. Chem. B 102*, 7492–7496 (1998).
M.H. Vos, M.R. Jones, J.L. Martin, *Chem. Phys. 233*, 179–190 (1998).
M.H. Vos, F. Rappaport, J.-C. Lambry, J. Breton, J.-L. Martin, *Nature 363*, 320–325 (1993).
A.G. Yakovlev, M.R. Jones, J.A. Potter, P.K. Fyfe, L.G. Vasilieva, A.Y. Shkuropatov, V.A. Shuvalov, *Chem. Phys. 319*, 297–307 (2005).

Antenna Systems

R.J. Cogdell, A.T. Gardiner, A.W. Roszak, C.J. Law, J. Southall, N.W. Isaacs, *Photosynth. Res.* *81*, 207–214 (2004).

J.L. Herek, N.J. Fraser, T. Pullerits, P. Martinsson, T. Polivka, H. Scheer, R.J. Cogdell, V. Sundström, *Biophys. J.* *78*, 2590–2596 (2000).

N. Ivashin, S. Larsson, *J. Phys. Chem. B* *109*, 23051–23060 (2005).

G. McDermott, S.M. Prince, A.A. Freer, A.M. Hawthornthwaite, M.Z. Lawless, M.Z. Papiz, R.J. Cogdell, *Nature* *374*, 517–521 (1995).

G. Trikunas, J.L. Herek, T. Polivka, V. Sundström, T. Pullerits, *Phys. Rev. Lett.* *86*, 4167–4170 (2001).

L. Valkunas, E. Åkesson, T. Pullerits, V. Sundström, *Biophys. J.* *70*, 2373–2379 (1996).

Photosystem 2

M.R.A. Blomberg, P.E.M. Siegbahn, S. Styring, G.T. Babcock, B. Åkermark, P. Korall, *J. Am. Chem. Soc.* *119*, 8285–8292 (1997).

K.N. Ferreira, T.M. Iverson, K. Maghlaoui, J. Barber, S. Iwata, *Science* *303*, 1831–1838 (2004).

C.T. Hoganson, G.T. Babcock, *Biochemistry* *31*, 11874–11880 (1992).

P.E.M. Siegbahn, M.R.A. Blomberg, *Biochim. Biophys. Acta* *1655*, 45–50 (2004).

Y. Umena, K. Kawakami, J.R. Shen, N. Kamiya, *Nature* *473*, 55–60 (2011).

A. Zouni, H.-T. Witt, J. Kern, P. Fromme, N. Krauss, W. Saenger, P. Orth, *Nature* *409*, 739–743 (2001).

Rubisco

B. Kannappan, J.E. Gready, *J. Am. Chem. Soc.* *130*, 15063–15080 (2008).

TEXT-BOOKS

R.E. Blankenship, *Molecular Mechanisms of Photosynthesis*, 1st edition (Blackwell Science, Oxford, 2002).

CHAPTER 16

PAPERS

P.W. Anderson, *Phys. Rev. Lett.* *34*, 953–955 (1975).

F. Bloch, *Z. Phys.* *52*, 555 (1928).

F. Bloch, *Z. Phys.* *57*, 545 (1929).

L. Brillouin, *Ann. Phys.* *17*, 88 (1922).

P. Drude, *Ann. Phys.* *306*, 566 (1900); *308*, 369 (1900).

R. Englman, J. Jortner, *J. Mol. Phys.* *18*, 145–164 (1970).

E. Grüneisen, E. Goens, *Z. Phys.* *44*, 615 (1927).

R.R. Heikes, W.D. Johnston, *J. Chem. Phys.* *582* (1957).

R.R. Heikes, A.A. Maradudin, R.C. Miller, *Ann. Phys.* *8*, 733–746 (1963).

C. Herring, *Phys. Rev.* *57*, 1169–1177 (1940).

T. Holstein, *Ann. Phys.* *8*, 325–342 (1959); *8*, 343–389 (1959); reprinted *Ann. Phys.* *281*, 706–724 (2000).

W.H. Houston, *Z. Phys.* *48*, 449 (1928); *Phys. Rev.* *34*, 279–283 (1928).

N.S. Hush, *Chem. Phys.* *10*, 361 (1975).

L.D. Landau, S.I. Pekar, *Zh. Eksp. Teor. Fiz.* *18*, 419–423 (1948).

S. Larsson, A. Klimkāns, *Mol. Cryst. Liq. Cryst.* *355*, 217–229 (2000).

M. Pai, J.M. Honig, *J. Solid State Chem. 40*, 59 (1981).
A.H. Romero, D.W. Brown, K. Lindenberg, *Phys. Rev. B 59*, 13728–13740 (1999).
J.C. Slater, *Phys. Rev. 35*, 509–529 (1930).
J.C. Slater, *Phys. Rev. 49*, 537–545 (1936).
J.C. Slater, *Phys. Rev. 51*, 846–851 (1937).
J.C. Slater, *Phys. Rev. 81*, 385–390 (1951); *82*, 538–541 (1951); *91*, 528–530 (1953).
J.C. Slater, G.F. Koster, *Phys. Rev. 94*, 1498 (1954).
A. Sommerfeld, *Z. Phys. 47*, 1 (1928).
E. Wigner, F. Seitz, *Phys. Rev. 46*, 509 (1934).

BOOKS

N.W. Ashcroft, N.D. Mermin, *Solid Stale Physics* (Saunders, Philadelphia, 1976).
C. Kittel, *Introduction to Solid State Physics*, 8th edition (Wiley, New York, 2005).
N.F. Mott, *Metal Insulator Transitions* (Taylor and Francis, London, 1974, 1990).
N.F. Mott, *Conduction in Non-Crystalline Materials* (Clarendon Press, Oxford, 1987).
R.E. Peierls, *Quantum Theory of Solids* (Oxford University Press, New York, 1955).
R.A. Smith, *Wave Mechanics of Crystalline Solids* (Chapman & Hall, London, 1961).

CHAPTER 17

PAPERS

J. Bardeen, L.N. Cooper, J.R. Schrieffer, *Phys. Rev. 108*, 1175–1204 (1957).
J.G. Bednorz, K.A. Müller, *Z. Phys. B 64*, 189–193 (1986).
C.W. Chu, P.H. Hor, R.L. Meng, L. Gao, Z.J. Huang, Y.Q. Wang, *Phys. Rev. Lett. 58*, 405 (1987).
A.J. Coleman, *Rev. Mod. Phys. 35*, 668–687 (1963).
A.J. Coleman, *Can. J. Phys. 45*, 1271 (1967).
L.N. Cooper, *Phys. Rev. 104*, 1189–1190 (1956).
H. Fröhlich, *Phys. Rev. 79*, 845 (1950).
A.Y. Ganin, Y. Takabayashi, Y.Z. Khimyak, S. Margadonna, A. Tamai, M.J. Rosseinsky, K. Prassides, *Nat. Mater. 7*, 367–371 (2008).
O. Gunnarsson, *Rev. Mod. Phys. 69*, 575–605 (1997).
L. van Hove, *Phys. Rev. 89*, 1189–1193 (1953).
J. Hubbard, *Proc. R. Soc. Lond. A 276*, 238–257 (1963).
N.S. Hush, in *Mixed-Valence Compounds in Chemistry, Physics and Biology*; NATO ASI Series, Editor D.B. Brown (D. Reidel Publishing Company, Dodrecht, The Netherlands, 1980), p. 151.
D.C. Johnston, *J. Low Temp. Phys. 25*, 145–175 (1976).
D.C. Johnston, H. Prakash, W.H. Zachariasen, R. Viwanathan, *Mat. Res. Bull. 8*, 777 (1973).
B. Josephson, *Phys. Lett. 1*, 251 (1962).
A. Klimkāns, S. Larsson, *J. Chem. Phys. 115*, 466–471 (2001).
S. Larsson, *Chem. Phys. Lett. 157*, 403–408 (1989).
S. Larsson, *Chem. Phys. 236*, 133–150 (1998).
S. Larsson, A. Klimkāns, *Int. J. Quantum Chem. 80*, 713–720 (2000).
F. London, *Phys. Rev. 54*, 947–954 (1938).
B.T. Matthias, *Phys. Rev. 97*, 74–76 (1955).
B.T. Matthias, G.R. Stewart, A.L. Giorgi, J.L. Smith, Z. Fisk, H. Barz, *Science 208*, 401–402 (1980).
N.F. Mott, *Proc. R. Soc. Lond. A 62*, 416 (1949).
N.F. Mott, *J. Non Cryst. Solids 164–166*, 1177–1178 (1993).

N.E. Phillips, *Phys. Rev. 114*, 676–685 (1959).

J.R. Reimers, N.S. Hush, *Chem. Phys. 299*, 79–82 (2004).

D.R. Rosseinsky, J.S. Tonge, *J. Chem. Soc. Faraday Trans. 1,78*, 3595–3603 (1982).

A.W. Sleight, *Science 242*, 1519–1527 (1988).

A.W. Sleight, J.L. Gillson, P.E. Bierstedt, *Solid State Commun. 17*, 27–28 (1975).

Y. Takabayashi, A.Y. Ganin, P. Jeglič, D. Arčon, T. Takano, Y. Iwasa, Y. Onishi, M. Takata, N. Takeshita, K. Prassides, M.J. Rosseinsky, *Science 323*, 1585–1590 (2009).

M.K. Wu, J.R. Ashburn, P.H. Torng, P.H. Hor, R.L. Meng, L. Gao, Z.J. Huang, Y.Q. Wang, C.W. Chu, *Phys. Rev. Lett. 58*, 908 (1987).

C.N. Yang, *Rev. Mod. Phys. 34*, 694 (1962).

Solvated Electrons

W. Bingel, *Ann. Phys. 12*, 57–83 (1953); Erratum: *Ann. Phys. 12*, 424 (1953).

D.A. Copeland, N.R. Kestner, *J. Chem. Phys. 58*, 3500–3503 (1973).

J.L. Dye, *Progr. Inorg. Chem. 32*, 327–441 (1984).

P.P. Edwards, *J. Supercond. 13*, 933–946 (2000).

C.A. Kraus, *J. Am. Chem. Soc. 30*, 1323 (1908).

R.A. Ogg, *J. Am. Chem. Soc. 68*, 155 (1946).

R.A. Ogg, *J. Chem. Phys. 14*, 114, 295 (1946).

R.A. Ogg, *Phys. Rev. 69*, 243–244 (1946); Erratum: *69*, 544 (1946); *69*, 668 (1946); *70*, 93 (1946).

J.C. Wasse, S. Hayama, N.T. Skipper, H.E. Fischer, *Phys. Rev. B 61*, 11993–11997 (2000).

Diamagnetic Currents

H. Fliegl, D. Sundholm, S. Taubert, J. Jusélius, W. Klopper, J. Gauss, *J. Phys. Chem. A 113*, 8668–8676 (2004).

J. Jusélius, D. Sundholm, J. Gauss, *J. Chem. Phys. 121*, 3952–3963 (2004).

A.F. Voter, W.A. Goddard III, *J. Am. Chem. Soc. 108*, 2830 (1986).

BOOKS

J.M. Blatt, *Theory of Superconductivity* (Academic Press, New York, London, 1964).

F. London, *Superfluids, vol. 1: Macroscopic Theory of Superconductivity*, 2nd edition, revised and with a new epilogue by M.J. Buckingham (Dover Publications, Inc., New York, 1961).

D.R. Tilley, *Superfluidity and Superconductivity*, 3rd Edition (Institute of Physics Publishing, Bristol and Philadelphia, 1990).

J.R. Waldram, *Superconductivity of Metals and Cuprates* (Institute of Physics Publishing, Bristol and Philadelphia, 1996).

CHAPTER 18

PAPERS

H. Akamatu, H. Inokuchi, Y. Matsunaga, *Nature 173*, 168–169 (1954).

R.H. Friend, D. Jérome, *J. Phys.C Solid State Phys. 12*, 1441–1477 (1979).

R.L. Green, G.B. Street, L.J. Suter, *Phys. Rev. Lett. 34*, 577–579 (1975).

A.J. Heeger, *Rev. Mod. Phys. 73*, 681–700 (2001).

D. Jérome, *Chem. Rev. 104*, 5665–5691 (2004).

E. Johansson, S. Larsson, *Synth. Met. 144*, 183–191 (2004).

M. Kertesz, H.C. Cheol, S.Y. Hong, *Synth. Met. 85*, 995–1778 (1997).

Y. Kuriyama, P.R. Ogilby, K.V. Mikkelsen, *J. Phys. Chem. 98*, 11918–11923 (1994).

S. Larsson, *Faraday Discuss. 131*, 69–77 (2006).

S. Larsson, L. Rodríguez-Monge, *Int. J. Quantum Chem. Symp. 27*, 655–665 (1993).

H.-C. Longuet-Higgins, L. Salem, *Proc. R. Soc. Lond. A 251*, 172 (1959).

A.G. MacDiarmid, *Rev. Mod. Phys. 73*, 701–713 (2001).

H. Naarman, N. Theophilou, *Synth. Met. 22*, 1 (1987).

J. Olofsson, S. Larsson, *J. Phys. Chem. B 105*, 10398–10496 (2001).

Y. Ooshika, *J. Phys. Soc. Jpn. 12*, 1246 (1957).

W.R. Salaneck, I. Lundström, W.-S. Huang, A.G. MacDiarmid, *Synth. Met. 13*, 291–297 (1986).

E.A. Silinsh, A. Klimkāns, S. Larsson, V. Čápec, *Chem. Phys. 198*, 311–331 (1995).

S. Stafström, J.-L. Brédas, *Phys. Rev. B 38*, 4180–4191 (1988).

T.-C. Tseng, C. Urban, Y. Wang, R. Otero, S.L. Tait, M. Alcamí, D. Écija, M. Trelka, J.M. Gallego, N. Lin, M. Konuma, U. Starke, A. Nefedov, A. Langner, C. Wöll, M.Á. Herranz, F. Martín, N. Martín, K. Kern, R. Miranda, *Nat. Chem. 2*, 374–379 (2010).

E.I. Zhilyaeva, O.N. Krasochka, R.N. Lyubovskaya, R.B. Lyubovskii, L.O. Atovmyan, M.L. Khidekel, *Russ. Chem. Bull. 35*, 2366–2368 (1986).

Conducting Polymers

W.J.D. Beenken, T. Pullerits, *J. Chem. Phys. 120*, 2490–2495 (2004).

R. McNeill, R. Siudak, J.H. Wardlaw, D.E. Weiss, *Aust. J. Chem. 16*, 1056–1075 (1963).

H. Shirakawa, *Rev. Mod. Phys. 73*, 713–718 (2001).

H. Shirakawa, E.J. Louis, A.G. MacDiarmid, C.K. Chiang, A.J. Heeger, *Chem. Commun.* 578 (1977).

E. Wang, Z. Ma, Z. Zhang, K. Vandewal, P. Henriksson, O. Inganäs, F. Zhang, M.R. Andersson, *J. Am. Chem. Soc. 133*, 14244–14247 (2011).

OLED

O. Inganäs, *Nat. Photon. 5*, 201–202 (2011).

D.H. O'Neil, A. Walsh, R.M.J. Jacobs, V.L. Kuznetsov, R.G. Egdell, P.P. Edwards, *Phys. Rev. B 81*, 085110 (1–8) (2010).

TTF, TCNQ

J. Ferraris, D.O. Cowan, V. Walatka Jr., J.H. Perlstein, *J. Am. Chem. Soc. 95*, 948–949 (1973).

L.R. Melby, R.J. Harder, W.R. Hertler, W. Mahler, R.E. Benson, W.E. Mochel, *J. Am. Chem. Soc. 84*, 3374–3387 (1962).

F. Wudl, D. Wobschall, E.J. Hufnagel, *J. Am. Chem. Soc. 94*, 670–672 (1972).

DNA

J.C. Genereux, A.K. Boal, J.K. Barton, *J. Am. Chem. Soc. 132*, 891–905 (2010).

Books

R. Peierls, *Quantum Theory of Solids* (Clarendon Press, Oxford, 1955).

M. Pope, C.E. Swenberg, *Electronic Processes in Organic Crystals* (Claendon Press, Oxford, New York, 1982).

S. Roth, D. Carroll, *One-Dimensional Metals* (Wiley-VCH Verlag GmbH & Co. KGaA, Weinheim, 2004).

L. Salem, *The Molecular Orbital Theory of Conjugated Systems* (W.A. Benjamin, New York, 1966).

E.A. Silinsh, V. Čápek, *Organic Molecular Crystals. Interaction, Localization, and Transport Phenomena* (AIP Press, New York, 1994).

Appendices

APPENDIX 1: INTEGRALS

Exponential functions were originally used by Slater as orbitals for a screened nucleus. They are referred to as *Slater orbitals*.

Exponential functions used as basis functions in actual calculations on atoms and molecules are called *Slater type orbitals* (STO), or simply *exponential type orbitals* (ETO), in quantum chemistry. In the case of molecules, exponential orbitals lead to very difficult and time-consuming many-center integrals, though there are some successful attempts to get beyond the bottleneck (J.D. Talman).

Simple integrals of the following type are easily evaluated by partial integration, for example,

$$W(n) = \int_0^\infty x^n \exp(-\alpha x)\,dx = \left| \frac{-x^n}{\alpha} \exp(-\alpha x) \right|_0^\infty + \frac{n}{\alpha} \int_0^\infty x^{n-1} \exp(-\alpha x)\,dx$$

$$= \frac{n}{\alpha} \int_0^\infty x^{n-1} \exp(-\alpha x)\,dx = \frac{n}{\alpha} W(n-1) \tag{A.1}$$

$$W(0) = \int_0^\infty \exp(-\alpha x)\,dx = \left| \frac{-1}{\alpha} \exp(-\alpha x) \right|_0^\infty = \frac{1}{\alpha}. \tag{A.2}$$

We obtain a recursion formula from which all $W(n)$ may be calculated.

Basis sets that contain functions of the type $\exp(-\alpha x^2)$ are called *Gaussian* basis sets. Gaussian many-center integrals are relatively easy to evaluate and therefore, Gaussian basis functions are used in quantum chemistry. A linear combination of Gaussians does not strictly satisfy the boundary conditions at $r \to 0$, since the derivative is equal to zero at the atomic nuclei, which is incorrect. The remedy is to include functions with a very high value of the exponent "a" in the basis set. We will not evaluate three-dimensional integrals here. Equations may be found in the original work by Boys and in a number of review articles.

Gaussian functions also appear in the vibrational time-independent SE, and the overlap integrals are very important here.

By partial integration, a recursion formula may be found for Gaussian indefinite integrals (for $n \geq 1$):

$$G_{n\alpha}(x) = \int x^n \exp(-\alpha x^2) dx = \int x^{n-1} x \exp(-\alpha x^2) dx$$

$$= x^{n-1} \frac{1}{-2\alpha} \exp(-\alpha x^2) + \frac{n-1}{2\alpha} \int x^{n-2} \exp(-\alpha x^2) dx \qquad (A.3)$$

$$= -\frac{x^{n-1}}{2\alpha} \exp(-\alpha x^2) + \frac{n-1}{2\alpha} G_{n-2,\alpha}.$$

For $n = 1$, the second part vanishes, so there is no need to define $G_{-1,\alpha}$. For $n \leq 0$, there is no useful equation. The definite (improper) integrals from $-\infty$ to ∞ may be easily evaluated from Equation A.1.

We need $G_{0,\alpha}$ integrated from $-\infty$ to ∞. If we try partial integration, we obtain:

$$I = \int_{-\infty}^{\infty} \exp(-\alpha x^2) dx = \left| x \exp(-\alpha x^2) \right|_{-\infty}^{\infty} + 2\alpha \int_{-\infty}^{\infty} x^2 \exp(-\alpha x^2) dx$$

$$= \frac{\alpha}{2\pi} \int_{-\infty}^{\infty} 4\pi r^2 \exp(-\alpha r^2) dr \qquad (A.4)$$

$$= \frac{\alpha}{\pi} \int_{0}^{\infty} \exp(-\alpha r^2) 4\pi r^2 dr.$$

This is, however, just the three-dimensional integral, integrated over all space with the integration element as spheres from 0 to ∞. Alternatively, since $r^2 = x^2 + y^2 + z^2$, we may evaluate this integral as:

$$I = \frac{\alpha}{\pi} \int_{0}^{\infty} \exp(-\alpha r^2) 4\pi r^2 dr$$

$$\qquad (A.5)$$

$$= \frac{\alpha}{\pi} \int_{-\infty}^{\infty} \exp(-\alpha x^2) dx \int_{-\infty}^{\infty} \exp(-\alpha y^2) dy \int_{-\infty}^{\infty} \exp(-\alpha z^2) dz = \frac{\alpha}{\pi} I^3.$$

We thus have to solve:

$$I = \frac{\alpha}{\pi} I^3 \qquad (A.6)$$

and obtain:

$$I = \int_{-\infty}^{\infty} \exp(-\alpha x^2) dx = \sqrt{\frac{\pi}{\alpha}}. \tag{A.7}$$

APPENDIX 2: COULOMB INTEGRALS

We will first calculate the $1/r_{12}$ integrals. Atomic integrals containing powers of r_{12} are most easily evaluated in the polar coordinate system (Figure 2.1). $r_{12} = |\vec{r}_1 - \vec{r}_2|$ is the distance between two points in space with the polar coordinates $(r_1, \theta_1, \varphi_1)$ and $(r_2, \theta_2, \varphi_2)$, respectively. The following expansion may be used:

$$\frac{1}{r_{12}} = \sum_{\ell=0}^{\infty} \frac{r_<^\ell}{r_>^{\ell+1}} \frac{4\pi}{2\ell+1} \sum_{m=-\ell}^{\ell} Y_{\ell m}^*(\theta_1, \varphi_1) Y_{\ell m}(\theta_2, \varphi_2) \tag{A.8}$$

(for a general power of r_{12}, see R.A. Sack, *J. Math. Phys.* 5, 245, 1964). The spherical harmonic functions, $Y_{\ell m}(\theta, \varphi)$, are given by:

$$Y_{\ell m}(\theta, \varphi) = (-1)^m \Theta_{\ell m}(\theta) \Phi_m(\varphi) \tag{A.9}$$

for $m \geq 0$; and

$$Y_{\ell m}(\theta, \varphi) = (-1)^m Y_{\ell-m}^*(\theta, \varphi) \tag{A.10}$$

for $m < 0$. The spherical harmonics, Θ and Φ, forming an orthonormal set, are defined in Chapter 2.

In Equation A.2, $r_<$ means the smaller of r_1 and r_2; and $r_>$ means the larger of r_1 and r_2:

$$
\begin{aligned}
r_> = r_1 \quad \text{and} \quad r_< = r_2 \quad \text{if } r_1 > r_2 \\
r_> = r_2 \quad \text{and} \quad r_< = r_1 \quad \text{if } r_2 > r_1.
\end{aligned}
\tag{A.11}
$$

We assume a charge distribution $\rho(r_2)$ and want to know the potential at a point at a distance r_1 from the center. Coulombs law is used and we integrate over all space:

$$V(r_1) = \int \frac{\rho(r_2)}{r_{12}} dv_2 = \int_0^\infty \frac{\rho(r_2)}{r_{12}} r_2^2 dr_2 \int_0^\pi \sin\theta d\theta \int_0^{2\pi} d\varphi \tag{A.12}$$

Equation A.8 for the expansion of $1/r_{12}$ leads to

$$V(r_1) = \sum_{\ell=0}^{\infty} \int_0^{\infty} \frac{r_<^\ell}{r_>^{\ell+1}} \rho(r_2) r_2^2 dr_2 \frac{4\pi}{2\ell+1} \int_0^{\pi} \sin\theta_2 d\theta_2 \int_0^{2\pi} d\varphi_2 \sum_{m=-\ell}^{\ell} Y_{\ell m}^*(\theta_1,\varphi_1) Y_{\ell m}(\theta_2,\varphi_2).$$

(A.13)

Since the density $\rho(r_2)$ has no angular dependence, we may carry out the integration over θ_2 and φ_2. Only the term with $l = 0$ and $m = 0$ remains. Using the expression for the spherical harmonics functions in Section 2.2.3, we obtain:

$$V(r_1) = \int_0^{\infty} \frac{1}{r_>} \rho(r_2) 4\pi r_2^2 dr_2 = \frac{1}{r_1} \int_0^{\eta} \rho(r_2) 4\pi r_2^2 dr_2 + \int_\eta^{\infty} \frac{1}{r_2} \rho(r_2) 4\pi r_2^2 dr_2.$$ (A.14)

Since $\rho(r_2)$ tends to zero exponentially as $r_2 \to \infty$, the second integral of the right member tends to zero as $r_1 \to \infty$. If r_1 is so large that the contribution from the second integral can be neglected, the potential V is the same as if all the charge was placed at the nucleus (the first term in Equation A.14).

We assume that the charge of a single electron is evenly distributed over a sphere with radius R. The density is ρ. The following equation has to be satisfied:

$$\frac{4\pi}{3} R^3 \cdot \rho = 1 \quad \Rightarrow \quad \rho = \frac{3}{4\pi R^3}.$$ (A.15)

At the center of this sphere, the repulsive potential may be obtained from Equation A.14 using $r_1 = 0$:

$$U = \int_0^R \frac{1}{r} 4\pi r^2 \rho(r) dr = \frac{4\pi R^2}{2} \frac{3}{4\pi R^3} = \frac{3}{2R}.$$ (A.16)

U is the self-energy of an electron with a constant density within a sphere with radius R. If the system is a metal, the self-energy is equal to zero, since $R \to \infty$. In the limit when the radius of the system tends to zero, the self-energy tends to infinity.

APPENDIX 3: PERTURBATION THEORY

We assume that the Hamiltonian can be written as

$$H = H_0 + V$$ (A.17)

The operator H_0 has known energy levels, E_n^0, and eigenstates, Ψ_n^0, of the time-independent SE. V represents a weak perturbation, which is so weak that the eigenfunctions of H_0 can be assumed to be a suitable basis set for the calculations of the eigenvalues and eigenfunctions of H:

$$\Psi = \sum_{i=0}^{n} C_i \Psi_i^0 \tag{A.18}$$

E and the coefficients are written in order of magnitude (upper index):

$$E = E_0^0 + E_0^1 + E_0^2 + \cdots; \quad C_k = C_k^0 + C_k^1 + C_k^2 + \cdots \tag{A.19}$$

where the terms of the summation may be expected to decrease as V, V^2, V^3, ... The zeroth order wave function of the ground state has the coefficients

$$C_0^0 = 1; \quad C_k^0 = 0, \quad \text{for } k \neq 0 \tag{A.20}$$

We assume that the eigenvalues of H_0 are nondegenerate and form a discrete spectrum. If H acts on Ψ in Equation A.16, we obtain:

$$H\Psi = \sum_{i=0}^{n} C_i H \Psi_i^0 = E\Psi \tag{A.21}$$

If Equation A.17 is inserted in Equation A.16, we obtain:

$$H\Psi = \sum_{i=0}^{n} (H_0 + V) C_i \Psi_i^0 = \sum_{i=0}^{n} (E_i^0 + V) C_i \Psi_i^0 = \sum_{i=0}^{n} EC_i \Psi_i^0 \tag{A.22}$$

If we multiply this equation by $\Psi_k^0{}^*$ and integrate, we obtain:

$$\sum_{i=0}^{n} (E_k^0 \delta_{ki} + V_{ki}) C_i = EC_k, \quad \text{for all } k \tag{A.23}$$

where $\langle \Psi_k^0 | \Psi_i^0 \rangle = \delta_{ki}$ and $V_{ki} = \langle \Psi_k^0 | V | \Psi_i^0 \rangle$.

In matrix form, we may alternatively write:

$$\begin{pmatrix} k=0 \\ k=1 \\ \vdots \\ k=n \end{pmatrix} \begin{pmatrix} E_0^0 + V_{00} & V_{01} & V_{02} & \cdots & V_{0n} \\ V_{10} & E_1^0 + V_{11} & V_{12} & \cdots & V_{1n} \\ \vdots & \vdots & \vdots & & \vdots \\ V_{n0} & V_{n1} & V_{n2} & \cdots & E_n^0 + V_{nn} \end{pmatrix} \begin{pmatrix} C_0 \\ C_1 \\ \vdots \\ C_n \end{pmatrix} = E \begin{pmatrix} C_0 \\ C_1 \\ \vdots \\ C_n \end{pmatrix}. \tag{A.24}$$

Insert the zeroth order wave function ($C_0 = 1$) in Equation A.23 and keep the terms to first order. Equation A.24 may be written as

$$
\begin{pmatrix}
E_0^0 + V_{00} & V_{01} & V_{02} & \cdots & V_{0n} \\
V_{10} & E_1^0 + V_{11} & V_{12} & \cdots & V_{1n} \\
\vdots & \vdots & \vdots & & \vdots \\
V_{n0} & V_{n1} & V_{n2} & \cdots & E_n^0 + V_{nn}
\end{pmatrix}
\begin{pmatrix}
1 \\ C_1 \\ \vdots \\ C_n
\end{pmatrix}
= E
\begin{pmatrix}
1 \\ C_1 \\ \vdots \\ C_n
\end{pmatrix}.
\tag{A.25}
$$

All products of first-order terms will be negligible. If we define $E = E_0^0 + E_0^1$, the first row ($k = 0$) is:

$$
E_0^0 + V_{00} = E \quad \Rightarrow \quad E_0^1 = V_{00}
\tag{A.26}
$$

Another row ($k > 0$) gives:

$$
V_{k0} + E_1^0 C_k^1 = \left(E_0^0 + E_0^1 \right) C_k^1.
\tag{A.27}
$$

E_0^1 may be assumed to be comparable to $E_0^0 - E_k^0$ in magnitude. Hence, we obtain:

$$
C_k^1 = \frac{V_{k0}}{E_0^0 - E_k^0}.
\tag{A.28}
$$

Equation A.26 therefore gives the first-order correction of the energy, while Equation A.28 gives the first-order correction to the wave function.

Subsequently, we may continue with the terms to second order in V, but now limit ourselves to energy corrections. With the first row of Equation A.24, we obtain:

$$
E_0^2 = \sum_{i \neq 0} \frac{V_{0i} V_{i0}}{E_0^0 - E_i^0} = \sum_{i \neq 0} \frac{V_{0i}^2}{E_0^0 - E_i^0}.
\tag{A.29}
$$

The approximations may be done in a different order. The derivation used here is due to Rayleigh and Schrödinger.

APPENDIX 4: PARTITIONING TECHNIQUE

We assume that the Schrödinger equation $H\Psi = E\Psi$ has been solved by expanding the wave function in relevant basis functions in a certain region of space (A). We want to know how the remaining part (B) of space influences the solution. For example, we may ask how far the wave function of an electron in a protein extends, to be able to judge if an electron can hop to another site in the same or a different protein molecule. Here we will present a theory, the *partitioning technique*, originally derived by Löwdin (*J. Mol. Spectr. 10*, 12, 1964), which may be helpful for that purpose. There exists a different version of this theory that is referred to as the Green function method.

The eigenvalue problem may be written as:

$$\begin{pmatrix} H_{aa} & H_{ab} \\ H_{ba} & H_{bb} \end{pmatrix} \begin{pmatrix} C_a \\ C_b \end{pmatrix} = \begin{pmatrix} E & 0 \\ 0 & E \end{pmatrix} \begin{pmatrix} C_a \\ C_b \end{pmatrix}. \tag{A.30}$$

The elements of the matrix, H_{xy} and C_x, are themselves matrices. Index a, and b denote the A and B spaces, respectively. The wave functions may be spin orbitals or N-electron wave functions. Equation A.30 consists of two matrix equations. The second of these equations may be written as:

$$H_{ba}C_a + H_{bb}C_b = EC_b. \tag{A.31}$$

Solving for C_b leads to:

$$C_b = -\left(H_{bb} - E\right)^{-1}\left(H_{ba}C_a\right). \tag{A.32}$$

E and C_a are the original solution in the A space that is now modified due to interaction with the B space. We assume that the latter interaction is so small that we may still use the original E and C_a in Equation A.32. Equation A.32, then, gives an approximate expression for C_b. We may subsequently solve for C_a in Equation A.31 and, after iterations, obtain the correct solution.

To get a more useful expression we diagonalize the H_{bb} matrix. The eigenvalue problem for H_{bb} is given by:

$$H_{bb}C_b = EC_b. \tag{A.33}$$

There are N eigenvectors, indexed with i. We form a square matrix U_b by letting the eigenvector coefficients of vector i be column elements of column i in this matrix. The complex conjugate of this matrix is $U_b^\dagger$ (the complex conjugate of the row vectors in this matrix are the eigenvector coefficients). A mathematical theorem states that the eigenvectors of a real matrix are orthogonal. They are also assumed to be normalized. Since H_{bb} is a real matrix, we have $U_b^\dagger U_b = 1$. If we multiply Equation A.33 from the left by $U_b^\dagger$, we obtain:

$$U_b^\dagger H_{bb} U_b - U_b^\dagger E U_b = 0. \tag{A.34}$$

E is a diagonal matrix that contains the eigenvalues of H_{bb} in the diagonal. Since E commutes with U_b and $U_b^\dagger U_b = 1$, Equation A.34 shows that the matrix $U_b^\dagger H_{bb} U_b$ is equal to E, a diagonal matrix with the eigenvalues of C_b in the diagonal.

We may rewrite Equation A.31, representing the interaction between the A and B space, as

$$U_b^\dagger H_{ba} C_a + (U_b^\dagger H_{bb} U_b - E) U_b^\dagger C_b = 0, \tag{A.35}$$

where E is now the "after interaction" matrix. In the same way as when Equation A.32 was derived from Equation A.31, we now obtain:

$$\left(U_b^\dagger H_{bb} U_b - E\right)^{-1} U_b^\dagger H_{ba} C_a + U_b^\dagger C_b = 0. \tag{A.36}$$

Here we realize that the inverse matrix is a diagonal matrix where the diagonal element is $(E_{bi} - E)^{-1}$. We multiply this equation by $U_b^\dagger U_b$ and obtain after some manipulations:

$$\left(U_b^\dagger H_{bb} U_b - E\right)^{-1} U_b^\dagger H_{ba} C_a + C_b = 0. \tag{A.37}$$

This gives an expression for C_b in the same way as we obtained Equation A.32. We see that the matrix $U_b^\dagger H_{ba} C_a$ has replaced $H_{ab} C_a$ and that the inverse matrix is significantly simplified. Again we have to insert for E an eigenvalue for H_a. Again several possibilities exist to express Equation A.37 in something that is reasonable (for example, first derive an expression for C_a).

We now assume that space A consists of both donor and acceptor, D and A. For example, in an electron transfer problem there is a donor, acceptor, and an intervening or bridging medium (B). The CI eigenvalue problem may be written as:

$$\begin{pmatrix} H_{dd} & H_{da} & H_{db} \\ H_{ad} & H_{aa} & H_{ab} \\ H_{bd} & H_{ba} & H_{bb} \end{pmatrix} \begin{pmatrix} C_d \\ C_a \\ C_b \end{pmatrix} = \begin{pmatrix} E & 0 & 0 \\ 0 & E & 0 \\ 0 & 0 & E \end{pmatrix} \begin{pmatrix} C_d \\ C_a \\ C_b \end{pmatrix}. \tag{A.38}$$

From the third equation, we obtain C_b as before. This expression for C_b is now inserted in the other two equations of Equation A.38:

$$\bar{H} \begin{pmatrix} C_d \\ C_a \end{pmatrix} = \begin{pmatrix} H_{dd} - H_{db}(H_{bb} - E)^{-1} H_{bd} & H_{da} - H_{db}(H_{bb} - E)^{-1} H_{ba} \\ H_{ad} - H_{ab}(H_{bb} - E)^{-1} H_{bd} & H_{aa} - H_{ab}(H_{bb} - E)^{-1} H_{ba} \end{pmatrix} \begin{pmatrix} C_d \\ C_a \end{pmatrix}$$

$$= \begin{pmatrix} E & 0 \\ 0 & E \end{pmatrix} \begin{pmatrix} C_d \\ C_a \end{pmatrix}. \tag{A.39}$$

$\bar{H}$ is an "effective" Hamiltonian matrix for donor and acceptor, which "contains" the interactions with the bridging molecular structure.

We now carry out the same derivation again but with two modifications. First, we denote by H_0 the Hamiltonian matrix for donor and acceptor together. Furthermore, we diagonalize the bridge matrix H_{bb} and apply the corresponding unitary transformation in Equation A.38:

$$\begin{pmatrix} 1 & 0 \\ 0 & U_b^\dagger \end{pmatrix}\begin{pmatrix} H_{00} & H_{0b} \\ H_{b0} & H_{bb} \end{pmatrix}\begin{pmatrix} 1 & 0 \\ 0 & U_b \end{pmatrix}\begin{pmatrix} 1 & 0 \\ 0 & U_b^\dagger \end{pmatrix}\begin{pmatrix} C_0 \\ C_b \end{pmatrix} = E\begin{pmatrix} 1 & 0 \\ 0 & U_b^\dagger \end{pmatrix}\begin{pmatrix} C_0 \\ C_b \end{pmatrix}. \quad \text{(A.40)}$$

If the matrix multiplications are carried out, we obtain:

$$\begin{pmatrix} H_0 & H_{0b}U_b \\ U_b^\dagger H_{b0} & U_b^\dagger H_{bb}U_b \end{pmatrix}\begin{pmatrix} C_0 \\ U_b^\dagger C_b \end{pmatrix} = E\begin{pmatrix} C_0 \\ U_b^\dagger C_b \end{pmatrix}. \quad \text{(A.41)}$$

$U_b^\dagger H_{bb}U_b$ is now a diagonal matrix, which we rename simply as H_{bb}. We do the following replacements:

$$H_{db} \rightarrow H_{db}U_b = \eta; \quad H_{ab} \rightarrow H_{ab}U_b = \theta \quad \text{(A.42)}$$

The η and θ matrices are thus Hamiltonian matrices where the expansion functions are the unchanged donor or acceptor functions on one side and the eigenfunctions of the bridge on the other side

$$H_{bb} \rightarrow U_b^\dagger H_{bb}U_b. \quad \text{(A.43)}$$

Finally, we rewrite the effective interaction matrix in Equation A.31 as:

$$\overline{H}_{da} = H_{da} - \eta(H_{bb} - E)^{-1}\theta^\dagger = H_{da} - \sum_k \frac{\eta_{ak}\theta_{kd}}{(E - E_k)}. \quad \text{(A.44)}$$

APPENDIX 5: EIGENVALUE PROBLEM

The most common way to calculate a wave function is to expand it in terms of N known functions $\{\chi_i\}$ in the following way:

$$\psi(x) = \sum_{i=1}^N c_i\chi_i(x) \quad \text{(A.45)}$$

The variational parameters are the coefficients $\{c_i\}$, which have to be calculated. If ψ is going to be a solution to the time-independent SE, it must hold that $H\psi = E\psi$:

$$H\psi(x) = \sum_{i=1}^N c_i H\chi_i(x) \approx E\psi \quad \text{(A.46)}$$

If we succeed in finding a good linear combination of these N functions, we may straighten out the sign $\approx$ in Equation A.46 to an equality sign and be in the possession of an eigenfunction to the Hamiltonian H. We multiply Equation A.46 with χ_j^* and integrate:

$$\sum_{i=1}^{N} \int \chi_j^* H \chi_i dx \cdot c_i = E \sum_{i=1}^{N} \int \chi_j^* \chi_i dx \cdot c_i = E \sum_{i=1}^{N} S_{ij} \cdot c_i, \quad \text{for all } j \qquad \text{(A.47)}$$

With $H_{ij} = \int \chi_j^* H \chi_i dx$ and $S_{ij} = \int \chi_j^* \chi_i dx$, we find that Equation A.47 is a matrix eigenvalue problem:

$$\begin{pmatrix} H_{11} & H_{12} & \cdots & H_{1N} \\ H_{21} & H_{22} & \cdots & H_{2N} \\ \vdots & \vdots & & \vdots \\ H_{N1} & H_{N2} & \cdots & H_{NN} \end{pmatrix} \begin{pmatrix} c_1 \\ c_2 \\ \vdots \\ c_N \end{pmatrix} = E \begin{pmatrix} S_{11} & S_{12} & \cdots & S_{1N} \\ S_{21} & S_{22} & \cdots & S_{2N} \\ \vdots & \vdots & & \vdots \\ S_{N1} & S_{N2} & \cdots & S_{NN} \end{pmatrix} \begin{pmatrix} c_1 \\ c_2 \\ \vdots \\ c_N \end{pmatrix}. \qquad \text{(A.48)}$$

Also written as:

$$(\mathbf{H} - E\mathbf{S})\mathbf{c} = \mathbf{0}, \qquad \text{(A.49)}$$

where $\mathbf{H}$ is the Hamiltonian matrix and $\mathbf{S}$ the overlap matrix. Equation A.48 is satisfied by the eigenvector $\mathbf{c}$ with the vector elements $c_1, c_2, \ldots$.

We will show that this solution is the same as the one obtained when the variation principle is used. E is defined as an expectation value to be optimized:

$$E = \frac{\langle \Psi | H | \Psi \rangle}{\langle \Psi | \Psi \rangle} = \frac{\displaystyle\sum_{i,j} c_i^* H_{ij} c_j}{\displaystyle\sum_{i,j} c_i^* S_{ij} c_j} \qquad \text{(A.50)}$$

E is now a function of the coefficients $\{c_i\}$ in Equation A.1. We multiply Equation A.50 by $\langle \Psi | \Psi \rangle$ and take the derivative:

$$\frac{\partial E}{\partial c_k} \sum_{i,j} c_i^* S_{ij} c_j + E \frac{\partial}{\partial c_k} \sum_{i,j} c_i^* S_{ij} c_j = \frac{\partial}{\partial c_k} \sum_{i,j} c_k^* H_{ij} c_j. \qquad \text{(A.51)}$$

The derivative of quadratic functions is simple. According to the variation principle, we set the derivatives of E with respect to the coefficients equal to zero. The first term in Equation A.51 disappears and we obtain:

$$\left(\sum_{i=1}^{N} H_{ji} c_i - E \sum_{i=1}^{N} S_{ji} \cdot c_i \right) = 0, \quad \text{for all } j. \qquad \text{(A.52)}$$

This equation is identical to Equation A.48.

A trivial solution of Equation A.49 is evident that all c_i are equal to zero. This solution is uninteresting, of course, since the wave function would be equal to zero. There are solutions different from zero, but, according to a theorem in linear analysis, such solutions exist only if the determinant of the matrix $(\mathbf{H} - E\mathbf{S})$ is equal to zero:

$$\begin{vmatrix} H_{11} - ES_{11} & H_{12} - ES_{12} & \cdots & H_{1N} - ES_{1N} \\ H_{21} - ES_{21} & H_{22} - ES_{22} & \cdots & H_{2N} - ES_{2N} \\ \vdots & \vdots & & \vdots \\ H_{N1} - ES_{N1} & H_{N2}ES_{N2} & \cdots & H_{NN} - ES_{NN} \end{vmatrix} = 0. \qquad (A.53)$$

Unfortunately, this determinant cannot be easily evaluated if $N > 3$, but it is easy to see that we would obtain a polynomial in E of degree N. Equation A.53, then, has N solutions, which are eigenvalues of the Hamiltonian matrix. Equation A.53 is referred to as the *secular equation*.

The solution of Equation A.53 are approximations to the eigenvalues of the time-independent SE that converge from above in order to the corresponding exact ones, when the number of basis functions (N) is increased. The solutions obtained of the secular Equation A.53 are the best ones in the sense that they give the best (lowest) energies in the chosen set of basis functions, Equation A.45. When N is allowed to increase, the energies are further improved, while at the same time, new solutions are added for the excited states. The latter are usually bad approximations.

There are a number of standard methods to solve the matrix eigenvalue problem. We refer to the MATLAB-programs. A number of scientists have developed effective methods, which are useful for practical purposes, such as large molecules. It should also be mentioned that reliable computer programs exist in most cases.

The explicit solution for $N = 2$ may be obtained analytically. This problem occurs if two states are interacting with the wave functions χ_1 and χ_2 in Equation A.45, for example, for chemical bonding and electron transfer. The two-dimensional eigenvalue problem reads:

$$\begin{pmatrix} H_{11} & H_{12} \\ H_{21} & H_{22} \end{pmatrix} \begin{pmatrix} c_1 \\ c_2 \end{pmatrix} = E \begin{pmatrix} S_{11} & S_{12} \\ S_{21} & S_{22} \end{pmatrix} \begin{pmatrix} c_1 \\ c_2 \end{pmatrix}. \qquad (A.54)$$

We assume that χ_1 and χ_2 are normalized and that $S_{12} = S_{21} = S$. It is possible to show that $H_{12} = H_{21}$. The eigenvalues may be obtained from the secular Equation A.53:

$$\begin{vmatrix} H_{11} - E & H_{12} - ES \\ H_{12} - ES & H_{22} - E \end{vmatrix} = 0 \qquad (A.55)$$

equivalent to the following polynomial equation in E:

$$(H_{11} - E)(H_{22} - E) = (H_{12} - ES)^2. \qquad (A.56)$$

Using $H = (H_{11} + H_{22})/2$:

$$E^2(1 - S^2) - 2E(H + SH_{12}) + H_{11}H_{12} - H_{12}^2 = 0. \qquad (A.57)$$

If we solve this second-degree equation, we obtain

$$E_{\pm} = \left[H - SH_{12} \pm \sqrt{\left(\frac{H_{11} - H_{22}}{2}\right)^2 (1 - S^2) + (H_{12} - HS)^2} \right] / (1 - S^2). \qquad (A.58)$$

In the special case $H_{11} = H_{22} = H$, we have

$$E_+ = \frac{H + H_{12}}{1 + S}; \quad E_- = \frac{H - H_{12}}{1 - S}. \qquad (A.59)$$

Since $H < 0$ and $H_{12} < 0$, E_+ is the solution with the lowest eigenvalue. The difference between the eigenvalues is

$$E_- - E_+ = \frac{2}{1 - S^2}(SH - H_{12}). \qquad (A.60)$$

Normally, the lowest eigenvalue decreases from H less than the upper one increases. If we let two electrons from different atoms occupy the wave function of the lower eigenvalue, we obtain a chemical bond between the atoms. However, if there are three or four electrons, for example, if the two atoms are noble gas atoms, the upper wave function has to be used. If both lower and upper orbitals are fully occupied, the chemical bond disappears. Noble gas atoms do not bond.

The calculation of the eigenvector s is left to the readers. It is recommended to start with the case $S = 0$ and $H_{11} = H_{22}$.

APPENDIX 6: VIBRATIONAL OVERLAP

To solve the vibrational overlap integrals we use Equations A.3 through A.7. First, we recall that the overlap between two Gaussians, centered at $x = a$ is:

$$\int_{-\infty}^{\infty} \exp\left[-\alpha(x - a)^2\right] \exp\left[-\alpha(x - a)^2\right] dx = \int_{-\infty}^{\infty} \exp\left[\left(-2\alpha(x - a)^2\right)\right] dx = \sqrt{\frac{\pi}{2\alpha}}. \qquad (A.61)$$

It is important to calculate the integral if only one of the Gaussians is displaced on the x-axis. We obtain:

$$\Omega_0 = \int_{-\infty}^{\infty} \exp\left(-\alpha x^2\right) \exp\left[-\alpha(x-a)^2\right] dx. \tag{A.62}$$

Here we use the easily proven theorem that a product of two Gaussians can always be written as a single Gaussian multiplied by a constant. The exponent in the integrand of Equation A.7 may thus be written as:

$$\alpha x^2 + \alpha(x-a)^2 = 2\alpha\left(x - \frac{a}{2}\right)^2 + \frac{\alpha}{2}a^2 \tag{A.63}$$

whereby the theorem follows. Using Equation A.7, we thus obtain:

$$\Omega_0 = \exp\left(-\alpha a^2/2\right)\sqrt{\frac{\pi}{2\alpha}}. \tag{A.64}$$

If the displaced function is multiplied by x, the following integral is obtained:

$$I_1 = \int_{-\infty}^{\infty} \exp\left(-\alpha x^2\right) x \exp\left[-\alpha(x-a)^2\right] dx = \exp\left(-\alpha a^2/2\right) \int_{-\infty}^{\infty} x \exp\left[-2\alpha(x-a/2)^2\right] dx$$

$$= \exp\left(-\alpha a^2/2\right) \left[\int_{-\infty}^{\infty} (x-a/2) \exp\left[-2\alpha(x-a/2)^2\right] dx + \int_{-\infty}^{\infty} (a/2) \exp\left[-2\alpha(x-a/2)^2 dx\right] \right]$$

$$= \exp\left(-\alpha a^2/2\right) \left[\int_{-\infty}^{\infty} (a/2) \exp\left[-2\alpha(x-a/2)^2 dx\right] \right] = \frac{a}{2}\exp\left(-\alpha a^2/2\right)\sqrt{\frac{\pi}{2\alpha}}.$$

$$\tag{A.65}$$

We may now evaluate the vibrational overlap:

$$S_{0,0} = \int_{-\infty}^{\infty} \Xi\Psi dQ = \left(\frac{\alpha}{\pi}\right)^{1/2} \int_{-\infty}^{\infty} \exp\left[-\alpha(Q-Q_e)^2/2\right] \exp\left[-\alpha(Q-Q_0)^2/2\right] \Psi dQ. \tag{A.66}$$

Substitute: $x = Q - Q_e$ and $x - a = Q - Q_0$, which implies $a = Q_e - Q_0$. Using Equation A.7 we obtain:

$$S_{0,0} = \left(\frac{\alpha}{\pi}\right)^{1/2} \int_{-\infty}^{\infty} \exp\left[-(\alpha/2)x^2\right] \exp\left[-(\alpha/2)(x-a)^2/2\right] dQ$$

(A.67)

$$= \exp\left(-\alpha a^2/4\right) = \exp\left(-\alpha(Q_e - Q_0)^2/4\right).$$

The same substitutions lead to:

$$S_{0,1} = \left(\frac{\alpha}{\pi}\right)^{1/2} (Q_e - Q_0) \exp\left(-\alpha(Q_e - Q_0)^2/4\right).$$

(A.68)

Index